HVAC

Level Three

Trainee Guide
Third Edition

Upper Saddle River, New Jersey
Columbus, Ohio

NCCER

President: Don Whyte
Director of Curriculum Revision and Development: Daniele Stacey
HVAC Project Manager: Carla Sly/Tania Domenech
Production Manager: Tim Davis
Quality Assurance Coordinator: Debie Ness
Editors: Rob Richardson and Matt Tischler
Desktop Publishing Coordinator: James McKay
Production Assistant: Brittany Ferguson

NCCER would like to acknowledge the contract service provider for this curriculum:
Topaz Publications, Liverpool, New York.

This information is general in nature and intended for training purposes only. Actual performance of activities described in this manual requires compliance with all applicable operating, service, maintenance, and safety procedures under the direction of qualified personnel. References in this manual to patented or proprietary devices do not constitute a recommendation of their use.

10 9
ISBN 0-13-604492-1
ISBN-13 978-0-13-604492-5

PREFACE

TO THE TRAINEE

Heating and air-conditioning systems (HVAC) regulate the temperature, humidity, and the total air quality in residential, commercial, industrial, and other buildings. This also extends to refrigeration systems used to transport food, medicine, and other perishable items. Other systems may include hydronics (water-based heating systems), solar panels, or commercial refrigeration. HVAC technicians and installers set up, maintain, and repair such systems. As a technician, you must be able to maintain, diagnose, and correct problems throughout the entire system. Diversity of skills and tasks is also significant to this field. You must know how to follow blueprints or other specifications to install any system. You may also need working knowledge of sheet metal practices for the installation of ducts, welding, basic pipefitting, and electrical practices.

Think about it! Nearly all buildings and homes in the United States alone use forms of heating, cooling and/or ventilation. The increasing development of HVAC technology causes employers to recognize the importance of continuous education and keeping up to speed with the latest equipment and skills. Hence, technical school training or apprenticeship programs often provide an advantage and a higher qualification for employment. NCCER's program has been designed by highly-qualified subject matter experts with this in mind. Our four levels present an apprentice approach to the HVAC field, including theoretical and practical skills essential to your success as an HVAC installer or technician.

As the population and the number of buildings grow in the near future, so will the demand for HVAC technicians. According to the U.S. Bureau of Labor Statistics, employment of HVAC technicians and installers is projected to increase 18 to 26 percent by 2014. We wish you the best as you begin an exciting and promising career.

WHAT'S NEW IN *HVAC LEVEL THREE*?

HVAC Level Three has four new modules: Refrigerants and Oils, Retail Refrigeration Systems, Commercial Hydronic Systems and Steam Systems. Two modules that were previously in Level Two – Compressors and Metering Devices – have now been moved to this level. Water Treatment, a module previously found in Level Four, was also modified and moved to Level Three.

The first two levels of HVAC now provide you with a broad knowledge of installation and service requirements for residential and commercial systems, while Levels Three and Four provide advanced maintenance, troubleshooting, design, and supervisory skills. Through this course design, you will enter the workforce with the critical knowledge and skills needed to perform productively in either the residential or commercial market.

CONTREN® LEARNING SERIES

The National Center for Construction Education and Research (NCCER) is a not-for-profit 501(c)(3) education foundation established in 1995 by the world's largest and most progressive construction companies and national construction associations. It was founded to address the severe workforce shortage facing the industry and to develop a standardized training process and curricula. Today, NCCER is supported by hundreds of leading construction and maintenance companies, manufacturers, and national associations. The Contren® Learning Series was developed by NCCER in partnership with Pearson Education, Inc., the world's largest educational publisher.

Some features of NCCER's Contren® Learning Series are as follows:

- An industry-proven record of success
- Curricula developed by the industry for the industry
- National standardization providing portability of learned job skills and educational credits
- Compliance with the Office of Apprenticeship requirements for related classroom training (CFR 29:29)
- Well-illustrated, up-to-date, and practical information

NCCER also maintains a National Registry that provides transcripts, certificates, and wallet cards to individuals who have successfully completed modules of NCCER's Contren® Learning Series. *Training programs must be delivered by an NCCER Accredited Training Sponsor in order to receive these credentials.*

Contents

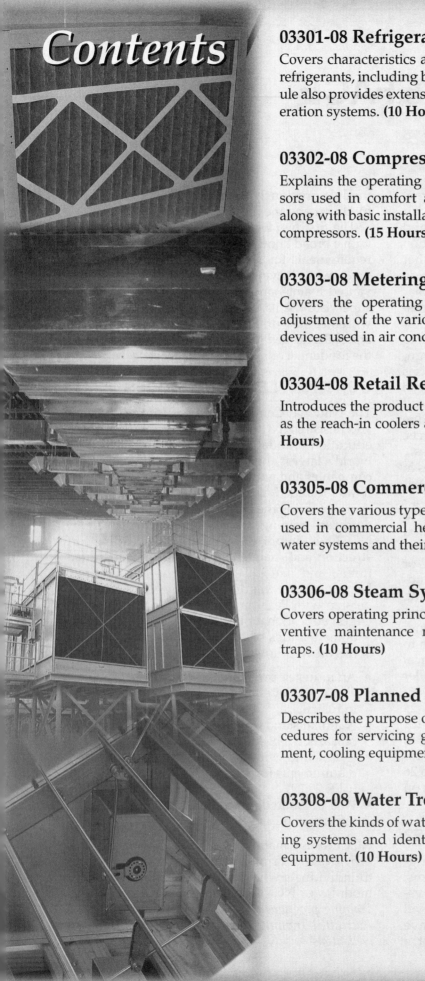

03301-08 Refrigerants and Oils NEW 1.i

Covers characteristics and applications of the current generation of refrigerants, including both pure and blended refrigerants. The module also provides extensive coverage of lubricating oils used in refrigeration systems. **(10 Hours)**

03302-08 Compressors. 2.i

Explains the operating principles of the different types of compressors used in comfort air conditioning and refrigeration systems, along with basic installation, service, and repair procedures for these compressors. **(15 Hours)**

03303-08 Metering Devices 3.i

Covers the operating principles, applications, installation, and adjustment of the various types of fixed and adjustable expansion devices used in air conditioning equipment. **(7.5 Hours)**

03304-08 Retail Refrigeration Systems 4.i

Introduces the product refrigeration components and systems, such as the reach-in coolers and freezers commonly used in markets. **(20 Hours)**

03305-08 Commercial Hydronic Systems 5.i

Covers the various types of boilers, components, and piping systems used in commercial heating applications. Also introduces chilled water systems and their components. **(12.5 Hours)**

03306-08 Steam Systems. 6.i

Covers operating principles, piping systems, components, and preventive maintenance requirements of steam systems and steam traps. **(10 Hours)**

03307-08 Planned Maintenance 7.i

Describes the purpose of planned maintenance and outlines the procedures for servicing gas and oil furnaces, electric heating equipment, cooling equipment, and heat pumps. **(20 Hours)**

03308-08 Water Treatment 8.i

Covers the kinds of water problems encountered in heating and cooling systems and identifies various water treatment methods and equipment. **(10 Hours)**

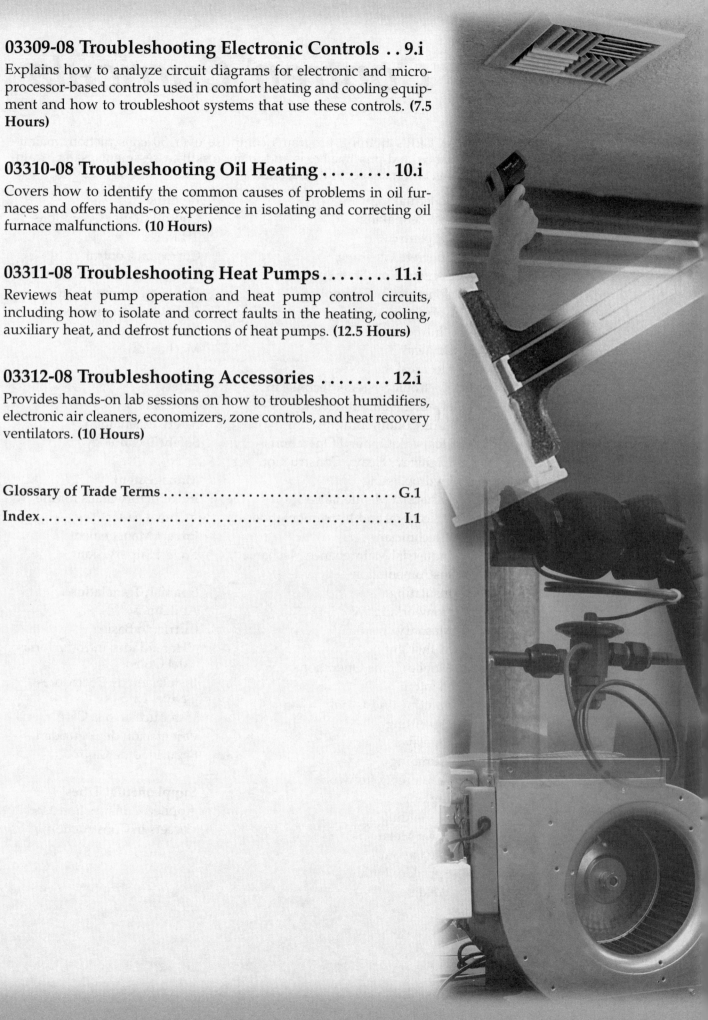

03309-08 Troubleshooting Electronic Controls .. 9.i

Explains how to analyze circuit diagrams for electronic and micro-processor-based controls used in comfort heating and cooling equipment and how to troubleshoot systems that use these controls. **(7.5 Hours)**

03310-08 Troubleshooting Oil Heating 10.i

Covers how to identify the common causes of problems in oil furnaces and offers hands-on experience in isolating and correcting oil furnace malfunctions. **(10 Hours)**

03311-08 Troubleshooting Heat Pumps 11.i

Reviews heat pump operation and heat pump control circuits, including how to isolate and correct faults in the heating, cooling, auxiliary heat, and defrost functions of heat pumps. **(12.5 Hours)**

03312-08 Troubleshooting Accessories 12.i

Provides hands-on lab sessions on how to troubleshoot humidifiers, electronic air cleaners, economizers, zone controls, and heat recovery ventilators. **(10 Hours)**

Glossary of Trade Terms G.1

Index I.1

Contren® Curricula

NCCER's training programs comprise over 50 construction, maintenance, and pipeline areas and include skills assessments, safety training, and management education.

Boilermaking
Cabinetmaking
Carpentry
Concrete Finishing
Construction Craft Laborer
Construction Technology
Core Curriculum:
  Introductory Craft Skills
Drywall
Electrical
Electronic Systems Technician
Heating, Ventilating, and
  Air Conditioning
Heavy Equipment Operations
Highway/Heavy Construction
Hydroblasting
Industrial Maintenance
  Electrical and Instrumentation
  Technician
Industrial Maintenance Mechanic
Instrumentation
Insulating
Ironworking
Masonry
Millwright
Mobile Crane Operations
Painting
Painting, Industrial
Pipefitting
Pipelayer
Plumbing
Reinforcing Ironwork
Rigging
Scaffolding
Sheet Metal
Site Layout
Sprinkler Fitting
Welding

Pipeline
Control Center Operations,
  Liquid
Corrosion Control
Electrical and Instrumentation
Field Operations, Liquid
Field Operations, Gas
Maintenance
Mechanical

Safety
Field Safety
Safety Orientation
Safety Technology

Management
Introductory Skills for the
  Crew Leader
Project Management
Project Supervision

Spanish Translations
Andamios
Currículo Básico
  Habilidades Introductorias
  del Oficio
Instalación de Rociadores
  Nivel Uno
Introducción a la Carpintería
Orientación de Seguridad
Seguridad de Campo

Supplemental Titles
Applied Construction Math
Careers in Construction

Acknowledgments

This curriculum was revised as a result of the
farsightedness and leadership of the following sponsors:

ABC of Wisconsin
Lincoln Technical Institute
W. B. Guimarin & Co., Inc.
Hunton Trane
Apex Technical School
Entek

This curriculum would not exist were it not for the dedication
and unselfish energy of those volunteers who served on the Authoring Team.
A sincere thanks is extended to the following:

Barry Burkan
Frank Kendall
Daniel Kerkman
Joe Moravek
Troy Staton
Mattew Todd

NCCER PARTNERING ASSOCIATIONS

American Fire Sprinkler Association
Associated Builders and Contractors, Inc.
Associated General Contractors of America
Association for Career and Technical Education
Association for Skilled and Technical Sciences
Carolinas AGC, Inc.
Carolinas Electrical Contractors Association
Center for the Improvement of Construction
   Management and Processes
Construction Industry Institute
Construction Users Roundtable
Design-Build Institute of America
Electronic Systems Industry Consortium
Merit Contractors Association of Canada
Metal Building Manufacturers Association
NACE International
National Association of Minority Contractors

National Association of Women in Construction
National Insulation Association
National Ready Mixed Concrete Association
National Systems Contractors Association
National Technical Honor Society
National Utility Contractors Association
NAWIC Education Foundation
North American Crane Bureau
North American Technician Excellence
Painting and Decorating Contractors of America
Portland Cement Association
SkillsUSA
Steel Erectors Association of America
Texas Gulf Coast Chapter ABC
U.S. Army Corps of Engineers
University of Florida
Women Construction Owners and Executives, USA

03301-08

Refrigerants and Oils

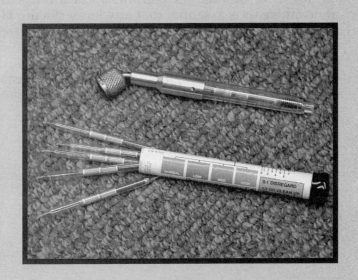

03301-08
Refrigerants and Oils

Topics to be presented in this module include:

1.0.0   Introduction 1.2
2.0.0   Refrigerant Structure 1.4
3.0.0   Refrigerant Identification 1.7
4.0.0   Refrigerant Composition 1.10
5.0.0   Refrigerant Leaks 1.14
6.0.0   Lubricating Oils 1.17
7.0.0   Oil and the Refrigeration System 1.19
8.0.0   Oil Handling Guidelines 1.23
9.0.0   System Conversion 1.26

Overview

In today's world, the HVAC technician must be thoroughly versed in the use and storage of refrigerants and oils used in the mechanical refrigeration cycle. Laws adopted in the 1990s mandate elimination of old refrigerants and their replacement with new, often more complex refrigerants. In addition to phasing out environmentally unsafe refrigerants, these laws provide harsh penalties for using banned refrigerants and for releasing refrigerants into the atmosphere. It is essential that anyone servicing HVAC and refrigeration equipment be thoroughly familiar with the rules governing refrigerants.

It is rare to find new refrigerants that are drop-in replacements for the banned refrigerants. Even if the two refrigerants are compatible, they are not likely to be compatible with the oil used to lubricate the compressor. So in addition to refrigerants, service technicians must know about oils.

Objectives

When you have completed this module, you will be able to do the following:

1. Identify the refrigerants in common use and state the types of applications in which each is used.
2. Explain the effects of releasing refrigerants into the atmosphere.
3. Explain how refrigerants are classified by their chemical composition.
4. Describe the color-coding scheme used to identify refrigerant cylinders.
5. Describe how azeotropes and near-azeotropes differ from each other and from so-called pure refrigerants.
6. Interpret a P-T chart for an azeotrope refrigerant.
7. Calculate superheat and subcooling.
8. Demonstrate refrigerant leak detecting methods.
9. Identify the different types of oils used in refrigeration systems and explain their relationships to the various refrigerants.
10. Explain how to add and remove oil from a system.
11. Describe how to test oil for contamination.
12. Perform a refrigerant retrofit.

Required Trainee Materials

1. Pencil and paper
2. Appropriate personal protective equipment

Prerequisites

Before you begin this module, it is recommended that you successfully complete *Core Curriculum*; *HVAC Level One*; and *HVAC Level Two*.

This course map shows all of the modules in the third level of the *HVAC* curriculum. The suggested training order begins at the bottom and proceeds up. Skill levels increase as you advance on the course map. The local Training Program Sponsor may adjust the training order.

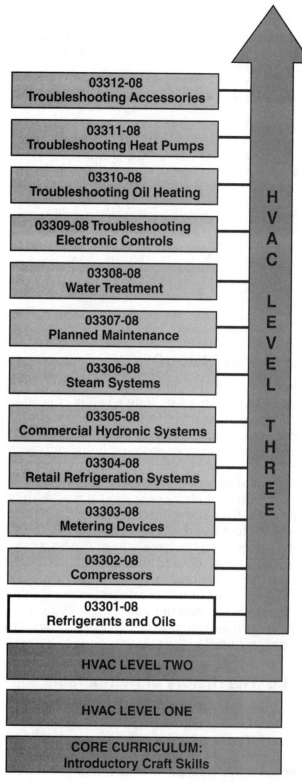

03312-08
Troubleshooting Accessories

03311-08
Troubleshooting Heat Pumps

03310-08
Troubleshooting Oil Heating

03309-08 Troubleshooting
Electronic Controls

03308-08
Water Treatment

03307-08
Planned Maintenance

03306-08
Steam Systems

03305-08
Commercial Hydronic Systems

03304-08
Retail Refrigeration Systems

03303-08
Metering Devices

03302-08
Compressors

03301-08
Refrigerants and Oils

HVAC LEVEL TWO

HVAC LEVEL ONE

CORE CURRICULUM:
Introductory Craft Skills

HVAC LEVEL THREE

301CMAP.EPS

Trade Terms

Alkylbenzene
Azeotrope
Binary blend
Bubble point
Chlorofluorocarbon (CFC)
Compound
Dew point
Dielectric strength
Flash point
Floc point
Fractionation
Glide
Global warming potential (GWP)
Greenhouse effect
Halogen
Hydrocarbon

Hydrogenated chlorofluorocarbon (HCFC)
Hydrogenated fluorocarbon (HFC)
Hygroscopic
Miscible
Mixture
Near-azeotrope
Ozone depletion potential (ODP)
Polyalkylene glycol (PAG)
Polyolester (POE)
Polyvinyl ether (PVE)
Pour point
Ternary blend
Viscosity
Zeotrope

1.0.0 ◆ INTRODUCTION

A refrigerant is the substance in the refrigeration system that changes its state when heat is added or taken away from it. In the condenser, gaseous refrigerant releases heat, allowing it to condense into a liquid. In the evaporator, liquid refrigerant boils into a gaseous state as it absorbs heat. This constant absorption and rejection of heat by the refrigerant as it changes state is the underlying principle that allows mechanical refrigeration systems to function. Heat is absorbed by the refrigerant from an area where the heat is not wanted and is rejected into an area where that heat is not objectionable.

The oils used in refrigeration systems are there to lubricate the moving parts of the compressor. The compressor oil must be compatible with the refrigerant used in the system.

1.1.0 The History of Refrigerants

The principles of mechanical refrigeration were put to practical use over 150 years ago when Dr. John Gorrie received a patent for a mechanical refrigeration system that compressed gas, condensed it into a liquid, and then changed the condensed gas to a vapor. He used the apparatus to keep his sick patients cool. Several years later, mechanical refrigeration systems using ammonia as the refrigerant were developed to produce ice.

In the early twentieth century, toxic refrigerants such as ammonia, sulfur dioxide, and methyl chloride were in widespread use in commercial applications. It was not until the 1920s that mod-ern, less toxic refrigerants were developed that used **halogens** such as fluorine and chlorine. These modern refrigerants enabled the evolution of domestic refrigerators and freezers, and the development and widespread use of air conditioning.

By the 1980s, scientists had come to realize that the chlorine used in refrigerants was a major contributor to atmospheric ozone depletion. International treaties were enacted to restrict and eventually phase out the use of these damaging refrigerants. New refrigerants that do not deplete atmospheric ozone have been developed and are now being phased in as the older refrigerants are phased out.

1.2.0 Desirable Refrigerant Characteristics

Many substances can be used as a refrigerant. However, they may possess certain characteristics, such as toxicity or explosiveness, that render them undesirable for use in that role. In order to be desirable as a refrigerant, a substance should be:

- Stable
- Nontoxic
- Nonflammable/nonexplosive
- Noncorrosive
- Easy to detect if leaks in the system occur

Most importantly, the substance should have characteristics that enable it to promote good heat transfer and perform satisfactorily under all normal conditions in the system in which it is used.

1.2.1 Refrigerant Stability

Refrigerants must perform under a wide variety of temperature and pressure conditions while being in contact with various metals and other nonmetallic substances commonly found in the refrigeration system. Under these normal conditions, the refrigerant must remain stable and not break down into its component elements or combine with other substances to form harmful byproducts.

1.2.2 Refrigerant Toxicity

Modern refrigerants, for the most part, are nontoxic. If leaks occur, a modern refrigerant is not toxic unless it is allowed to accumulate in a confined space such as an equipment vault where it may displace oxygen and cause suffocation. Ammonia, which is still used in large commercial refrigeration systems, is not toxic but is extremely irritating. However, ammonia systems are designed with special safety and alarm systems that alert personnel if a leak exists.

1.2.3 Refrigerant Flammability

Refrigerants should not be flammable or explosive for obvious reasons. The normal operation of HVAC electrical equipment can produce arcing. Servicing or repair procedures involving the use of a flame could ignite any leaking refrigerant that is flammable or explosive.

1.2.4 Corrosive Properties of Refrigerants

Refrigerant used in a system should not corrode the materials the system is made of. Some refrigerants, notably ammonia, are corrosive to copper and bronze in the presence of moisture. For that reason, ammonia refrigerant piping is made of iron or steel, materials it does not corrode.

1.2.5 Leak Detection

Refrigerants should be easily detectable if they leak. Modern leak detectors are often electronic devices that draw in a sample of air and check for the presence of a refrigerant (*Figure 1*). When using a modern leak detector, ensure that it is capable of detecting the type of refrigerant in the system being checked.

1.2.6 Individual Refrigerant Characteristics

The refrigerant used for a particular application must have characteristics that enable it to perform well in that application. For example, the low boiling point of ammonia (−28°F) makes it ideal for use in very low-temperature applications. At these temperatures, the evaporator pressure is still positive. Typically, a refrigerant is selected based on the conditions under which the system must operate; the physical design of the system, including the materials used in the piping, coils, and accessories; and the cost and availability of the refrigerant.

301F01.EPS

Figure 1 ◆ Electronic leak detector.

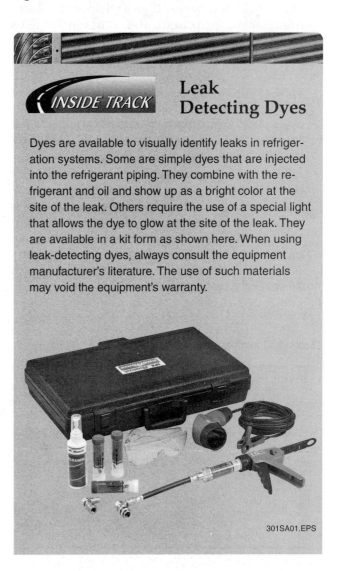

INSIDE TRACK   **Leak Detecting Dyes**

Dyes are available to visually identify leaks in refrigeration systems. Some are simple dyes that are injected into the refrigerant piping. They combine with the refrigerant and oil and show up as a bright color at the site of the leak. Others require the use of a special light that allows the dye to glow at the site of the leak. They are available in a kit form as shown here. When using leak-detecting dyes, always consult the equipment manufacturer's literature. The use of such materials may void the equipment's warranty.

301SA01.EPS

1.3.0 Common Refrigerant Applications

Different refrigerants are used in different applications with dozens of different refrigerants available. The more common refrigerants and their applications are shown in *Table 1*.

- *HCFC-22* – Widely used in residential and light commercial air conditioning applications and in some refrigeration applications. This refrigerant contains ozone-depleting chlorine. After December 31, 2009, this refrigerant can no longer be produced or imported into the United States for use in new equipment. Existing inventory and reclaimed refrigerant can be used for service applications after that date.

- *HFC-410A* – Widely regarded as the refrigerant most likely to be used in air conditioning systems that currently use HCFC-22 refrigerant. Several major equipment manufacturers now offer products that use this refrigerant. However, it cannot be used as a drop-in replacement for equipment that currently uses HCFC-22 because of pressure and oil incompatibility, as well as metering device selection.

- *HFC-407C* – This refrigerant is currently available as an easy conversion replacement for HCFC-22 because it has similar operating characteristics. Equipment modification such as the replacement of the compressor lubricating oil may be necessary.

- *HCFC-123* – This refrigerant contains chlorine but has a much lower ozone depleting potential than CFC-11, the refrigerant that it replaced. It is widely used in low-pressure centrifugal chillers. Equipment modification may be required if used to retrofit a CFC-11 system. After 2020, the availability of HCFC-123 will be severely restricted.

- *HFC-404A* – This refrigerant is used in refrigeration applications and is a drop-in replacement for CFC-502, a phased-out refrigerant, and HCFC-22. Its characteristics allow it to replace both. Equipment modification such as the replacement of the compressor lubricating oil may be necessary.

- *HFC-134a* – This refrigerant is widely used in domestic and commercial refrigeration units as well as in automotive air conditioning systems. It is also a drop-in replacement for CFC-12, a phased-out refrigerant, because of its similar characteristics. Equipment modification such as the replacement of the compressor lubricating oil may be necessary when retrofitting a CFC-12 system.

Ammonia is a **compound** of nitrogen and hydrogen, so it is chemically different from the refrigerants just described. It is used in low-temperature applications such as refrigerated warehouses and ice-making plants. Ammonia does not damage the environment, but the refrigeration system components must be made of iron or steel because ammonia reacts with copper and bronze.

2.0.0 ◆ REFRIGERANT STRUCTURE

Refrigerants may consist of or contain **hydrocarbons**, which are compounds composed only of carbon and hydrogen. When some or all of the hydrogen atoms in a hydrocarbon are combined with a halogen such as chlorine, fluorine, or bromine, the end result is a compound known as a halogenated hydrocarbon or halocarbon. A fully halogenated hydrocarbon has all the hydrogen atoms replaced with a halogen atom. A partially halogenated hydrocarbon has at least one hydrogen atom remaining and the rest replaced with a halogen atom (*Figure 2*).

Table 1 Common Refrigerants

Refrigerant	Use	Comments
HCFC–22	Air conditioning and some refrigeration	Can no longer be produced or imported into the U.S. for use in new equipment after 12-30-2009.
HFC–410A	HCFC–22 replacement	Cannot be used in "drop-in" retrofits.
HFC–407C	HCFC–22 "drop-in" replacement	Equipment modification required.
HCFC–123	CFC–11 replacement	Use restricted after the year 2020.
HFC–404A	CFC–502 and HCFC–22 "drop-in" replacement	Equipment modification required.
HFC–134a	CFC–12 replacement	Used in domestic refrigeration and automotive air conditioning. Equipment modification required for "drop-in" replacement.

301T01.EPS

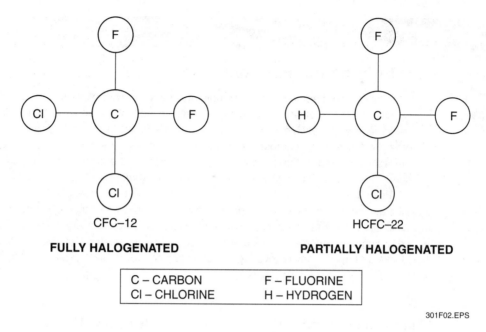

Figure 2 ◆ Halogenated hydrocarbons.

2.1.0 Refrigerant Classifications

Refrigerants fall into the following three major classifications of halocarbons:

- **Chlorofluorocarbons (CFCs)**, consisting of atoms of chlorine, fluorine, and carbon.
- **Hydrogenated chlorofluorocarbons (HCFCs)**, consisting of hydrogen, chlorine, fluorine, and carbon atoms.
- **Hydrogenated fluorocarbons (HFCs)**, do not contain any chlorine atoms, only hydrogen, fluorine, and carbon atoms.

Refrigerants can exist in their pure form or they can be mixed with other refrigerants to obtain the desired characteristics. For example, HCFC-22 could be considered a pure compound because it is not mixed with any other refrigerant. On the other hand, HFC-410A, a replacement for HCFC-22, consists of a 50-50 blend by weight of HFC-32 and HFC-125. Different types of refrigerant blends and **mixtures** will be discussed later in this module.

2.1.1 CFC Refrigerants

Chlorofluorocarbon (CFC) refrigerants are the most damaging to the environment because they contain more chlorine atoms in their structure. Production of CFCs in the United States ended in 1995. Equipment using CFCs still exists. Obtaining CFCs to service this obsolete equipment has forced owners to pay exorbitant costs for the refrigerant, if it can even be found. This has resulted in the scrapping and replacement of old equipment or the retrofit to a compatible HFC refrigerant that does not damage the environment.

2.1.2 HCFC Refrigerants

Hydrogenated chlorofluorocarbon (HCFC) refrigerants contain chlorine atoms in their structure, but fewer than found in CFCs. They are kinder to the environment than CFCs, allowing them to be phased out over a longer period of time. By 2030, all HCFC refrigerants will no longer be manufactured.

2.1.3 HFC Refrigerants

Hydrogenated fluorocarbon (HFC) refrigerants such as HFC-123 contain no chlorine atoms in their structure. That makes this group of refrigerants the most ozone-friendly of the three classifications of halocarbon refrigerants. Generally speaking, the replacement refrigerants for CFCs and HCFCs have tended to be HFCs. Equipment using HFCs have been available since the early 1990s.

2.2.0 Organic Refrigerants

Some refrigerants touted as organic and environmentally friendly have been marketed for several years. They often consist of different blends of flammable hydrocarbons such as propane or butane in liquid form, and are marketed as

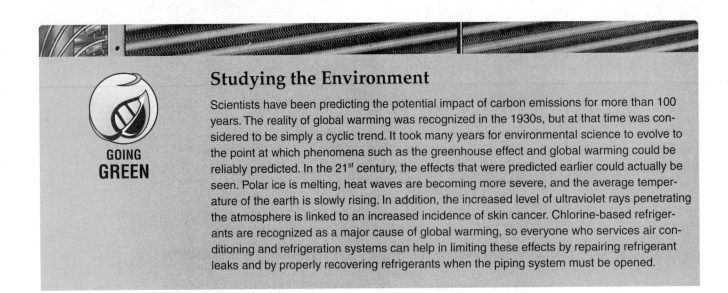

Studying the Environment

Scientists have been predicting the potential impact of carbon emissions for more than 100 years. The reality of global warming was recognized in the 1930s, but at that time was considered to be simply a cyclic trend. It took many years for environmental science to evolve to the point at which phenomena such as the greenhouse effect and global warming could be reliably predicted. In the 21st century, the effects that were predicted earlier could actually be seen. Polar ice is melting, heat waves are becoming more severe, and the average temperature of the earth is slowly rising. In addition, the increased level of ultraviolet rays penetrating the atmosphere is linked to an increased incidence of skin cancer. Chlorine-based refrigerants are recognized as a major cause of global warming, so everyone who services air conditioning and refrigeration systems can help in limiting these effects by repairing refrigerant leaks and by properly recovering refrigerants when the piping system must be opened.

GOING **GREEN**

replacements for CFC-12 in automotive and other applications. While they do not deplete the ozone layer, safety issues related to flammability have led the United States Environmental Protection Agency (EPA) and some state governments to ban their use as refrigerants.

2.3.0 Refrigerants and the Environment – Ozone Depletion

Ozone in the atmosphere prevents excessive ultraviolet radiation from the sun from reaching the Earth where it can damage plants and cause skin cancer in animals and humans. Refrigerants containing chlorine are stable and eventually rise into the upper levels of the atmosphere where ultraviolet radiation causes the refrigerant to break down, releasing chlorine atoms. The chlorine atoms then combine with one of the three oxygen atoms that make up an ozone molecule. This results in the breakdown of the ozone molecule. This sets forth a series of other chemical reactions that further break down ozone molecules.

CFC refrigerants are the biggest offender in ozone depletion because they contain more atoms of chlorine in their makeup. For that reason, CFCs were the first refrigerants to have their use restricted and eventually eliminated in the United States. HCFC refrigerants have less **ozone depletion potential (ODP)** than CFCs, hence their continued use and slower phase-out schedule. HFCs contain no chlorine atoms and are considered the most environmentally friendly of the three halocarbon refrigerant classifications.

Refrigerants are listed according to their ozone depletion potential (ODP). The lower the number, the lower the ozone depletion potential (*Table*

2). CFCs such as CFC-11 and CFC-12 have ozone depletion potentials of 1.0. HCFC-22 has an ozone depletion potential of 0.05, while HCFC-123 has an ozone depletion potential of 0.02. All HFCs have ozone depletion potentials of zero.

2.4.0 Refrigerants and the Environment – Global Warming

Another environmental concern is the **greenhouse effect**, said to contribute to global warming (*Figure 3*). In this process, various pollutants in the atmosphere, notably carbon dioxide, trap the heat of the sun in the atmosphere in a manner similar to how glass in a greenhouse traps the sun's heat. Compared to other substances that are emitted into the atmosphere in huge quantities, refrigerants are not big contributors to global warming even though they may have a high **global warming potential (GWP)** number. Here's why:

- Compared to other major pollutants such as carbon dioxide, their quantity in the atmosphere is very low. HFC-23, a byproduct of the production of HCFC-22, has increased its concentration in the atmosphere over time. This pollutant should decline as HCFC-22 production is phased out.

Table 2 Ozone Depletion Potential of Refrigerants

Refrigerant	ODP
CFC–12	1.0
HCFC–22	0.05
HCFC–123	0.02
HFC–134a	0.0

301T02.EPS

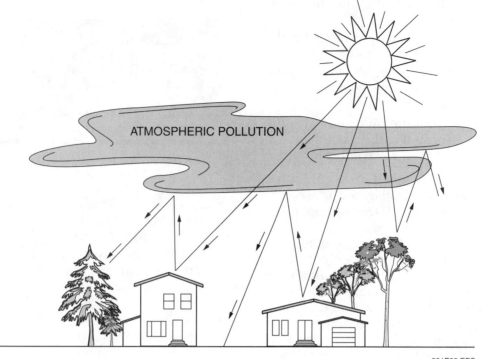

Figure 3 ◆ Greenhouse effect.

301F03.EPS

- Refrigerants in the atmosphere tend to break down over time, limiting their effect. For example, HFC-134a has an atmospheric life of about 14 years, while HFC-152A, with a global warming potential number of 140, has an atmospheric life of less than 1.5 years.
- Laws prohibiting the discharge of refrigerants into the atmosphere minimize refrigerant pollution.

2.4.1 Working with Refrigerants

As an HVAC technician working with refrigerants, it is your responsibility to prevent refrigerant from escaping into the atmosphere when servicing or repairing equipment. In the United States, all HVAC technicians must be EPA certified to handle refrigerants (*Figure 4*).

Four technician certification types are available. Type I allows technicians to work on small appliances containing less than 5 pounds of refrigerant. Type II allows technicians to work on high-pressure and very-high pressure appliances such as residential and commercial air conditioners. Type III allows technicians to work on low-pressure chillers. Universal certification allows technician to work on any systems covered by Type I, II, or III certifications.

Figure 4 ◆ EPA certification card.

3.0.0 ◆ REFRIGERANT IDENTIFICATION

The American Society of Heating, Refrigeration, and Air Conditioning Engineers (ASHRAE) has developed a standardized method of numbering refrigerants based on the number of atoms of the various elements contained in the refrigerant. Note that the numbering system begins on the right.

- The first digit on the right indicates the number of fluorine atoms.
- The second digit from the right indicates the number of hydrogen atoms plus one.
- The third digit from the right indicates the number of carbon atoms minus one. It is not used when equal to zero.
- The fourth digit from the right indicates the number of unsaturated carbon bonds in the compound. It is not used when equal to zero.

When carbon bonds are not all occupied by fluorine or hydrogen atoms, they are attached to chlorine atoms. There are a total of four carbon bonds available per carbon atom.

Let's use the refrigerant $CHClF_2$, to determine its number. The first number on the right (number of fluorine atoms) is 2. The second digit from the right (number of hydrogen atoms plus one) is 2 (1 hydrogen atom + 1 = 2). The third digit from the right (number of carbon atom minus one) is zero (there is only one carbon atom). Since all the carbon atoms are bonded to other elements, there are zero unsaturated carbon bonds available, so the fourth number is not used. The one carbon atom is bonded to two fluorine atoms, one hydrogen atom, and one chlorine atom. The two numbers resulting (2 and 2) provide the refrigerant number. In this case, $CHClF_2$ is HCFC-22. In the past, refrigerants were designated with the letter R followed by a number. For example, the CFC refrigerant 12 was called R-12. Today, letters that better represent the chemical makeup of the refrigerant are more commonly used (*Table 3*). What was once called R-12 is now called CFC-12.

3.1.0 Refrigerant Safety Classifications

Refrigerants are placed in four broad safety classifications. *Table 3* contains the classifications of common refrigerants.

Table 3 Old Versus New Refrigerants

Old	New	Classification
R–22	HCFC–22	II
R–410A	HFC–410A	II
R–407C	HFC–407C	II
R–123	HCFC–123	I
R–404A	HFC–404A	II
R–134a	HFC–134a	II

301T03.EPS

- *Class I, Liquid Refrigerants* – Refrigerants in this class have a normal boiling point above 68°F and are typically packaged in drums.
- *Class II, Low-Pressure Refrigerants* – Refrigerants in this class are compressed gases with a minimum cylinder pressure that does not exceed 500 psig.
- *Class III, High-Pressure Refrigerants* – Refrigerants in this class are compressed gases with a minimum cylinder pressure that exceed 500 psig.
- *Class IV, Flammable Refrigerants* – Refrigerants in this class have a flammability rating of 2 or 3 in *ANSI/ASHRAE Standard 34*. Flammable refrigerant cylinders have a red band on the top of the cylinder to indicate the flammability hazard.

3.1.1 Refrigerant Cylinder Identification

Refrigerant cylinders are color-coded to avoid confusion when selecting a refrigerant. *Table 4* contains the color codes used on cylinders containing some of the more popular refrigerants. For example, cylinders of the refrigerant HCFC-22 are color-coded light green, while HFC-410A cylinders are rose (pink) colored. Cylinders used to recover refrigerant have their own unique color code—gray with a yellow shoulder.

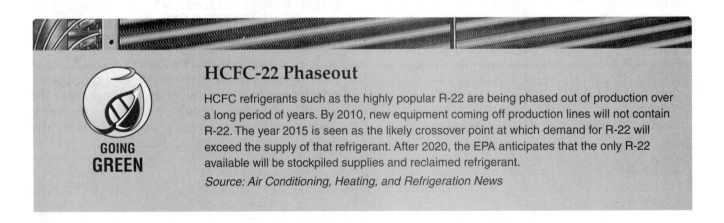

HCFC-22 Phaseout

HCFC refrigerants such as the highly popular R-22 are being phased out of production over a long period of years. By 2010, new equipment coming off production lines will not contain R-22. The year 2015 is seen as the likely crossover point at which demand for R-22 will exceed the supply of that refrigerant. After 2020, the EPA anticipates that the only R-22 available will be stockpiled supplies and reclaimed refrigerant.

Source: Air Conditioning, Heating, and Refrigeration News

GOING GREEN

Table 4 Refrigerant Cylinder Color Codes

Refrigerant	Cylinder Color
HCFC–22	Light green
HFC–410A	Pink
HFC–407C	Brown
HCFC–123	Blue-gray
HFC–404A	Orange
HFC–134a	Light blue
Recovery cylinder	Gray/yellow top

301T04.EPS

3.2.0 Refrigerant Cylinder Safety

Technicians will encounter three types of refrigerant cylinders: disposable, returnable/reusable, and recovery. Each type has its own unique handling requirement to ensure safety.

Disposable cylinders (*Figure 5*) are single-use cylinders. They are discarded after the refrigerant is used. It is against the law to reuse them in any way, including as a recovery cylinder. Break off the valve on the cylinder or puncture the rupture disk before disposing of the cylinder. To prevent damaging the cylinder during transport, it is recommended that it be kept in its cardboard shipping container to prevent it from rolling around. If the cardboard container is not available, restrain the cylinder in the upright position with light chains or bungee straps.

Returnable/reusable cylinders are to be returned to the distributor and/or refrigerant manufacturer for refilling after they become empty (*Figure 6*). Generally, they are too heavy for field service work and are usually used in fixed locations where they can be moved short distances to equipment with a special handcart. When not in use, these cylinders must be secured in an upright position using light chains. The protective cap must be screwed on top of the cylinder to protect the shutoff and relief valves.

Recovery cylinders are special cylinders designed for use with refrigerant recovery machines (*Figure 7*). Once a repair to the equipment is done,

301F06.EPS

Figure 6 ◆ Reusable cylinder.

the recovery cylinder can be used to recharge the recovered refrigerant back into the repaired equipment. Recovery cylinders must be dedicated for use with a specific refrigerant and have the refrigerant clearly identified on the cylinder. Never overfill a recovery cylinder. Generally, the

301F05.EPS

Figure 5 ◆ Disposable cylinders.

301F07.EPS

Figure 7 ◆ Recovery unit and cylinder.

maximum capacity of a recovery cylinder is 80 percent of the cylinder's gross weight (weight of cylinder and refrigerant). Some recovery cylinders are equipped with a float switch that shuts off the recovery machine when the maximum safe level in the cylinder is reached.

Manufacturers of all types of refrigerant cylinders design them so that when the gross weight is reached, about 20 percent of the volume in the cylinder remains for refrigerant expansion. Even though room for expansion is provided, technicians should avoid exposing refrigerant cylinders, especially full cylinders, to high temperatures. In desert daytime conditions, temperatures in a closed service truck can exceed 160°F. These high temperatures could cause the relief valve or rupture disc on the cylinder to open, releasing the refrigerant.

NOTE

The U.S. Department of Transportation (DOT) governs the construction and labeling of cylinders used for refrigerant storage and recovery. Recovery cylinders are colored gray with a yellow top. However, cylinders used to recover HFC-410A, even though they use that color scheme, must be specifically manufactured and labeled for use with that refrigerant because of its higher pressures.

4.0.0 ◆ REFRIGERANT COMPOSITION

Refrigerants can be made in two ways. Chemists can develop a pure refrigerant that consists of a single compound. HCFC-22 is an example of a single compound. In other cases, chemists will blend two different refrigerant compounds together to obtain a new refrigerant mixture with characteristics that are somewhat different from the characteristics of the two individual refrigerants. HFC-410A, a replacement for HCFC-22, is an example of a blended refrigerant. It consists of a 50-50 blend by weight of HFC-32 and HFC-125. Blends containing two refrigerants are called **binary blends**. Blends of three refrigerants are called **ternary blends**. Blended refrigerants fall into three categories: **azeotropes**, **zeotropes**, and **near-azeotropes**.

4.1.0 Azeotropes

An azeotropic blend behaves like a pure refrigerant. The refrigerant evaporates and condenses at one given pressure and temperature. From the technician's standpoint, working with an azeotrope is like working with a pure refrigerant. Many CFC refrigerants are azeotropes.

4.2.0 Zeotropes

A zeotropic blend never mixes chemically. As a result, the refrigerant evaporates and condenses over a temperature range, called the **glide**, at a constant pressure (*Figure 8*). The refrigerant HFC-407C is a zeotropic blend. When looking at a pressure-temperature (P-T) chart for a zeotropic blend, two points are given: the **bubble point** and the **dew point** (*Figure 9*). These are the points at which refrigerant starts evaporating (bubble point) and ends evaporating (dew point) when heat is added to the refrigerant, and the point that refrigerant starts condensing (dew point) and finishes condensing (bubble point) when heat is removed from the refrigerant.

When heat is added to the refrigerant, the dew point can be looked at as the point at which the blend stops evaporating, and superheating of the vapor begins. Any vapor temperature measured above the dew point would indicate superheat.

When heat is removed from the refrigerant, the dew point can also be looked at as the point at which liquid refrigerant first begins to condense from vapor.

When heat is added to refrigerant, bubble point can be looked at as the point at which refrigerant begins to evaporate. Any temperature measured below the bubble point would indicate subcooling. When heat is removed from the refrigerant, the bubble point can be looked at as the point where all vapor has condensed to a liquid. When calculating superheat with a zeotropic blend, use the dew point. When calculating subcooling, use the bubble point.

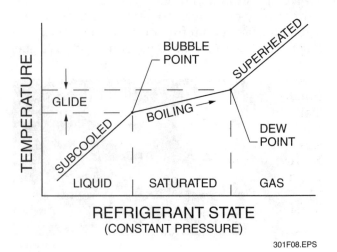

301F08.EPS

Figure 8 ◆ Temperature glide.

°F	°C	124	134a	12†	MP39 Bubble	MP39 Dew	500†	MP66 Bubble	MP66 Dew	409A Bubble	409A Dew	22
−40	−40.0	22.1	14.8	11.0	8.1	13.2	7.6	6.5	11.8	5.7	13.6	6
−35	−37.2	20.9	12.5	8.4	5.1	10.7	4.5	3.3	9.1	2.4	11.1	6
−30	−34.4	19.4	9.8	5.5	1.7	7.9	1.1	0.2	6.1	0.6	8.4	9
−25	−31.7	17.8	6.9	2.4	1.0	4.8	1.3	2.1	2.8	2.6	5.4	4
−20	−28.9	16.0	3.7	0.5	3.0	1.4	3.3	4.3	0.5	4.7	2.0	0.2
−15	−26.1	14.0	0.1	2.4	5.2	1.2	5.5	6.6	2.5	7.1	0.8	3.2
−10	−23.3	11.8	1.9	4.5	7.7	3.3	7.9	9.2	4.7	9.6	2.8	6.5
−5	−20.6	9.3	4.1	6.7	10.3	5.5	10.6	12.0	7.1	12.4	5.0	20.1
0	−17.8	6.6	6.5	9.1	13.2	8.0	13.4	15.1	9.7	15.5	7.5	24.0
5	−15.0	3.6	9.1	11.7	16.3	10.7	16.5	18.4	12.6	18.7	10.1	28.3
10	−12.2	0.3	11.9	14.6	19.7	13.7	19.9	22.0	15.8	22.3	13.0	32.8
15	−9.4	1.6	15.0	17.7	23.4	16.9	23.5	25.9	19.2	26.1	16.1	37.8
20	−6.7	3.6	18.4	21.0	27.4	20.4	27.4	30.1	23.0	30.3	19.5	43.1
25	−3.9	5.7	22.1	24.6	31.7	24.2	31.7	34.6	27.0	34.7	23.2	48.8
30	−1.1	8.0	26.1	28.4	36.4	28.3	36.2	39.5	31.4	39.5	27.1	55.0
35	1.7	10.5	30.4	32.5	41.3	32.8	41.0	44.8	36.1	44.6	31.4	61.5
40	4.4	13.2	35.0	36.9	46.6	37.6	46.2	50.4	41.1	50.1	36.1	68.6
45	7.2	16.1	40.1	41.6	52.4	42.7	51.8	56.4	46.6	56.0	41.0	76.1
50	10.0	19.3	45.4	46.6	58.5	48.2	57.7	62.8	52.4	62.2	46.3	84.1
55	12.8	22.7	51.2	51.9	65.0	54.1	64.0	69.6	58.7	68.9	52.0	92.6
60	15.6	26.3	57.4	57.6	71.9	60.4	70.7	76.9	65.4	75.9	58.1	101.6
65	18.3	30.2	64.0	63.7	79.3	67.2	77.8	84.7	72.5	83.4	64.7	111.2
70	21.1	34.4	71.1	70.1	87.1	74.4	85.4	92.9	80.1	91.4	71.6	121.4
75	23.9	38.9	78.7	76.8	95.4	82.1	93.4	101.6	88.2	99.8	79.0	132.2
80	26.7	43.7	86.7	84.0	104.2	90.2	101.9	110.9	96.8	108.7	86.9	143.6
85	29.4	48.8	95.2	91.6	113.6	98.9	110.8	120.7	106.0	118.1	95.2	155.7
90	32.2	54.3	104.3	99.6	123.4	108.1	120.3	131.0	115.6	128.0	104.1	168.4
95	35.0	60.1	113.9	108.0	133.8	117.9	130.3	141.9	125.9	138.4	113.4	181.8
100	37.8	66.2	124.2	116.9	144.8	128.2	140.8	153.4	136.8	149.4	123.4	195.9
105	40.6	72.7	135.0	126.3	156.4	139.1	151.9	165.5	148.2	161.0	133.8	210.8
100	43.3	79.6	146.4	136.1	168.5	150.6	163.5	178.3	160.4	173.1	144.9	226.4
115	46.1	86.9	158.4	146.4	181.3	162.8	175.7	191.6	173.1	185.9	156.6	242.8
120	48.9	94.6	171.2	157.3	194.7	175.6	188.6	205.7	186.6	199.2	168.9	260.0
125	51.7	102.8	184.6	168.6	208.8	189.0	202.1	220.4	200.7	213.2	181.8	278.0
130	54.4	111.3	198.7	180.5	223.6	203.2	216.2	235.9	215.6	227.8	195.5	296.9
135	57.2	120.4	213.6	193.0	239.1	218.1	231.0	252.0	231.2	243.1	209.8	316.7
140	60.0	129.9	229.2	206.0	255.3	233.8	246.5	268.9	247.6	259.1	224.8	337.4
145	62.8	139.9	245.7	219.7	272.2	250.2	262.7	286.6	264.8	275.7	240.5	359.0
150	65.6	150.4	262.9	233.9	289.8	267.4	279.7	305.0	282.8	293.1	257.0	381.7

301F09.EPS

Figure 9 ◆ P-T chart for azeotropes.

Another problem with zeotropic blends is **fractionation**. Fractionation becomes a problem if a leak occurs in the system. The different refrigerants in the blend boil off at different rates and leak at different rates, upsetting the blend proportions in the remaining refrigerant. When a technician approaches a leaking system that contains a zeotropic blend, he or she does not know the state of the remaining refrigerant.

For that reason, any remaining charge should be recovered, the leak repaired, and the unit evacuated and recharged. When adding charge, zeotropic blends must always be in the liquid state. If charge is added to an operating system through the suction service port, the liquid must be passed through a metering orifice before it enters the unit to ensure it is in vapor form. For best results in charging zeotropic blends, always follow the equipment manufacturer's charging instructions.

4.3.0 Near-Azeotropes

Near-azeotropes can best be described as zeotropes with azeotropic properties. The refrigerant HFC-410A is an example of a near-azeotrope. Like zeotropes, near-azeotropes exhibit glide characteristics, but they are usually minimal to the point that they are not a factor for technicians to consider when servicing a unit containing such a refrigerant. For example, the charging charts that one equipment manufacture provides for its line of residential products using R-410A look identical (except for pressures) to charging charts for similar equipment using HCFC-22, an azeotrope. The same holds true for P-T charts for the two refrigerants (*Figure 10*). On the chart shown, there are no bubble points or dew points as seen on a P-T chart for a zeotrope like HFC-407C. Like zeotropes, near-azeotrope refrigerants must be charged into a system as a liquid.

4.4.0 Using Pressure-Temperature (P-T) Charts

A pressure-temperature (P-T) chart shows the pressure and temperature relationships at saturation for a specific refrigerant. Saturation is defined as the point at which a refrigerant holds as much vapor as it is capable of holding. For any given temperature there is a corresponding pressure, and vice versa. If a temperature does not correspond to its pressure, the refrigerant may be in a superheated or subcooled state. Superheat is defined as heat added to a refrigerant after it is fully vaporized. Subcooling is defined as heat removed from a refrigerant after it has fully condensed to a liquid. A P-T chart can be used to calculate both. Use the P-T chart in *Figure 10* to perform the following calculations.

4.4.1 Calculating Superheat

The tools and instruments needed to calculate superheat include a gauge manifold, an accurate electronic thermometer, and a P-T chart for the refrigerant in the system.

Step 1 With power off, connect the gauge manifold to the suction line and liquid line gauge ports. Attach the thermometer probe to the suction line as close to the compressor as possible. Insulate the probe for a more accurate reading.

Step 2 Turn the system on and let it run for about 10 minutes to allow temperatures

and pressures to stabilize. Record the suction pressure and the suction line temperature. On this HCFC-22 unit, suction pressure is slightly over 71 psig and the suction line temperature is 50°F.

Step 3 Consult the HCFC-22 P-T chart and find the pressure as close to 71 psig as possible. In this case, it is 71.5 psig. To the left of the pressure reading, read the corresponding saturated temperature, 42°F in this example.

Step 4 Calculate superheat by subtracting the saturated temperature from the measured temperature. The difference is superheat.

50°F (measured temp.) – 42°F
(saturated temp.) = 8° superheat

Superheat can determine if a system is charged properly if other conditions, such as temperature, humidity, and the required superheat for the equipment are known. Based on the known conditions and the measured superheat, refrigerant may have to be added or removed from the system to obtain a correct charge.

4.4.2 Calculating Subcooling

The tools and instruments needed to calculate subcooling are the same as those used to calculate superheat; a gauge manifold, an accurate electronic thermometer, and a P-T chart for the refrigerant in the system.

Step 1 With power off, connect the gauge manifold to the suction line and liquid line gauge ports. Attach the thermometer probe to the liquid line close to the exit of the condenser coil and near the liquid line gauge port. Insulate the probe for a more accurate reading.

Step 2 Turn the system on and let it run for about 10 minutes to allow temperatures and pressures to stabilize. Record the liquid line pressure and the liquid line temperature. On this HCFC-22 unit, liquid line pressure is 250 psig and the liquid line temperature is 107°F.

Step 3 Consult the HCFC-22 P-T chart and find the pressure as close to 250 psig as possible. In this case, it is 249.5 psig. To the left of the pressure reading, read the corresponding saturated temperature, 117°F in this example.

°F	R-410A	R-22	°F	R-410A	R-22	°F	R-410A	R-22	°F	R-410A	R-22
−40°	10.8	0.6	10	62.2	32.8	60	169.6	101.6	110	364.1	226.4
−39°	11.5	1.0	11	63.7	33.8	61	172.5	103.5	111	369.1	229.6
−38°	12.1	1.4	12	65.2	34.8	62	175.4	105.4	112	374.2	232.8
−37°	12.8	1.8	13	66.8	35.8	63	178.4	107.3	113	379.4	236.1
−36°	13.5	2.2	14	68.3	36.8	64	181.5	109.3	114	384.6	239.4
−35°	14.2	2.6	15	69.9	37.8	65	184.5	111.2	115	389.9	242.8
−34°	14.9	3.1	16	71.5	38.8	66	187.6	113.2	116	395.2	246.1
−33°	15.6	3.5	17	73.2	39.9	67	190.7	115.3	117	400.5	249.5
−32°	16.3	4.0	18	74.9	40.9	68	193.9	117.3	118	405.9	253.0
−31°	17.1	4.5	19	76.6	42.0	69	197.1	119.4	119	411.4	256.5
−30°	17.8	4.9	20	78.3	43.1	70	200.4	121.4	120	416.9	260.0
−29°	18.6	5.4	21	80.0	44.2	71	203.6	123.5	121	422.5	263.5
−28°	19.4	5.9	22	81.8	45.3	72	207.0	125.7	122	428.2	267.1
−27°	20.2	6.4	23	83.6	46.5	73	210.3	127.8	123	433.9	270.7
−26°	21.1	6.9	24	85.4	47.6	74	213.7	130.0	124	439.6	274.3
−25°	21.9	7.4	25	87.2	48.8	75	217.1	132.2	125	445.4	278.0
−24°	22.7	8.0	26	89.1	50.0	76	220.6	134.5	126	451.3	281.7
−23°	23.6	8.5	27	91.0	51.2	77	224.1	136.7	127	457.3	285.4
−22°	24.5	9.1	28	92.9	52.4	78	227.7	139.0	128	463.2	289.2
−21°	25.4	9.6	29	94.9	53.7	79	231.3	141.3	129	469.3	293.0
−20°	26.3	10.2	30	96.8	55.0	80	234.9	143.6	130	475.4	296.9
−19°	27.2	10.8	31	98.8	56.2	81	238.6	146.0	131	481.6	300.8
−18°	28.2	11.4	32	100.9	57.5	82	242.3	148.4	132	487.8	304.7
−17°	29.2	12.0	33	102.9	58.8	83	246.0	150.8	133	494.1	308.7
−16°	30.1	12.6	34	105.0	60.2	84	249.8	153.2	134	500.5	312.6
−15°	31.1	13.2	35	107.1	61.5	85	253.7	155.7	135	506.9	316.7
−14°	32.2	13.9	36	109.2	62.9	86	257.5	158.2	136	513.4	320.7
−13°	33.2	14.5	37	111.4	64.3	87	261.4	160.7	137	520.0	324.8
−12°	34.2	15.2	38	113.6	65.7	88	265.4	163.2	138	526.6	329.0
−11°	35.3	15.9	39	115.8	67.1	89	269.4	165.8	139	533.3	333.2
−10°	36.4	16.5	40	118.1	68.6	90	273.5	168.4	140	540.1	337.4
−9°	37.5	17.2	41	120.3	70.0	91	277.6	171.0	141	547.0	341.6
−8°	38.6	17.9	42	122.7	71.5	92	281.7	173.7	142	553.9	345.9
−7°	39.8	18.7	43	125.0	73.0	93	285.9	176.4	143	560.9	350.3
−6°	40.9	19.4	44	127.4	74.5	94	290.1	179.1	144	567.9	354.6
−5°	42.1	20.1	45	129.8	76.1	95	294.4	181.8	145	575.1	359.0
−4°	43.3	20.9	46	132.2	77.6	96	298.7	184.6	146	582.3	363.5
−3°	44.5	21.7	47	134.7	79.2	97	303.0	187.4	147	589.6	368.0
−2°	45.7	22.4	48	137.2	80.8	98	307.5	190.2	148	596.9	372.5
−1°	47.0	23.2	49	139.7	82.4	99	311.9	193.0	149	604.4	377.1
0	48.3	24.0	50	142.2	84.1	100	316.4	195.9	150	611.9	381.7
1	49.6	24.9	51	144.8	85.7	101	321.0	198.8			
2	50.9	25.7	52	147.4	87.4	102	325.6	201.8			
3	52.2	26.5	53	150.1	89.1	103	330.2	204.7			
4	53.6	27.4	54	152.8	90.8	104	334.9	207.7			
5	55.0	28.3	55	155.5	92.6	105	339.6	210.8			
6	56.3	29.2	56	158.2	94.4	106	344.4	213.8			
7	57.8	30.1	57	161.0	96.1	107	349.3	216.9			
8	59.2	31.0	58	163.8	98.0	108	354.2	220.0			
9	60.7	31.9	59	166.7	99.8	109	359.1	223.2			

Source: Honeywell. The above data, including recommendations for application and use of R-410A (Genetron® AZ-20®) are available at http://www.genetron.com.

301F10.EPS

Figure 10 ◆ Comparison of R-410A and R-22.

Step 4 Calculate subcooling by subtracting the measured temperature from the saturated temperature. The difference is subcooling.

$$117°F \text{ (saturated temp.)} - 107°F \text{ (measured temp.)} = 10° \text{ subcooling}$$

Subcooling can determine if a system is charged properly if other conditions, such as temperature, and the required subcooling for the equipment are known. Based on the known conditions and the measured subcooling, refrigerant may have to be added or removed from the system to obtain a correct charge.

5.0.0 ◆ REFRIGERANT LEAKS

Refrigerants cannot be allowed to enter the atmosphere. They damage the environment, and it is against the law to do so in the United States. The Clean Air Act of 1990 has significantly impacted the HVAC industry in the following ways:

- Refrigerants cannot be released into the atmosphere, either voluntarily or involuntarily. Persons who voluntarily release refrigerant into the atmosphere are subject to fines and/or imprisonment if convicted. Minor refrigerant releases that occur with service procedures such as disconnecting gauge hoses (called de-minimus releases) are allowed.
- Persons handling refrigerants must be EPA certified.
- All transactions related to refrigerants (sale, use, disposal) must be recorded.
- Leaks in large systems containing over 50 pounds of refrigerant must be repaired.

5.1.0 Finding Leaks

Poor equipment performance is often a clue that a system has lost refrigerant. This can be confirmed by checking system temperatures and pressures. If it is confirmed that the charge is low, the leak must first be found and repaired before the system can be recharged. Some leaks can be located using the physical senses.

- Large leaks can often be heard. The escaping refrigerant will hiss like a leaking tire.
- Some large leaks emit a stream of refrigerant that can be felt. If combined with oil, the stream may be visible.
- Oil in the system will leak out with the refrigerant, leaving an oil stain at the site of the leak.

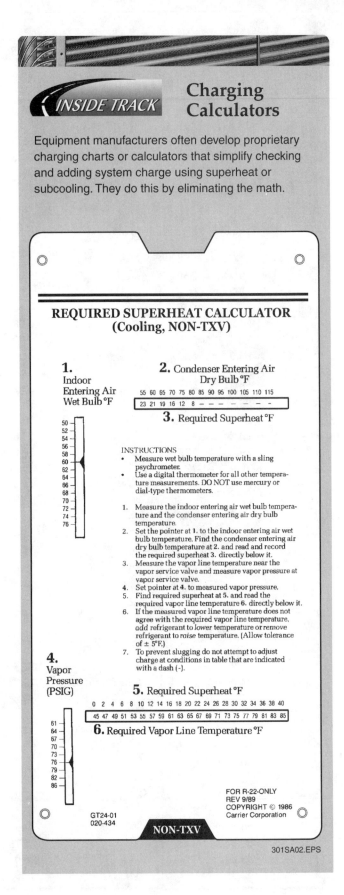

- The refrigerant may have a distinct odor. Refrigerant from a system that has experienced a compressor motor burnout has a distinct rotten-egg smell. Ammonia has its own distinct and pungent odor.

Leaks can occur anywhere in the refrigeration system, but some places are more susceptible to leaks than others. The following are some common leak sites:

- Factory- or field-made brazed joints. Leaks at factory-made joints are usually caught before the unit leaves the production line, but leaks occasionally slip through. Sometimes the jostling a unit encounters in shipping from the factory can cause a marginal joint to leak. Poorly made field repair joints or poorly made joints on interconnecting tubing used with spit systems are common leak sites.

- Mechanical fittings such as flare or compression fittings used on factory- or field-installed filter-driers, solenoid valves, and other refrigerant piping accessories are prime leak sites. Any bolted or gasketed joint such as found on semi-hermetic compressors can leak (*Figure 11*).

- Unsecured refrigerant lines that are allowed to move or vibrate can eventually crack or break from metal fatigue. These are often the result of poor field piping practices.

- Coils can leak due to factory defects, but are more likely to leak from corrosion or vandalism. Condenser coils in seaside environments are prone to corrosion. Dirty evaporator coils can corrode in the presence of the water that condenses on their fins during normal operation. Vandalism often occurs on units installed at ground level or in other easily accessible areas.

5.2.0 Isolating Leaks

Once a leak is suspected, there are several methods and devices that can isolate the leak if the physical senses are unable to find it. They all work on the assumption that the system has enough pressure to cause the refrigerant to leak. Some of the common methods used to isolate leaks include:

- *Leak-detecting fluids* – These fluids are applied to a suspected leak site with an applicator and form bubbles in the presence of a leak. Their formulation allows them to bubble even in the presence of a very small leak if the fluid is left on long enough (*Figure 12*).

301F11.EPS

Figure 11 ◆ Service valves.

301F12.EPS

Figure 12 ◆ Leak detecting fluid.

- *Halide torch* – This leak detector has been around for decades. It samples air from around a suspected leak and passes it through a torch flame surrounding a copper element. If a halogen such as chlorine is present, the flame will turn green. It is not as sensitive as electronic leak detectors and is not widely used any more.
- *Leak-detecting dyes* – When added to the refrigerant or oil in an operating system, they display a bright color at the site of the leak. Some dyes require an ultraviolet light that causes the dye to glow at the site of the leak. Check the equipment manufacturer's literature before using a leak-detecting dye.
- *Electronic leak detectors* – These leak detectors (*Figure 13*) draw in a sample of air surrounding a suspected leak site. If refrigerant is present in

the sample, the detector will give an audible signal such as a buzz or squeal and/or a visual signal such as a blinking light. The intensity of the indicating signal often indicates the severity of the leak. Electronic leak detectors must be compatible with the refrigerant they are attempting to detect. Other electronic leak detectors are basically electronic amplifiers that use a microphone to pick up the sound of the leaking refrigerant. They are not especially effective for field service work.

5.2.1 Pressurizing for Leak Detection

If there is insufficient refrigerant in the system to cause the refrigerant to leak, the system must be pressurized to a level that will allow a leak. Use the following procedure to pressurize a system:

Step 1   Recover all remaining refrigerant from the system until the pressure gauge reads zero psig.

Step 2   Pressurize the system to about 10 psig of HCFC-22 refrigerant vapor. (This does not apply to ammonia systems.) Then, increase the system pressure by adding regulated dry nitrogen (*Figure 14*) to about 125 psig. The nitrogen provides pressure, and the refrigerant in the nitrogen/refrigerant mixture is at a level sufficient to be detected with an electronic leak detector.

Step 3   Use an electronic leak detector to probe suspected areas for the leak.

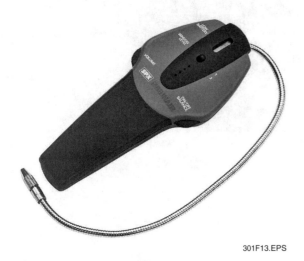

301F13.EPS

Figure 13 ◆ Electronic leak detector.

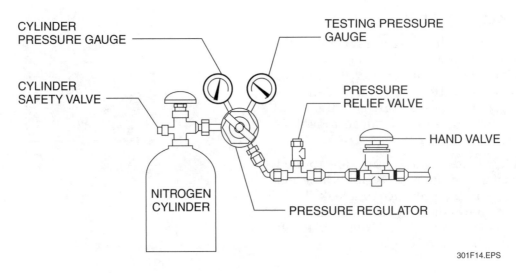

301F14.EPS

Figure 14 ◆ Nitrogen setup.

HCFC-22 Availability

Effective January 10, 2010, HCFC-22 can only be produced for use in systems manufactured before that date. Following that date, and until the complete phaseout in 2020, the Clean Air Act allows HCFC-22 to be produced only for the purpose of servicing existing equipment. However, the production level is limited, and is scheduled to shrink every year. Thus, it is possible that the demand for this refrigerant may exceed the available supply. In that case, there will not be enough of the refrigerant to go around.

Step 4 If a leak detecting fluid is being used instead of an electronic leak detector, simply pressurize the system with regulated dry nitrogen to 125 psig and look for bubbles at the leak site. No refrigerant is needed.

WARNING!

Never pressurize a system with unregulated nitrogen. The cylinder pressure (over 2,000 psig) can rupture system components and cause personal injury. Never pressurize a system with oxygen, compressed air, or any flammable gas. Such a mixture can explode on contact with oil or other flammable material, causing personal injury.

6.0.0 ◆ LUBRICATING OILS

Oil is required in a refrigeration system compressor to lubricate the moving parts. Oil also helps provide a fluid seal that helps to separate the high and low sides of the system, and helps dampen mechanical noises within the compressor. It acts as a coolant, and even flushes tiny metal particles off moving parts and into the oil sump in the bottom of the compressor. Lubrication, the primary purpose of oil in a refrigeration system, is a process in which a material, called a lubricant, is placed between two moving surfaces to reduce friction. In the case of a moving compressor crankshaft and its fixed bearings, a microscopically thin film of oil between these two surfaces reduces friction and allows the crankshaft to rotate freely without wearing the bearing surface. The most common lubricants take the form of liquids (oil) and solids (grease). Specialized lubricants are also available in gaseous form.

For many years, mineral oils were used with systems that used CFC and HCFC refrigerants. Additives used to increase the lubricating qualities and stability of mineral oils made them adequate for most applications. The new HFC refrigerants that are replacing CFC and HCFC refrigerants are not compatible with mineral oils. As a result, different types of synthetic oils have been developed that are compatible with this new family of refrigerants. All refrigerant oils must have good lubricating qualities and be **miscible** with the refrigerant with which it is being used. Miscibility allows the oil and refrigerant to dissolve within each other, allowing the mixture to move around the system and ensuring oil return to the compressor. Additionally, oil must have a **viscosity** that allows it to flow under all the temperature conditions in which the compressor must operate. Compressor oil must also be free of water.

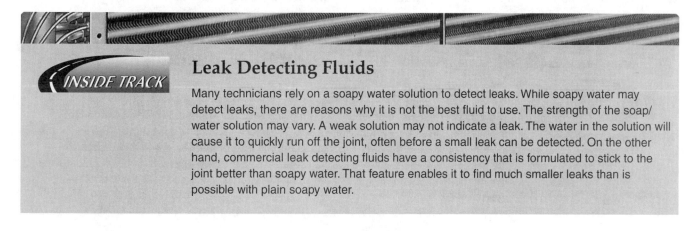

INSIDE TRACK

Leak Detecting Fluids

Many technicians rely on a soapy water solution to detect leaks. While soapy water may detect leaks, there are reasons why it is not the best fluid to use. The strength of the soap/water solution may vary. A weak solution may not indicate a leak. The water in the solution will cause it to quickly run off the joint, often before a small leak can be detected. On the other hand, commercial leak detecting fluids have a consistency that is formulated to stick to the joint better than soapy water. That feature enables it to find much smaller leaks than is possible with plain soapy water.

6.1.0 Lubricating Oil Properties

All lubricating oils, whether mineral or synthetic, must possess properties that make them useful as a lubricant in a refrigeration system. Desirable properties include the following:

- Stability
- Viscosity
- **Pour point**
- **Dielectric strength**
- **Floc point**
- **Flash point**
- Compatibility
- Foaming

6.1.1 Stability

Refrigerant oil, like refrigerant, must be stable over the normal range of temperature and pressure conditions in which it must function. It must not break down or combine with refrigerant or other materials in the compressor or refrigerant piping to form harmful byproducts.

6.1.2 Viscosity

Viscosity can be described as the resistance of a fluid to flow. Honey has a higher viscosity than water. As a result, it has a harder time flowing from a container than water does. Viscosity of a fluid can vary, depending on the temperature of the fluid. Warm a jar of honey and it will flow from the jar much faster because it is less viscous. When applied to refrigerant oil, viscosity must be such that the oil will flow and provide proper lubrication over the wide range of temperatures encountered in refrigeration and air conditioning systems.

6.1.3 Pour Point

Pour point is an oil quality related to viscosity. It can be defined as the temperature at which oil

first starts to flow. In low-temperature refrigeration systems, it is important to have oil with a low pour point to ensure that lubrication takes place at the low operating temperatures.

6.1.4 Dielectric Strength

Dielectric strength is the ability of oil or any material to resist breaking down in the presence of voltage. Voltages of varying values exist inside the shells of operating compressors. The compressor oil must have a dielectric strength high enough so that it does not break down or arc under normal operating voltage conditions. Contaminants such as water can lower the dielectric strength of the oil and cause it to break down.

6.1.5 Floc Point

The floc point of oil is the temperature at which a 90/10 mixture of oil/refrigerant forms a cloudy or flocculent precipitate of wax in the mixture. Wax is an undesirable material to have in a refrigeration system. It can plug metering orifices and prevent check valves from functioning properly. Wax can form because there are impurities in mineral oils that cannot be refined out. The floc point of oil should be low enough that wax does not form at the normal operating temperatures encountered in the system in which the oil is used.

6.1.6 Flash Point

The flash point of oil is the temperature at which heated oil vapors burst into flame. It is a measure of the flammability of the lubricant. Ideally, compressor lubricant oils should have a high flash point for safety reasons.

6.1.7 Material Compatibility

Compressor lubricating oils must be compatible with the various materials the compressor is

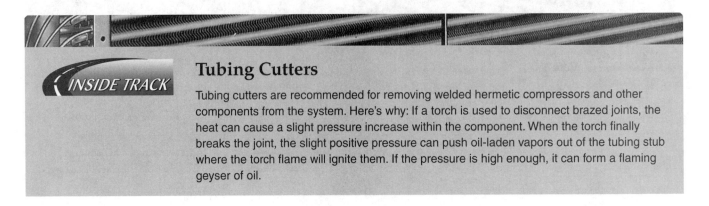

INSIDE TRACK

Tubing Cutters

Tubing cutters are recommended for removing welded hermetic compressors and other components from the system. Here's why: If a torch is used to disconnect brazed joints, the heat can cause a slight pressure increase within the component. When the torch finally breaks the joint, the slight positive pressure can push oil-laden vapors out of the tubing stub where the torch flame will ignite them. If the pressure is high enough, it can form a flaming geyser of oil.

made of, the metals and other materials used in the refrigeration system, as well as the refrigerant and other chemicals in the system. Under normal operating conditions, lubricating oils should not cause corrosion or react with materials such as plastics, desiccants, and refrigerant to form contaminants.

6.1.8 Foaming

Foaming of the compressor oil is a normal occurrence caused by mechanical movement in the compressor and/or dissolved refrigerant boiling out of the oil. If foaming becomes excessive, the foam can reduce the lubricating qualities of the oil, leading to mechanical wear or failure of the compressor. Foaming can be reduced by maintaining oil in the compressor at the correct level, and by using a crankcase heater to prevent refrigerant from migrating to the compressor crankcase during off cycles. Refrigerant oils typically contain antifoaming additives.

6.2.0 Oil Types

Compressors in air conditioning and refrigeration systems have several different types of oils available. They include mineral oils and synthetic oils.

6.2.1 Mineral Oils

Mineral oils have traditionally been used in systems using CFC and HFC refrigerants. Mineral oils have a paraffin base, meaning they are derived from the same crude oil from which gasoline and diesel fuel are refined. Mineral oils are available in a wide range of viscosities and contain specific additives to make them compatible with the refrigerant and the application in which they are being used.

6.2.2 Synthetic Oils

The predominant synthetic oils are glycol-, ester-, ether-, or **alkylbenzene**-based. The glycol-, ester-, and ether-based oils are widely used with the newer HFC refrigerants. Alkylbenzene oil is closely related to mineral oil and used in similar applications.

Polyalkylene glycol (PAG) oil is widely used in automotive air conditioning systems where HFC-134a is the refrigerant. When converting CFC-12 automotive air conditioners to HFC-134a, the compressor oil must be drained and new PAG oil added because HFC-134a is not compatible with mineral oils.

Polyolester (POE) oil is very compatible with most of the newer HFC refrigerants and is widely used in those applications. Like PAG oils, it is not compatible with mineral or alkylbenzene oils.

Polyvinyl ether (PVE) oil has properties similar to mineral oils but can be used with HFC refrigerants. It has an advantage over POE and PAG oils in that it is not **hygroscopic.**

Alkylbenzene oil is compatible with mineral oil, but has better oil return properties. It is widely used in low-temperature applications. It is not compatible with synthetic oils.

Refrigerant compatibility with the various refrigerant oils varies. The equipment manufacturer or refrigerant manufacturer should be consulted for recommendations on what refrigerant and oil are best suited for the application (*Table 5*). For example, HCFC-123 can use mineral oil, alkylbenzene oil, or synthetic oils such as POE or PVE. On the other hand, most HFCs such as HFC-134a and HFC-410A must be used with a synthetic such as POE or PVE only.

7.0.0 ◆ OIL AND THE REFRIGERATION SYSTEM

Lubricating oil is located in the sump of the compressor shell where it is used to lubricate the moving parts of the compressor. Two lubrication methods are used with reciprocating compressors (*Figure 15*). In the splash lubrication system, oil from the sump is splashed onto the cylinder walls and bearing surfaces of the compressor during each revolution of the crankshaft while the compressor is running. Some compressor connecting rods have little dips or scoops attached to the lower end that scoop up the oil and splash it around to the other parts. The second and most popular method of lubrication is the pressure (force-feed) lubrication system. Arrows in the figure show the direction of oil flow. This method is used in all sizes of compressors. It uses an internal oil pump mounted on the end of the crankshaft that forces oil through the

Table 5 Refrigerants and Oils

Refrigerant	Oil Requirement
HCFC–22	Mineral oil
HFC–410A	Synthetic (POE, PVE)
HFC–407C	Synthetic (POE, PVE)
HCFC–123	Mineral, alkylbenzene, or synthetic (POE, PVE)
HFC–404A	Synthetic (POE, PVE)
HFC–134a	Synthetic (PAG, POE)

301T05.EPS

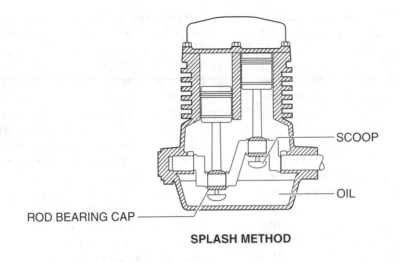

SCOOP

OIL

ROD BEARING CAP

SPLASH METHOD

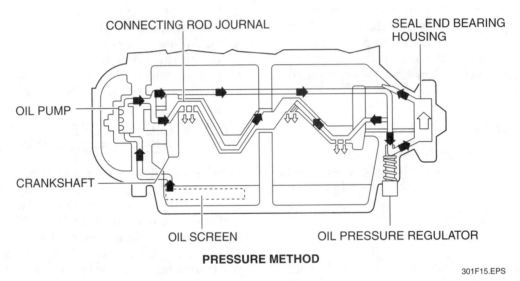

CONNECTING ROD JOURNAL

SEAL END BEARING HOUSING

OIL PUMP

CRANKSHAFT

OIL SCREEN

OIL PRESSURE REGULATOR

PRESSURE METHOD

301F15.EPS

Figure 15 ◆ Compressor lubrication systems.

crankshaft to the bearing surfaces. In some cases, the connecting rods are drilled so that pressurized oil is also supplied to the piston pins. An oil screen may be located in the bottom of the sump to filter the oil taken from there before it is pumped to the bearing surfaces. Some systems use an oil pressure regulator to prevent excessive oil pressure that could result in high power consumption, loss of oil, and damage to the compressor.

7.1.0 Refrigeration System Piping

As a part of normal operation, the compressor discharges a small amount of oil with the refrigerant. During cold starts, the refrigerant/oil mixture in the crankcase can foam as refrigerant boils out. This results in more oil being lost. To ensure that this oil returns to the compressor, the suction and liquid lines must be properly sized to maintain enough velocity to move the oil up vertical piping sections, called risers. Problems occur when the lines are oversized, resulting in the oil settling in low spots. Suction lines should be pitched in the direction of flow (toward the compressor) to maintain flow back to the compressor and to avoid any backflow of oil during shutdown. If a split-system has the evaporator installed above the condenser, install a trap in the suction line piping as it leaves the evaporator (*Figure 16*). The velocity of the suction gas through the trap will help carry the oil out for return to the compressor. When installing refrigeration system piping, always follow the equipment manufacturer's installation instructions to avoid any oil return problems. Other components in the system such as receivers and accumulators can trap oil. Both of these components have features designed to aid in oil return, and unless the component is defective or improperly sized or installed, these components should not impede oil return.

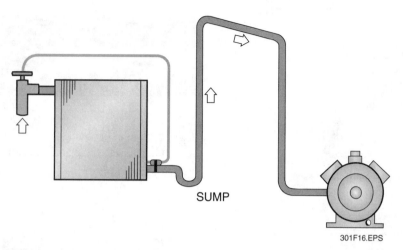

Figure 16 ◆ Trapped suction line.

7.2.0 Refrigeration System Contamination

Ideally, the only two substances that should be inside a refrigeration system are refrigerant and refrigerant oil. Anything else has to be considered a contaminant.

Typical contaminants include air, water, acid, and solid contaminants such as metal particles. These contaminants can enter a system in a variety of ways and end up in both the oil and the refrigerant.

Air usually enters the system during installation when the installer neglects to properly evacuate the system. Air is a non-condensable, which causes head pressure to increase and may cause erratic operation. A bigger problem with air is that it may contain moisture, a harmful contaminant. The best way to avoid air contamination is to thoroughly evacuate the system during installation.

Water can enter the system in many ways, including through the air, through the use of incorrect piping materials, careless installation practices, or through leaks in refrigerant-to-water heat exchangers. The use of ACR-grade refrigerant tubing and proper system evacuation can help prevent water from entering the system. Evacuation and the use of filter-driers can remove water from a contaminated system.

Acid forms within the system as the result of a compressor motor burnout or by contaminants combining under high temperature with refrigerant, oil, and water in the system. Proper evacuation can prevent moisture from entering the system. Problems causing abnormally high temperatures should be corrected. Once acid is in the system, it must be cleaned from the system using filter-driers and/or by replacing the compressor oil.

Solids can take the form of metal particles formed during the normal running and wear of a compressor or leftovers from the compressor manufacturing process. Other solids particles can come from improper repair procedures such as cutting tubing with a hacksaw, or from components such as a filter drier core. Solid particles can plug orifices and cause all kinds of damage within the refrigeration system. Some compressors are equipped with magnetic drain plugs in the compressor sump that can trap any ferrous particles. Draining the compressor oil (if possible) can remove some solid contaminants. Filter screens in the compressor or other parts of the system, as well as filter-driers (*Figure 17*) can trap solid contaminants.

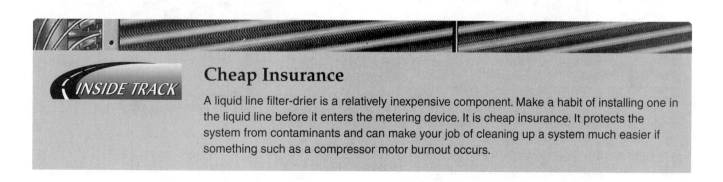

Cheap Insurance

A liquid line filter-drier is a relatively inexpensive component. Make a habit of installing one in the liquid line before it enters the metering device. It is cheap insurance. It protects the system from contaminants and can make your job of cleaning up a system much easier if something such as a compressor motor burnout occurs.

301F17.EPS

Figure 17 ◆ Liquid line filter-drier cutaway.

7.2.1 Oil Testing

The oil and/or refrigerant can be tested for the presence of refrigeration system contaminants. An acid/moisture tester (*Figure 18*) can be used to sample the refrigerant in a gaseous state. Crystals in a sealed tube change color in the presence of acid and/or moisture. Acid test kits (*Figure 19*) require that a sample of the compressor oil be obtained. When mixed with other chemicals, the sample will turn color if contaminants are present. In some cases where very large compressors are used, oil samples are periodically taken and sent to a laboratory for a thorough analysis.

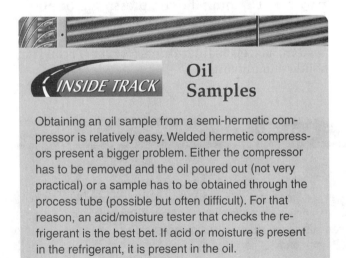
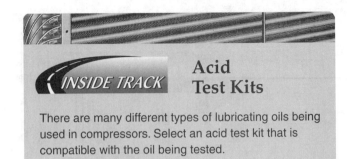
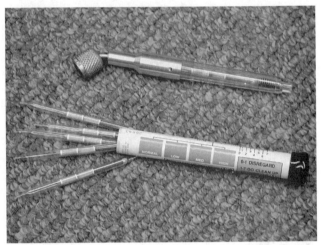

301F18.EPS

Figure 18 ◆ Acid/moisture test kit.

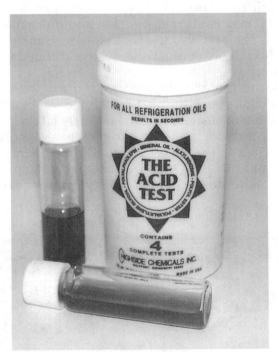

301F19.EPS

Figure 19 ◆ Acid test kit for refrigerant oil.

7.2.2 Oil Additives and Treatments

There are a number of different aftermarket oil additives available that claim to cure various compressor problems or to increase compressor performance. Some of these products may do what they claim to do; others may do nothing but are harmless; and still others may damage the compressor or other system components. Before using any additive or oil treatment, always consult the compressor manufacturer for guidance. Because some additives may damage the compressor or other components, the equipment manufacturer may void the compressor warranty if additives are used.

8.0.0 ◆ OIL HANDLING GUIDELINES

Refrigerant oils must be handled properly to avoid personal injury and to avoid damaging materials and the environment. When in doubt about the potential hazards of a specific refrigerant oil, consult the oil supplier's material safety data sheet (MSDS). This sheet lists all potential hazards and states what to do to protect against them. It will also state any special handling requirements. *Figure 20* shows an MSDS for refrigerant oil.

8.1.0 Personal Protective Equipment

When handling refrigerant oils, wear oil-resistant gloves and eye protection. Use caution when handling oil from a compressor that has experienced a motor burnout because the oil may contain acid. Synthetic oils such as POE will irritate exposed skin. Use soap and water to wash the affected area if contact occurs. If a spill occurs, clean it up to prevent slips or falls.

8.2.0 Working with Oils

All refrigerant oils are hygroscopic. That is, they absorb moisture from the air. Because moisture is a major refrigeration system contaminant, it must be kept from entering the system. Synthetic oils such as POE and PAG are much more hygroscopic than other oils. Synthetic oil also can damage other materials, especially certain types of roofing materials. When working with synthetic oils, follow these guidelines:

- If a system must be opened for a repair, seal any open tubing ends if the system will be exposed to the air longer than 15 minutes.
- Never open a system or attempt to add or remove oil from a system on a rainy or foggy day.

- Always use fresh oil from a new, unopened container. Obtain a container of oil sized close to the oil requirement. If one quart is needed, obtain a one-quart container. Discard any unused oil.
- Pump oil into the system. Do not pour it in, as this exposes the oil to air.
- When retrofitting a system that used mineral oil, purge all oil from the system using regulated nitrogen. Synthetic oils can tolerate up to 5 percent mineral oil in the system. Because purging oil from a system can be difficult and time consuming, some manufacturers recommend replacing the system to avoid potential problems.
- If the oil becomes contaminated with moisture, it cannot be removed with a vacuum pump. Instead, use a refrigerant-specific filter-drier to remove the moisture.
- Follow the equipment manufacturer's instructions for applying protection to prevent damage to roofing materials. If protective measures are not taken, roofs can still be damaged, even if spilled synthetic oil is cleaned off the roof.

CAUTION

Synthetic oil can act as a mild solvent. When used in a retrofit installation, the oil can clean the inside of pipes and other components, possibly releasing harmful contaminants into the system. For that reason, the installation of a filter-drier is recommended when replacing oil.

8.2.1 Adding and Removing Oil from a System

Before adding or removing oil from a compressor, it first must be determined why there is a lack of or excess of oil in the compressor, especially if the compressor had the correct amount of oil to begin with. If the compressor is low on oil, it may indicate a piping design problem that is preventing oil from returning, or there may be a defective component in the system, such an accumulator with a plugged oil return orifice that is trapping oil. If the oil level is excessive, it may indicate that oil that was trapped in the piping has now returned to the compressor. The too-little, too-much oil problem is often encountered when a compressor has failed and a replacement compressor installed. Never blindly add or remove compressor oil. Always determine the reason for the lack of or excess of oil and correct the problem before adding or removing oil from the compressor.

LUBRIPLATE®
MATERIAL SAFETY DATA SHEET

Section 1

PRODUCT NAME OR NUMBER
LUBRIPLATE Refrigeration Compressor Oil 68-P

GENERIC/CHEMICAL NAME:
Petroleum Lubricating Oil

Manufacturer's Name
Fiske Brothers Refining Co., d/b/a LUBRIPLATE Lubricants Co.

Address
1500 Oakdale Ave., Toledo, Ohio 43605 - 129 Lockwood St., Newark, NJ 07105

FORMULA
Severely Hydrotreated Paraffinic Mineral Oil

NSF Registration No:
N/A

Emergency Telephone Number
1-800-255-3924 - CHEM-TEL (24 hour)

Telephone Number for Information
419-691-2491 - Toledo Office

Section 2 – Composition/Information on Ingredients

Ingredients with CAS#'s	CAS #	Percentage
Severely Hydrotreated Heavy Paraffinic Distillate	726-23-87-1 & 726-23-85-9	100%

Hazardous Material Identification System (HMIS): Health - 1, Flammability - 1, Reactivity – 0

Not a Controlled Product under (WHMIS) - Canada   Special Protection: See Section 9

Section 3 - Health Hazard Data

Threshold Limit Value   5 mg/m^3 for oil mist in air. OSHA Regulation 29 CFR 1910.1000

Effects of Overexposure   Prolonged or repeated skin contact may cause skin irritation. Product contacting the eyes may cause eye irritation. Human health risks vary from person to person. As a precaution, exposure to liquids, vapors, mists and fumes should be minimized. This product has a low order of acute oral toxicity, but minute amounts aspirated into the lungs during ingestion may cause mild to severe pulmonary injury.

Carcinogenicity: NTP? No   IARC Monographs? No   OSHA Regulated? No

Section 4 - Emergency and First Aid Procedures

EYE CONTACT: Flush with clear water for 15 minutes or until irritation subsides. If irritation persists, consult a physician.

SKIN CONTACT: Remove any contaminated clothing and wash with soap and warm water. If injected by high pressure under skin, regardless of the appearance or its size, contact a physician IMMEDIATELY. Delay may cause loss of affected part of the body.

INHALATION: Vapor pressure is very low and inhalation at room temperature is not a problem. If overcome by vapor from hot product, immediately remove from exposure and call a physician.

INGESTION: If ingested, call a physician immediately. Do not induce vomiting.

Section 5 - Fire and Explosion Hazard Data

Flash Point (Method Used)	COC - 468°F **Flammable Limits** **LEL** 0.9% **UEL** 7.0%
Extinguishing Media	Foam, Dry Chemical, Carbon Dioxide or Water Spray (Fog)
Special Fire Fighting Procedures	Cool exposed containers with water. Use air-supplied breathing equipment for enclosed or confined spaces.
Unusual Fire and Explosion Hazards	Do not store or mix with strong oxidants. Empty containers retain residue. Do not cut, drill, grind, or weld, as they may explode.

301F20A.EPS

Figure 20 ◆ MSDS for refrigerant oil (1 of 2).

PRODUCT NAME OR NUMBER - LUBRIPLATE Refrigeration Compressor Oil 68-P

Section 6 - Physical/Chemical Characteristics

Boiling Point	>550°F	**Specific Gravity (H_2O = 1)**	0.9
Vapor Pressure (mm Hg.)	<0.01	**Melting Point**	Liquid
Vapor Density (AIR = 1)	>5	**Evaporation Rate (Butyl Acetate = 1)**	<0.01
Solubility in Water	Negligible		
Appearance and Odor	Pale liquid with mineral oil odor.		

Section 7 - Reactivity Data

Stability Unstable
Stable X
Conditions to Avoid N/A

Incompatibility (Materials to Avoid) Avoid contact with strong oxidants like liquid chlorine, concentrated oxygen, strong acids.

Hazardous Decomposition or Byproducts May form SO_2. If incomplete combustion, Carbon Monoxide.

Hazardous Polymerization May Occur
Will Not Occur X
Conditions to Avoid N/A

Section 8 - Spill or Leak Procedures

Steps to be taken in case material is released or spilled
Recover liquid, wash remainder with suitable petroleum solvent or add absorbent. Keep petroleum products out of sewers and water courses. Advise authorities if product has entered or may enter sewers and water courses.

Waste disposal method
Assure conformity with applicable disposal regulations. Dispose of absorbed material at an approved waste disposal facility or site.

Section 9 - Special Protection Information

Respiratory Protection (Specify type) Normally not needed

Ventilation **Local Exhaust** Used to capture fumes and vapors **Special** N/A
Mechanical (General) **Other** N/A

Protective Gloves Use oil-resistant gloves, if needed. **Eye Protection** If chance of eye contact, wear goggles.

Other Protective Equipment Use oil-resistant apron, if needed.

Section 10 - Special Precautions

Precautions to be taken in handling and storing
Keep containers closed when not in use. Do not handle or store near heat, sparks, flame, or strong oxidants.

Other Precautions
Avoid breathing oil mist. Remove oil-soaked clothing and launder before reuse. Cleanse skin thoroughly after contact.

The above information is furnished without warranty, expressed or implied, except that it is accurate to the best knowledge of Fiske Brothers Refining Co., d/b/a LUBRIPLATE Lubricants Co. The data on these sheets relates only to the specific material designated herein. Fiske Brothers Refining Co., d/b/a LUBRIPLATE Lubricants Co. assumes no legal responsibility for use or reliance upon this data.

Date Prepared: September 24, 2007 Prepared by: James R. Kontak

301F20B.EPS

Figure 20 ◆ MSDS for refrigerant oil (2 of 2).

WARNING!
Wear rubber gloves and eye protection when handling refrigerant oils.

Generally, oil is added or removed from semi-hermetic compressors that have an oil level indicator such as a sight glass. Oil is not usually added to welded hermetic compressors because there is no reliable way of measuring their oil level short of removing the compressor and pouring the oil out, which is an impractical procedure. It is best to add new oil to a compressor using a pump (*Figure 21*) and to remove oil using a pump, siphon, or gravity drain.

To add oil to a semi-hermetic compressor, reduce crankcase pressure to zero. To do this, close the suction service valve of an operating compressor, drawing crankcase pressure to zero. Then close the discharge service valve. Another way is to recover all refrigerant with a recovery machine. Oil is then pumped into the crankcase through an angled service valve. Some oil pumps are capable of pumping oil into a compressor against as much as 250 psig system pressure.

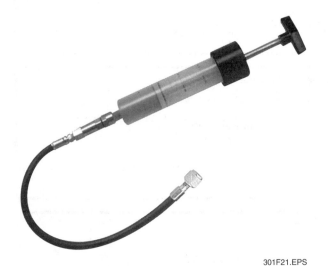

301F21.EPS

Figure 21 ◆ Oil pump.

To remove oil from a semi-hermetic compressor, reduce crankcase pressure to zero as previously described. If the compressor is equipped with an oil drain plug, open the drain and allow gravity to drain the oil into a suitable container. If the compressor shell has a removable plug or service port above the oil level, this can be opened and a pump hose or siphon tube inserted into the bottom of the oil sump to remove the oil. Always follow the compressor manufacturer's instructions when adding or removing oil from a compressor.

8.3.0 Waste Oil Disposal

Waste oil contaminated with CFCs is not considered a hazardous waste unless it is mixed with other oils. Oil contaminated with acid, lead, or other heavy metals or contaminants that would make it flammable or corrosive would qualify it as a hazardous waste. Take uncontaminated used oil to a used oil recycler. Oils deemed hazardous must be taken to a hazardous waste disposal facility. Federal and state regulations governing the handling of hazardous wastes are constantly evolving. For that reason, technicians should check with local and national agencies for the most up-to-date regulations.

WARNING!
Wear rubber gloves and eye protection when handling waste oil.

9.0.0 ◆ SYSTEM CONVERSION

With CFC and HCFC refrigerants being phased out, refrigerant manufacturers have developed replacement refrigerants that allow equipment designed to operate with those older refrigerants to operate with newer, less environmentally damaging refrigerants. This extends the life of the equipment, resulting in savings to the end user. Issues to be addressed during the conversion include the following:

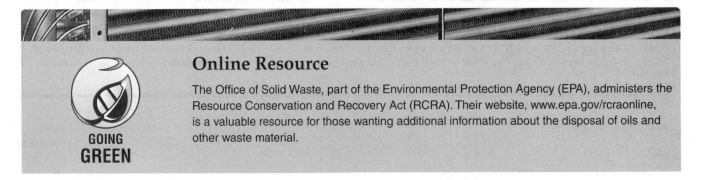

Online Resource

The Office of Solid Waste, part of the Environmental Protection Agency (EPA), administers the Resource Conservation and Recovery Act (RCRA). Their website, www.epa.gov/rcraonline, is a valuable resource for those wanting additional information about the disposal of oils and other waste material.

GOING GREEN

- *Material compatibility* – Materials used in the system with the original refrigerant may not be compatible with the new refrigerant or oil, requiring replacement. Synthetic oil may attack certain gasket or O-ring materials and may be incompatible with the existing compressor oil. Replacement refrigerants may require filter-driers with different core materials.
- *Refrigerant characteristics* – Compressor oil may have to be changed. Some replacement refrigerants may change equipment capacity or have better or worse performance under certain conditions. If certain combinations of approved oils and accumulators and receivers are used, oil return may be affected. Metering devices often have to be changed.
- *Lubricant compatibility* – Synthetic oils and mineral oils do not mix, requiring a thorough purge of all old oil from the system. The hygroscopic nature of some synthetic oils introduces new service requirements.

9.1.0 Common Refrigerant Conversions

There are a variety of refrigerants available for retrofitting systems from CFC and HCFC refrigerants to more environmentally friendly refrigerants (*Figure 22*). Some of the more common ones include the following:

- *CFC-11 retrofits* – Use HCFC-123 as it is compatible with all oil types. It is due to be phased out by 2030.
- *CFC-12 retrofits* – There are several refrigerants available. HCFC-401A is compatible with all oil types, but its use is limited to medium-temperature applications. HCFC-401B is compatible with all oil types and is best suited for low-temperature applications. HCFC-409A is compatible with all oil types and is considered a good broad-range substitute for CFC-12. All of the above refrigerants are blends that have moderate to high glides.
- *HCFC-22 retrofits* – HFC-407C requires synthetic oil and has a high glide. It is considered the best retrofit alternative to HCFC-22. HFC-417A can use various oils depending on the application. It has a high glide and a lower cooling capacity than HCFC-22. There also can be problems with oil return if accumulators and/or receivers are in the system and mineral oil is used.

HFC-410A is considered to be the likely replacement refrigerant for HCFC-22. However, it cannot be used as a retrofit refrigerant because of its much higher operating pressures. It can only be used in new equipment specifically designed to use it.

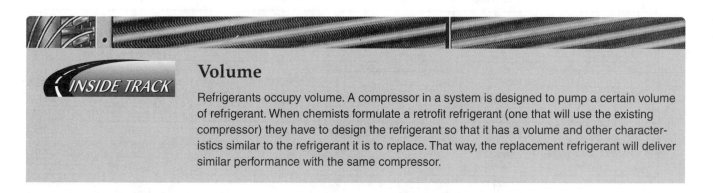

Volume

Refrigerants occupy volume. A compressor in a system is designed to pump a certain volume of refrigerant. When chemists formulate a retrofit refrigerant (one that will use the existing compressor) they have to design the refrigerant so that it has a volume and other characteristics similar to the refrigerant it is to replace. That way, the replacement refrigerant will deliver similar performance with the same compressor.

Guide to Alternative Refrigerants

Commercial Refrigeration
Long Term Refrigerants

Honeywell

Genetron® Refrigerants

ASHRAE #	Trade Name	Manufacturer	Replaces	Type[b][e]	Typical Lubricant[a]	Applications	Comments
R-404A 125/143a/134a (44%/52%/4%)	Genetron 404A Suva HP62 Forane 404A	Honeywell DuPont Arkema	R-502 R-22 HP-80 R-408A	Blend HFC (low guide)	Synthetic (POE, PVE etc.)	New Equipment Retrofits	Most widely used low and medium temperature replacement.
R-507 125/143a (50%/50%)	Genetron AZ-50 Suva 507	Honeywell DuPont	R-502 R-22 HP-80 R-408A	Azeotrope (no guide) HFC	Synthetic (POE, PVE etc.)	New Equipment Retrofits	Slightly higher pressure and efficiency than R404A Best choice for systems with flooded evaporators.
R-422A 125/134a/600a (85.1%/11.5%/3.4%)	One Shot Isceon 79	ICOR DuPont	R-502 R-22	Blend HFC	Synthetic (POE, PVE etc.)	New Equipment Retrofits	Similar performance to R-404A. Equipment with suction line accumulators and receivers should use synthetic oils to avoid oil return issues.
R-422D 125/134a/600a (65.1%/31.5%/3.4%)	Genetron 422D Isceon MO 29	Honeywell Dupont	R-22	Blend (moderate glide) HFC	Mineral Oil POE	New Equipment Retrofits	Low capacity Use of POE will enhance oil return, if required.
R-407C 32/125/134a (23%/25%/52%)	Genetron 407C Suva 9000 Forane 407C	Honeywell Dupont Arkema	R-22	Blend (high glide) HFC	Synthetic (POE, PVE etc.)	New Equipment Retrofits	Reasonable performance match to R-22 in medium temperature refrigeration. Lower capacity in low temperature refrigeration system.
R-134a	Genetron 134a Suva 134a Forane 134a Klea 134a	Honeywell Dupont Arkema INEOS	R-12	Single Component Fluid HFC	Synthetic (POE, PVE etc.)	New Equipment	Performs well in small hermetic systems.

Interim HCFC Based Refrigerants

ASHRAE #	Trade Name	Manufacturer	Replaces	Type[b][e]	Typical Lubricant[a]	Applications[c]	Comments
R-401A 22/152a/124 (53%/13%/34%)	Genetron MP39 Suva MP39	Honeywell Dupont	R-12	Blend (moderate glide)HCFC/HFC	Alkylbenzene Synthetic (POE, PVE etc.) Mineral Oil	Med Temp Retrofits[c]	In most cases no oil change is needed. Best for applications with >0 degrees F suction.
R-401B 22/152a/124 (61%/11%/28%)	Genetron MP66 Suva MP66	Honeywell Dupont	R-12 R-500	Blend (moderate glide) HCFC/HFC	Alkylbenzene Synthetic (POE, PVE etc.) Mineral Oil	Transport refrigeration Low Temp Retrofits[c] Retrofits including air conditioners and dehumidifiers	In most cases no oil change is needed. Best for low temp R-12 and R-500 retrofit applications
R-409A 22/124/142b (60%/25%/15%)	Genetron 409A Suva 409A Forane FX-56	Honeywell Dupont Arkema	R-12 R-500	Blend (high glide) HCFC	Alkylbenzene Synthetic (POE, PVE etc.) Mineral Oil	Retrofits[c] Low and Med Temp	In most cases no oil change is needed. Good broad range R-12 substitute.
R-402A 125/290/22 (60%/2%/38%)	Genetron HP80 Suva HP80	Honeywell Dupont	R-502	Blend (low glide) HFC/HC/HCFC	Alkylbenzene Synthetic (POE, PVE etc.)	Retrofits[c] Low and Med Temp	Most widely used R-502 retrofit substitute. Higher discharge pressure than R-502. Use either synthetic oil or blend of AB/MO with AB>50%.
R-402B 125/290/22 (38%/2%/60%)	Genetron HP81 Suva HP81	Honeywell Dupont	R-502	Blend (low glide) HFC/HC/HCFC	Alkylbenzene Synthetic (POE, PVE etc.) Mineral Oil	Ice Machines	Niche refrigerant used in some ice machines.
R-408A 125/143a/22 (7%/46%/47%)	Genetron 408A Suva 408A Forane FX-10	Honeywell Dupont Arkema	R-502	Blend (low glide) HFC/HCFC	Alkylbenzene Synthetic (POE, PVE etc.)	Retrofits[c] Low and Med Temp	Works well as R-502 substitute. Higher discharge temperatures than R-502. Use either synthetic oil or blend of AB/MO with AB>50%.

301F22A.EPS

Figure 22 ◆ Guide to alternative refrigerants (1 of 2).

Ultra Low Temp Refrigerants

ASHRAE #	Trade Name	Manufacturer	Replaces	Type[b] [e]	Typical Lubricant[a]	Applications	Comments
R-23	Genetron 23 Suva 23	Honeywell DuPont	R-13 R-503	Single component Fluid HFC	Synthetic (POE, PVE etc.)	New Equipment Retrofits	Higher discharge temperatures than R-13 or R-508B
R-508B 23/116 (46%/54%)	Genetron 508B Suva 95	Honeywell DuPont	R-13 R-503	Azeotrope HFC	Synthetic (POE, PVE etc.)	New Equipment Retrofits	Lower discharge temperatures than R-13 or R-23. Good Performance match to R-503.

Air Conditioning
Centrifugal Chiller Refrigerants

ASHRAE #	Trade Name	Manufacturer	Replaces	Type[b] [e]	Typical Lubricant[a]	Applications	Comments
R-123	Genetron 123 Suva 123 Forane 123	Honeywell DuPont Arkema	R-11	Single component Fluid HFC	Alkylbenzene Mineral Oil Synthetic (POE, PVE etc.)	New Equipment Retrofits	Due for phase out in 2030
R-245fa	Genetron 245fa	Honeywell	R-11	Single component Fluid HFC	Synthetic (POE, PVE etc.)	New Equipment	Equipment redesign Organic Rankine Cycle & as Heat Transfer Fluid
R-134a	Genetron 134a Suva 134a Forane 134a Klea 134a	Honeywell DuPont Arkema INEOS	R-12 R-500	Single component Fluid HFC	Synthetic (POE, PVE etc.)	New Equipment Retrofits	Used in many new chiller designs.

Long Term Refrigerants — Air Conditioning and Heat Pumps

ASHRAE #	Trade Name	Manufacturer	Replaces	Type[b] [e]	Typical Lubricant[a]	Applications	Comments
R-407C 32/125/134a (23%/25%/52%)	Genetron 407C Suva 407C Forane 407C	Honeywell DuPont Arkema	R-22 R-500	Blend (high glide) HFC	Synthetic (POE, PVE etc.)	New Equipment Retrofits	Best retrofits alternative to R-22. Close performance match with slightly higher operating pressures.
R-410A 32/125 (50%/50%)	Genetron AZ-20 Suva 410A Forane 410A Puron	Honeywell DuPont Arkema	R-22	Azeotropic Mixture (near zero glide) HFC	Synthetic (POE, PVE etc.)	New Equipment	High pressure, high efficiency refrigerant designed for new equipment. NOT FOR RETROFITTING.
R-422D 125/134a/600a (65.1%/31.5%/3.4%)	Genetron 422D Isceon MO 29	Honeywell DuPont	R-22	Blend (moderate glide) HFC	Mineral Oil POE	New Equipment Retrofits	Low Capacity. Use of POE will enhance oil return, if required.
R-134a	Genetron 134a Isceon 134a Forane 134a Klea 134a	Honeywell DuPont Arkema INEOS	R-12 R-500	Single Component Fluid (no glide) HFC	Synthetic (POE, PVE etc.)	New Equipment Retrofits	Used in large screw chillers

R22 Phase-out Schedule for Air Conditioning and Refrigeration

January 1, 2004:	35% reduction in HCFC consumption required by Montreal Protocol.
January 1, 2010:	No new R-22 can be manufactured for new equipment manufactured or imported in US or Canada. 65% reduction HCFC consumption required by Montreal Protocol.
January 1, 2015:	90% reduction in HCFC consumption required by Montreal Protocol.
January 1, 2020:	No new R-22 refrigerant can be manufactured or imported into the US or Canada. Any servicing must be done using stockpiled or reclaimed material. 100% reduction in HCFC consumption required by Montreal Protocol.

(d) The refrigerant R-600 is butane. The refrigerant R-600 is isobutane.
(e) CFC=Chlorofluorocarbon: HCFC=Hydrochlorofluorocarbon: HFC=Hydrofluorocarbon HC=Hydrocarbon: FC=Fluorcarbon

Trademarks
Genetron, AZ-20, AZ-50 are registered trademarks of Honeywell International
Puron is a registered trademark of Carrier Corporation
Klea is a registered trademark of INEOS
Suva is a registered trademark of DuPont
Forane is a registered trademark of Arkema
Isceon is a registered trademark of Dupont

Notes:
(a) Check with the compressor manufacturer for their recommended lubricant.
(b) Interim replacements contain HCFCs that are scheduled for phase out.
(c) Not recommended for automotive air conditioning.

Disclaimer
All statements, information and data given herein are believed to be accurate and reliable but are presented without guaranty, warranty or responsibility of any kind, expressed or implied. Statements or suggestions concerning possible use of our products are made without representation or warranty that any such use is free of patent infringement and are not recommendations to infringe any patent. The user should not assume that all safety measures are indicated, or that other measures may not be required.

Honeywell

Genetron® Refrigerants
G-525-043 04-07 15K
Printed in USA © 2005 Honeywell International

301F22B.EPS

Figure 22 ◆ Guide to alternative refrigerants (2 of 2).

1. HFC-410A is an ideal drop-in replacement for HCFC-22.

    a. True   b. False

2. A hydrocarbon that has some or all of its hydrogen atoms replaced with chlorine atoms is called a(n) _____.

    a. hydrogenated hydrocarbon
    b. HFC-410A
    c. halocarbon
    d. HFC

3. The refrigerants considered the friendliest to the environment are _____.

    a. halocarbons
    b. CFCs
    c. HCFCs
    d. HFCs

4. Which of the following refrigerants has the highest ozone depletion potential?

    a. CFC-12
    b. HCFC-22
    c. HCFC-123
    d. HFC-134a

5. Class _____ refrigerants are typically supplied in drums.

    a. I
    b. II
    c. III
    d. IV

6. If you had a light green disposable refrigerant cylinder, you would know that it contained _____.

    a. HFC-410A
    b. CFC-12
    c. HCFC-22
    d. HCFC-123

Use the P-T chart shown in *Figure 1* to answer Question 7.

7. On an HCFC-22 system, suction pressure is 78 psig and the measured suction line temperature is 50°F. Superheat is _____.

    a. 1°F
    b. 2°F
    c. 4°F
    d. 8°F

8. Liquid line temperature and suction pressure are required to calculate subcooling.

    a. True   b. False

9. When checking a system for leaks using an electronic leak detector, always _____.

    a. add a small amount of refrigerant backed up with compressed air
    b. add a small amount of refrigerant backed up with regulated nitrogen
    c. pressurize the system with oxygen
    d. pressurize the system with regulated dry nitrogen

10. The temperature at which oil first begins to flow is known as the _____.

    a. floc point
    b. viscosity
    c. flash point
    d. pour point

11. Which of the following refrigerants is compatible with mineral oil?

    a. HFC-410A
    b. HFC-407A
    c. HFC-134a
    d. HCFC-22

12. You need to add 12 ounces of POE oil to a compressor. For best results, obtain the required oil from a _____.

    a. 55-gallon drum
    b. one-gallon container
    c. pint container
    d. quart container

13. One of the possible causes of a low oil level in a system is a plugged accumulator.

    a. True   b. False

14. Refrigerant oil that contains a slight amount of diesel fuel is not considered a hazardous material.

    a. True   b. False

15. Which of the following refrigerant retrofits is *not* valid?

    a. HFCF-123 for CFC-11
    b. HCFC-409A for CFC-12
    c. HFC-407C for HCFC-22
    d. HFC-410A for HCFC-22

°F	R-410A	R-22	°F	R-410A	R-22	°F	R-410A	R-22	°F	R-410A	R-22
−40°	10.8	0.6	10	62.2	32.8	60	169.6	101.6	110	364.1	226.4
−39°	11.5	1.0	11	63.7	33.8	61	172.5	103.5	111	369.1	229.6
−38°	12.1	1.4	12	65.2	34.8	62	175.4	105.4	112	374.2	232.8
−37°	12.8	1.8	13	66.8	35.8	63	178.4	107.3	113	379.4	236.1
−36°	13.5	2.2	14	68.3	36.8	64	181.5	109.3	114	384.6	239.4
−35°	14.2	2.6	15	69.9	37.8	65	184.5	111.2	115	389.9	242.8
−34°	14.9	3.1	16	71.5	38.8	66	187.6	113.2	116	395.2	246.1
−33°	15.6	3.5	17	73.2	39.9	67	190.7	115.3	117	400.5	249.5
−32°	16.3	4.0	18	74.9	40.9	68	193.9	117.3	118	405.9	253.0
−31°	17.1	4.5	19	76.6	42.0	69	197.1	119.4	119	411.4	256.5
−30°	17.8	4.9	20	78.3	43.1	70	200.4	121.4	120	416.9	260.0
−29°	18.6	5.4	21	80.0	44.2	71	203.6	123.5	121	422.5	263.5
−28°	19.4	5.9	22	81.8	45.3	72	207.0	125.7	122	428.2	267.1
−27°	20.2	6.4	23	83.6	46.5	73	210.3	127.8	123	433.9	270.7
−26°	21.1	6.9	24	85.4	47.6	74	213.7	130.0	124	439.6	274.3
−25°	21.9	7.4	25	87.2	48.8	75	217.1	132.2	125	445.4	278.0
−24°	22.7	8.0	26	89.1	50.0	76	220.6	134.5	126	451.3	281.7
−23°	23.6	8.5	27	91.0	51.2	77	224.1	136.7	127	457.3	285.4
−22°	24.5	9.1	28	92.9	52.4	78	227.7	139.0	128	463.2	289.2
−21°	25.4	9.6	29	94.9	53.7	79	231.3	141.3	129	469.3	293.0
−20°	26.3	10.2	30	96.8	55.0	80	234.9	143.6	130	475.4	296.9
−19°	27.2	10.8	31	98.8	56.2	81	238.6	146.0	131	481.6	300.8
−18°	28.2	11.4	32	100.9	57.5	82	242.3	148.4	132	487.8	304.7
−17°	29.2	12.0	33	102.9	58.8	83	246.0	150.8	133	494.1	308.7
−16°	30.1	12.6	34	105.0	60.2	84	249.8	153.2	134	500.5	312.6
−15°	31.1	13.2	35	107.1	61.5	85	253.7	155.7	135	506.9	316.7
−14°	32.2	13.9	36	109.2	62.9	86	257.5	158.2	136	513.4	320.7
−13°	33.2	14.5	37	111.4	64.3	87	261.4	160.7	137	520.0	324.8
−12°	34.2	15.2	38	113.6	65.7	88	265.4	163.2	138	526.6	329.0
−11°	35.3	15.9	39	115.8	67.1	89	269.4	165.8	139	533.3	333.2
−10°	36.4	16.5	40	118.1	68.6	90	273.5	168.4	140	540.1	337.4
−9°	37.5	17.2	41	120.3	70.0	91	277.6	171.0	141	547.0	341.6
−8°	38.6	17.9	42	122.7	71.5	92	281.7	173.7	142	553.9	345.9
−7°	39.8	18.7	43	125.0	73.0	93	285.9	176.4	143	560.9	350.3
−6°	40.9	19.4	44	127.4	74.5	94	290.1	179.1	144	567.9	354.6
−5°	42.1	20.1	45	129.8	76.1	95	294.4	181.8	145	575.1	359.0
−4°	43.3	20.9	46	132.2	77.6	96	298.7	184.6	146	582.3	363.5
−3°	44.5	21.7	47	134.7	79.2	97	303.0	187.4	147	589.6	368.0
−2°	45.7	22.4	48	137.2	80.8	98	307.5	190.2	148	596.9	372.5
−1°	47.0	23.2	49	139.7	82.4	99	311.9	193.0	149	604.4	377.1
0	48.3	24.0	50	142.2	84.1	100	316.4	195.9	150	611.9	381.7
1	49.6	24.9	51	144.8	85.7	101	321.0	198.8			
2	50.9	25.7	52	147.4	87.4	102	325.6	201.8			
3	52.2	26.5	53	150.1	89.1	103	330.2	204.7			
4	53.6	27.4	54	152.8	90.8	104	334.9	207.7			
5	55.0	28.3	55	155.5	92.6	105	339.6	210.8			
6	56.3	29.2	56	158.2	94.4	106	344.4	213.8			
7	57.8	30.1	57	161.0	96.1	107	349.3	216.9			
8	59.2	31.0	58	163.8	98.0	108	354.2	220.0			
9	60.7	31.9	59	166.7	99.8	109	359.1	223.2			

Source: Honeywell. The above data, including recommendations for application and use of R-410A (Genetron® AZ-20®) are available at http://www.genetron.com.

301RQ01.EPS

Figure 1

Summary

A refrigerant is the substance in the refrigeration system that changes state when heat is added or taken away from it. In the condenser, gaseous refrigerant releases heat, allowing it to condense into a liquid. In the evaporator, liquid refrigerant boils into a gaseous state as it absorbs heat. This constant absorption and rejection of heat by the refrigerant as it changes state is the underlying principle that allows mechanical refrigeration systems to function. Until the 1990s, chlorine-based CFCs and HCFCs were the predominant refrigerants. As the result of international treaties and federal legislation, CFCs have been banned and the use of HCFCs has been restricted because the chlorine they contain contributes to atmospheric ozone depletion. Alternatives to CFCs and HCFCs have been developed. Most are HFCs, which contain no chlorine. When retrofitting a CFC or HCFC system to use the replacement refrigerants, system modifications are often required.

HFC refrigerants are, for the most part, not compatible with the mineral oils used as lubricants in the compressors of CFC and HCFC systems. As a result, new synthetic oils have been developed that are compatible with HFC refrigerants. The synthetic oils are usually not compatible with mineral oils, so in retrofit situations, mineral oils often have to be thoroughly purged from the compressor, refrigerant piping, and other components before synthetic oil can be added. Due to the hygroscopic nature of synthetic oils, special handling requirements must be followed to prevent moisture from entering the refrigeration system when synthetic oils are used.

Notes

Trade Terms Introduced in This Module

Alkylbenzene: A type of synthetic hydrocarbon oil that is compatible with mineral oil but has better oil return properties. It is widely used in low-temperature applications. It is not compatible with synthetic oils.

Azeotrope: A blended refrigerant that behaves like a pure refrigerant. The refrigerant evaporates and condenses at one given pressure and temperature. From the technician's standpoint, working with an azeotropic blend is like working with a pure refrigerant.

Binary blend: A blended refrigerant consisting of two refrigerants.

Bubble point: The point at which refrigerant starts evaporating when heat is added to a zeotropic refrigerant. It is also the point that refrigerant finishes condensing when heat is removed from the refrigerant. See *dew point*.

Chlorofluorocarbons (CFCs): The most damaging refrigerants to the environment since they contain more chlorine atoms in their structure. Production of CFCs in the United States ended in 1995.

Compound: A substance made up of different elements. A refrigerant that is not a blend is a compound.

Dew point: The point at which refrigerant stops evaporating when heat is added to a zeotropic refrigerant. It is also the point that refrigerant starts condensing when heat is removed from the refrigerant. See *bubble point*.

Dielectric strength: The ability of refrigerant oil or any material to resist breaking down in the presence of voltage.

Flash point: The temperature at which heated oil vapors burst into flame. It is a measure of the flammability of the lubricant. Ideally, compressor lubricant oils should have a high flash point for safety reasons.

Floc point: The temperature at which a 90/10 mixture of oil/refrigerant forms a cloudy or flocculent precipitate of wax in the mixture.

Fractionation: A process in which the component refrigerants of a blended refrigerant boil off into a vapor state at different temperatures.

Glide: The temperature range in which a zeotropic refrigerant blend evaporates and condenses.

Global warming potential (GWP): A measure of a substance's ability to contribute to global warming that is expressed as a number. The higher the number, the greater the warming potential.

Greenhouse effect: An effect in which atmospheric gases such as carbon monoxide trap solar heat in the atmosphere in the same way that a greenhouse captures and holds solar heat.

Halogen: A class of elements that include chlorine and fluorine that are used in the manufacture of refrigerants. A fully halogenated refrigerant is one in which all hydrogen atoms of a hydrocarbon molecule are replaced by halogen atoms.

Hydrogenated chlorofluorocarbon (HCFC): A refrigerant that contain chlorines atoms in its structure, but fewer than found in CFCs. They are kinder to the environment than CFCs, allowing them to be phased out over a longer period of time. By 2030, all HCFC refrigerants will no longer be manufactured.

Hydrogenated fluorocarbon (HFC): A refrigerant that contains no chlorine atoms in its structure, which makes this group of refrigerants the most environmentally friendly of the three classifications of halocarbon refrigerants.

Hydrocarbon: A compound composed only of carbon and hydrogen atoms. When some or all of the hydrogen atoms in a hydrocarbon are combined with a halogen such as chlorine or fluorine, the end result is a compound known as a halogenated hydrocarbon or halocarbon. Most common refrigerants are halocarbons.

Hygroscopic: The ability of a substance to absorb moisture from the air. Many synthetic refrigerant oils are very hygroscopic.

Miscible: The desirable property that allows oil to dissolve in refrigerant and refrigerant to dissolve in oil.

Mixture: The combination of two or more compounds. Blended refrigerants are mixtures.

Near-azeotrope: A zeotrope with azeotropic properties. The refrigerant HFC-410A is an example of a near-azeotrope. Like zeotropes, near-azeotropes exhibit glide characteristics but they are usually minimal to the point that they are not a factor for technicians to consider when servicing a unit containing such a refrigerant.

Ozone depletion potential (ODP): A measure of a substance's ability to deplete atmospheric ozone that is expressed as a number. CFC refrigerants have greater ozone depletion potentials than HFC refrigerants.

Polyalkylene glycol (PAG): Synthetic refrigerant oil used with HFC refrigerants. It is very hygroscopic.

Polyolester (POE): Synthetic refrigerant oil used with HFC refrigerants. It is very hygroscopic.

Polyvinyl ether (PVE): A synthetic refrigerant oil with properties similar to mineral oils which can be used with HFC refrigerants. It has an advantage over POE and PAG oils in that it is not hygroscopic.

Pour point: An oil quality related to viscosity. It can be defined as the temperature at which oil first starts to flow. See *viscosity.*

Ternary blend: A blended refrigerant consisting of three refrigerants.

Viscosity: The resistance of a fluid to flow. Viscosity of a fluid can vary, depending on the temperature of the fluid. When applied to refrigerant oil, viscosity must be such that the oil will flow and provide proper lubrication over the wide range of temperatures encountered in refrigeration and air conditioning systems. See *pour point.*

Zeotrope: A blended refrigerant that never mixes chemically. As a result, it evaporates and condenses over a temperature range, called the glide.

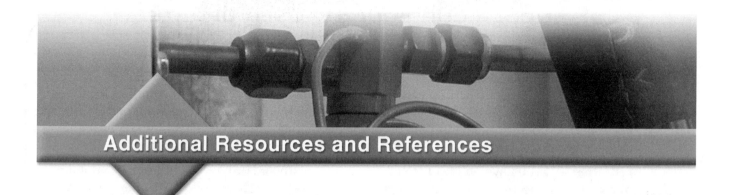

Additional Resources and References

Additional Resources

This module is intended to be a thorough resource for task training. The following reference work is suggested for further study. This is optional material for continued education rather than for task training.

Guide to the E.P.A. Refrigerant Handling Certification Exam, Prentice Hall, Upper Saddle River, NJ.

Figure Credits

NCCER CURRICULA — USER UPDATE

NCCER makes every effort to keep its textbooks up-to-date and free of technical errors. We appreciate your help in this process. If you find an error, a typographical mistake, or an inaccuracy in NCCER's curricula, please fill out this form (or a photocopy), or complete the online form at **www.nccer.org/olf**. Be sure to include the exact module ID number, page number, a detailed description, and your recommended correction. Your input will be brought to the attention of the Authoring Team. Thank you for your assistance.

Instructors – If you have an idea for improving this textbook, or have found that additional materials were necessary to teach this module effectively, please let us know so that we may present your suggestions to the Authoring Team.

NCCER Product Development and Revision

13614 Progress Blvd., Alachua, FL 32615

Email: curriculum@nccer.org
Online: www.nccer.org/olf

❏ Trainee Guide   ❏ AIG   ❏ Exam   ❏ PowerPoints   Other _____

Craft / Level: _____   Copyright Date: _____

Module ID Number / Title: _____

Section Number(s): _____

Description: _____

_____

_____

_____

_____

Recommended Correction: _____

_____

_____

_____

_____

Your Name: _____

Address: _____

_____

Email: _____   Phone: _____

03302-08

Compressors

03302-08
Compressors

Topics to be presented in this module include:

1.0.0   Introduction 2.2

2.0.0   The Role of the Compressor 2.2

3.0.0   Open, Hermetic, and Semi-Hermetic Compressors 2.4

4.0.0   Types of Compressors 2.4

5.0.0   Capacity Control of Compressors 2.13

6.0.0   Compressor Electric Motors .2.18

7.0.0   Other Compressor Protection Devices 2.24

8.0.0   Reduced-Voltage Motor Starting 2.28

9.0.0   Causes of Compressor Failure 2.29

10.0.0   System Checkout Following Compressor Failure 2.36

11.0.0   Compressor Changeout 2.39

Overview

The compressor is considered to be the heart of the HVAC system. It provides the force that moves the refrigerant through the cycle and it raises the pressure of the refrigerant so that heat absorbed in the evaporator can be expelled at the condenser. There are several types of compressors, ranging from the small rotary compressors used in appliances and room air conditioners to the giant centrifugal and screw compressors used to power chillers. In between are hermetic reciprocal and scroll compressor and the semi-hermitic reciprocating compressors often used in commercial systems. It is important to understand compressors and know how to service and troubleshoot them because the compressor is the most costly part of the system. Improper servicing can result in serious damage to the compressor, while incorrect diagnosis of a problem can result in the unnecessary replacement of a compressor.

Objectives

When you have completed this module, you will be able to do the following:

1. Identify the different types of compressors.
2. Demonstrate or describe the mechanical operation for each type of compressor.
3. Demonstrate or explain compressor lubrication methods.
4. Demonstrate or explain methods used to control compressor capacity.
5. Demonstrate or describe how compressor protection devices operate.
6. Perform the common procedures used when field servicing open and semi-hermetic compressors, including:
   - Shaft seal removal and installation
   - Valve plate removal and installation
   - Unloader adjustment
7. Demonstrate the procedures used to identify system problems that cause compressor failures.
8. Demonstrate the system checkout procedure performed following a compressor failure.
9. Demonstrate or describe the procedures used to remove and install a compressor.
10. Demonstrate or describe the procedures used to clean up a system after a compressor burnout.

Trade Terms

Burnout
Capacity control
Clearance volume
Compliant scroll
  compressor
Compression
Flooded starts
Flooding
Line duty device

Pilot duty device
Positive-displacement
  compressor
Pressure (force-feed)
  lubrication system
Short cycling
Single phasing
Splash lubrication system

Required Trainee Materials

1. Pencil and paper
2. Appropriate personal protective equipment

Prerequisites

Before you begin this module, it is recommended that you successfully complete *Core Curriculum*; *HVAC Level One*; *HVAC Level Two*; and *HVAC Level Three*, Module 03301-08.

This course map shows all of the modules in the third level of the *HVAC* curriculum. The suggested training order begins at the bottom and proceeds up. Skill levels increase as you advance on the course map. The local Training Program Sponsor may adjust the training order.

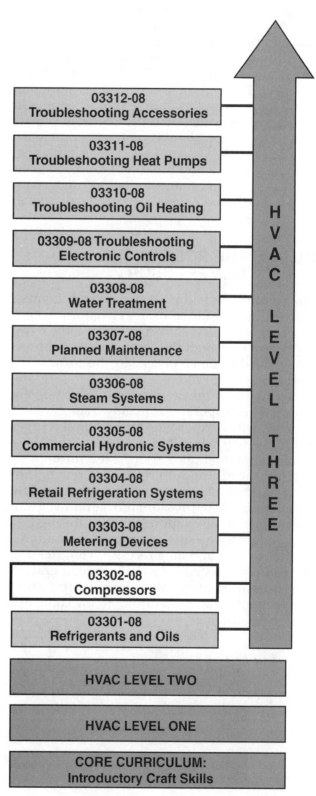

03312-08
Troubleshooting Accessories

03311-08
Troubleshooting Heat Pumps

03310-08
Troubleshooting Oil Heating

03309-08 Troubleshooting
Electronic Controls

03308-08
Water Treatment

03307-08
Planned Maintenance

03306-08
Steam Systems

03305-08
Commercial Hydronic Systems

03304-08
Retail Refrigeration Systems

03303-08
Metering Devices

03302-08
Compressors

03301-08
Refrigerants and Oils

HVAC LEVEL TWO

HVAC LEVEL ONE

CORE CURRICULUM:
Introductory Craft Skills

H V A C   L E V E L   T H R E E

302CMAP.EPS

1.0.0 ◆ INTRODUCTION

Compressors typically are the most expensive and, functionally, the most important part of a complete refrigeration system. Replacing a compressor is one of the most critical service procedures performed on refrigeration systems. Many compressors that are replaced are not defective. In many cases, this is because compressor operation and its relationship to the whole system are not fully understood by the HVAC technician. The result is needless service time and customer expense, as well as the failure to repair the equipment. Because understanding how a compressor works is so important, this module reviews some basic information about compressors previously covered in *HVAC Level One*. This is enhanced with new information that will help you accurately diagnose a compressor failure, understand why it failed, and correctly install a replacement compressor.

2.0.0 ◆ THE ROLE OF THE COMPRESSOR

The compressor is the heart of the refrigeration system. Its role in the basic refrigeration cycle deserves review. As shown in *Figure 1*, the refrigerant flows through the components of the system in the direction indicated by the arrows.

The evaporator receives low-temperature, low-pressure liquid refrigerant from the expansion device. The evaporator is mainly a series of tubing coils which expose the cooler liquid refrigerant to the warmer air passing over them. Heat from the warmer air is transferred through the evaporator tubing into the cooler refrigerant, causing it to boil or vaporize. Superheated, low-temperature, low-pressure refrigerant vapor at the output of the evaporator flows through the suction line to the suction input of the compressor. There, through the process of **compression**, the compressor converts the low-temperature, low-pressure vapor into a high-temperature, high-pressure vapor that flows to the condenser via the hot gas line.

Like the evaporator, the condenser is mainly a series of tubing coils through which the refrigerant flows. As cooler outside air moves across the condenser tubing, the hot refrigerant vapor gives up superheat and cools. As the refrigerant continues to give up heat to the outside air, it cools further, and the temperature drops to the saturated refrigerant vapor point, where the refrigerant begins to change from a vapor into a liquid. As more cooling takes place, called subcooling, the refrigerant is cooled below the saturation temperature. This medium-temperature, high-pressure liquid flows through the liquid line to the input of the expansion device.

The expansion or metering device regulates the flow of refrigerant to the evaporator. It also decreases the pressure, and therefore the temperature, of the refrigerant applied to the evaporator. Through the use of a built-in restriction, such as a small hole or orifice, it converts the medium-temperature, high-pressure refrigerant from the condenser into the low-temperature, low-pressure refrigerant needed to absorb heat in the evaporator.

The compressor performs two operations: it draws the refrigerant from the evaporator coil (suction cycle) and it forces the refrigerant into the condenser (discharge cycle). During this process, certain conditions are created:

- The pressure and temperature of the refrigerant in the evaporator are lowered, allowing the refrigerant to boil and absorb heat from its surroundings.

- The pressure and temperature of the refrigerant in the condenser are raised, allowing the refrigerant to give up heat at existing temperatures to whatever medium (air or water) is used to absorb the heat.

- A pressure difference is maintained between the high- and low-pressure sides of the system. It is this pressure difference that causes the refrigerant to flow through the system.

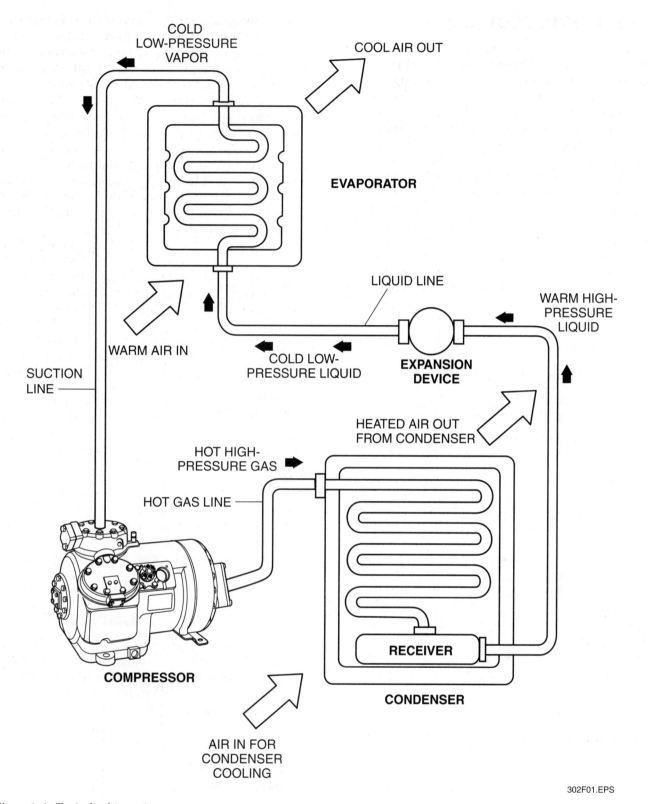

Figure 1 ◆ Typical refrigeration system.

302F01.EPS

3.0.0 ◆ OPEN, HERMETIC, AND SEMI-HERMETIC COMPRESSORS

Compressors are usually driven by an electric motor. Very large compressors can be driven by internal combustion engines or steam turbines. Compressors are divided into three groups based on the way they are joined to their motors or engines (*Figure 2*).

- *Open-drive compressor* – This compressor is separate from its motor. One end of its horizontally mounted shaft extends outside the case.

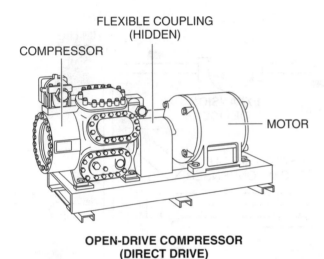

**OPEN-DRIVE COMPRESSOR
(DIRECT DRIVE)**

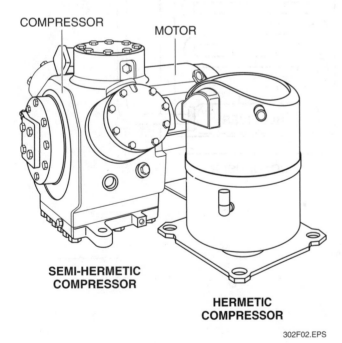

**SEMI-HERMETIC
COMPRESSOR**

**HERMETIC
COMPRESSOR**

302F02.EPS

Figure 2 ◆ Types of compressors.

A mechanical seal on the rotating shaft prevents leakage of the refrigerant. The compressor motor drives the compressor by a belt (belt drive) or a flexible coupling (direct drive). Belt-driven arrangements allow the motor to run at one speed, while the compressor can run at another. The proper combination of pulleys, also called drives, produces the desired speed of the compressor. Most direct-drive systems use an electric motor to drive the compressor. This means that the compressor also runs at the speed of the drive motor.

- *Hermetic (welded hermetic) compressor* – In this unit, the compressor and motor have a common drive shaft. Hermetic reciprocating, rotary, and scroll compressors have the motor and drive shaft mounted vertically in the compressor casing. The reciprocating and rotary compressors usually have the motor mounted above the compressor. In the scroll compressor, the motor is usually mounted below the compressor. They are sealed in a welded steel enclosure or shell. Hermetic compressors (sometimes called tin cans) are more compact, less noisy, and require less maintenance than open-type compressors because they have no belts or couplings to break or wear out. Because they are sealed, the entire unit must be replaced when it fails.

- *Semi-hermetic (serviceable hermetic) compressor* – Similar to the hermetic compressor, the compressor and motor share the same housing and a common, horizontally mounted drive shaft. When they fail, access to the compressors or motors for limited repair is possible by removing the heads and/or bottom and end plates.

4.0.0 ◆ TYPES OF COMPRESSORS

The variety of refrigerants, and the size, location, and application of the systems are some of the factors that create the need for many types of compressors. The following compressors are commonly used in mechanical refrigeration systems:

- Reciprocating compressors
- Rotary compressors
- Scroll compressors
- Screw compressors
- Centrifugal compressors

Table 1 summarizes the uses and characteristics of each type of compressor.

Table 1 Compressor Comparison Chart

	Reciprocating	Rotary	Scroll	Screw	Centrifugal
Use	Refrigeration and air conditioning, heat pumps, and transportation	Refrigerators, room air conditioners	Central systems for refrigeration, air conditioning, and heat pumps	Refrigeration, air conditioning, and heat pumps	Refrigeration, air conditioning, and heat pumps
Size/Range	Fractional tonnage through 150 tons	Smallest: 5 tons or less	1.5 to 70 tons	50 to 750 tons	Largest: 100 to over 10,000 tons
Types	Open, serviceable hermetic, and hermetic	Sliding vane and rolling piston type, hermetic only	Compliant and noncompliant hermetic	Rolling rotor, open, and hermetic	Single and multi-stage open and hermetic
Displacement	Positive	Positive	Positive	Positive	Positive
Typical Capacity Control	Two-speed, on-off, and cylinder unloaders	On-off	On-off, unloaders, variable from 10 to 100 percent	Variable speed and intake slide valve	Variable speed and inlet guide vanes
Suction Valves	Yes	No	No	No	No
Discharge Valves	Yes	Yes	No	No	No

Open-Drive Compressors

Open-drive compressors are commonly used in many larger stationary air conditioning and refrigeration systems. Open-drive compressors of various sizes are also widely used in mobile applications such as in air conditioned vehicles and refrigerated trailers. In these applications, the belt-driven open-drive compressor is coupled to the vehicle's engine through a magnetic clutch mechanism located in the compressor drive pulley hub. The magnetic clutch, controlled by the vehicle's thermostat, engages or disengages the compressor belt-drive pulley and the compressor shaft. The clutch is operated by forcing a clutch disk, mounted to the compressor shaft, against the belt pulley. When the thermostat in the passenger compartment or refrigerated space of the vehicle calls for cooling, the clutch is engaged and causes the compressor to run. When the temperature reaches the desired cooling level, the thermostat signal releases the clutch. This disengages the compressor from the vehicle's motor and causes the compressor to stop running.

Welded Hermetic Compressors

In the past, the lower capacity of welded hermetic compressors limited their use. In air conditioning applications, they were typically found in systems of less than 10 tons capacity. Systems over that capacity tended to use semi-hermetic compressors. Today, welded hermetic compressors are widely used in higher-capacity air conditioning systems by installing them together in multiple-compressor configurations.

4.1.0 Reciprocating Compressors

Reciprocating compressors come in many types and can have from one to as many as ten pistons moving back and forth within a cylinder or cylinders. The main parts include a cylinder, piston, connecting rod, crankshaft, cylinder head, and suction and discharge valves enclosed in a crankcase. Piston movement is synchronized with the opening and closing of the suction and discharge valves. These valves control the intake and discharge of the refrigerant into and from the cylinder. The crankcase contains the crankshaft and stores oil that is used to lubricate the moving compressor components.

Figure 3 shows the intake (suction) and compression (discharge) strokes for one cylinder of a reciprocating compressor. The following events take place inside a reciprocating compressor cylinder during the intake stroke of operation:

1. At the start of the intake stroke, both the discharge and suction valves are closed. As the piston starts down, a low pressure is formed under the suction valve. When this low pressure becomes less than the suction line pressure, the suction valve opens and the cylinder begins to fill with gas supplied from the suction line.

2. As the piston continues to the bottom of the intake stroke, the cylinder becomes nearly full. At the bottom of the stroke there is a very slight time lag as the crankshaft rotates the connecting rod to complete the stroke. During this time lag, the cylinder continues to fill with suction gas.

3. When the piston reaches its lowest point of travel, called bottom dead-center, a spring on the suction valve closes the valve.

The following events take place inside a reciprocating compressor cylinder during the compression stroke of operation (*Figure 3*):

1. At the start of the compression stroke, both the suction and discharge valves are closed. The piston begins its upward stroke, compressing the gas in the cylinder. This increases the pressure and saturation temperature of the gas. The superheat of the gas is also increased. The suction valve remains held shut by the pressure in the cylinder while the discharge valve remains shut by the pressure in the discharge line.

2. As the piston continues its upward travel, the discharge valve remains held shut by the pressure in the discharge line until the piston gets close to the top of the cylinder. At this point in travel, the pressure in the cylinder becomes

greater than the pressure in the discharge line, and the discharge valve opens. This allows the high-temperature, high-pressure gas to pass into the discharge line on its way to the condenser.

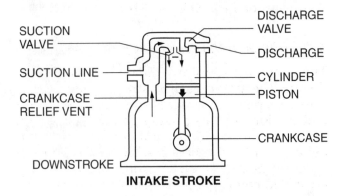

INTAKE STROKE

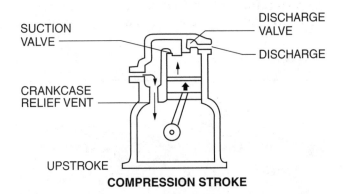

COMPRESSION STROKE

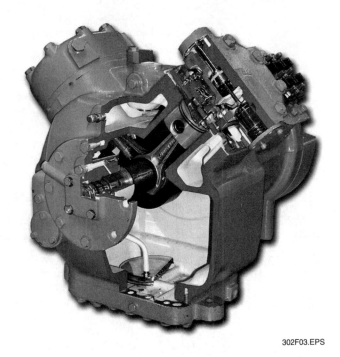

302F03.EPS

Figure 3 ◆ Reciprocating compressor.

3. At the top of its travel, called top dead-center, the piston stops moving and leaves gas in a clearance area at the top of the cylinder. This area is called the **clearance volume**. The gas, which is at the discharge pressure, will re-expand when the piston starts back down at the start of the next intake stroke.

4.1.1 Piston and Piston Rings

The piston is exposed to the high-pressure gas during the compression stroke. During the upstroke, pistons have high-pressure gas on top and suction or low-pressure gas on the bottom. The piston must slide up and down in the cylinder in order to pump the gas. Two types of piston rings are mounted on the piston. The upper rings are called compression rings. They prevent the high-pressure discharge gas from leaking past the piston into the crankcase. Rings mounted on the piston below the compression rings are called

oil rings. They function to control the oil flow past the piston. Some small hermetic compressors have no rings. In these compressors, the refrigerant oil acts as the seal.

4.1.2 Compressor Cylinder Valves

Two types of compressor valves are in common use. They are the reed (flapper) valve and the ring valve (*Figure 4*). Both types can be used either as a suction or discharge port valve. The reed valve is a thin, flexible piece of spring steel mounted on a valve plate. The valve plate mounts between the head of the compressor and the top of the cylinder wall. Some portion of the valve covers the suction or discharge port. The natural spring tension of the reed valve material works to keep the valve closed. It is forced open by pressures that exceed the valve spring tension. Reed valves have many different shapes, with each manufacturer having an individual version.

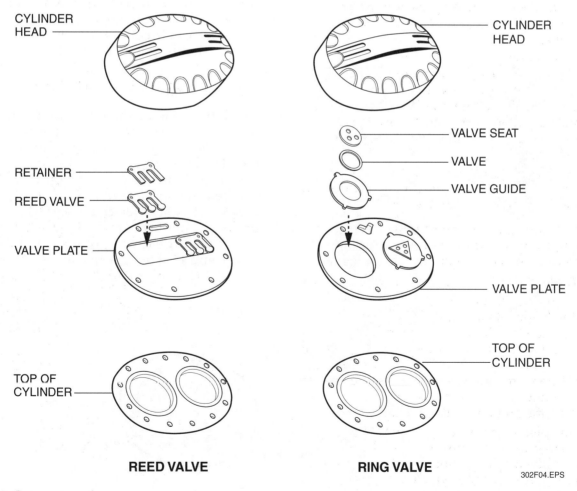

Figure 4 ◆ Compressor valves.

302F04.EPS

Ring valves are also mounted on valve plates. The assembly consists of valve seats, circular ring valves, and valve guides. The valve plate is mounted at the top of the cylinder, with the suction valve below the plate and the discharge valve above the plate. The ring valves are held in the closed position by a number of small springs installed in equally spaced holes around the guide. When the pressure exerted under either the discharge or suction valve is greater than the tension of the valve springs, the valve lifts from its seat, allowing gas to flow.

The operation, seating, and tightness of the valves is important. Broken valves, bent valves, or leaking valve plate gaskets affect the seal between the high-pressure and low-pressure sides of the compressor. High-pressure, high-temperature gas will leak into the suction side of the compressor. If this happens, the suction pressure will rise and the discharge pressure will be reduced. This can result in the compressor overheating. If the leak is small, such as might occur with a slightly bent or cracked valve, the pressure differences from normal operation will also be small. Continuous running of the compressor, low capacity, or upper cylinder head overheating may indicate that a compressor has worn or broken valves, or blown gaskets.

NOTE

Under normal operation, all refrigerant applied to a compressor is vaporized. However, if a problem occurs, such as a faulty metering device, a slug of liquid refrigerant may be drawn into the compressor. This can have the same disastrous effect in a compressor as when birds are sucked into a jet engine. Liquid refrigerant does not compress. Trying to compress a slug of refrigerant in a compressor generates pressures of over 1,000 psi. When applied to the compressor's valves or connecting rods, this pressure can damage (and even break) the valves or rods.

Overheating of the compressor resulting from high suction superheat, high compression ratios, or air trapped in the refrigeration system can also damage the compressor by carbonizing the discharge valves and/or valve guides.

4.1.3 Open Compressor Crankshaft Seals

Open compressors must use a leakproof seal where the crankshaft exits the compressor crankcase. This seal prevents the refrigerant from leaking out of the compressor, regardless of whether or not the shaft is rotating. In certain low-temperature systems, such as those using R134a refrigerant, the compressor crankcase pressure can be in a vacuum (below 0 psig). With this condition, a leaking seal can allow air and water vapor to enter the refrigerant system. A leaking seal usually causes the loss of the system refrigerant, resulting in constant running of the compressor and poor cooling.

All seals use two rubbing surfaces. One surface turns with the crankshaft and is sealed to the shaft with an O-ring or synthetic material. The other surface is stationary and is mounted on the housing with leakproof gaskets. Most compressors use a bellows seal or a variation of the bellows seal. The bellows seal (*Figure 5*) consists of a bellows assembly, carbon ring, and cover plate installed on the compressor crankshaft in the same order as listed. The sealing surface is between the nose of the bellows and a shoulder on the crankshaft. Other seals in use are the packing gland, diaphragm, and rotary.

If installed correctly, crankcase seals can last for years without wearing out. For long seal life, it is important that the compressor shaft be properly aligned with the mating drive motor (or engine) shaft. In a direct-drive unit, the shafts should be aligned according to the manufacturer's instructions. In a belt-drive unit, the belts must be adjusted for the proper tension.

4.1.4 Lubrication Systems

Two lubrication methods are used with reciprocating compressors to lubricate the moving parts. Some compressors use a **splash lubrication system**. In the splash lubrication system (*Figure 6*), crankcase oil is splashed onto the cylinder walls and bearing surfaces of the compressor during each revolution of the crankshaft while the compressor is running. Some compressor connecting rods have little dips or scoops attached to the lower end that scoop up the oil and splash it around to the other parts.

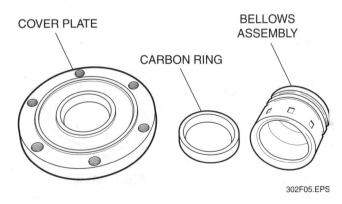

Figure 5 ◆ Bellows-type crankshaft seal.

Oil-Free Semi-Hermetic Compressor

Improved compressor efficiency translates into reduced consumption of electricity. In the HVACR compressor world, the introduction of the oil-free semi-hermetic compressor represents a significant improvement over the standard oil-flooded semi-hermetic compressor at full and part load.

GOING GREEN

302SA01.EPS

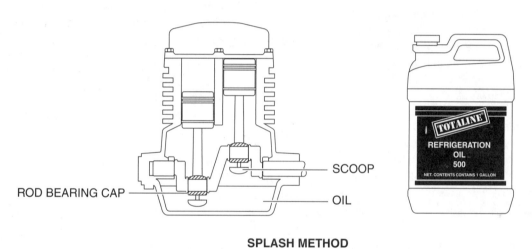

ROD BEARING CAP

SCOOP

OIL

SPLASH METHOD

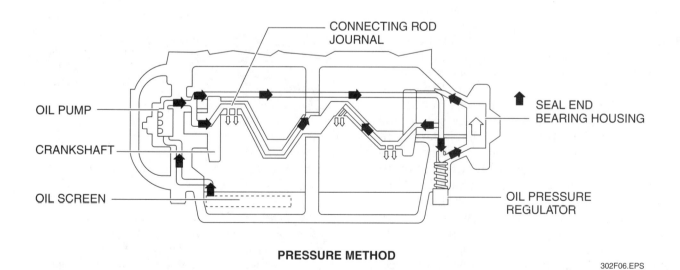

CONNECTING ROD JOURNAL

OIL PUMP

CRANKSHAFT

OIL SCREEN

SEAL END BEARING HOUSING

OIL PRESSURE REGULATOR

PRESSURE METHOD

302F06.EPS

Figure 6 ◆ Compressor lubrication systems.

The second and most popular method of lubrication is the **pressure (force-feed) lubrication system**. This method is used in all sizes of compressors. It uses an internal oil pump mounted on the end of the crankshaft. This pump forces oil through the crankshaft to the bearing surfaces. In some cases, the connecting rods are drilled so the oil under pressure is also supplied to the piston pins. An oil screen may be located in the bottom of the crankcase to filter the oil supplied from the crankcase to the oil pump. An oil pressure regulator in some systems prevents the buildup of excessive oil pressure, which could result in high power consumption, loss of oil, and damage to the compressor. *Figure 6* shows typical pressure (force-feed) lubrication systems. The arrows show the direction of oil flow through the compressor.

NOTE

A certain amount of compressor lubricating oil is normally entrained in the refrigerant being circulated in a system. The uniform movement of this oil through the system piping and back to the compressor is dependent on a reasonably high refrigerant velocity. However, at minimum load, the movement of refrigerant within the system can be greatly reduced, resulting in the trapping of oil within the system evaporator and discharge and suction piping. Excessive trapping of oil can reduce the oil level in the compressor, eventually resulting in bearing failure.

4.2.0 Rotary Compressors

Rotary compressors are usually welded hermetic compressors. The motor and drive shaft are mounted vertically in the compressor housing with the motor above the compressor. Rotary compressors are usually of two types: stationary vane and rotary vane. In both types, the vanes, under spring tension, slide in and out of their retaining slots to provide a continuous seal for the refrigerant vapor within the cylinder. Rotary compressors do not have suction valves, but use a discharge valve to prevent backflow of refrigerant into the compressor when it is turned off. Normally, they have a check valve installed in the intake passage to prevent the compressor oil from being forced back into the suction line and into the evaporator. Lubrication in the rotary compressor is provided either by a splash system or force-feed system. Regardless of the method, proper operation of the rotary compressor depends on maintaining a continuous film of oil on the cylinder, roller, and vane surfaces.

4.2.1 Stationary Vane Compressors

In the stationary vane compressor (*Figure 7*), a shaft with an attached off-center (eccentric) rotor rotates or rolls around the cylinder. As it rotates, one point on its circumference is always in contact with the cylinder wall. A stationary vane that is under spring tension is mounted in the compressor housing. It slides in and out and follows the out-of-round motion of the rotor as the rotor moves within the cylinder. This vane also isolates the suction and discharge sides of the cylinder. As the shaft turns, the rotor rolls around the cylinder, drawing suction gas in the intake opening while at the same time compressing the gas against the cylinder wall on the discharge or compression side. This process continues as long as the compressor is running. An exhaust valve mounted on the discharge port keeps the compressed gas from leaking back into the cylinder and into the suction side during the off cycle.

4.2.2 Rotary Vane Compressors

Rotary vane compressors (*Figure 8*) have a rotor centered on the drive shaft. However, the drive shaft is positioned off-center in the cylinder. Mounted on the rotor are two or more vanes that slide in and out to follow the shape of the cylinder. As the rotor turns, low-pressure suction gas from the suction line is drawn into the cylinder behind the vanes. The trapped vapor ahead of the vanes is compressed against the cylinder wall until it is forced out of the discharge opening. The vanes also keep the compressed gas from mixing with the incoming low-pressure gas.

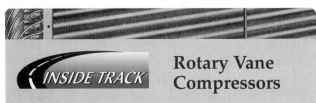

INSIDE TRACK

Rotary Vane Compressors

Rotary vane compressors must be installed in clean systems. If small particles of debris are present, they can jam the rotor and stop the compressor. For this reason, rotary compressors are mainly used in packaged room air conditioners and packaged terminal air conditioners (PTACs) where the cleanliness of the sealed system can be controlled at the factory. In the past, rotary compressors were used in some split air conditioner systems. However, dirt and contamination introduced during installation caused a high rate of compressor failure, prompting manufacturers to discontinue their use in those systems.

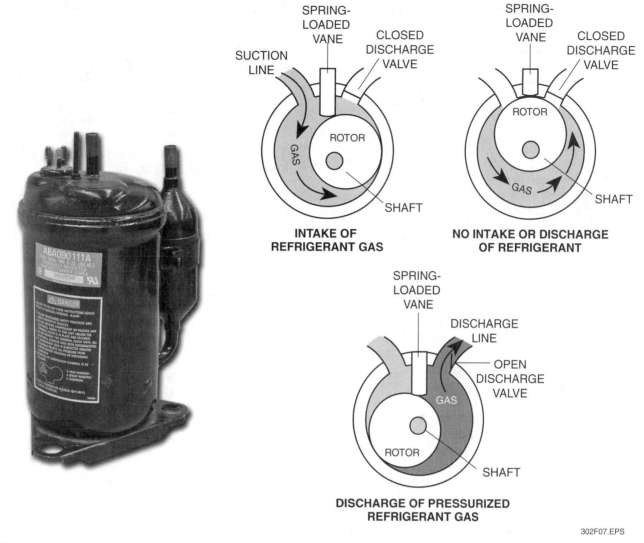

INTAKE OF REFRIGERANT GAS

NO INTAKE OR DISCHARGE OF REFRIGERANT

DISCHARGE OF PRESSURIZED REFRIGERANT GAS

302F07.EPS

Figure 7 ◆ Rotary compressor.

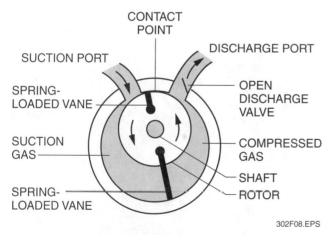

302F08.EPS

Figure 8 ◆ Rotary vane compressor operation.

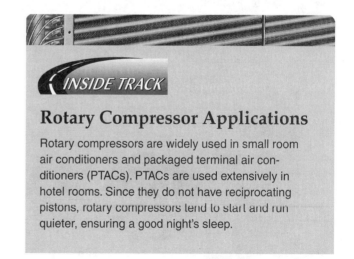

Rotary Compressor Applications

Rotary compressors are widely used in small room air conditioners and packaged terminal air conditioners (PTACs). PTACs are used extensively in hotel rooms. Since they do not have reciprocating pistons, rotary compressors tend to start and run quieter, ensuring a good night's sleep.

4.3.0 Scroll Compressors

Scroll compressors are usually welded hermetic compressors. The motor and drive shaft are mounted vertically in the compressor housing with the motor below the compressor. This compressor has neither suction nor discharge valves because its construction resists refrigerant backflow.

Of all the compressor types, the scroll compressor has the fewest working parts. It operates efficiently even in applications that have large changes in refrigerant pressures, such as with commercial refrigeration and heat pumps.

The scroll compressor achieves compression through the use of two spiral-shaped parts, called scrolls (*Figure 9*). The upper scroll is fixed; the lower scroll is driven by the motor and moves with an orbiting action inside the fixed scroll. There is contact between the two. Refrigerant gas enters the suction port at the outer edge of the scroll and, after compression, is squeezed out a separate discharge port at the center of the stationary scroll. The orbiting action draws gas into pockets between the two spirals. As this action continues, the gas opening is sealed off and the gas is compressed and forced into smaller pockets as it progresses toward the center.

A version of the scroll compressor, called a **compliant scroll compressor**, allows the orbiting scroll to temporarily shift from its normal operating position if liquid refrigerant enters the compressor.

> **WARNING!**
> The top shell of an operating scroll compressor is very hot.

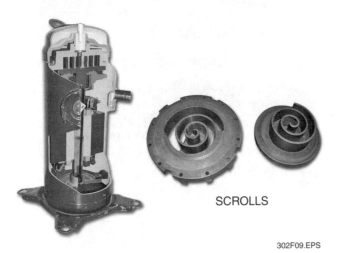

SCROLLS

302F09.EPS

Figure 9 ◆ Scroll compressor.

INSIDE TRACK

Scroll Compressors

Scroll compressors make unique sounds when starting up, running, and shutting down. If you are not aware that these sounds are normal, you may think there is something wrong with the compressor. Various compressor manufacturers and equipment manufacturers have training programs and materials available that can help you recognize the normal and abnormal sounds made by scroll compressors.

4.4.0 Screw Compressors

Screw compressors are used in large commercial and industrial applications requiring capacities from 50 to 750 tons. They are made in both open and hermetic styles.

Screw compressors use a matched set of screw-shaped rotors (*Figure 10*), one male and one female, enclosed within a cylinder. The male rotor is driven by the compressor motor. In turn,

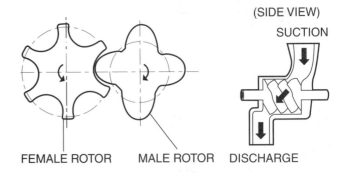

FEMALE ROTOR   MALE ROTOR   DISCHARGE

(SIDE VIEW)
SUCTION

SCREW COMPRESSOR ROTORS

302F10.EPS

Figure 10 ◆ Screw compressor.

it drives the female rotor. Normally, the driven male rotor turns faster than the female rotor because it has fewer lobes than the female rotor. Typically, the male has four lobes, and the female has six. As these rotors turn, they mesh with each other and compress the gas between them. Oil keeps them from actually touching. The screw threads form the boundaries separating several compression chambers, which move down the compressor at the same time. In this way, the gas entering the compressor is moved through a series of progressively smaller compression stages until the gas exits at the compressor discharge in its fully compressed state.

4.5.0 Centrifugal Compressors

Centrifugal compressors are made in open and hermetic designs. They are typically used on commercial and industrial refrigeration and air conditioning systems using compressors larger than 100 tons. Standard models range up to 10,000 tons of capacity; custom-built models exceed 20,000 tons.

Centrifugal compressors use a high-speed impeller with many blades that rotate in a spiral-shaped housing (*Figure 11*). The impeller is driven at high speeds (typically 10,000 rpm) inside the compressor housing. Refrigerant vapor is fed into the housing at the center of the impeller. The impeller throws this incoming vapor in a circular path outward from between the blades and into the compressor housing. This action, called centrifugal force, creates pressure on the high-velocity gas and forces it out the discharge port. Often, several impellers are put in series to create a greater pressure difference and to pump a sufficient volume of vapor. A compressor that uses one impeller is a single-stage machine; one which uses two impellers is a two-stage machine, and so on. When more than one stage is used, the discharge from the first stage is fed into the inlet of the next stage.

5.0.0 ◆ CAPACITY CONTROL OF COMPRESSORS

When a cooling system is operating properly, it undergoes a reduction in load whenever there is less heat for the system to dissipate. Loads handled by commercial cooling systems change rapidly and vary more than those usually encoun-

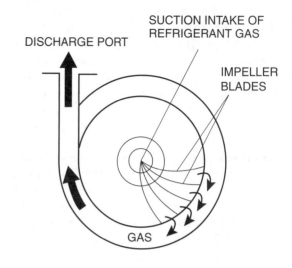

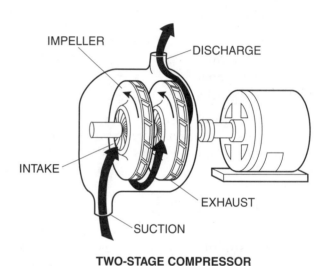

TWO-STAGE COMPRESSOR

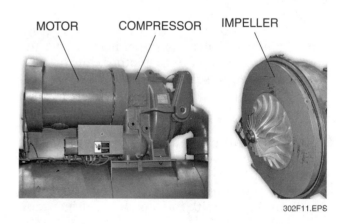

302F11.EPS

Figure 11 ◆ Centrifugal compressor.

tered by residential systems. Commercial systems typically operate at 50 percent of the design (peak) cooling load. If the system compressor operated at its full capacity under these conditions, it would lower the evaporator pressure and the dew point of the evaporator coil, causing the relative humidity and temperature within the building to drop below a comfortable level. If the dew point dropped low enough, frost would build up on the evaporator coil, restricting airflow and resulting in compressor **short cycling** and erratic system operation. If the condition continued, it might even cause a compressor failure.

Common methods of controlling compressor capacity include the following:

- On/off cycling
- Multiple compressors
- Cylinder unloading
- Hot gas bypass
- Intake slide valves
- Inlet guide vanes
- Multiple-speed motors

Capacity control of compressors is one of several methods used to change the pumping capacity of a compressor in order to match changes in the system load.

5.1.0 On/Off Cycling

The simplest form of capacity control is the cycling on and off of the compressor. In a standard air conditioning system, a room thermostat supplies a call-for-cooling signal to the equipment. This energizes the compressor, allowing it to run until the temperature in the room reaches the thermostat's setpoint. At that time, the thermostat circuit opens and the call-for-cooling signal is removed. Once the signal is removed, the compressor cycles off until another call-for-cooling signal energizes it.

5.2.0 Multiple Compressors

The use of multiple compressors is an effective way to achieve capacity control. The two most common methods include multiple compressors with a common refrigeration system and multiple compressors with independent refrigeration systems.

When installed with a common refrigeration system, multiple compressors share a common evaporator, condenser, and refrigerant piping. The compressors are essentially connected in par-

allel, often by use of a manifold. When demand is low, only a single compressor may be required to handle the load. As the load increases, more compressors are brought on line. One problem that must be addressed in this piping arrangement is maintaining equal oil levels in the multiple compressors. If one compressor runs more often than others, most of the oil in the system may be returned to that compressor, starving the other compressors of oil. To maintain oil levels, an oil equalizer line is installed between the compressors and located at the normal oil level. When all compressors are off, oil in all the compressors will equalize at the common level.

Other unique design features can help maintain equal oil levels. When two compressors are connected in parallel and the upstream compressor runs, most of the oil will be returned to the upstream compressor. The use of a restrictor in a common suction line provides a small pressure drop when the downstream compressor is running. This keeps the oil pressure in the sump of the downstream compressor slightly lower than the oil pressure in the upstream compressor. This pressure differential allows excess oil in the upstream compressor to move through the oil equalizer line to the downstream compressor (*Figure 12*).

Another common way to employ multiple compressors for capacity controls is to design a system that has independent circuits for each compressor. In a system using two compressors, the evaporator and condenser coils contain two independent circuits completely isolated from each other. Each compressor has a separate piping system and each system has its own refrigerant charge. When demand is low, a single compressor runs, using its part of each coil to function.

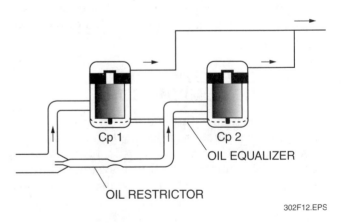

Figure 12 ◆ Oil management.

As additional capacity is needed, the other compressor comes in, using its portion of both coils. At that point, both coils are fully utilized. In such an arrangement, condenser and evaporator fan speeds are adjusted up and down to match system capacity. This piping arrangement eliminates the problem of oil level management between the two compressors.

5.3.0 Cylinder Unloading

Cylinder unloading is a popular method of capacity control used with open and semi-hermetic reciprocating compressors, usually above 20 tons of capacity. A control valve internal to the compressor blocks the flow of refrigerant gas to some of the cylinders of a multi-cylinder compressor. The cylinders continue to move up and down but do not pump refrigerant into the system. Four types of cylinder unloading methods are in common use:

• Suction bypass unloading
• Suction cutoff unloading
• Hydraulic unloading
• Cylinder bypass

The suction bypass unloading method uses a special cylinder head and a pressure actuated or electrically actuated unloader control valve. When the system cooling load drops, the pressure and temperature at the compressor suction input also drops. The unloader control device senses this decrease in pressure (pressure actuated valve) or temperature (electrically actuated valve) and operates to actuate the unloader. A piston in the unloader retracts at a preset point and unloads the cylinder. This causes the suction refrigerant gas to be recirculated between the discharge and suction sides of the unloaded cylinder. *Figure 13* shows a pressure actuated bypass unloader operating in the unloaded position. The suction cutoff unloading method is actuated in a similar manner. With a suction cutoff unloader, the suction gas is prevented from entering the unloaded cylinder, rather than being recirculated through it.

Normally, both the suction bypass and suction cutoff types of unloader valves have adjustments to set the operating point at which the control valve will load and unload the cylinders. One adjustment, called the control setpoint, adjusts the point where the cylinder loads. Another adjustment is made to provide a pressure differential between the cylinder load and unload point.

The hydraulic unloader achieves unloading in compressors by holding open the suction valve, thus preventing compression of the gas. The unloader valve lifting mechanism is powered by the compressor lubricating system. When capacity reduction is needed, oil pressure is relieved from a piston located in the capacity reduction unit. This piston operates a mechanical linkage to lift the suction valve from its seat. The compressor piston is then no longer able to compress refrigerant within the cylinder and just travels up and down within the cylinder.

Like the other unloaders, the hydraulic unloader has adjustments to set the operating point at which the cylinders are loaded and unloaded.

Another type of capacity control is used in welded hermetic compressors in which one or both pistons in a two-cylinder compressor are engaged. To accomplish this, one of the pistons is equipped with a mechanism that only allows that piston to engage if the compressor crankshaft is rotating in a certain direction. With both pistons engaged, the compressor delivers full capacity. When the compressor motor reverses rotation, the crankshaft reverses rotation. The piston then disengages and the compressor delivers 50 percent capacity. The second cylinder of the compressor is effectively bypassed.

5.4.0 Hot Gas Bypass

The hot gas bypass method reduces compressor capacity by routing some of the compressor hot gas discharge through a bypass line back into the suction line. The flow of hot gas through the bypass line is controlled by a solenoid stop valve in the line.

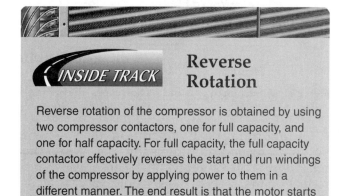

INSIDE TRACK

Reverse Rotation

Reverse rotation of the compressor is obtained by using two compressor contactors, one for full capacity, and one for half capacity. For full capacity, the full capacity contactor effectively reverses the start and run windings of the compressor by applying power to them in a different manner. The end result is that the motor starts and runs in the opposite direction.

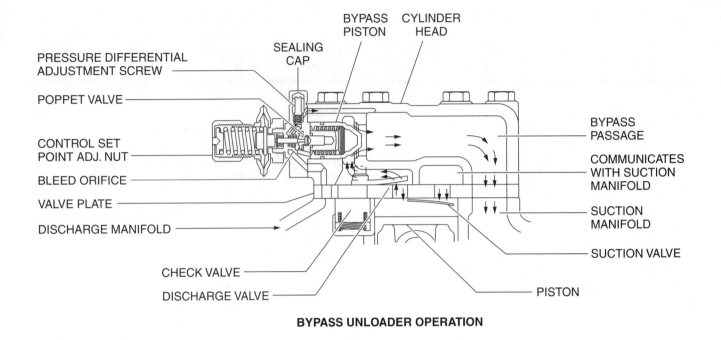

BYPASS UNLOADER OPERATION

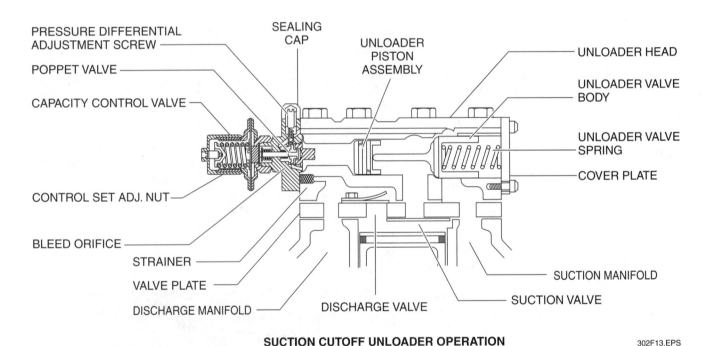

SUCTION CUTOFF UNLOADER OPERATION

302F13.EPS

Figure 13 ◆ Suction bypass and suction cutoff unloaders.

During full capacity operation, the solenoid valve is closed and blocks the line. When the related pressure or temperature sensor calls for a reduction in capacity, the solenoid valve is actuated and the bypass line is opened. This allows some of the hot gas discharged from the compressor to be returned to the suction line. In this method, the amount of capacity reduction is determined by the amount of gas bypassed.

Scroll compressors use a form of bypass to achieve capacity control. An internal solenoid operates bypass ports. At low capacity, the solenoid is de-energized and the bypass ports are open (*Figure 14*). At high capacity, the solenoid is energized and the bypass ports are closed. The opening and closing of the bypass ports effectively changes the volume in which the refrigerant is compressed.

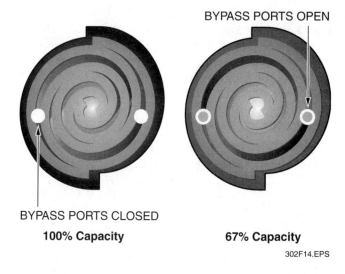

BYPASS PORTS OPEN

BYPASS PORTS CLOSED

100% Capacity

67% Capacity

302F14.EPS

Figure 14 ◆ Scroll compressor capacity control.

5.5.0 Intake Slide Valve

The intake slide valve method of capacity control is used in screw compressors. It uses a sliding seal mechanism that varies the compressor capacity by changing the point where the suction entraps the vapor and starts to compress it as it moves through the rotors.

5.6.0 Inlet Guide Vane

The inlet guide vane method of capacity control is used in centrifugal compressors. It uses inlet guide vanes mounted in front of the impeller inlet. These vanes open and close while directing the refrigerant vapor into the impeller, changing compressor performance. The more the vanes are closed, the lower the capacity of the compressor.

5.7.0 Compressor Speed Control

All compressors, except the centrifugal compressor, are **positive-displacement compressors**. Positive-displacement compressors are those in which the pumping action is created by moving pistons or moving chambers. When they are operated at a constant speed, they pump a constant volume of gas. The higher the compressor displacement, the higher its capacity. For a reciprocating compressor, the displacement is calculated as follows:

Compressor displacement =
     piston displacement × rpm

Compressor displacement (cubic inches) =
     $(pD^2Ln \div 4) \times rpm$

Where:

     p   = 3.1416

     D   = cylinder bore (in inches)

     L   = length of stroke (in inches)

     n   = number of cylinders

     rpm = revolutions per minute

The displacement, and therefore the capacity, of a compressor is proportional to the speed of its drive motor. For this purpose, multiple-speed motors are often used to drive compressors, especially reciprocating, scroll, and rotary compressors. Electronic controls that automatically select the proper motor speed to match load conditions make their use practical.

Variable-speed motors have been used to drive compressors. However, the complexity and expense of variable-speed motors have limited their use.

> **NOTE**
> One equipment manufacturer has patented technology that promises increased use of variable-speed motors to drive compressors.

Multiple-speed motors contain dedicated sets of motor windings used for specific motor speeds. Care must be taken in the design of the circuits controlling the motor to ensure that windings for different speeds are not energized simultaneously. Doing so would damage the motor. Often, separate contactors are used for the different motor windings. They are typically mechanically and electrically interlocked to prevent simultaneous energizing.

5.8.0 Scroll Capacity Modulation

One manufacturer of scroll compressors has introduced a means of modulating the capacity of the scroll compressor. The scroll compressor uses a fixed upper plate and an orbiting lower plate. Scrolls considered compliant are able to lift the upper plate (an axial movement, because it moves in the same direction as the drive motor shaft) when placed under stress by liquid refrigerant, which cannot be compressed. Capacity control in the scroll compressor is achieved by taking this feature one step further. In 20-second cycles, the two plates are either separated to cease compression (unloaded) or brought together for normal compression (loaded). Over a period of time, the number of 20-second cycles in the loaded or unloaded position are managed to precisely track the load. With this strategy, scroll compressor capacity can be infinitely modulated between 10 and 100 percent, resulting in reduced operating costs and consistent load control with fewer compressor starts required.

6.0.0 ◆ COMPRESSOR ELECTRIC MOTORS

The electric motors used to drive compressors operate in the same way as other motors used in refrigeration systems. You have studied the principles of operation for electric motors previously in the *HVAC Level Two* module, *Alternating Current*. This section describes cooling, mechanical, and electrical considerations specific to compressor motors. Also described are methods commonly used to protect compressors and compressor motors.

6.1.0 Compressor Motor Cooling

Unless specially designed, open compressors and their drive motors must be located in a well-ventilated area to aid in motor cooling. Electric motors used to drive open compressors are cooled by the surrounding air. Some have ventilation openings on their end bells or sides to pass external cooling air over and around the windings. This type of motor must be used in relatively clean locations to prevent the entrance of dust and other foreign material into the motor housing. Other motors used to drive open compressors are totally enclosed to prevent free air exchange between the inside and outside. They depend on radiation for cooling the windings. Often, the enclosed motor has external fins and/or a fan connected to its shaft to aid in cooling.

In hermetic compressors, the electric motor is enclosed in the same housing as the compressor. As the motor turns the compressor, electric energy is changed into mechanical energy and heat energy. This heat must be removed or it will build up, causing compressor and motor damage. In serviceable-hermetic and welded-hermetic compressors, cool refrigerant from the evaporator passes over and cools the motor windings before the refrigerant enters the suction side of the compressor.

6.2.0 Compressor and Drive Motor Shaft Alignment

Electric motors used to drive open compressors are coupled to the compressor crankshaft through pulleys and belts or a flexible coupling. Proper bearing wear depends on achieving the best possible alignment of the compressor and the drive motor pulleys or flexible couplings. The compressor and motor mountings should be firm and rigid to eliminate vibration and bearing wear. Refer to the *HVAC Level Two* module, *Basic Installation and Maintenance Practices* for the methods used to align compressor and motor pulleys and couplings.

6.3.0 Input Power

Depending on the size of the motor and the application, electric motors used to drive compressors usually operate on either single-phase or three-phase AC power. Wiring to a motor must be done in accordance with the latest edition of the *National Electrical Code*® (*NEC*®) and local code requirements. Wire size should be based on the motor nameplate full-load amperage (FLA) or rated-load amperage (RLA), any listed NEC de-rating factor, and any increase in wire size needed to prevent voltage drop on long runs.

To prevent damage to motor windings, make sure the operating voltage, frequency, and phase stamped on the motor nameplate are compatible with the input electric power source. In three-phase systems, also make sure that the input voltage phase imbalance is no more than 2 percent. Should the imbalance exceed 2 percent, contact your local power company.

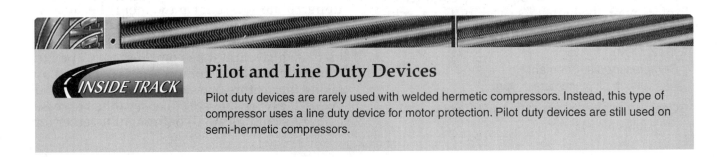

Pilot and Line Duty Devices

Pilot duty devices are rarely used with welded hermetic compressors. Instead, this type of compressor uses a line duty device for motor protection. Pilot duty devices are still used on semi-hermetic compressors.

6.4.0 Compressor Motor Overload Protection

The most common causes of motor failure are overloads and overheating. An overload condition is produced when current to the motor exceeds the motor's normal operating current flow. An overload condition may result in melted conductors or burned insulation on the motor's wiring. Considerable damage may result if the compressor motor overheats. Overheating can occur without the current draw becoming excessive. Heat can be caused by a defective start relay, excessive load, or loss of refrigerant gas cooling in a hermetic compressor. Another cause is operation at too high or too low a voltage. In three-phase motors, a leg-to-leg voltage imbalance exceeding 2 percent will shorten the life of the motor.

Another problem common only to three-phase motors is **single phasing**. Single phasing is when the motor continues to run after one of the three input phases is lost while the motor is operating. The resulting imbalance increases the temperature in the remaining two windings, causing the motor to overheat. Devices designed to prevent single phasing of a three-phase motor consist of an electronic module and current sensors that monitor current flowing in each of the three motor windings. If current flow ceases in one of the three windings, the sensor sends that information to the electronic module. The module, in turn, opens the circuit to the compressor contactor, removing power from the motor to prevent any damage from occurring.

It is important to protect a motor from both current overloads and overheating. To accomplish this protection, current-sensing and/or heat-sensing devices are used to open the circuit before damage is caused to the motor. Electrical protection devices used with compressors are classified as either **pilot duty devices** or **line duty devices** (*Figure 15*).

- Pilot duty devices sense current overload or temperature within the motor and open the motor contactor control circuit to remove power from the motor.
- Line duty devices sense current flow and temperature in the motor winding and open the winding circuit to remove the line voltage when an overload occurs.

Once an overload device has opened or tripped, it must be reset manually or automatically before the motor can operate again. A manual-reset overload device must be physically reset once it has tripped. An automatic-reset overload device will automatically reconnect the power to the motor after the overcurrent or over-temperature condition has passed. Reset time of automatic devices is based on recovery from the out-of-tolerance condition, and can vary from seconds to hours. Many types of motor protection devices are used, including:

- External line break overloads
- Internal line break overloads
- Motor thermostat overloads
- Electronic overloads
- Three-phase overloads
- Current monitoring devices

The operation of the devices described in this section are typical of those you will encounter in the field.

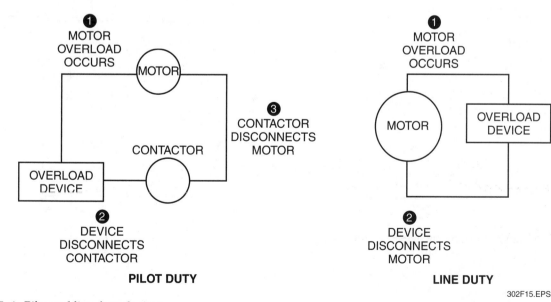

302F15.EPS

Figure 15 ◆ Pilot and line duty devices.

6.4.1 External Line Break Overloads

The external line break overload, often called a klixon, is a common single-phase motor overload protector. Some types provide current protection, while others are sensitive to both temperature and current. These devices use a bimetal warped-disc switch. Generally, the external line break is located in a motor terminal box and connected for line duty in series with the motor's common terminal. Some versions are used in pilot duty applications. In this case, the contacts are connected in series with the contactor circuit. Both devices reset automatically after recovery from the overload condition. *Figure 16* shows two examples of an external line break overload.

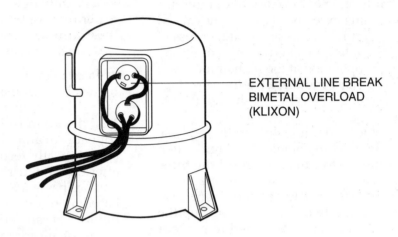

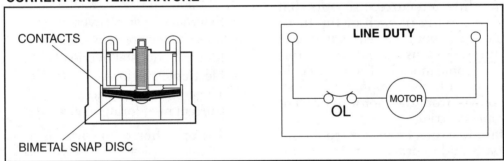

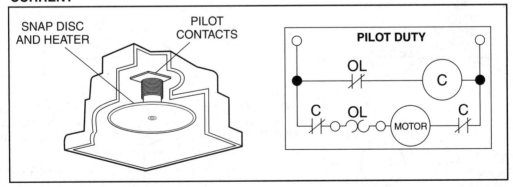

302F16.EPS

Figure 16 ◆ Examples of external line break overloads.

INSIDE TRACK

Line Duty External Overloads

Line duty external overloads are commonly used on the compressors used in room air conditioners.

6.4.2 Internal Line Break Overloads

The internal line break overload is used often in single-phase hermetic compressors. This protector is a line duty device located inside the motor (*Figure 17*). It is placed in series with the terminal motor common and trips on either winding current or temperature, or a combination of both. It resets automatically after recovery of the overload condition. Because this type of protector is

INTERNAL CURRENT AND TEMPERATURE

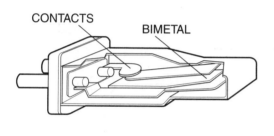

CONTACTS

BIMETAL

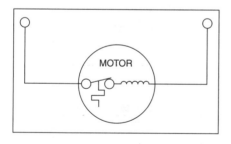

MOTOR

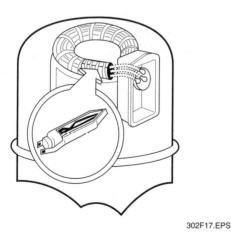

302F17.EPS

Figure 17 ◆ Example of an internal line break overload.

located inside the compressor motor, it is impossible to remove and inspect. The complete motor must be tested in order to isolate a faulty protector.

6.4.3 Motor Thermostat Overloads

Motor thermostats protect motors from overheating. They are either externally mounted or internally mounted directly on the motor windings. Three thermostat overloads are shown in *Figure 18*. The external shell device is mounted on the motor shell. When the shell overheats, the disc warps upward to open the control circuit. The internal device is wound into the motor windings and will open the control circuit when an over-temperature condition occurs. The hermetic motor thermostat shown on the bottom in *Figure 18* has rate-of-rise compensation. The case and internal strip expand at different rates. It trips early if the temperature rises rapidly, such as in a locked rotor condition. If the temperature rise is gradual, it will trip at its normal setting. This method prevents a rapidly rising temperature from overshooting the trip-out point and damaging the motor. The reset of all types is automatic.

6.4.4 Electronic Overloads

Thermistors are resistors that vary their resistance in response to temperature changes. They can decrease resistance as temperature increases (negative temperature coefficient thermistor, or NTC), or increase resistance as temperature increases (positive temperature coefficient thermistor, or PTC). Either type can be embedded in a compressor motor winding to sense temperature. As motor temperature increases, the resistance of the thermistor changes. The resistance change becomes an input to an electronic control that reacts to the temperature increase by opening the control circuit of the compressor motor power circuit. Manufacturers typically supply resistance versus temperature charts that are useful in determining if the thermistor is within range. Heat is the main enemy of a hermetic compressor motor.

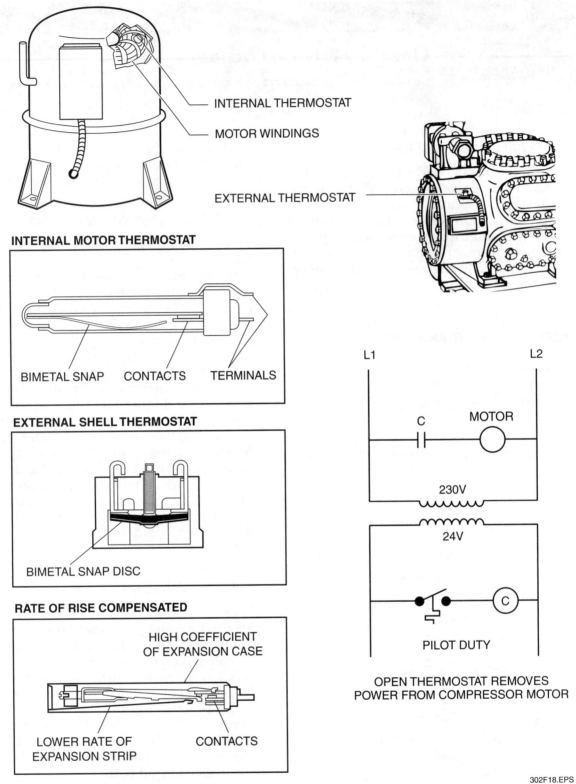

INTERNAL THERMOSTAT

MOTOR WINDINGS

EXTERNAL THERMOSTAT

INTERNAL MOTOR THERMOSTAT

BIMETAL SNAP CONTACTS TERMINALS

EXTERNAL SHELL THERMOSTAT

BIMETAL SNAP DISC

RATE OF RISE COMPENSATED

HIGH COEFFICIENT
OF EXPANSION CASE

LOWER RATE OF
EXPANSION STRIP

CONTACTS

L1   L2

C   MOTOR

230V

24V

PILOT DUTY

OPEN THERMOSTAT REMOVES
POWER FROM COMPRESSOR MOTOR

302F18.EPS

Figure 18 ◆ Open thermostat removes power from compressor motor.

Some of the input to an electric motor is converted to heat as well as to useful work, but excessive heat can damage the motor. Therefore, it is essential that the hermetic motor be effectively protected. One type of overload protection uses a single-module, three-sensor system. The sensors are inserted in the hermetic motor winding (*Figure 19*). The resistance of the sensors increases as the temperature rises. This action causes the module to react and break a set of contacts to open the control circuit of the compressor when the temperature reaches a preset level.

An overload module is shown in *Figure 20*. The module requires a power supply, sensor connections, control circuit contacts, and manual reset connections. *Figure 21* shows the connections that are required between the module and other components in the system.

6.4.5 Three-Phase Motor Overloads

Three-phase connected motors in the smaller horsepower range can also be protected by line duty current-temperature devices like the ones shown in *Figure 22*. They can be internal or external to the motor when used with open compressors, but are always internal when used with hermetic compressors. When an overload occurs, the bimetal disc warps, breaking the electrical circuit. Reset is automatic when the disc cools.

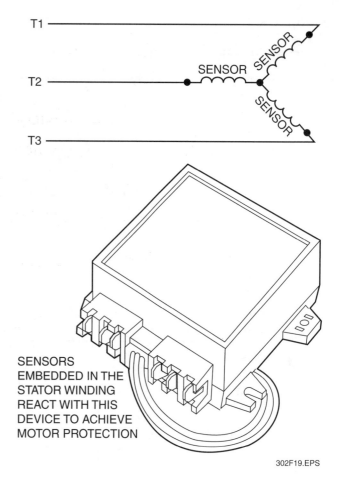

SENSORS EMBEDDED IN THE STATOR WINDING REACT WITH THIS DEVICE TO ACHIEVE MOTOR PROTECTION

302F19.EPS

Figure 19 ◆ Sensor location in a three-phase motor.

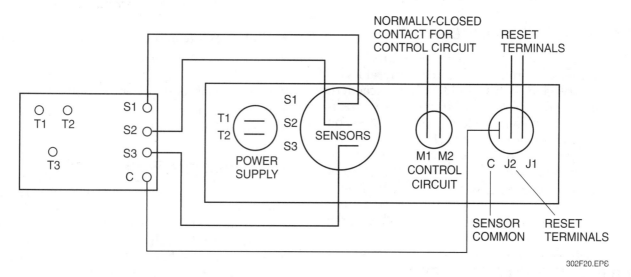

302F20.EPS

Figure 20 ◆ Wiring connections for an overload module.

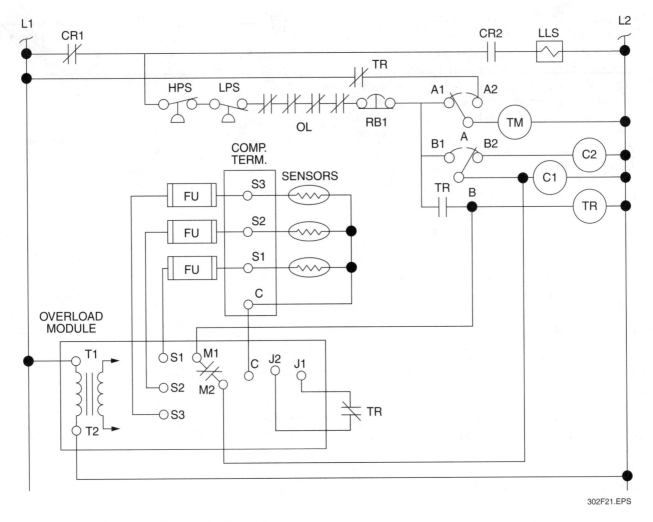

Figure 21 ◆ Control circuit with an overload module.

6.4.6 *Current Monitoring Devices*

Electronic controls have been developed that monitor compressor current to ensure that preset levels are not exceeded. If current levels are exceeded, the control shuts the compressor off. These devices use either current-sensing loops (transformers) on the compressor power leads (*Figure 23*), or are wired in series with the compressor leads. Controls for single-phase compressors monitor current only in the common lead while three-phase controls monitor current in all three legs. Similar current monitoring devices are used in the single-phasing protectors described earlier.

7.0.0 ◆ OTHER COMPRESSOR PROTECTION DEVICES

In addition to its electric drive motor overload devices, a compressor is often protected from damage by other devices. Typically, these protection devices include the following:

- High-pressure and low-pressure protection
- Loss of evaporator or condenser airflow protection
- Operational sequence protection
- Short cycling protection
- Head pressure control

7.1.0 Pressure Protection

The use of high-pressure and low-pressure safety cutout switches is the most common method of protecting the compressor when a pressure problem exists. High discharge or condensing pressure is one of the most harmful conditions affecting the compressor. High condensing temperature raises the temperature of the refrigerant vapor and oil moving across the compressor discharge valves. This could result in oil and refrigerant breakdown. In hermetic compressors, motor cooling depends on the amount and temperature of the returning refrigerant in the suction line. If suction

EXTERNAL CURRENT
AND TEMPERATURE

INTERNAL CURRENT
AND TEMPERATURE

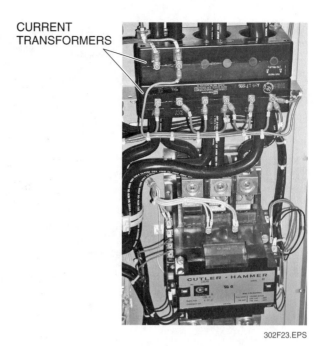

CURRENT
TRANSFORMERS

LINE DUTY

SNAP DISC   HEATER
LINE DUTY

302F23.EPS

Figure 23 ◆ Current-sensing transformer.

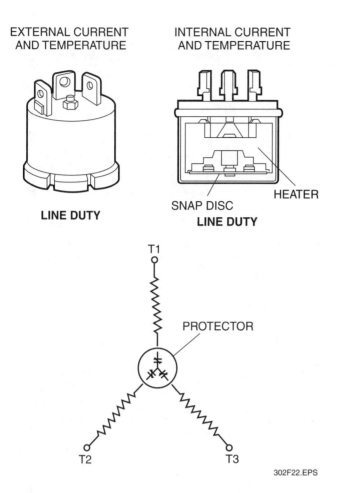

302F22.EPS

Figure 22 ◆ Line duty three-phase overloads.

pressure is too low, the motor may overheat and cause compressor failure.

As shown in *Figure 24*, the high-pressure and low-pressure switches are usually wired in series with the compressor contactor coil and any other protective devices. A sensor tube from the high-pressure switch is attached to the discharge line. The pressure in the discharge line acts against a diaphragm in the switch. The contacts in the

switch are kept closed by a calibrated spring. If the pressure in the system exceeds the switch cutout setting, as determined by the spring tension, the contacts open and stop the compressor. Operation of the low-pressure switch is similar to that of the high-pressure switch, except the sensor tube is connected to the suction line. If the pressure in the suction line falls below the suction cutout level, the spring opens the contacts and stops the compressor. The cutout points for the switches are normally preset at the factory.

7.2.0 Oil Pressure Protection

Larger compressors, primarily semi-hermetic and open-drive units, often incorporate protection against the loss of lubricating oil pressure. Oil

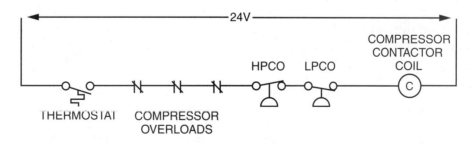

PRESSURE SWITCHES

HPCO = HIGH PRESSURE CUTOUT
LPCO = LOW PRESSURE CUTOUT

302F24.EPS

Figure 24 ◆ Compressor pressure protection devices.

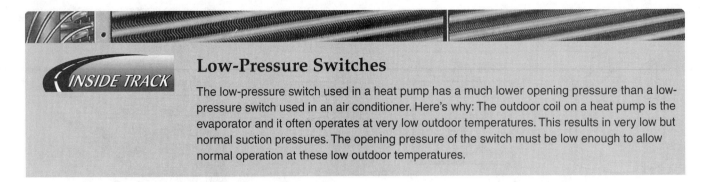
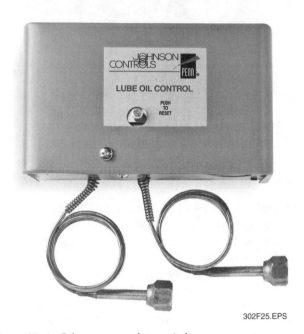

302F25.EPS

Figure 25 ◆ Oil pressure safety switch.

pressure safety switches, such as the one shown in *Figure 25* are generally of the differential pressure style, which requires two pressure inputs to the control. One input is connected to a point on the compressor that provides a suction pressure value, while the other is connected to the discharge pressure of the oil pump. Actual, or net, oil pressure is calculated by subtracting the suction pressure from the oil pump discharge pressure. This is due to the fact that the oil pump inlet is subjected to the system suction pressure.

The safety control incorporates a time delay to allow the compressor time to build proper oil pressure after startup. A set of normally-closed contacts in the control circuit allows the compressor to start. These contacts open after the delay period (often 90 seconds), which begins when the control circuit is energized and the compressor starts. A second set of contacts in the switch, in parallel with the first set, closes when the oil pressure reaches the setpoint. This keeps the control circuit intact after the time delay period has elapsed when proper oil pressure is present. If

proper oil pressure has not been achieved before the end of the time delay period, the control circuit will be broken and the compressor will shut down. The control must then be manually reset using the button on the front before the compressor can start again.

Some oil pressure safety switches are available with fixed differentials to satisfy the design requirements of specific compressor manufacturers. Others are adjustable and can be applied to a variety of compressors. When using the adjustable models, the appropriate minimum oil pressure for the compressor being serviced must be known in order to properly set the control. It is recommended that the pressure connections on the compressor be made using tees with service access valves. This allows a refrigeration gauge manifold to be connected simultaneously to the same points for troubleshooting or recording the net oil pressure during periodic maintenance inspections.

For compressors fitted with the necessary pressure transducers, electronic oil pressure safety switches can be used. This eliminates the copper capillary tubes, replacing them with sensor leads and eliminating the potential from leaks due to broken or chafed tubes.

7.3.0 Lockout Protection

Compressor lockout protection is used to prevent the compressor from operating out of sequence due to the opening and closing of the control circuit by any of the automatic reset safety controls. *Figure 26* shows a lockout relay (sometimes called an impedance relay) used for this purpose. The relay coil, due to its high resistance, is not energized during normal operation. However, when any one of the safety controls opens the circuit to the compressor contactor coil, current flows through the lockout relay coil, causing it to energize and open its contacts. The relay remains energized, keeping the compressor contactor circuit open until the power is interrupted from the control circuit (either by the thermostat or the

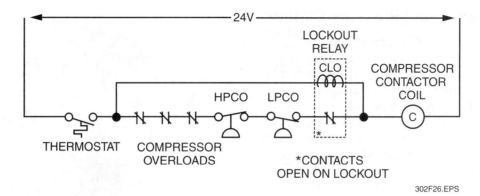

Figure 26 ◆ Compressor lockout relay protection.

main power switch) after the safety control has been reset.

Proper operation of this circuit depends on the resistance of the lockout relay coil being much greater than the resistance of the compressor contactor coil. If the lockout relay becomes faulty, it should be replaced with an exact replacement to maintain the proper circuit balance.

Another type of compressor lockout device uses a current-sensing loop to monitor current in the common leg of a single-phase compressor. This device must detect a compressor current within a few seconds of the compressor contactor energizing or the lockout will open the control circuit and lock it out until manually reset.

7.4.0 Short Cycling Protection

Before restarting a compressor after it has been turned off, enough time should be allowed to pass so that the pressures in the system can equalize. Short cycling is a condition in which the compressor is restarted immediately after it has been turned off. This causes the compressor to restart against a high discharge (head) pressure, which can cause damage to the compressor or motor windings. It can also cause the circuit

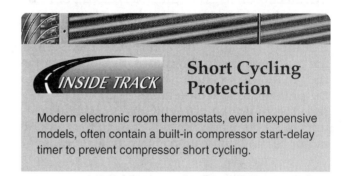

Modern electronic room thermostats, even inexpensive models, often contain a built-in compressor start-delay timer to prevent compressor short cycling.

breaker or overload to open. Short cycling can result when there is a momentary interruption of power to the compressor, such as might occur during severe thunder storms, or when the thermostat setting is manually changed in a manner which first opens then closes its contacts. Short cycling can also be caused by erratic operation of a marginal system. Short cycling protection can be provided by placing a solid-state electronic timer in the compressor contactor control circuit (*Figure 27*). When power is applied, the anti-short cycling device must first time out before power can be applied to the compressor contactor coil. Timers typically have delays ranging from 30 seconds to 5 minutes. The specific time delay used depends on the application.

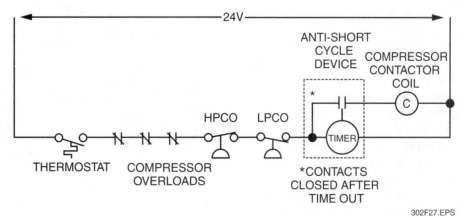

Figure 27 ◆ Compressor anti-short cycle simplified circuit.

7.5.0 Electronic Head Pressure Control

Commercial air conditioners are required to operate over a much wider outdoor temperature range than residential equipment. For example, a cooling unit may need to run in a crowded restaurant even if the outdoor temperature is below freezing. Operating the unit at this low ambient temperature will result in low head pressure and possible liquid flood-back to the compressor.

To maintain a workable head pressure, electronic low ambient temperature head pressure controls (*Figure 28*) are available which control condenser fan motor speed. These controls use thermistors attached to pre-determined points on the condenser coil to monitor the coil temperature. If the thermistors detect the coil is getting too cold, that input is read by the control that then signals the condenser fan to slow down. By slowing down the fan motor, less heat is rejected by the coil and the pressures and temperatures in the coil rise as a result. A motor equipped with special ball bearings must be used with this control to maintain motor bearing lubrication at low motor speeds.

8.0.0 ◆ REDUCED-VOLTAGE MOTOR STARTING

When starting larger compressor motors, the starting current can sometimes be almost six times the rated load current. The motors are built to handle this current, but there may be a large voltage drop in the power system. In these cases, reduced-voltage starting methods are used to control the starting current and limit the voltage drop to a tolerable value. All reduced starting methods use specially designed motor starters and are controlled by adjustable timers between the start and run functions.

Common methods of reduced-voltage starting include the following:

- *Primary resistor or reactor* – This method uses a series resistance or reactance to reduce the current on the first step. After a preset interval, the motor is connected directly across the line. This method can be used with any standard motor.
- *Autotransformer* – An autotransformer is used to directly reduce voltage and current on the first step. After a preset time interval, the motor is connected directly across the line. This method can be used with any standard motor.
- *Wye-delta* – This method induces the voltage across the wye-connection to reduce voltage and current on the first step. After a preset time interval, the motor is connected in delta, allowing full current. This method requires a motor capable of wye-delta connection.

302F28.EPS

Figure 28 ◆ Electronic low ambient temperature head pressure control.

- *Part-winding* – This method uses a motor with two separate winding circuits. When starting, only one winding circuit is energized, allowing the current to be reduced. After a preset time interval, both winding circuits of the motor are connected directly across the line. To avoid overheating and possible damage to the windings, the time between the connection of the first and second windings must be limited. Typically, this time is about four seconds.
- *Solid-state reduced-voltage starters* – Today, solid-state reduced-voltage starters (*Figure 29*) are widely used with large compressor motors and other types of motors in new industrial and commercial HVAC equipment installations. They are also being used to replace older electromechanical-type reduced-voltage starters in retrofit work. Solid-state reduced-voltage starters provide for smooth start and acceleration. For this reason, they are referred to as soft-start controllers.

302F29.EPS

Figure 29 ◆ Solid-state reduced-voltage starters.

9.0.0 ◆ CAUSES OF COMPRESSOR FAILURE

If not corrected, most mechanical refrigeration system problems will result in compressor failure. If you assume the cause of the compressor failure lies within the compressor, it is likely you will make the first of many unnecessary compressor replacements. If the compressor fails and the cause has not been found and eliminated, the replacement compressor will also fail.

Problems that can cause compressor failure include the following:

- Slugging of liquid refrigerant and/or oil in the compression area of the compressor
- **Flooding** of liquid refrigerant into the crankcase of the compressor
- **Flooded starts** when the oil in the crankcase is mixed with a quantity of liquid refrigerant when it migrates back to the compressor at shutdown
- Loss of lubrication by loss of oil or by refrigerant diluting the oil
- Contamination of the refrigeration system with air, moisture, and dirt
- Overheating of the compressor cylinder components and/or the hermetic motor windings
- Electrical irregularities in voltage and current
- Incorrect installation of any system components, piping, or accessories

9.1.0 Slugging

Slugging occurs when a compressor tries to compress liquid refrigerant, oil, or both instead of superheated gas. If slugging occurs, it will occur at startup or during rapid changes in system operating conditions. It can sometimes be detected by a periodic knocking noise at the compressor. When a compressor tries to compress a liquid refrigerant or oil, extremely high pressures that may exceed 1,000 psi can be reached in the cylinder. These pressures can result in blown cylinder heads and/or valve plate gaskets and damage to the compressor pistons and discharge valves. The following conditions may cause slugging:

- There is an overcharge of refrigerant.
- The crankcase heater has failed.
- There is an oversized or damaged thermostatic expansion valve or loose sensing bulb.
- Condensed refrigerant in any cold part of the system, such as the evaporator, during the off cycle. Buried refrigerant lines or lines passing through cold spots can allow the refrigerant to condense back into a liquid at shutdown.
- Slugs of oil are trapped in the suction line because the suction gas does not have enough velocity to return the oil to the compressor. Normally, the oil and refrigerant mix. The oil is circulated through the system in very small drops as it is being swept along by the velocity of the refrigerant vapor. If it gets trapped in the system piping and returns all at once, it can cause slugging. This condition tends to be found in the suction line of built-up systems. It also occurs in systems that use compressors with unloaders, especially when the compressor runs unloaded for long periods of time.
- The system has an overcharge of oil.

9.2.0 Flooding

Flooding is the continuous return of liquid refrigerant in the suction vapor to the compressor during operation. Flooding usually dilutes the oil, resulting in crankcase foaming and overheating of bearing surfaces. If severe enough, it can result

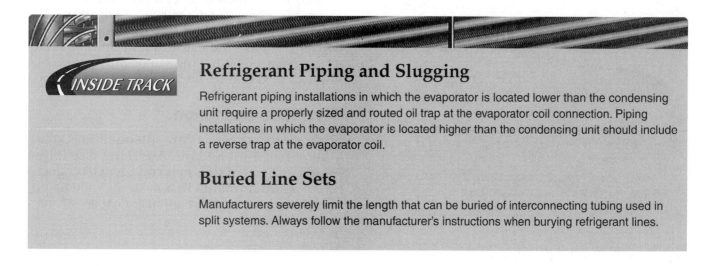

Refrigerant Piping and Slugging

Refrigerant piping installations in which the evaporator is located lower than the condensing unit require a properly sized and routed oil trap at the evaporator coil connection. Piping installations in which the evaporator is located higher than the condensing unit should include a reverse trap at the evaporator coil.

Buried Line Sets

Manufacturers severely limit the length that can be buried of interconnecting tubing used in split systems. Always follow the manufacturer's instructions when burying refrigerant lines.

in damage to the pistons, rings, and valves because the refrigerant washes the oil off the bearing surfaces. The following conditions may result in flooding:

- The thermostatic expansion valve is oversized.
- The thermostatic expansion valve sensing element is broken, mislocated, in poor contact, or improperly insulated.
- The superheat setting is too low.
- There is a low load on the evaporator caused by low airflow. Reduced airflow often causes frosting of the coil, which adds to the problem. Possible causes of restricted airflow are dirty filters, air restriction, and dirty fan wheels.
- There is an overcharge of refrigerant in systems that use fixed-orifice metering devices. Since fixed metering devices do not react to load change, an overcharge of refrigerant can raise the head pressure, which can increase the flow rate to a point where there is more flow than available heat transfer.

9.3.0 Flooded Starts

Flooded starts are a result of the oil in the compressor crankcase absorbing refrigerant. This condition usually occurs during shutdown. Oil will absorb refrigerant under most conditions. The amount absorbed depends on the temperature of the oil and the pressure in the crankcase. The lower the temperature and the higher the pressure, the more refrigerant is absorbed.

On startup, this refrigerant-rich oil mixture is pumped through the oil pump of the compressor, resulting in marginal or inadequate lubrication of the bearings. This results in the bearing surfaces being overheated and scored because of the inadequate lubrication. As the crankcase pressure drops after start-up, the refrigerant will flash from a liquid to a gas, causing foaming. This

INSIDE TRACK

Metering Devices

Because of the risk to equipment reliability and the need to meet higher energy efficiency standards, one major equipment manufacturer has abandoned the use of fixed-orifice metering devices on the indoor coils of residential split-system products and has switched entirely to TXVs.

foaming can restrict the oil passages and cause pressure to build. It can also cause a hydraulic slug of the oil and liquid mixture to enter the cylinder, resulting in compressor damage.

The problem of flooded starts can be minimized by following the manufacturer's specifications. It is important to maintain the correct refrigerant-to-oil ratio by making sure the system has a proper refrigerant charge and the correct amount of oil in the crankcase. Crankcase heaters are normally used to raise the temperature of the oil during shutdown and to prevent refrigerant from migrating to the compressor crankcase during the off cycle.

Crankcase heaters are often thermostatically controlled. When the outdoor temperature is warm enough (usually above 65°F) to where refrigerant migration to the compressor crankcase is not a problem, the thermostat opens and the crankcase heater is de-energized. Below 65°F, the thermostat closes and power is applied to the crankcase heater. This feature provides energy savings while still protecting the compressor.

Crankcase heaters can prevent refrigerant from condensing directly in the crankcase, but they cannot prevent migration and condensation in the suction line leading up to the compressor. In cases where refrigerant migration may occur, a liquid line solenoid valve should be used to prevent movement during the off cycle. The valve can be controlled along with the compressor itself, opening at startup and closing immediately upon compressor shutdown.

Proper trapping of the suction line, often done to ensure that oil is entrained in the refrigerant vapor and returned to the compressor, can also help prevent oil migration. During the off cycle, oil often fills the trap, blocking refrigerant vapor from moving freely through the system. Because the management and return of oil to the compressor is usually the primary reason for traps installed in the suction line, liquid line solenoids cannot be considered a substitute for traps. On the other hand, suction line traps in some systems may make the use of liquid line solenoids unnecessary by blocking vapor movement, but this can be very difficult to predict.

9.4.0 Contamination

Refrigeration systems are intended to contain only refrigerant and oil. Anything else in the closed refrigerant system is considered a contaminant. Contaminants in a system, such as air, moisture, acid, and dirt are major causes of compressor failure.

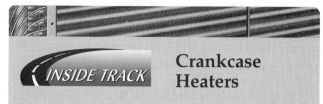
Crankcase Heaters

Compressor crankcase heaters are used to prevent liquid refrigerant from migrating into the compressor and mixing with the oil when the compressor is off. All heaters evaporate refrigerant from the oil. Heaters used with semi-hermetic compressors are typically fastened to the bottom of the crankcase. Another type, called an immersion heater, is inserted directly into the compressor crankcase. Wrap-around (bellyband) crankcase heaters like the one shown here are widely used to encircle the outside shell of welded hermetic compressors.

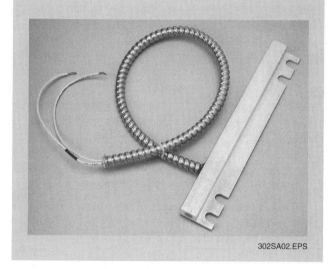

302SA02.EPS

- Air, moisture, and dirt usually enter the system accidentally during installation or servicing.
- Acid, soot, varnish, hard carbon, and copper plating are created in the system either by chemical reactions with contaminants or as a result of compressor motor **burnout**.

9.4.1 Air Contamination

Air not only contains moisture but is noncondensable. Since air is noncondensable, it can accumulate in the condenser, taking up space needed for condensing the refrigerant. This increases the condensing temperature and makes the system work harder. It also promotes the creation of acids by chemical reaction with the oil and refrigerant mixture. These acids erode machined surfaces, which can create copper plating, causing copper from the system to be deposited on the heated bearing surfaces in the compressor.

Installing Split Systems

Split systems are at risk for dirt contamination because the two sections of the system have to be connected with field-installed refrigerant tubing. Here is how to minimize the chances of system contamination when installing a split system:

- Use only ACR or refrigeration grade tubing, or use the refrigerant line sets specified by the equipment manufacturer.
- Keep the ends of the tubing plugged until just before making the connections to the equipment. Install a filter-drier in the liquid line.
- Evacuate the interconnecting tubing between the units to 500 microns before opening the service valves.

9.4.2 Moisture Contamination

Moisture is a common contaminant. Under the heat of compression, it will react with the refrigerant to form hydrochloric and hydrofluoric acid. These acids cause corrosion of metals and breakdown of the insulation on the motor windings and other motor wiring. Moisture in the refrigerant can also cause oil sludge, which reduces the lubricating properties of the oil and plugs oil passages and screens in the compressor. Moisture can also freeze at the expansion device, especially in heat pumps where they operate at lower temperatures than those found in cooling-only systems. The presence of moisture in a system can be determined by testing the refrigerant with an acid/moisture test kit. The POE and PAG oils used with newer HFC refrigerants such as R-410A is very hygroscopic, meaning it readily absorbs moisture from the air. For that reason, great care must be taken to prevent moisture from entering the refrigeration system during repair procedures on equipment using these refrigerants.

9.4.3 Acid Contamination

Acid is not a contaminant that is introduced into a system. It is formed inside an improperly operating system by the reaction of air or moisture with the refrigerant. Acid is produced in great quantities in a system that has experienced a severe motor burnout. It creates sludge and varnish, which plug oil passages and restrict the strainers in the lubrication system. The presence of acid in a system can be determined by testing the refrigerant with an acid/moisture test kit or the oil with an oil acid test kit. There are simple, inexpensive acid testers available today. The tester shown in *Figure 30* is designed to take a tiny refrigerant sample from the suction side service valve by depressing the Schrader core. If acid is present, the strip inside the tube will turn red. There are also acid test kits made specifically for testing refrigerant oils for the presence of acid (*Figure 31*).

9.4.4 Dirt Contamination

Foreign material such as filings, chips, or lint can get into the system during installation or

servicing. These particles can easily clog small oiling passages in the compressor, preventing its normal lubrication and forming restrictions in the system that cause pressures to increase or decrease.

Copper oxide is a black, flaky, solid contaminant that is formed around brazed joints. It forms when the copper tubing is heated to over 1,000°F in the presence of oxygen. It is easily washed from the tubing surfaces when the system is operated, allowing it to circulate with the refrigerant and oil. These solid particles of copper oxide circulating through the system can cause many problems that eventually lead to compressor failure. It is critical to prevent the formation of

302F30.EPS

Figure 30 ◆ Refrigerant acid tester.

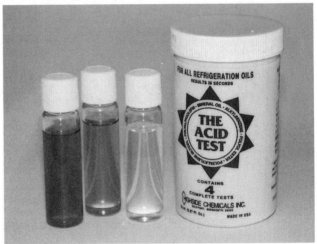

302F31.EPS

Figure 31 ◆ Acid test kit for oil.

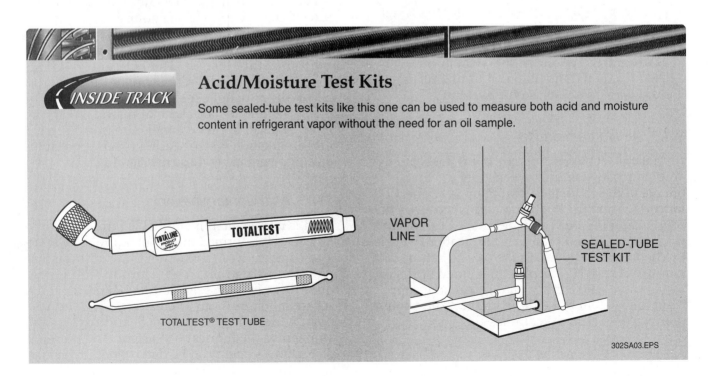

copper oxides within the tubing while brazing. You must practice proper brazing techniques when installing or servicing a system. This includes purging the refrigerant lines with nitrogen gas in order to displace the oxygen in the system when brazing the joints.

9.4.5 Eliminating Contaminants

Contaminants must be eliminated when installing new systems or when servicing existing equipment. Contaminants can be eliminated as follows:

- *Air* – Evacuate the system.
- *Moisture* – Evacuate and dehydrate the system and install a filter-drier.
- *Acid* – Recover and replace or recycle the system refrigerant, replace system filter-driers, and replace the compressor oil.
- *Chips and dirt* – Work carefully and install strainers and filters in the system.

9.5.0 Electrical

Compressor motor electric failures typically occur as a result of other compressor problems such as contamination, overheating, and the loss of lubrication. Some electrical problems in compressors are caused by a failed component in the electrical start or run circuit, such as the capacitors, potential relay, or start switch. Review the material on electric motors, their components, and the procedures used to test them previously studied in the *HVAC Level Two* module, *Alternating Current*. The remainder of this section describes electrical problems external to the compressor motor that will result in compressor failure if not corrected.

9.5.1 Nameplate Information

Valuable information about a hermetic compressor's operating frequency, voltage, and current can be obtained from the unit information plate (*Figure 32*). This information plate is attached to the unit. In addition to specific electrical information, it also gives information such as refrigerant type and model number, which may be needed if the compressor must be replaced. Unfortunately, many compressor nameplates are hard to see because the compressor is mounted deep within the unit. Fortunately, the important compressor electrical information is normally also included in a visible location on the unit nameplate.

An important piece of information given on the nameplate is the RLA or FLA value. RLA stands for rated load amps and FLA for full load amps. Nameplates for hermetic compressors built after 1972 are marked with the RLA. The RLA for a hermetic compressor is determined by the manufacturer by placing the compressor under actual operation at rated refrigerant pressure and temperature, voltage, and frequency. To determine a compressor's RLA, the manufacturer first determines the compressor's maximum continuous current (MCC). This is the maximum current value that the compressor's motor protector will carry without opening, thus stopping the compressor. Any additional current draw above the MCC will cause the protector to open. Sometimes the MCC value is also marked on the unit nameplate.

Once the MCC is determined, the RLA is then calculated. According to the Underwriters' Laboratory (UL), this is done by dividing the MCC by 1.56 to establish the minimum RLA; however, some compressor manufacturers may use a dif-

SERIAL 9370052		MODEL PAC036010		
		OUTDOOR FAN MOTOR		
FACTORY CHARGED R-22		VOLTS AC 208/230		
LBS 7.0	Kg 3.2	HP 1/4		FLA 1.4
		PH 1		HZ 60
POWER SUPPLY 208/230 VOLTS		INDOOR FAN MOTOR		
PH 1	HZ 60	VOLTS AC 208/230		
PERMISSIBLE VOLTAGE AT UNIT		HP 1/3		FLA 2.8
MAX 253	MIN 187	PH 1		HZ 60
SUITABLE FOR OUTDOOR USE		DESIGN/TEST PRESSURE GAUGE		
COMPRESSOR		HI PSI 300		kPA 2068
VOLTS AC 208/230		LO PSI 250		kPA 1034
PH 1	HZ60	MINIMUM CIRCUIT AMPS 26.7		
RLA 18.0	LRA 96	MAX FUSE40 40MAX CKT-BKR (*)		
		HEATER PACKAGE		
		VOLTS AC 209/230		
		HEATER AMPS 26.5		

302F32.EPS

Figure 32 ◆ Typical unit nameplate.

ferent divisor than 1.56 (such as 1.4). The RLA value is important because it can be used to size wire and contactors used with the compressor. In refrigeration low-temperature compressor applications, compressors are typically equipped with a crankcase pressure regulator (CPR) to prevent the pressure in the compressor crankcase from rising to a level that will overload the compressor. In this case, the RLA value is also used when adjusting the CPR so that the compressor amp draw does not exceed its RLA value.

9.5.2 Motor Operating Voltage Ranges

Too high or too low an operating voltage can cause overheating of a compressor motor and may cause compressor failure. Operating voltages must be maintained within minimum-maximum limits from the voltage value given on the motor's nameplate. If the operating voltage falls outside these limits, the system should be turned off and the problem corrected before restarting the system. The problem may be with the building distribution system or the electric utility supply to the building. The voltage tolerances used for motors are as follows:

- *Single-voltage rated motors* – The input supply voltage should be within ±10 percent of the nameplate voltage. For example, a motor with a nameplate single voltage rating of 230V should have an input voltage that ranges between 207V and 253V (±10 percent of 230V).
- *Dual-voltage rated motors* – The input supply voltage should be within ±10 percent of the nameplate voltage. For example, a motor with a nameplate dual voltage rating of 208/230V should have an input voltage that ranges between 187V (–10 percent of 208V) and 253V (+10 percent of 230V).

9.5.3 Voltage Imbalance

The voltage imbalance between any two legs of the supply voltage applied to a three-phase motor should not exceed 2 percent. A small imbalance in the input voltage results in a considerable amount of heat being generated in the motor windings. For example, if the voltage imbalance were to exceed 2 percent, the temperature rise generated in the motor windings would increase to 8 percent over the safe level. With only a 5 percent imbalance, the winding temperature can increase to 50 percent over the safe level. *Figure 33* shows an example of how to calculate the voltage imbalance in a three-phase system.

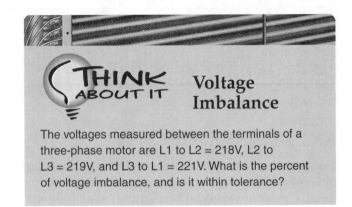

THINK ABOUT IT — **Voltage Imbalance**

The voltages measured between the terminals of a three-phase motor are L1 to L2 = 218V, L2 to L3 = 219V, and L3 to L1 = 221V. What is the percent of voltage imbalance, and is it within tolerance?

The percent of voltage imbalance is calculated using the following formula:

% voltage imbalance =

$$\frac{\text{maximum voltage imbalance}}{\text{average voltage}} \times 100$$

9.5.4 Current Imbalance

Current imbalance between any two legs of a three-phase motor should not exceed 10 percent. Voltage imbalance will always produce current imbalance, but a current imbalance may occur without a voltage imbalance. This can happen when an electrical terminal or contact becomes loose or corroded, causing a high resistance in the leg. Since current follows the path of least resistance, the current in the other two legs will increase, causing more heat to be generated in the windings. Current imbalance is calculated using the following formula:

% current imbalance =

$$\frac{\text{maximum current imbalance}}{\text{average voltage}} \times 100$$

Single-phasing of a three-phase motor is an example of a severe case of current imbalance. If the compressor is operating and one phase opens,

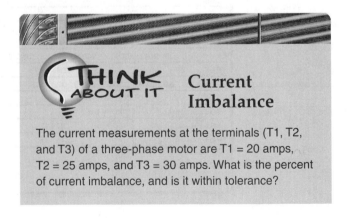

THINK ABOUT IT — **Current Imbalance**

The current measurements at the terminals (T1, T2, and T3) of a three-phase motor are T1 = 20 amps, T2 = 25 amps, and T3 = 30 amps. What is the percent of current imbalance, and is it within tolerance?

	PHASE	MEASURED READING
(1)	L1 TO L2	215V
	L2 TO L3	221V
	L3 TO L1	224V

$$\text{(2)} \quad \text{AVERAGE VOLTAGE} = \frac{215 + 221 + 224}{3} = 220V$$

(3) INDIVIDUAL PHASE IMBALANCE FROM AVERAGE

$$\text{L1 TO L2} = 220 - 215 = 5V$$
$$\text{L2 TO L3} = 221 - 220 = 1V$$
$$\text{L3 TO L1} = 224 - 220 = 4V$$

(4) $5V$ = MAXIMUM IMBALANCE

$$\text{(5)} \quad \text{\% IMBALANCE} = \frac{\text{MAXIMUM VOLTAGE IMBALANCE}}{\text{AVERAGE VOLTAGE}} \times 100$$

$$\text{(6)} \quad \text{\% IMBALANCE} = \frac{5V}{220V} \times 100 = 2.27\%$$

302F33.EPS

Figure 33 ◆ How to check voltage imbalance in a three-phase system.

the motor can continue to run. The other two phases will attempt to carry the load, resulting in a current increase of about 1½ times the normal running current. If the compressor is loaded, this will probably cause an overload device to trip. If unloaded, the overload might not trip, allowing the motor to continue running and causing the motor windings to overheat. Normally, once the motor is stopped, it cannot be restarted again because it will continuously trip the overload protectors. This can eventually cause a compressor failure.

9.6.0 Compressor Heating

Compressors normally generate heat and are designed to handle it. One indication of possible compressor overheating is a high compression ratio. When the compressor is overheated, the cause must be determined. As discussed previously, high superheat, lubrication problems, condenser problems, or electrical problems are all possible causes. High temperatures between 275°F and 300°F in the hot gas discharge line cause oil and refrigerant to break down with the potential for compressor failure. Laboratory tests have shown that for each 18°F rise above normal in the discharge temperature, the chemical reaction between the refrigerant-oil mixture and moisture and acid in the system doubles.

 WARNING!
Certain parts of operating compressors can be very hot under normal operating conditions. Wear gloves and other personal protective equipment to avoid being burned when working around operating compressors.

10.0.0 ◆ SYSTEM CHECKOUT FOLLOWING COMPRESSOR FAILURE

Before condemning the compressor, system checkout procedures should be performed to make sure the compressor is bad. Even if the compressor proves to be bad, these checks must be made to find out why it failed. Remember that if a compressor fails and the cause has not been determined and corrected, the replacement compressor will also fail. The system checkout procedure consists of three phases:

- Preliminary inspection
- Analyzing system operating conditions
- Final compressor tests

10.1.0 Preliminary Inspection

Preliminary inspection of the system is performed using your senses of sight, sound, touch, and smell to identify problems.

WARNING!

Be sure all electrical power to the equipment is turned off. Open, lock, and tag disconnects. Watch out for pressurized or hot components. Follow all safety instructions labeled on the equipment and given in the manufacturer's service manual for the equipment.

The preliminary inspection is performed as follows:

Step 1 Turn off the equipment. Lockout and tag equipment so that no one can start it.

Step 2 Look for an evaporator or condenser mounted above the compressor which might dump liquid refrigerant into the compressor.

Step 3 Look for the following piping problems:
- Long or uninsulated suction line which might develop excessive superheat
- Liquid line running through an unconditioned space (hot or cold) which might affect subcooling
- Buried lines which might cause refrigerant to condense
- Extremely long liquid line which might hold an excessive amount of refrigerant

Step 4 At the evaporator and condenser, check for the following:
- Fin collars corroded
- Fins or coils dirty or damaged
- Supply plenum dirty
- Filters dirty or missing
- Fan belts at improper tension
- Blowers and fans dirty
- Evaporator shows signs of freezing up

Step 5 As applicable, at the compressor:
- Check that the service valves are fully open.
- Check that the hold-down bolts are loosened or unloosened per the manufacturer's instructions.
- Check that the crankcase heater is working.
- Inspect the cylinder heads to see if they are scorched or blistered from excessive heat.
- Inspect for rust streaks, indicating condensation from cold return gas.
- Check that the oil level is at the proper height in the sight glass.

10.2.0 Analyzing System Operating Conditions

In order to tell if the system or compressor has a problem, the actual conditions that exist in the system must be known.

10.2.1 System Operation Checks

If the compressor is operable, system operation should be monitored and the critical parameters measured. This is necessary so the actual conditions can be compared against a set of previously recorded normal system operating parameters. Check the system operating conditions and make sure to record the values for system parameters as described in this section. It cannot be stressed enough how important it is to properly record the operating conditions and measured values for a system. The record may serve as a historical document that can be used for comparison purposes by a technician when troubleshooting in the future. The recorded data can also serve as a system commissioning record in order to verify compliance with contractual system specifications.

WARNING!

Watch out for rotating, pressurized, or hot components. Follow all safety instructions labeled on the equipment and given in the manufacturer's service manual for the equipment. Failure to do so could result in personal injury.

Step 1 Install a gauge manifold set on the system gauge ports. Install thermometers on the suction line at the compressor input, on the hot gas discharge line, and on the liquid line at the input to the expansion device.

Step 2 Turn on the system and adjust the thermostat to call for cooling.

Step 3 Start the system and monitor its operating characteristics.

- Listen for excessive vibration of the compressor, piping, motors, and fans.
- Listen for compressor knocks or rattles, which may indicate liquid refrigerant is being drawn into the cylinders. If this condition continues for more than a few seconds, shut down the system and look for the cause of excess liquid return.

Step 4 Check the compressor oil sight glass (if applicable). Heavy foaming at the sight glass should clear 5 to 10 minutes after startup. If not, there may be excessive refrigerant in the oil.

Step 5 Measure and record the input voltage and current at the compressor contactor as follows. Current in the compressor motor leads can be measured with a clamp-on ammeter.

- The measured voltage should be within ±10 percent of the motor nameplate value.
- In a three-phase motor, the voltage imbalance between phases should not exceed 2 percent. Any current imbalance between legs should not exceed 10 percent.

WARNING!

Danger exists if the compressor terminals are damaged and the system is pressurized. Disturbing the terminals to make measurements could cause them to blow out, causing injury. When making voltage, current, or continuity checks on a hermetic or semi-hermetic compressor in a pressurized system, always make measurements at terminal boards and points of test away from the compressor. Once the refrigerant has been recovered and the system is no longer under pressure, measurements can be made at the compressor terminals.

Step 6 Check that all fans, motors, and pumps are operational and moving the proper amounts of air or water.

Step 7 As applicable, measure or calculate and record the following operating parameters:

- Suction pressure
- Saturated suction temperature
- Suction line temperature
- Superheat
- Discharge pressure
- Saturated discharge temperature
- Discharge line temperature
- Subcooling
- Liquid line temperature entering the expansion device
- Oil pressure
- Hot gas discharge line temperature differential above the saturated condensing temperature (head temperature)
- Compressor temperature at the bottom of the cylinder heads
- Compressor temperature at top and bottom of the motor barrel
- Crankcase temperature
- Compression ratio
- Evaporator capacity
- Condenser capacity

10.2.2 Analyzing System Conditions

Once the actual conditions that exist in a system are known, they can be compared against a set of normal system operating parameters in order to determine if there is a problem and where it is. Remember, readings vary because of equipment application, operating conditions, and type of refrigerant used. The data in *Table 2* is typical of an HCFC-22 air conditioning system at 90°F outdoor temperature. Always refer to the manufacturer's service manual or system maintenance log to find typical readings for the specific equipment you are servicing.

Based on the analysis of system operation, repair the system, repair or replace the compressor, or both. Some common problems and their causes are as follows.

- Failure to start:
  - Thermostat setting too high
  - Power circuit voltage inadequate or missing
  - Control circuit voltage inadequate or missing
  - Safety switches open or defective

- Overheated compressor:
  - High superheat caused by low charge or defective metering device
  - Electrical problems such as low voltage or voltage imbalance

  - Lubrication problems such as low oil or poor oil return
  - Condenser problems such as dirty or plugged coil, or failed condenser fan motor

- Reduced system capacity:
  - Check load
  - Check evaporator flow
  - Check refrigerant flow
  - Check refrigerant charge

10.3.0 Final Compressor Checks

Before replacing a compressor, some final checks should be made on the compressor to be sure it is defective. Compressor failures result from either an electrical failure or a mechanical failure.

10.3.1 Electrical Reasons for Failure

The methods used to troubleshoot motors were studied in the *HVAC Level Two* module, *Alternating Current*. Review the material about electric motors, their components, and the procedures used to test them. The most likely causes of electrical failure include:

- A grounded, open, or shorted motor
- Open internal overload (stuck open)
- Defective start relay (stuck open or closed)
- Open or shorted start or run capacitor
- Electrical connections
- Relays
- Improper input voltage
- Malfunctioning contactor

Table 2 Typical Operating Conditions for HCFC-22 System at 90°F Outdoors

System Operation	Parameter
Suction pressure	68 psig
Saturated suction temperature	40°F
Suction line temperature	50°F
Superheat	12°F
Discharge pressure	160 psig
Saturated discharge temperature	120°F
Subcooling	10°F
Liquid line temperature entering the expansion device	110°F
Hot gas discharge line temperature differential above the saturated condensing temperature (head temperature)	70°F
Oil pressure	40 psig
Compressor temperature at the bottom of the cylinder heads	80°F to 120°F (max)
Compressor temperature at the top and bottom of cylinder heads	90°F
Crankcase temperature	100°F
Compression ratio	3.32 to 1

Internal Line Break Overloads

Internal line break overloads can take an hour or more to reset, under some conditions. If you are on a service call, you don't want to waste your valuable time and the customer's money waiting for the overload to reset. Here's how you can speed up the reset time:

- Shut off all power to the unit.
- Protect any exposed electrical components and motors from water entry.
- Run water from a garden hose over the compressor to help cool it down.

10.3.2 Mechanical Reasons for Failure

Mechanical reasons for compressor failure include loss of charge or overcharge, physical damage, broken valves or rings, and loss of lubrication.

Leaks that exist at the weld or terminals of welded hermetic compressors cannot be repaired and the compressor must be replaced. If a leak exists at the stubs, recover all refrigerant and repair the leak with silver brazing alloy using the necessary precautions. If the leak cannot be repaired, the compressor must be replaced. System leaks can also contribute to compressor failure and should be repaired.

If a base unit or compressor is mishandled, internal damage can occur. A broken suspension spring or bent line can cause excessive running noise. Compressors with these problems should be replaced.

Compressor valves and rings provide a seal between the high-pressure and low-pressure sides. If damaged, the compressor must be replaced or repaired if it is an open or serviceable hermetic. Suction and discharge pressure checks can be used to test for this condition. Bad valves or rings may exist if the suction will not pull down or discharge builds up with the system properly charged.

Another check can be made by measuring the running current with an ammeter. If, under loaded conditions, the running current is considerably lower than normal, faulty valves or rings should be suspected. If the compressor has unloaders, make sure they are not activated. On heat pumps, make sure the reversing valve and check valves are not stuck open.

Many compressors that are replaced are incorrectly judged to be seized. Seized or tight motors usually hum but will not run. Also, when the current is measured, the motor is drawing locked rotor current. Before replacing a compressor with these symptoms, make sure the following conditions do not exist:

- Unequal system pressures (common with PSC motors)
- Low supply voltage
- Contactor not making good contact on all poles
- Defective (open) start relay
- Start or run capacitor open or shorted

If a compressor will not start and it checks out electrically, attempt to start the compressor with a temporary capacitance boost.

11.0.0 ◆ COMPRESSOR CHANGEOUT

The procedures used to change out a compressor differ depending on whether the cause of the compressor failure is electrical or mechanical. There are many types and variations of compressors. The procedures in this section describe the methods used to replace a welded hermetic-type compressor. Regardless of the type of compressor being replaced, the guidelines given in the procedures normally apply. You must always consult the system and replacement compressor manufacturer's service literature for the specific system being serviced and the compressor being used.

11.1.0 Compressor Replacement Due to Mechanical Failure

When replacing a compressor because of a mechanical problem, use the following guidelines.

 WARNING!

Be sure all electrical power to the equipment is turned off. Open, lock, and tag disconnects. Watch out for pressurized or hot components. Follow all safety instructions labeled on the equipment and given in the manufacturer's service manual. Failure to do so may result in personal injury.

Step 1   Turn off the equipment. Lockout and tag equipment so that no one can start it.

Step 2   Recover the refrigerant, regardless of its condition. Do not vent the refrigerant to the atmosphere.

Step 3   Disconnect and tag all wiring from the compressor.

 WARNING!

Danger exists if the compressor terminals are damaged and the system is pressurized. Disturbing the terminals to make measurements could cause them to blow out, causing injury. When making voltage, current, or continuity checks on a hermetic or semi-hermetic compressor in a pressurized system, always make measurements at terminal boards and points of test away from the compressor. Once refrigerant has been recovered and the system is no longer under pressure, measurements can be made at the compressor terminals.

Step 4 Use a tubing cutter to cut the system refrigerant tubing connected to the compressor. Do not use a hacksaw; it can introduce harmful chips into the system. Tape the open lines in the equipment to prevent dirt or moisture from entering.

WARNING!
Never use a torch to cut the compressor lines because oil vapor in the lines can flare up and cause severe burns.

Step 5 Remove the holddown bolts, then remove the old compressor. Be sure to get help to avoid injury caused by heavy lifting.

Step 6 Unpack the new compressor and read any manufacturer's literature that accompanies it. Replacement compressors often have new features or devices that should be used. Compare nameplates between the old and new compressors to be sure the new compressor is the correct type.

Step 7 Test the new compressor motor for open windings, grounds, or shorts.

Step 8 Install stubs on the new compressor.

Step 9 Remove the mounting grommets from the old compressor and install them on the new one.

Step 10 Mount the new compressor in the equipment and bolt it down. Be sure to get adequate help to avoid injury caused by heavy lifting.

Step 11 Braze the system refrigerant lines to the compressor using sweat couplings. Clean any flux from the joints and paint them for protection.

Step 12 Remove the existing liquid line filter-drier and replace it with one that is one size larger. If the system is not equipped with a filter-drier, install one.

Step 13 Leak test the system, then deep evacuate to a level of 500 microns.

Step 14 While evacuating the system, consult the equipment wiring diagram and connect all wiring to the compressor.

Step 15 Recharge the system with the correct type and weight of refrigerant.

Step 16 Start the system and allow it to run in order to stabilize the system pressures. Refer to the equipment manufacturer's service instructions and follow them to make any adjustments in the refrigerant charge as deemed necessary.

11.2.0 Compressor Replacement Due to Electrical Failure

Electrical failures involving motor windings are often called burnouts. A burnout is the breakdown of the motor winding insulation, which causes the motor to short out or ground electrically. Burnouts are classified either as mild or severe.

- A mild burnout usually occurs suddenly, causing the motor to stop before the contaminants created by the burnout leave the compressor. Few contaminants, if any, are produced and little or no chemical reaction of the refrigerant and oil has occurred.

- Severe burnouts (cookouts) usually occur over a long period of time. Considerable contaminants are produced and may be pumped through the system while the compressor is still running.

WARNING!
When working on a system suspected of having a compressor burnout, wear appropriate personal protective equipment, including rubber gloves and eye protection. Contaminated refrigerant oil may contain heavy concentrations of acid. Do not allow contact with the skin or eyes as severe burns may result.

Before replacing a compressor that has failed because of a burnout, you must determine if the burnout was mild or severe. To do this, you must check the system for acid or moisture, using one of several acid/moisture test kits or oil test kits that are readily available. Acid/moisture test kits typically can be connected to a system service port to obtain a sample of the refrigerant. Oil test kits require that a sample of the system oil be taken. For systems with hermetic compressors, an oil sample may be obtained by inserting an oil trap in the suction line. In either case, follow the test kit manufacturer's instructions for using the kit and for determining the amount of contamination in the system.

> **WARNING!**
>
> Do not cut into or attempt to unbraze a line with any system pressure in it. There is a potential for explosion, as well as production of deadly phosgene gas. The system pressure must be 0 psig before repairs are attempted.

Another way to check out the type of burnout in a hermetic compressor is to cut the suction and discharge lines and check for carbon by running a clean, lint-free swab into the lines. In a semi-hermetic compressor, the easiest way to tell the type of burnout is to remove the cylinder head. If you find carbon using either method, it is a good indication that a severe burnout has occurred. With a severe burnout, the refrigerant will have a very strong rotten-egg smell.

11.2.1 Replacement Procedure After a Mild Burnout

Since most contaminants produced by a mild burnout are contained within the compressor, the procedure for changing the compressor is the same as that used with a mechanical failure, including the installation of an oversized filter-drier. One exception is that after the compressor installation is completed, the system should be triple evacuated before charging it with refrigerant. The triple evacuation method is described in the *HVAC Level Two* module, *Leak Detection, Evacuation, Recovery, and Charging*.

11.2.2 Replacement Procedure After a Severe Burnout

Servicing a system after a motor burnout requires not only compressor replacement, but also a thorough cleanup of the entire system to remove all harmful contaminants left by the burnout. Successive failures on the same system can generally be traced to improper cleanup. Unless the cleanup is performed correctly, a repeat failure usually occurs in a short period of time. The following is an overview of the procedure for replacing a compressor after a severe burnout:

Step 1   Recover the system refrigerant and remove the compressor.

Step 2   Remove the liquid line filter-drier.

Step 3   Purge the system piping with dry nitrogen in the direction opposite to normal refrigerant flow.

Step 4   Remove, clean, and/or replace metering devices, accumulators, reversing valves, and related components, if contaminated.

Step 5   Install the new compressor.

Step 6   Add or replace the liquid line filter-drier. Use the next larger size.

Step 7   Add a suction line filter-drier with input and output pressure taps.

Step 8   Triple evacuate the system.

Step 9   Recharge the system with new refrigerant.

Step 10   Run the system for one hour. Stop the system and change the suction line filter-drier any time the pressure drop across it is just below or at 3 psig.

Step 11   After one hour has elapsed, stop the system and change the liquid and suction line filter-driers. If using a semi-hermetic compressor, replace the oil.

Step 12   Run the system for two more hours.

Step 13   After two hours have elapsed, stop the system and test for acid or moisture contamination to make sure the system is clean. If clean, change the liquid line filter-drier. Remove the suction line filter-drier from the system. Never leave a suction line filter-drier in a system for an extended period of time. Once the system is clean, remove it. Liquid line filter-driers may be left in a system once the system is clean.

Step 14   If acid or moisture is still present in the system at Step 13, change the oil (semi-hermetic) and repeat Steps 12 and 13 as necessary to achieve a clean system.

1. A compressor that has a piston that travels back and forth in a cylinder is a _____ compressor.
   a. reciprocating
   b. rotary
   c. centrifugal
   d. screw

2. In a scroll compressor, _____.
   a. both scrolls move
   b. only the upper scroll moves
   c. only the lower scroll moves
   d. neither scroll moves

3. A primary concern when multiple compressors share a common refrigeration circuit is _____.
   a. oil level management
   b. compressor noise
   c. refrigerant charge
   d. constant operation

4. The cylinder unloader method of capacity control is used mainly with _____ compressors.
   a. centrifugal
   b. screw
   c. rotary
   d. open and semi-hermetic reciprocating

5. The capacity of a compressor is related to the horsepower of the driving motor.
   a. True
   b. False

6. A compressor electrical motor pilot duty protection device _____.
   a. resets automatically
   b. senses current flow and/or temperature in the motor winding and opens the winding circuit to remove line voltage from the motor
   c. senses a current overload and/or temperature within the motor and opens the motor contactor control circuit to remove power from the motor
   d. must be reset manually

7. A compressor protection device that prevents the compressor from operating out of sequence is a _____.
   a. high-pressure switch
   b. short-cycle timer
   c. lockout relay
   d. airflow switch

8. Reduced-voltage starting of a compressor is used to _____.
   a. boost the motor starting current
   b. limit the supply voltage drop when the motor starts
   c. control the capacity of the compressor
   d. decrease the compressor motor rpm

9. Compressing a slug of liquid refrigerant can result in pressures applied to the compressor valves and pistons that exceed _____ psi.
   a. 40
   b. 100
   c. 1,000
   d. 2,000

10. Excessive foaming in the oil sight glass shows the oil is mixed with _____.
    a. refrigerant
    b. air
    c. acid
    d. moisture

11. A clean refrigeration system may contain _____.
    a. moisture
    b. air
    c. acid
    d. oil

12. The nameplate on a three-phase compressor motor is marked 460 volts. For proper motor operation, the input supply voltage must range between _____ volts.
    a. 414 and 506
    b. 430 and 590
    c. 437 and 483
    d. 451 and 469

13. Voltage imbalance for a three-phase electrical supply must not exceed _____ percent.

    a. 1
    b. 2
    c. 5
    d. 10

14. When determining the type of burnout in a system with a hermetic compressor, the first thing you would do is _____.

    a. turn off the power and remove the wiring from the compressor
    b. recover the refrigerant charge
    c. check the moisture and acid content of the system
    d. evacuate the system

15. A metering device should be cleaned or replaced after a _____.

    a. mild compressor burnout
    b. capacitor failure
    c. severe compressor burnout
    d. compressor mechanical failure

Summary

Compressors used in refrigeration systems fall into three groups: open compressors, welded hermetic compressors, and semi-hermetic compressors. Within each group, five different types of compressors can be found depending on the application. These are: reciprocating, rotary, scroll, screw, and centrifugal. Welded hermetic compressors are used in the majority of smaller commercial and residential systems, while open and semi-hermetic compressors are typically used for large commercial or industrial applications.

Because of the expense and time involved, compressor changeouts are considered to be one of the critical service procedures performed on cooling systems. The best way to prevent problems is to do a clean, professional installation. Ideally, the installation should be followed by periodic routine maintenance and service checks that enable early detection and correction of potential compressor problems. If not corrected, most mechanical refrigeration system problems will result in compressor failure. If you assume the cause of the compressor failure lies within the compressor, it is likely you will make the first of many unnecessary compressor replacements. If a compressor fails and the cause has not been found and corrected, the replacement compressor will also fail.

When a compressor fails, a systematic approach must be followed to correctly diagnose the cause of the failure. This includes examining the rest of the system for abnormal system conditions which may have caused the compressor to fail. These conditions include:

- Incorrect installation
- Refrigerant flooding
- Loss of lubrication
- System contamination
- Overheating
- Electrical problems
- Compressor incorrectly sized for the load

Notes

Burnout: A condition in which the breakdown of the motor winding insulation causes the motor to short out or ground electrically.

Capacity control: Methods used in cooling systems to adjust system operation to match changes in the system cooling load.

Clearance volume: The amount of clearance between a piston at the top dead-center position of travel and the cylinder head.

Compliant scroll compressor: A version of the scroll compressor that allows the orbiting scroll to temporarily shift from its normal operating position if liquid refrigerant enters the compressor.

Compression: The reduction in volume of a vapor or gas by mechanical means.

Flooded starts: A condition in which slugging, foaming, and inadequate lubrication occur at compressor startup as a result of the oil in the compressor crankcase having absorbed refrigerant during shutdown.

Flooding: A condition in which there is a continuous return of liquid refrigerant in the suction vapor being returned to the compressor during operation.

Line duty device: A motor protection device that senses current flow and temperature in the motor winding. If an overload occurs, it opens the motor winding circuit to remove the line voltage.

Pilot duty device: A motor protection device that senses current overload or temperature within the motor. If an overload occurs, it opens the motor contactor control circuit to remove power from the motor.

Positive-displacement compressor: Any compressor where the pumping action is created by pistons or moving chambers.

Pressure (force-feed) lubrication system: A method of compressor lubrication that uses an oil pump mounted on the end of the crankshaft to force oil to the compressor main bearings, lower connecting rod bearings, and piston pins.

Short cycling: A condition in which the compressor is restarted immediately after it has been turned off.

Single phasing: A condition in which a three-phase motor continues to run after losing one of the three input phases while operating.

Splash lubrication system: Method of compressor lubrication in which the crankcase oil is splashed onto the cylinder walls and bearing surfaces during each revolution of the crankshaft while the compressor is running.

Additional Resources and References

Additional Resources

This module is intended to be a thorough resource for task training. The following reference works are suggested for further study. These are optional materials for continued education rather than for task training.

Capacity Control–General Training Compressors, 1990. Syracuse, NY: Carrier Corporation.

Clean-Up After Burnout–General Training Compressors, 1985. Syracuse, NY: Carrier Corporation.

Compressors–General Training Air Conditioning, 1991. Syracuse, NY: Carrier Corporation.

Figure Credits

Topaz Publications, Inc., 302F03 (photo), 302F07 (photo), 302F09, 302F10 (photo), 302F11 (photo), 302F23

High-efficiency, variable speed, oil-free centrifugal compressor by Danfoss Turbocor Compressors, Inc., 302SA01

Carrier Corporation, 302F06, 302F12, 302F14, 302F19, 302SA03

W.W. Grainger, Inc., 302F25

ICM Controls, 302F28

Eaton's Electrical Group, manufacturer of Cutler-Hammer electrical control and distribution products, 302F29

Emerson Climate Technologies, 302SA02

Mainstream Engineering Corporation, 302F30

Highside Chemicals, Inc., 302F31

NCCER CURRICULA — USER UPDATE

NCCER makes every effort to keep its textbooks up-to-date and free of technical errors. We appreciate your help in this process. If you find an error, a typographical mistake, or an inaccuracy in NCCER's curricula, please fill out this form (or a photocopy), or complete the online form at **www.nccer.org/olf**. Be sure to include the exact module ID number, page number, a detailed description, and your recommended correction. Your input will be brought to the attention of the Authoring Team. Thank you for your assistance.

Instructors – If you have an idea for improving this textbook, or have found that additional materials were necessary to teach this module effectively, please let us know so that we may present your suggestions to the Authoring Team.

NCCER Product Development and Revision
13614 Progress Blvd., Alachua, FL 32615

Email: curriculum@nccer.org
Online: www.nccer.org/olf

❏ Trainee Guide ❏ AIG ❏ Exam ❏ PowerPoints Other _____

Craft / Level: _____ Copyright Date: _____

Module ID Number / Title: _____

Section Number(s): _____

Description: _____

_____

_____

_____

_____

Recommended Correction: _____

_____

_____

_____

_____

Your Name: _____

Address: _____

_____

Email: _____ Phone: _____

03303-08

Metering Devices

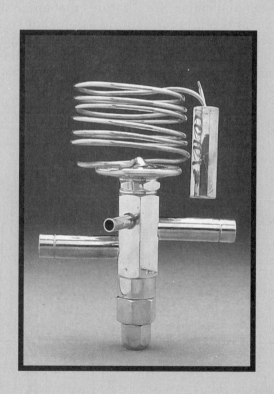

03303-08
Metering Devices

Topics to be presented in this module include:

1.0.0	Introduction	3.1
2.0.0	Basic Operation	3.2
3.0.0	Fixed Metering Devices	3.4
4.0.0	Expansion Valves	3.8
5.0.0	Distributors	3.17
6.0.0	TXV Replacement	3.18
7.0.0	Metering Device Problems	3.21

Overview

The metering device provides the pressure drop that the refrigerant needs in order to boil in the evaporator and absorb heat from the conditioned space. There are several types of metering devices. Small refrigeration appliances often use a capillary tube, while low-tonnage residential systems commonly use a fixed metering device with a selectable orifice. The thermostatic expansion valve—or TEV—is among the most common types, and is widely used in commercial packaged units. It is also used in residential heat pumps in order to obtain the higher SEER ratings that are in demand today. The TEV offers the advantage of maintaining a constant superheat as system conditions change. Proper installation of thermostatic expansion valves is essential. If the sensing bulb is not properly located and is not securely attached to the suction line and correctly insulated, the system will not operate properly.

Objectives

When you have completed this module, you will be able to do the following:

1. Explain the function of metering devices.
2. Describe the operation of selected fixed-orifice and expansion valves.
3. Identify types of expansion valves.
4. Describe problems associated with replacement of expansion valves.
5. Describe the procedure for installing and adjusting selected expansion valves.

Trade Terms

Capillary tube
Design load
Direct-expansion (DX)
  evaporator
Distributor
Distributor line
Electronic expansion
  valve (EEV)
Fixed-orifice metering
  device
Flash gas

Flood-back
Flooded evaporator
Hunting
Orifice
Superheat
Surge chamber
Thermal expansion valve
  (TEV or TXV)
Thermal-electric
  expansion valve
  (TEEV or THEV)

Required Trainee Materials

1. Pencil and paper
2. Appropriate personal protective equipment

Prerequisites

Before you begin this module, it is recommended that you successfully complete *Core Curriculum*; *HVAC Level One*; *HVAC Level Two*; and *HVAC Level Three*, Modules 03301-08 and 03302-08.

This course map shows all of the modules in the third level of the *HVAC* curriculum. The suggested training order begins at the bottom and proceeds up. Skill levels increase as you advance on the course map. The local Training Program Sponsor may adjust the training order.

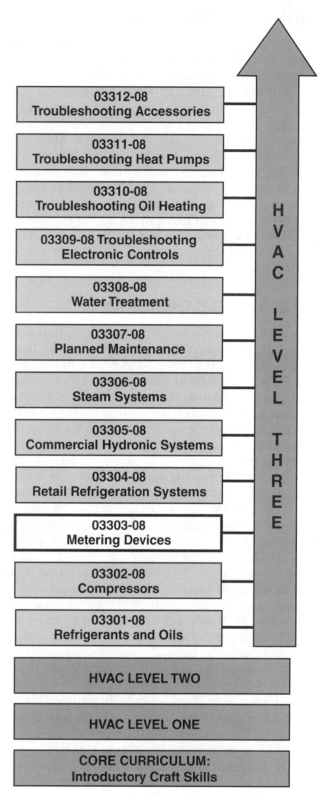

303CMAP.EPS

1.0.0 ◆ INTRODUCTION

The metering device in a refrigeration system performs two important functions. First, it matches the rate of refrigerant flow entering the evaporator with the rate at which the evaporator will boil liquid refrigerant into vapor. Second, it provides a pressure drop that separates the high side of the system from the low side, where the actual cooling occurs. A variety of metering devices are used in modern systems to control the flow of refrigerant. These devices include **capillary tubes**, precision **orifices**, pressure-operated valves, **thermal expansion valves (TEVs or TXVs)**, float valves, **electronic expansion valves (EEVs)**, and **distributors**. The metering device is also called a flow control device or refrigerant control.

2.0.0 ◆ BASIC OPERATION

The metering device exists because of the need for a refrigeration system to operate at both high and low pressures. The refrigerant entering the condenser must be at a high pressure, because its temperature must be higher than that of the outdoor air in order for heat transfer to take place. The refrigerant entering the evaporator, on the other hand, must have a low pressure and temperature, because the temperature of the refrigerant entering the evaporator must be lower than that of the air or water entering the evaporator in order for heat transfer to take place. To accomplish this, a metering device is installed between the condenser and the evaporator, usually at the evaporator input (*Figure 1*). In order for a metering device to operate, there must be a minimum pressure drop across the orifice.

In a metering device, high-temperature, high-pressure, subcooled refrigerant from the condenser is forced through a tiny opening or orifice, which reduces the pressure at the output (see *Figure 2*). In addition, some of the refrigerant turns into vapor, known as **flash gas**, which helps to cool the remaining refrigerant. If the pressure of the refrigerant entering the metering device is 300 psig, the pressure at the evaporator entry might be 69 psig. If the temperature at the metering device input is 110°F, the leaving temperature might be about 40°F. This is the temperature at which HCFC-22 boils at 69 psig.

In many systems, the flash gas enters the evaporator along with the liquid refrigerant. Thus, the refrigerant entering the evaporator is a mix of about 75 to 80 percent liquid and 20 to 25 percent vapor. This applies to systems of 100 tons or less that use **direct-expansion (DX) evaporators**. Large systems containing centrifugal or absorption chillers use **flooded evaporators**, in which only liquid refrigerant flows through the evaporator. In these systems, a **surge chamber** at the evaporator inlet is used to separate the liquid from the flash gas. The liquid then reenters the evaporator, while the vapor is bypassed directly to the compressor suction line.

This module focuses on DX systems; large commercial systems using flooded evaporators are covered later in your training.

2.1.0 Function

The metering device controls (meters) the amount of liquid refrigerant that enters the evaporator. Ideally, only as much liquid as is needed for system operation is allowed to pass through the metering device. It must pass enough liquid to provide the cooling requirements of the evaporator (the liquid needed for evaporation), while at the same time preventing a liquid surplus.

If not enough liquid enters the evaporator, it evaporates too quickly and much of the coil surface becomes ineffective. In some cases, the coil may freeze. Conversely, if too much liquid enters the evaporator, some of it will not boil into a gas and will pass in liquid form through the coil and into the suction line. If this excess liquid flows back through the suction line to the compressor, it is known as **flood-back**. Flood-back can wash out or dilute the lubricating oil in the crankcase, causing increased wear and shortening the life of the compressor. In addition, flood-back often results in slugging, a condition in which excess liquid finds its way into the compressor cylinder. Because liquids cannot be compressed, slugging often results in severe mechanical damage to the compressor in the form of broken valves, pistons, connecting rods, and in some cases, the crankshaft. Thus, the metering device must perform these two functions:

- Restrict the refrigerant flow to maintain the pressure drop needed to produce the low-temperature refrigerant required for cooling.
- Regulate the quantity of refrigerant that can pass into the evaporator according to the cooling demand, without allowing excess liquid to pass that can cause mechanical damage to the compressor.

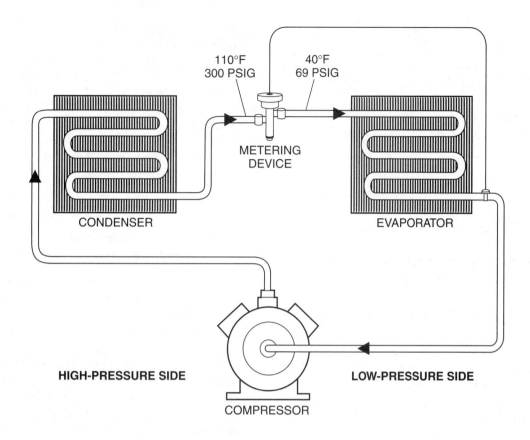

110°F
300 PSIG

40°F
69 PSIG

METERING
DEVICE

CONDENSER

EVAPORATOR

HIGH-PRESSURE SIDE

LOW-PRESSURE SIDE

COMPRESSOR

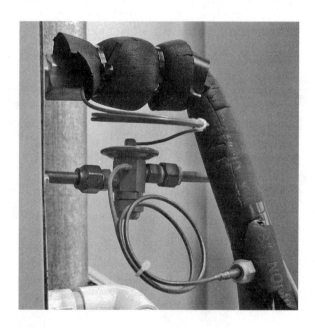

303F01.EPS

Figure 1 ◆ Metering device location.

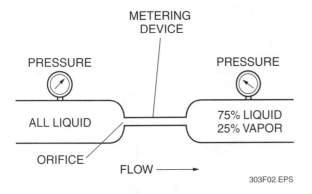

Figure 2 ◆ Metering device operation.

2.2.0 Adapting to Load Changes

The expansion valve metering device adapts to changing loads by increasing or decreasing the amount of refrigerant flowing to the evaporator. Fixed metering devices, on the other hand, are selected to meet the maximum cooling load or **design load**. This means that the flow capacity of the device is equal to the pumping capacity of the compressor.

Fixed metering devices are suitable for small systems that operate under a fairly constant load (such as refrigerators, freezers, and air conditioners of five tons or less). Such systems can, to some extent, adapt themselves to changing loads because system pressures vary in response to changes in the amount of heat absorbed by the refrigerant as it flows through the evaporator.

As the cooling load reduces, for example, the compressor pumps less refrigerant through the system. The refrigerant flow through a fixed metering device depends on the condenser pressure and the size of the metering device's orifice.

In the case of a capillary tube, the refrigerant flow also depends on the length of the tubing.

3.0.0 ◆ FIXED METERING DEVICES

This category includes capillary tubes and **fixed-orifice metering devices**. These devices have a fixed opening through which the refrigerant from the condenser passes on its way to the evaporator. The outside temperature affects the condensing pressure achieved with fixed metering devices. On a hot day, the condensing pressure will be high, driving more refrigerant through the fixed metering device. On a mild day, the condenser pressure will be lower, reducing the amount of refrigerant flow. **Superheat** will be lower on hot days and higher on mild days. It is important to note that fixed metering devices have no capability to adjust to changes in load or other external variables. The volume of refrigerant that passes is based on the pressure difference

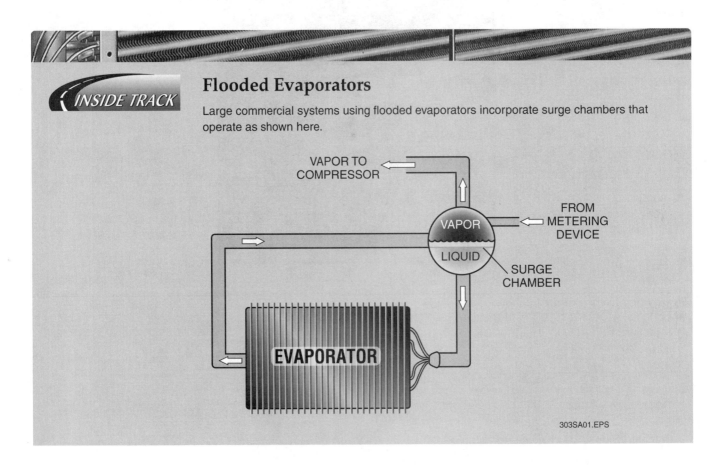

Metering Devices

Because of the risk to equipment reliability and the need to meet higher energy efficiency standards, one major equipment manufacturer has abandoned the use of fixed-orifice metering devices on the indoor coils of residential split-system products and has switched entirely to TXVs.

between its inlet and outlet, and the size of the opening. For this reason, great care must be taken in the engineering and selection of fixed-orifice devices to ensure proper performance.

One of the drawbacks of fixed metering devices is that they do not stop refrigerant flow during the off-cycle. Thus, liquid refrigerant from the condenser can migrate to the evaporator. When the system cycles on, a slug of liquid will be drawn into the compressor. If the liquid migrates to the compressor crankcase during the off-cycle, it will mix with the compressor oil and affect lubrication. Over time, this could seriously damage the compressor. At minimum, system performance will be affected. To prevent liquid from condensing in the oil, a crankcase heater can be used to keep the compressor oil warm.

Although such devices cannot stop off-cycle flow, the advantage is rapid equalization of refrigerant pressure as the compressor cycles off. This allows the compressor to start more easily, reducing motor wear. This reduces the need for starting components such as start capacitors and potential relays.

3.1.0 Capillary Tubes

The capillary tube, or cap tube, (*Figure 3*) is a copper tube with an inside diameter ranging from .026" up to .01". Capillary tubes vary widely in length, from roughly 12" to 140". They are usually coiled to save space and protect it from damage. The combination of diameter and length are selected to match the pressure drop with the pumping capacity of the compressor at design load.

> **NOTE**
>
> Capillary tubes should be secured so they do not rub against each other or the case. Leaks can occur over time.

Because capillary tubes are small and relatively fragile, they must be handled with extreme care. Because of the risk of damage, cap tubes must not be cut with regular tubing cutters or hacksaws.

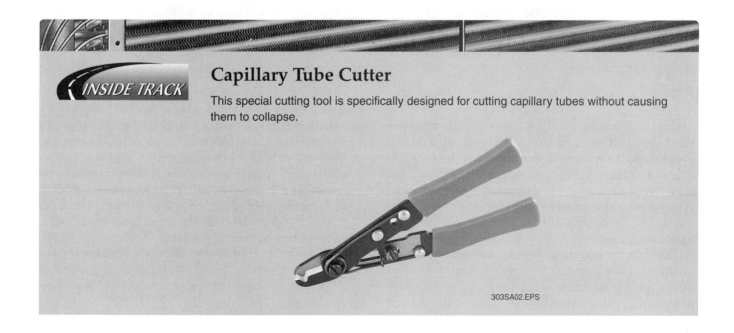

INSIDE TRACK

Capillary Tube Cutter

This special cutting tool is specifically designed for cutting capillary tubes without causing them to collapse.

303SA02.EPS

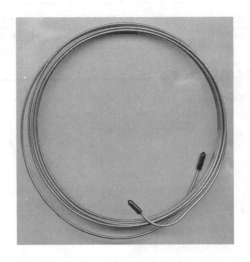

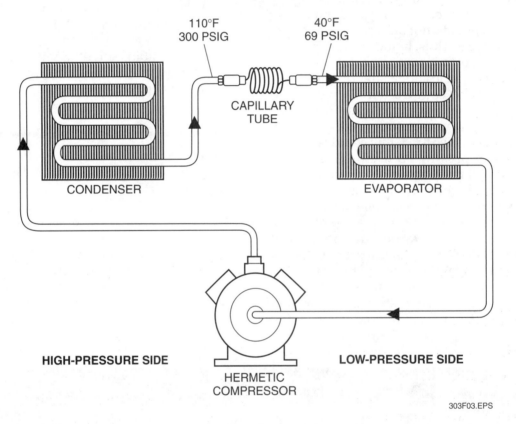

110°F
300 PSIG

40°F
69 PSIG

CAPILLARY
TUBE

CONDENSER

EVAPORATOR

HIGH-PRESSURE SIDE

LOW-PRESSURE SIDE

HERMETIC
COMPRESSOR

303F03.EPS

Figure 3 ◆ Capillary tube.

Cap tubes may be cut by using a file to score the tube, and then filing the tube until it easily breaks apart. Be careful not to pinch the tube, as this may change the inside diameter (ID) of the tube, which will affect system performance. Always remember to examine the tube ends for any burrs or debris before performing maintenance or repair.

The cap tube is usually selected and installed at the factory because its length and diameter are critical. If the tube is too long or the diameter too small, the evaporator will be starved for refrigerant, and excess liquid will build up in the condenser. The effect will be high head pressure, high superheat, high subcooling, and inadequate cooling. The same effect would be caused by a restricted tube. Therefore, a liquid strainer is sometimes provided at the cap tube input. A liquid line filter drier is always recommended to protect any metering device.

If the tube is too short or the diameter too great, excess liquid will be fed to the evaporator. This could cause refrigerant flood-back and/or slugging.

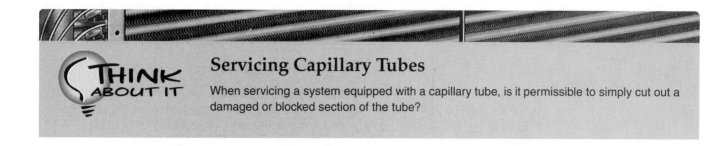

3.2.0 Fixed-Orifice Metering Devices

One popular metering device uses a removable piston that contains a fixed orifice (*Figure 4*). It has the advantage of being much smaller and more rugged than the cap tube and is less prone to damage.

In a split system, the condenser capacity determines the orifice size. Therefore, it is common for the manufacturer to ship a metering piston with the condensing unit.

The evaporator is also shipped with a piston installed in a metering device. In most cases, the piston shipped with the evaporator should be replaced with the piston supplied with the condensing unit. For long line applications, refer to the installation instructions to determine the correct piston to use.

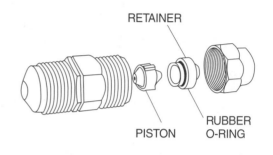

303F04.EPS

Figure 4 ◆ Fixed-orifice device.

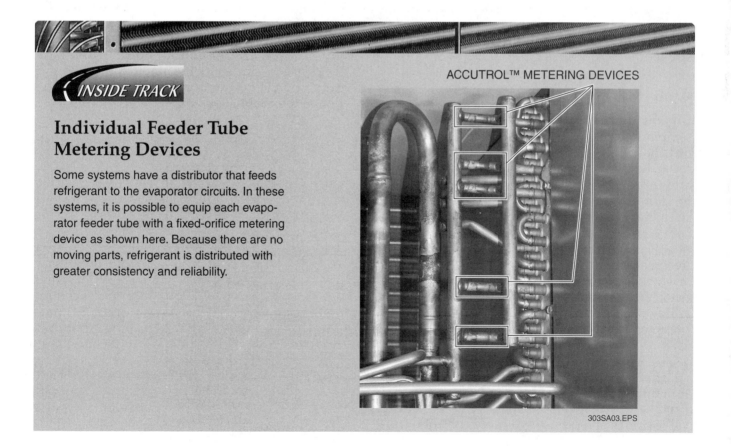

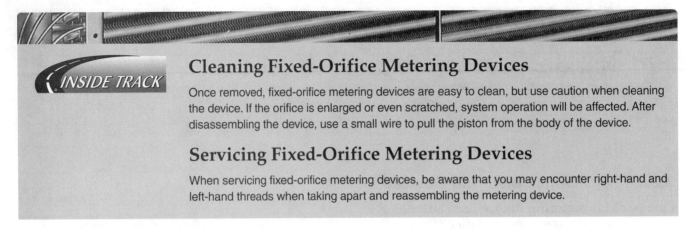
Fixed-orifice metering devices have been widely used in heat pumps because their design eliminates the need for check valves. The piston can slide within its housing. Movement in one direction creates a seal around the perimeter of the device as two machined surfaces meet, forcing all refrigerant to pass through the center orifice. In this position, the piston acts as a metering device. When flow is reversed, the piston slides to the opposite end of the housing that has no matching machined surface. In this position, refrigerant can flow not only through the orifice, but around the perimeter of the orifice itself. Minor turbulence is created, but there is a relatively insignificant pressure drop.

An advantage of the fixed-orifice metering device is that it can be removed easily from the system and replaced with a device of a different size to fine-tune system performance.

4.0.0 ◆ EXPANSION VALVES

Cooling demand varies according to changes in temperature and the quantity of indoor and/or outdoor air. When the air in the conditioned space becomes warmer, it increases in quantity and more heat is available to be absorbed. The ideal metering device reacts and allows more refrigerant to flow into the evaporator. When the conditioned air becomes cooler or decreases in quantity, less heat is available to be absorbed and the metering device allows less refrigerant flow.

There are several types of automatic expansion devices. There are also manual expansion valves, which are adjusted by the installer to match the pressure drop to the design load.

4.1.0 Manual Expansion Valves

A manual expansion valve (*Figure 5*) is a special hand valve with needle-pointed valve stems. The valve orifice and adjusting needle control refrigerant flow.

Manual expansion valves are adjusted by hand to match the design load, so they are only suitable in systems with a fairly constant load. Even then, constant monitoring is required by the system operator to adjust for changing conditions, such as condensing temperatures, to ensure the appropriate amount of refrigerant is admitted to the evaporator. If they are used in systems with variable loads, and not readjusted as the load changes, system performance will be affected (for example, the evaporator will be either starved or flooded as the load increases or decreases).

4.2.0 High-Side Float Valves

The high-side float valve (*Figure 6*) reacts to the level of liquid from the condenser, opening and closing the orifice in response to changes in the liquid level. The amount of liquid leaving the condenser is driven by the amount of vapor produced by the evaporator. This, in turn, is a function of the cooling load on the evaporator.

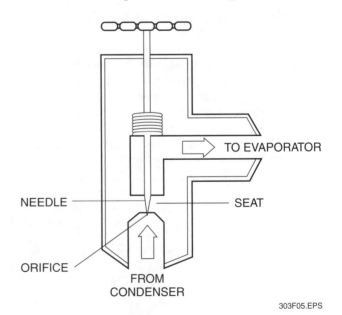

303F05.EPS

Figure 5 ◆ Manual expansion valve.

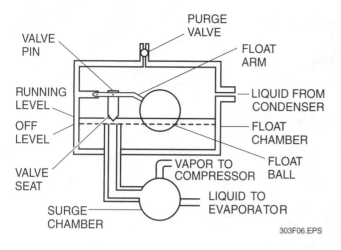

Figure 6 ◆ High-side float valve used with flooded evaporator.

As the cooling load increases, more liquid leaves the condenser. This extra liquid raises the float, and in doing so, increases the opening in the orifice to send more refrigerant to the evaporator.

In a system with a high-side float valve, most of the refrigerant stays in the evaporator when the compressor is not running. System charging is therefore critical. If the system is overcharged, the excess charge in the evaporator will flood the compressor with liquid when the system cycles on. High-side float valves are commonly used in the flooded cooler of a centrifugal chiller; however, they are also used with DX evaporators. When used with a flooded evaporator, the metering device output is sent through a surge chamber. This bypasses flash gas to the compressor, leaving only liquid at the evaporator input. High-side float valves are used on systems with a constant load.

4.3.0 Low-Side Float Valves

The low-side float valve (*Figure 7*) is the best control device for use with flooded evaporators.

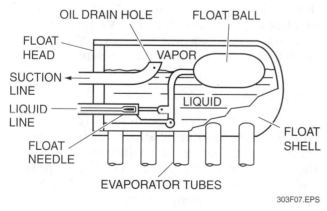

Figure 7 ◆ Low-side float valve.

It may be installed directly in the evaporator or in a separate float chamber outside the evaporator. Note that it is installed on the low side of the metering orifice, while the high-side float valve is installed on the high side. When the load on the evaporator increases, more liquid is boiled into vapor, reducing the amount of liquid in the float chamber. The lowered liquid level lowers the float, opening the metering orifice, and supplying more refrigerant to the evaporator.

4.4.0 Automatic Expansion Valves

The automatic expansion valve (*Figure 8*) maintains a constant evaporator pressure. The pressure maintains the temperature in the evaporator. It relies on a spring-loaded diaphragm and valve that react to pressure in the evaporator.

At a preset pressure, the force applied by the spring and the force provided by the evaporator cancel each other out. If the evaporator pressure changes in response to a changing load, the diaphragm will raise or lower, depending on whether the pressure is increased or decreased. Movement of the diaphragm causes the valve to expand or reduce the orifice opening. Automatic expansion valves are best suited for small capacity equipment with a fairly constant load, such as refrigerators and freezers.

The automatic expansion valve responds to load changes in a manner opposite of that you might expect. When there is an increase in the heat load being handled by the related evaporator coil, the suction pressure starts to rise. This causes the automatic expansion valve to start to reduce, or throttle, the amount of refrigerant being fed to the evaporator by closing the valve enough to maintain the constant suction pressure setpoint. This has the effect of starving the coil slightly. A larger increase in the heat load will cause even more starving of the coil.

Similarly, when there is a decrease in the heat load being handled by the evaporator, the suction pressure goes down. This causes the automatic expansion valve to open and feed more refrigerant to the coil. If the load is reduced too much, it is possible for liquid refrigerant to leave the evaporator coil and flood into the suction line going to the compressor. It is important to note that this valve has no way of preventing inadequate or excessive refrigerant flow, because it responds only to pressure changes and is not influenced by temperature. It also cannot differentiate between refrigerant liquid or vapor states.

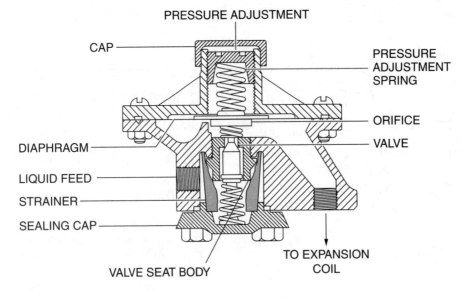

PRESSURE ADJUSTMENT

CAP

PRESSURE
ADJUSTMENT
SPRING

ORIFICE

VALVE

DIAPHRAGM

LIQUID FEED

STRAINER

SEALING CAP

VALVE SEAT BODY

TO EXPANSION
COIL

(A) AUTOMATIC EXPANSION VALVE COMPONENT

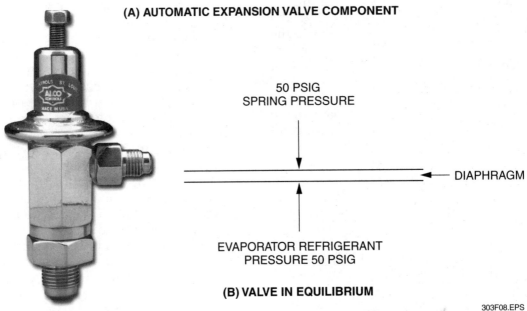

50 PSIG
SPRING PRESSURE

DIAPHRAGM

EVAPORATOR REFRIGERANT
PRESSURE 50 PSIG

(B) VALVE IN EQUILIBRIUM

303F08.EPS

Figure 8 ◆ Automatic expansion valve.

4.5.0 Thermal Expansion Valves

The thermal expansion valve is a metering device with an external sensing bulb that senses the refrigerant temperature at the evaporator outlet. The bulb sends this information, which represents the amount of superheat produced by the evaporator, back to the metering device, where it is used to adjust the flow of refrigerant to match the load. This type of device is alternately referred to as a thermal expansion valve or thermostatic expansion valve, and may be abbreviated TEV or TXV. The terms are interchangeable, as are the abbreviations. It tries to maintain a constant superheat at the evaporator outlet. This device should not be confused with the **thermal-electric expansion valve (TEEV/ THEV)** or the electronic expansion valve (EEV), both of which use a thermistor as a temperature sensor.

4.5.1 Operating Principles

The TXV is a device in which control is based chiefly on the amount of superheat in the refrigerant when it leaves the cooling coil and enters the suction line (see *Figure 9*). The DX evaporator coil in a system running at design conditions carries a mix of low-temperature liquid and vapor refrigerant. This refrigerant is boiling, and the air

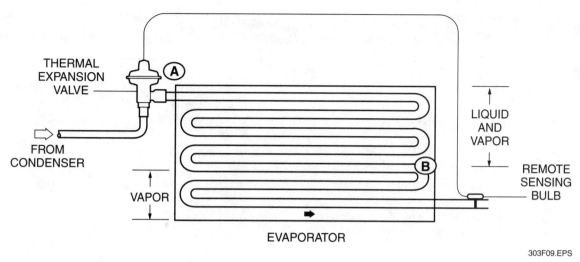

Figure 9 ◆ Refrigerant flow in the evaporator.

303F09.EPS

flowing through the evaporator is supplying the heat to change the refrigerant from a liquid to a vapor. When the refrigerant becomes a saturated vapor, its sensible heat begins to increase, creating superheat.

As the cooling load fluctuates, the point at which the liquid becomes a gas will move back and forth in the evaporator. The expansion valve must open and close automatically to adjust refrigerant flow to meet the load, thereby keeping the superheat at a consistent value and minimizing movement of this point in the evaporator.

When the load is constant, the diaphragm is balanced by the sensing bulb pressure at the top and a combination of spring pressure and evaporator inlet pressure at the bottom (see *Figure 10*). When the cooling load changes, the pressure at the evaporator outlet will also change. The diaphragm will move up or down, opening or closing the orifice to change the amount of refrigerant applied to the evaporator.

If the cooling load increases, the evaporator will be starved for refrigerant and superheat will begin earlier. The suction line temperature will increase because of the increased evaporator superheat. The increased superheat causes an increase in the pressure of the refrigerant in the sensing bulb.

The pressure increase is transmitted back to a diaphragm (or bellows). This opens the refrigerant needle valve to a greater degree. The thermostatic fluid used in the sensing bulb is often the same refrigerant used in the system. This refrigerant does not mix with the system refrigerant. For air conditioning, a TXV is commonly selected for 15°F to 20°F superheat at the factory. In most cases, the TXV does not need to be adjusted.

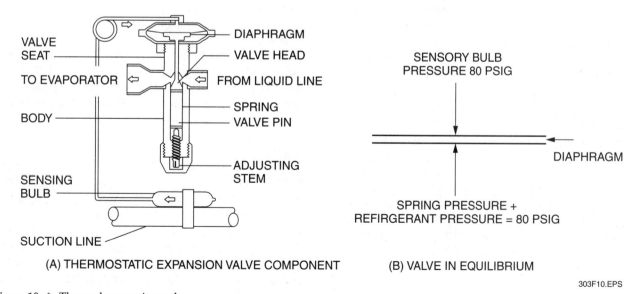

Figure 10 ◆ Thermal expansion valve.

303F10.EPS

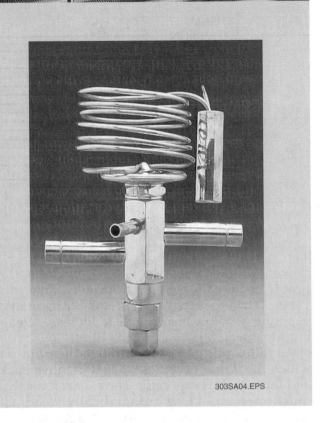
In a conventional TXV, high head pressure pushing against the metering piston in the orifice can overcome the spring force holding the piston closed, forcing it open. This affects the superheat setting. Today's TXVs use a balanced port design that keeps excessively high head pressures from acting as an opening force on the piston. One design uses a hole drilled through the piston. This allows pressure on the top of the piston to be sensed on the bottom of the piston, canceling the effects of the pressure. Another method uses identical cross-sectional areas for the port and pushrod. This also has the effect of canceling the respective pressure drops.

4.5.2 Equalizers

Along with spring pressure, the evaporator inlet pressure, which is also the pressure at the outlet of the TXV, acts on the bottom of the diaphragm. The sensing bulb pressure acts on the top of the diaphragm. If there is a significant refrigerant pressure drop through the evaporator, the pressure on the bottom of the valve diaphragm (a closing force) will be higher than the pressure exerted on the top (an opening force). This results in increased superheat.

In systems with little or no pressure drop across the evaporator, an internal equalizer is used (see *Figure 11*). The allowable pressure drop will vary with the refrigerant and the evaporating temperature, and may range from 0.75 psig to 3 psig. In an internally equalized TXV, evaporator inlet pressure is fed to the bottom of the diaphragm through an internal passage. Internal equalization is used only with evaporators having a very small pressure drop. These include refrigerators, food freezers, and small air conditioners.

If there is a larger pressure drop across the evaporator, an external equalizer is used. The use of a distributor or insufficiently sized **distributor lines** are factors that can cause an excessive pressure drop. When a distributor is used, an externally equalized expansion valve will also be used.

The external equalizer (*Figure 12*) isolates the TXV from the evaporator inlet pressure. The evaporator outlet pressure is sensed at a location immediately after the TXV sensing bulb. It is then fed back through the equalizer fitting to the bottom of the diaphragm. Because the pressures at the top and bottom of the diaphragm are sensed at the same point, the TXV is equalized.

Low Ambient Control

When air conditioners operate at low outdoor air temperatures, low head pressure causes a low-pressure differential across the TXV. This results in the TXV being starved of refrigerant. Head pressure can be maintained by controlling operation of the condenser fan motor or it can be maintained by modifying the refrigeration system. The refrigeration system is modified so as to restrict liquid flow from the condenser to the receiver, and at the same time divert hot gas to the inlet of the receiver. This causes liquid refrigerant to back up into the condenser, reducing its capacity, which in turn increases pressure in the condenser. Concurrently, the hot gas raises liquid pressure in the receiver, allowing the system to operate normally. The device used to accomplish this is called a low-ambient control. In operation, discharge pressure bleeds around the pushrod to the underside of the diaphragm where it opposes the dome pressure. When the outdoor temperature falls, the condensing pressure falls, causing the discharge pressure to fall. When the discharge pressure falls below the dome pressure, the valve modulates open to the discharge port, allowing discharge gas to bypass the condenser. Mixing discharge gas with liquid creates a high pressure at the condenser outlet, reducing the flow and causing liquid to back up in the condenser. Flooding the condenser reduces its surface, allowing head pressure to rise.

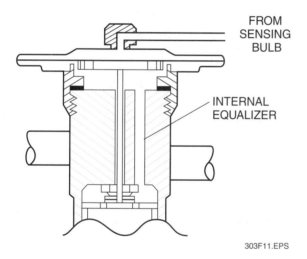

303F11.EPS

Figure 11 ◆ Expansion valve with internal equalizer.

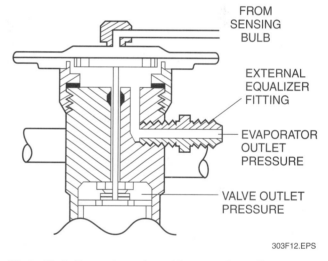

303F12.EPS

Figure 12 ◆ Expansion valve with external equalizer.

Refer to *Figure 13* for an example of a typical R-22 system. Assume that you want to maintain 10°F of superheat at the evaporator outlet to achieve optimum system performance. If the pressure at the TXV outlet is 76 psig and the spring pressure is equal to 18 psig, the combined pressure on the underside of the diaphragm is 94 psig. With zero pressure drop across the evaporator, the superheated vapor at the evaporator outlet will produce a suction line temperature of 55°F. In the TXV sensing bulb, the 55°F temperature corresponds to a pressure of 94 psig, which is fed back to the top of the diaphragm. Since the pressures on the top and bottom of the dia-

phragm are equal, the valve maintains a constant refrigerant flow. With a small pressure drop, the TXV may fluctuate for a while, but will eventually stabilize. In this situation, the internally equalized TXV is suitable for the conditions.

If the evaporator had a 6 psig pressure drop, the effect with an internally equalized TXV would be very different. The evaporator outlet pressure would be 6 psig less than the inlet pressure. When the bulb sensed this pressure drop, the valve would modulate, reducing the amount of refrigerant to the evaporator. This would cause the evaporator to operate at about 15°F of superheat, which would adversely affect system performance.

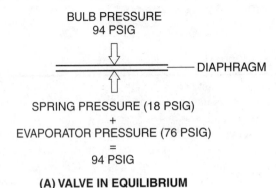

(A) VALVE IN EQUILIBRIUM

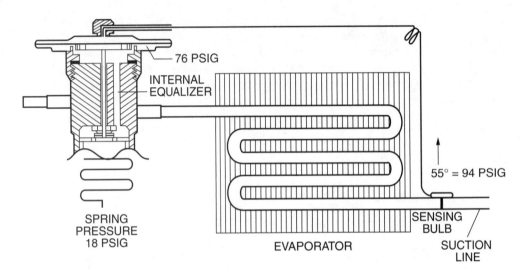

(B) INTERNALLY EQUALIZED TXV COMPONENTS

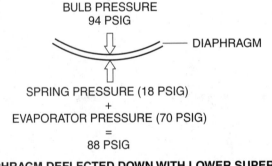

(C) VALVE DIAPHRAGM DEFLECTED DOWN WITH LOWER SUPERHEAT

303F13.EPS

Figure 13 ◆ Internally equalized TEV.

With an externally equalized TXV, this problem is eliminated (*Figure 14*). The pressure in the suction line is tapped and fed back to the bottom of the diaphragm. Since the pressure at the top of the diaphragm is also a function of the suction line pressure, the forces on the two sides of the diaphragm remain equal as long as the load remains the same.

Many valve manufacturers recommend externally equalized valves for improved performance. When replacing a defective internally equalized valve, it may be necessary to replace the valve with an externally equalized valve. The equalizer line is usually ¼" soft copper tubing run from the valve to the suction line near the evaporator outlet. In these cases, an external equalizer

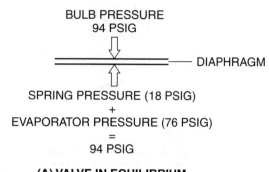

BULB PRESSURE
94 PSIG

⇓

DIAPHRAGM

⇑

SPRING PRESSURE (18 PSIG)
+
EVAPORATOR PRESSURE (76 PSIG)
=
94 PSIG

(A) VALVE IN EQUILIBRIUM

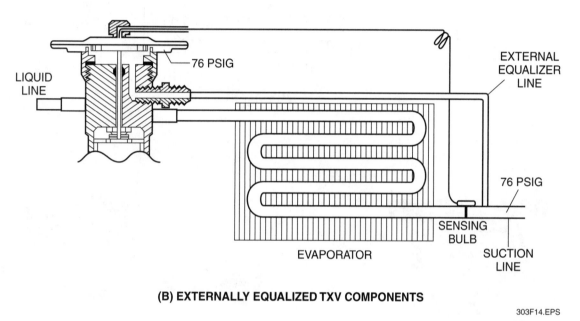

(B) EXTERNALLY EQUALIZED TXV COMPONENTS

303F14.EPS

Figure 14 ◆ Externally equalized TXV.

line must be installed on the system. Make sure not to cap off the external equalizer fitting.

4.5.3 Sensing Bulb Refrigerant Charge

The sensing bulb of a TXV contains refrigerant that expands and contracts in response to temperature changes in the suction line. The refrigerant in the bulb must be the same refrigerant that is in the system. This enables the valve to accurately respond to changes in temperature because the refrigerant in the bulb has the same characteristics as the refrigerant in the system. If the bulb loses its charge, the pressure on the internal diaphragm will be lost, and the spring on the metering piston will force the valve closed. In this situation, the valve must be replaced. Some TXVs allow the sensing bulb portion of the valve to be replaced while leaving the rest of the valve in place.

4.6.0 Thermal-Electric Expansion Valves

The thermal-electric expansion valve (it may be abbreviated as either TEEV or THEV) senses the temperature of the refrigerant leaving the evaporator (see *Figure 15*). A thermistor placed in the suction line acts as the sensor. As the cooling load changes, the resulting changes in the refrigerant temperature will reduce or increase the resistance offered by the thermistor. Inside the valve, there is a bimetal needle that reacts to the change in current by modulating the valve to reduce or increase refrigerant flow. The valve will maintain a constant superheat in the system.

4.7.0 Electronic Expansion Valves

An electronic expansion valve (EEV) (*Figure 16*) is a microprocessor-controlled device that is driven by a precision DC motor known as a stepper

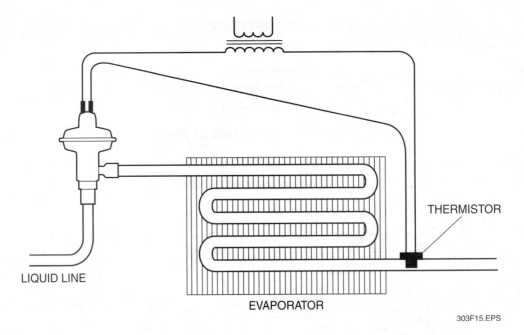

Figure 15 ◆ Thermal-electric expansion valve location.

LIQUID LINE

THERMISTOR

EVAPORATOR

303F15.EPS

303F16.EPS

Figure 16 ◆ Electronic expansion valve (EEV).

303F17.EPS

Figure 17 ◆ Pulsing solenoid EEV.

motor. This motor can provide 760 discrete steps of orifice control in the metering device. Like the TEEV, the EEV responds to temperature changes sensed by a thermistor and reacts to these changes to maintain a constant superheat at the evaporator outlet. Some EEVs are driven by a pulse system instead of a stepper motor.

Another type of electronic expansion valve used with compressors with capacity control is the pulsing solenoid (*Figure 17*). It contains a fixed-orifice sized for about half capacity through which refrigerant always flows. Another parallel orifice is pulsed open and shut in response to signals from an electronic control module. The more time the solenoid is open, the more refrigerant flows. This results in increased capacity. When the solenoid is open 100 percent of the time, the metering device is delivering full capacity. Electronic expansion valves often employ sensors in the form of thermistors that supply ongoing temperature change information to the electronic module that controls valve operation.

4.8.0 Hunting

Hunting is a term commonly used to describe the changes in refrigerant flow as the expansion device adapts to changing conditions. As the load increases, for example, the valve may open too wide, allowing too much refrigerant into the evaporator. When this happens, the system may overcompensate in the other direction. Hunting occurs because there is a time lag between when

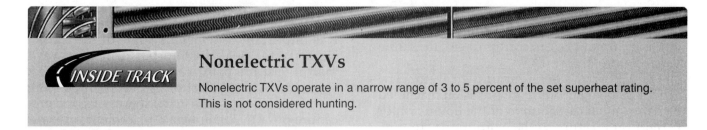

Nonelectric TXVs

Nonelectric TXVs operate in a narrow range of 3 to 5 percent of the set superheat rating. This is not considered hunting.

the expansion valve modulates and when the sensing bulb senses the change. Proper selection of the expansion valve and adjusting for the correct superheat can minimize hunting. Oversized expansion valves are particularly susceptible to hunting. The position of the sensing bulb and how firmly the bulb is clamped to the suction line can also affect hunting.

Some devices include anti-hunting features such as thermal ballast. The thermal ballast is included inside the sensing bulb. It delays the response of the sensing tube, and thus prevents the valve from overfeeding or underfeeding before the system can balance. It also acts as a safety device by quickly closing the valve when it senses a liquid slug in the suction line.

Hunting can only be verified by making several suction line temperature or pressure measurements over a period of time. If there is a repetitive pattern such as that shown in *Figure 18*, hunting is occurring. It may be necessary to select a smaller valve to correct the situation.

5.0.0 ◆ DISTRIBUTORS

For best system performance, the mixture of liquid and gas coming from the TXV must be distributed evenly throughout the evaporator coil. It is critical that the pressure drop within the evaporator be held within reasonable limits. Additionally, refrigerant must be regulated to guarantee proper oil return to the compressor. These needs are usually met by adding multiple refrigerant circuits in the cooling coil.

For some applications, each circuit can be served by its own TXV. However, normal practice is to employ a **distributor** that evenly proportions refrigerant flow from a single expansion

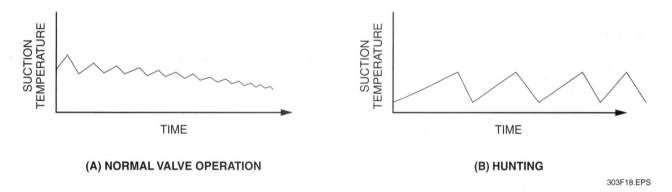

(A) NORMAL VALVE OPERATION

(B) HUNTING

303F18.EPS

Figure 18 ◆ Suction temperature variations indicate hunting.

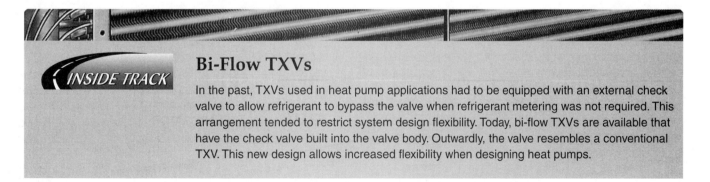

Bi-Flow TXVs

In the past, TXVs used in heat pump applications had to be equipped with an external check valve to allow refrigerant to bypass the valve when refrigerant metering was not required. This arrangement tended to restrict system design flexibility. Today, bi-flow TXVs are available that have the check valve built into the valve body. Outwardly, the valve resembles a conventional TXV. This new design allows increased flexibility when designing heat pumps.

valve to all circuits (*Figure 19*). Ideally, a distributor should provide even refrigerant distribution with a minimal pressure drop to ensure stable system control by the TXV.

Figure 20 shows one type of distributor. The orifices in the distributor are precision machined to ensure equal distribution of the liquid/vapor refrigerant mixture to the evaporator circuits. Because the pressure drop through distributors is often significant, they are sometimes considered extensions of the metering device.

6.0.0 ◆ TXV REPLACEMENT

This section covers TXV replacement, including selection and placement of the new TXV. Sensing bulb location and TXV adjustment are also discussed. Proper installation and insulation of the TXV sensing bulb are critical to correct operation.

6.1.0 Selection

It is important to select a TXV with the correct capacity for peak system performance. Most manufacturers print selection tables. Correct expansion valve selection is vitally important; therefore, it is advisable to review a selection procedure prior to obtaining a replacement valve.

Valve capacity must be compatible with system capacity. For example, if a load is calculated at 17.5 tons and the only valves available have capacities of 12 tons and 18 tons, the 18-ton valve should be selected. The smaller-capacity valve would starve the evaporator at full load, whereas the slightly oversized valve would be acceptable. TXV selection tables usually contain data in relation to system capacity in tons, condensing temperature, evaporator temperature, and pressure drop across the valve.

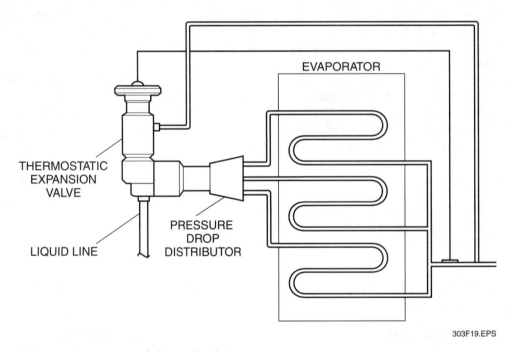

303F19.EPS

Figure 19 ◆ Multi-circuit evaporator fed by a distributor.

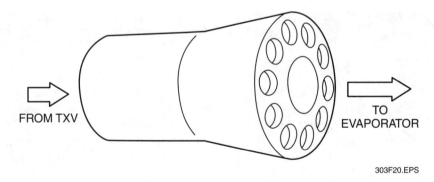

303F20.EPS

Figure 20 ◆ Distributor.

If an extremely oversized valve is selected, it will not be able to control closely at part-load conditions and may cause slugging of the compressor.

During selection, consider such factors as correct capacity, selective charge of the sensing bulb, and whether it should have an external or internal equalizer. According to some manufacturers, TXVs may be mounted in any position; however, it is advisable to place them as close to the evaporator inlet as possible and to follow each manufacturer's installation instructions. If a distributor is used, it should be mounted directly to the valve outlet. Common valve mountings are brazed, soldered, flared, or flanged.

6.2.0 Placement

The best performance is obtained if the valve feeds vertically up or down into the distributor. If a hand valve is located on the outlet side of the expansion valve, it should not have a restricted port, nor should any restrictions appear between the TXV and the evaporator (refrigerant distributor excepted). To minimize refrigerant migration on some types of valves, it is recommended that the valve diaphragm case be insulated so that it remains warmer than the bulb.

If TXVs are located in corrosive atmospheres, they must be protected with appropriate materials to prevent early failure.

It is usually not necessary to disassemble solder-type valves when soldering to the connecting lines. Any of the commonly used solders such as 95-5, Sil-Fos, Easy-Flow, or Phos-Copper may be used. However, exercise caution to keep the flame of the torch directed away from the valve body to avoid excessive heat on the valve diaphragm. A wet cloth wrapped around the valve during the soldering process is an extra precaution that is well worth the effort.

A nitrogen purge will prevent copper oxide contamination during the brazing process. Nitrogen will also help keep the valve cool. Replace the liquid line filter drier after installing the new valve.

6.3.0 Sensing Bulb

Locate the sensing bulb with care and follow accepted piping standards to allow for precise valve control. The sensing bulb should be attached to a horizontal suction line, but it may be attached to a vertical line if absolutely necessary. The bulb should not be mounted in a trap or pocket in the suction line because refrigerant boiling out of the trap will falsely influence the temperature of the bulb, resulting in improper valve control. The bulb should never be installed on unions or fittings because the thermal conductance of these materials will delay response to changes in suction line temperature.

On suction lines up to and including ⅝" outside diameter (OD), the bulb should be installed on top of the suction line (see *Figure 21*). On lines with a ⅞" OD or larger, the bulb should be clamped near the bottom of this line. Locating the bulb on the bottom of the line is not recommended because oil-refrigerant mix is usually present at this point.

For difficult installations, the proper bulb location may be determined by trial and evaluation. Good thermal contact between the bulb and the suction line is essential; therefore, the bulb should be fastened securely to a clean, straight section of the suction line with two bulb straps. The straps should be a copper alloy to prevent reaction with the copper. The sensing bulb should be insulated to improve performance. Good piping practices usually include attaching the TXV to a horizontal suction line leaving the evaporator. This line is pitched downward.

When a vertical riser follows the horizontal line, a short trap (*Figure 22*) is placed immediately ahead of the vertical line. This will collect any liquid refrigerant and/or oil passing through the suction line that might influence the temperature of the sensing bulb. On multiple evaporator installations, the piping and sensing bulb should be arranged so the flow from any one valve cannot influence the others.

BULB
⅝" O.D. OR SMALLER

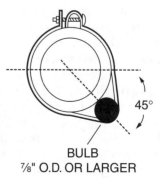

45°

BULB
⅞" O.D. OR LARGER

303F21.EPS

Figure 21 ◆ Sensing bulb positioning.

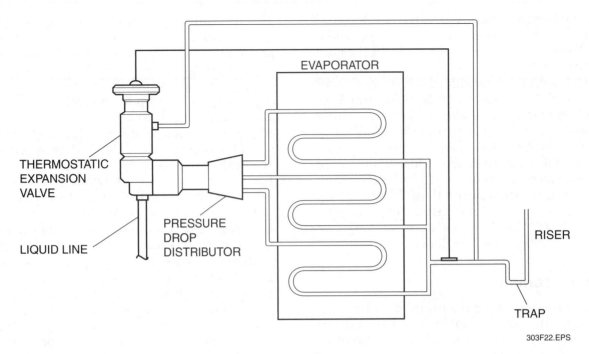

Figure 22 ◆ Trap in a suction line riser.

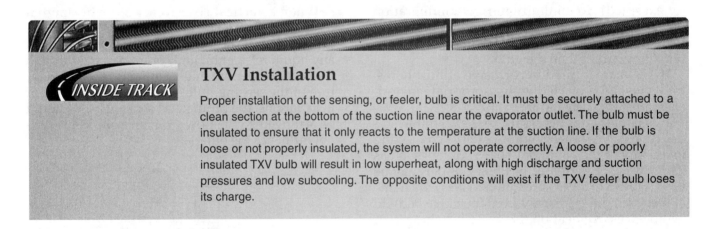

TXV Installation

Proper installation of the sensing, or feeler, bulb is critical. It must be securely attached to a clean section at the bottom of the suction line near the evaporator outlet. The bulb must be insulated to ensure that it only reacts to the temperature at the suction line. If the bulb is loose or not properly insulated, the system will not operate correctly. A loose or poorly insulated TXV bulb will result in low superheat, along with high discharge and suction pressures and low subcooling. The opposite conditions will exist if the TXV feeler bulb loses its charge.

On commercial and low-temperature applications, the bulb should be clamped on the suction line at a point where the bulb temperature will be the same as the evaporator temperature during the off-cycle. The insulation used to wrap the sensing bulb must not be water absorbent.

On brine tanks and water coolers, the bulb should be attached below the surface of the liquid. If the bulb is located in a brine tank, the bulb and cap tubing must be painted with corrosion-resistant paint or pitch. If the bulb has to be located where its temperature will be higher than the evaporator during the off-cycle, a solenoid valve must be used ahead (upstream) of the TXV.

When an externally equalized TXV is installed in the system, the equalizer connection should be made at a point immediately downstream from the sensing bulb. If evaporator pressure or temperature control valves are located in the suction line near the evaporator outlet, the equalizer line must be connected on the evaporator side of these valves.

Many valve failures are due to the presence of dirt, sludge, and moisture. Therefore, filter-driers should be installed in the system whenever the expansion valve is replaced. It may also be desirable to install a sight glass and moisture indicator at the same time. As an added precaution, most replacement TXVs are equipped with built-in inlet screens for removing scale or other particles that could obstruct the closure of the valve and seat.

6.4.0 TXV Adjustment

All TXVs are tested and set at the factory prior to shipment. The factory superheat setting will be correct in relation to the size of the valve; no further adjustment is required in most cases. If operating conditions require a different setting, consult the equipment manufacturer's literature and install the specified replacement valve.

The amount of superheat for a given evaporator may be obtained using the following procedure:

Step 1   Record the temperature of the suction line at the position where the sensing bulb is attached.

Step 2   Obtain the suction pressure in the suction line at the bulb location. Place a gauge in the external equalizer line (if the unit is so equipped) or record the gauge pressure at the suction valve of the compressor. If the pressure is obtained at the compressor, an estimated line pressure drop must be added to the gauge reading. The sum of the gauge reading and estimated pressure drop should be about equal to the suction line pressure at the sensing bulb.

Step 3   Convert the recorded pressure to saturated refrigerant temperature using a pressure-temperature chart. Some compound gauges have the corresponding temperatures for this conversion.

Step 4   Subtract the two temperatures (bulb location temperature and converted suction line pressure-temperature). The difference is the superheat.

To find the system superheat rather than the evaporator superheat, measure the suction line temperature near the compressor and follow Steps 2 through 4. Most manufacturers require the system superheat reading.

Acceptable superheat readings may range from as low as 10°F to as high as 20°F. The manufacturer's recommendations should be followed for the correct valve size on factory equipment.

7.0.0 ◆ METERING DEVICE PROBLEMS

Metering devices are often blamed for system problems that can be caused by any number of other faults. For example, if the temperature leaving the evaporator is too high, the metering device may be blamed for not feeding enough refrigerant. The symptoms are obvious (not cooling properly), but the cause may be more difficult to find.

NOTE

Before troubleshooting any system, make sure the system is properly charged and has the correct level of subcooling.

Causes of poor system performance include:

• Moisture due to water and oil frozen in the valve port or working parts when operating below 32°F

• Dirt or other contaminants that may not be trapped by the strainer and obstruct the flow of refrigerant through the valve port

INSIDE TRACK

Before Troubleshooting a TXV

Before troubleshooting a TXV, establish a baseline superheat reading. The operation of a TXV can then be checked by removing the sensing bulb and holding it, or holding the bulb up to a hot light source. The system should respond to the increase in the bulb's temperature by lowering the superheat.

Next, place the sensing bulb in a glass of ice water. The system superheat should increase, because the drop in temperature of the sensing bulb will cause the TXV to close. If the TXV does not respond, it is defective and must be replaced.

If the sensing bulb completely loses its refrigerant charge, this removes a primary opening force and the valve will often close very tight. Many TXVs incorporate replaceable thermostatic elements. The valve remains in place and only the element is replaced.

- Wax that precipitates at low temperature and builds up on the needle valve and seat
- Refrigerant shortage due to a low state of charge
- Gas in the liquid line due to long or undersized lines or improperly piped systems
- Misapplication of an internally equalized valve in place of an externally equalized valve, or improper location of the externally equalized valve line
- Restricted, plugged, or capped external equalizer tube
- Undersized valve
- High superheat adjustment
- High pressure drop through the evaporator

Other problems, such as refrigerant overfeeding, feeding too much on startup, improper valve feeding, or hunting and cycling could be due to some of the same causes.

Keep in mind that many system malfunctions may be eliminated if the system is kept clean, dry, and in a proper state of charge. Check these conditions before changing the metering device.

Neither TXVs nor distributors function well at very low loads. Low-load conditions may be caused by oversizing. If a distributor load falls below 50 percent, the liquid-vapor mixture entering the evaporator may separate. The liquid will go to the bottom circuits and the vapor will go to the top circuits. Excessive superheat and liquid flood-back to the compressor will result. An easy way to check for this problem is to look for high temperatures at the top of the evaporator and low temperatures at the bottom. If this problem occurs, consult the equipment manufacturer to determine the proper corrective action.

1. A metering device is a refrigeration system control that _____.
   a. converts high-pressure, high-temperature liquid refrigerant to a low-temperature, low-pressure pure vapor refrigerant
   b. converts high-pressure, high-temperature liquid refrigerant to a low-temperature, low-pressure mix of liquid and vapor refrigerant
   c. is located at the evaporator outlet
   d. always contains a feature to automatically adjust the refrigerant flow

2. A significant difference between a DX evaporator and a flooded evaporator is that _____.
   a. the DX evaporator is used primarily on large commercial systems, while the flooded evaporator is used more on small systems
   b. in the DX evaporator, the refrigerant is converted to vapor in the evaporator, whereas it remains in liquid form in the flooded evaporator
   c. DX evaporators are always used with a surge chamber, while flooded evaporators seldom are
   d. there is more flash gas at the inlet of a flooded evaporator

3. Which of the following is true of a capillary tube?
   a. It is always the same length.
   b. It is made of PVC pipe.
   c. It has no sensing bulb.
   d. It is used only with flooded evaporators.

4. In which of these systems would a manual expansion valve be a good choice?
   a. A system with a fairly constant load.
   b. A system that experiences frequent load changes.
   c. A lobby in a busy hotel.
   d. A movie theater.

5. The main difference between high-side and low-side float valves is that the _____.
   a. high-side float valve is located at the compressor outlet, while the low-side float valve is installed at the evaporator inlet
   b. low-side float valve is always used with DX evaporators, while the high-side float valve is used with flooded evaporators
   c. low-side float valve is installed at the bottom of the evaporator, while the high-side float valve is installed at the top of the evaporator
   d. low-side float valve is installed on the low side of the metering orifice, while the high-side float valve is installed on the high side of the metering orifice

6. A distinguishing feature of the TXV is that it _____.
   a. has a sensing bulb filled with refrigerant or other thermostatic fluid
   b. needs to be manually adjusted as the cooling load changes
   c. uses a thermistor as a sensing device
   d. uses a float to maintain a constant liquid level in the evaporator

7. Which of the following is best suited for a system with a large pressure drop across the evaporator?
   a. An internally equalized TXV.
   b. An externally equalized TXV.
   c. A high-side float valve.
   d. A manual expansion valve.

8. The primary purpose of a distributor in an air conditioning system is to _____.
   a. provide high voltage to the spark plugs
   b. feed equal amounts of refrigerant to the evaporator circuits
   c. make sure the evaporator gets enough refrigerant
   d. distribute refrigerant from the condenser to the metering devices

9. Selective charge of the sensing bulb is an important consideration in selecting a TXV.

   a. True
   b. False

10. The correct location for a TXV sensing bulb is _____.

   a. at the condenser outlet
   b. on the suction riser
   c. on the compressor side of the external equalizer tube
   d. at the evaporator outlet

Summary

The purpose of the TXV or any other refrigerant metering device is to maintain the pressure drop needed to produce low-temperature liquid refrigeration for cooling and to regulate the quantity of refrigerant that can pass into the evaporator in response to the cooling load. The importance of the TXV cannot be overemphasized. If it is to operate properly, it must be carefully selected and precisely installed in a clean, dry, properly charged system.

Notes

Trade Terms Introduced in This Module

Capillary tube: A copper tube with a fixed length and fixed diameter, usually with an inside diameter of $\frac{1}{16}$" to $\frac{1}{8}$". Used as a metering device.

Design load: The maximum load at which a system is designed to operate.

Direct-expansion (DX) evaporator: An evaporator in which liquid is completely converted to vapor; also known as dry-expansion.

Distributor: A special fitting, generally machined, containing multiple passageways that distribute liquid refrigerant evenly to the evaporator circuits.

Distributor line: Tubing between the distributor and evaporator or the line between the metering device and evaporator.

Electronic expansion valve (EEV): A microprocessor-controlled expansion valve in which the orifice is controlled by a precision DC motor. The EEV maintains a constant superheat.

Fixed-orifice metering device: A device in which the metering orifice is located in a replaceable piston. The term fixed-orifice metering device may also be used to refer to a capillary tube.

Flash gas: Vapor refrigerant that is formed as the liquid refrigerant is squeezed through the metering orifice. Flash gas is produced when some of the liquid refrigerant boils off to cool the remaining liquid as the liquid passes through the metering device.

Flood-back: Liquid refrigerant that makes its way past the evaporator and returns to the compressor. It often causes dilution of the oil in the crankcase and a more severe condition known as slugging.

Flooded evaporator: An evaporator in which the refrigerant remains in liquid form as it leaves the evaporator.

Hunting: A condition in which an expansion valve alternately underfeeds and overfeeds the evaporator.

Orifice: A tiny opening designed to pass liquid refrigerant.

Superheat: Heat added to the refrigerant above the refrigerant's boiling point.

Surge chamber: A device that separates liquid and vapor refrigerant. Used with flooded evaporators. Liquid is recirculated back to the chiller or evaporator and vapor returns to the compressor.

Thermal expansion valve (TEV or TXV): A metering device with an external sensing bulb that senses the refrigerant temperature at the evaporator outlet. Also referred to as a thermostatic expansion valve. The valve maintains superheat around a setpoint. The terms TEV and TXV are used interchangeably.

Thermal-electric expansion valve (TEEV or THEV): An expansion valve in which a thermistor senses liquid line temperature and adjusts the orifice by changing the current flow through a bimetal needle. The terms TEEV and THEV are used interchangeably.

Additional Resources and References

Additional Resources

This module is intended to be a thorough resource for task training. The following reference work is suggested for further study. This is optional material for continued education rather than for task training.

Air Conditioning Systems: Principles, Equipment, and Service, 2001. Joseph Moravek. Upper Saddle River, NJ: Prentice Hall, Inc.

Figure Credits

NCCER CURRICULA — USER UPDATE

NCCER makes every effort to keep its textbooks up-to-date and free of technical errors. We appreciate your help in this process. If you find an error, a typographical mistake, or an inaccuracy in NCCER's curricula, please fill out this form (or a photocopy), or complete the online form at **www.nccer.org/olf**. Be sure to include the exact module ID number, page number, a detailed description, and your recommended correction. Your input will be brought to the attention of the Authoring Team. Thank you for your assistance.

Instructors – If you have an idea for improving this textbook, or have found that additional materials were necessary to teach this module effectively, please let us know so that we may present your suggestions to the Authoring Team.

NCCER Product Development and Revision

13614 Progress Blvd., Alachua, FL 32615

Email: curriculum@nccer.org
Online: www.nccer.org/olf

❏ Trainee Guide   ❏ AIG   ❏ Exam   ❏ PowerPoints   Other _____

Craft / Level: _____ Copyright Date: _____

Module ID Number / Title: _____

Section Number(s): _____

Description: _____

_____

_____

_____

_____

Recommended Correction: _____

_____

_____

_____

_____

Your Name: _____

Address: _____

_____

Email: _____ Phone: _____

03304-08

Retail Refrigeration Systems

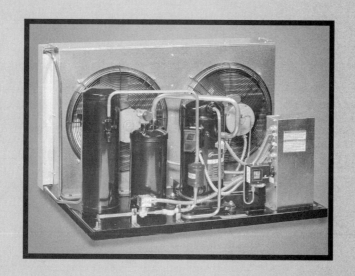

03304-08
Retail Refrigeration Systems

Topics to be presented in this module include:

1.0.0   Introduction 4.2
2.0.0   Mechanical Refrigeration Systems 4.2
3.0.0   Defrost Systems 4.17
4.0.0   Retail Refrigeration Equipment and Fixtures 4.20
5.0.0   Common Refrigeration System Controls 4.26
6.0.0   Troubleshooting 4.30

Overview

Think about how many convenience stores, supermarkets, restaurants, hotels, and clubs there are in your hometown. Then think about how many reach-in coolers, freezers, ice machines, water coolers, and other refrigeration machines are needed in all those places. The existence of all this refrigerated equipment creates a huge demand for people to maintain them. When a supermarket freezer fails, for example, a service technician will get an urgent call to repair it before the contents spoil. All this boils down to the fact that a properly trained refrigeration service tech has the opportunity for a busy and rewarding career in the trade.

Objectives

When you have completed this module, you will be able to do the following:

1. Describe the mechanical refrigeration cycle as it applies to retail refrigeration systems.
2. Explain the differences in refrigerants and applications in low-, medium-, and high-temperature refrigeration systems.
3. Identify and describe the primary refrigeration cycle components used in retail refrigeration systems.
4. Identify and describe the supporting components and accessories used in retail refrigeration systems.
5. Describe the various methods of defrost used in retail refrigeration systems.
6. Identify and describe the applications for the various types of retail refrigeration systems.
7. Describe the control system components used in retail refrigeration systems.
8. Explain the operating sequence of a retail refrigeration system.
9. Interpret wiring diagrams and troubleshooting charts to isolate malfunctions in retail refrigeration systems.

Trade Terms

Phase change   Unit cooler
Slinger ring

Required Trainee Materials

1. Pencil and paper
2. Appropriate personal protective equipment

Prerequisites

Before you begin this module, it is recommended that you successfully complete *Core Curriculum*; *HVAC Level One*; *HVAC Level Two*; and *HVAC Level Three*, Modules 03301-08 through 03303-08.

This course map shows all of the modules in the third level of the *HVAC* curriculum. The suggested training order begins at the bottom and proceeds up. Skill levels increase as you advance on the course map. The local Training Program Sponsor may adjust the training order.

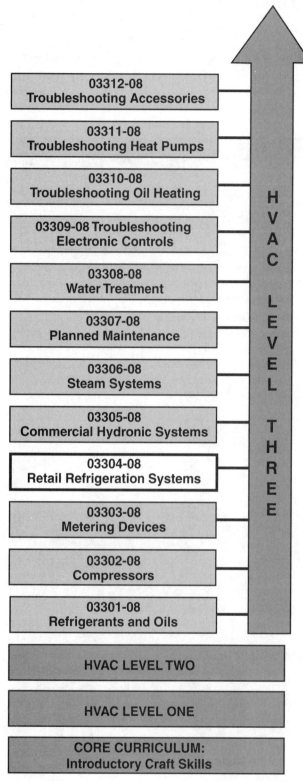

304CMAP.EPS

1.0.0 ◆ INTRODUCTION

How many times each day does refrigeration impact your life? How different and difficult might it be without mechanical refrigeration? In earlier times, ways of preserving foods were limited to curing, dehydrating, or using natural resources such as snow and winter ice to avoid spoilage.

If not for bacteria, many foods could remain edible for long periods of time. Once it was understood that bacteria were responsible for most food degradation, and later discovered that bacteria were unable to grow at lower temperatures, development of modern approaches to refrigeration began in earnest. Throughout the 1800s, advances continued in methods to protect and preserve foods. In 1880, a ship carrying meat from Australia to England was fitted with the necessary apparatus to cool the meat for the voyage. However, the meat was accidentally frozen, and the unexpected result led to the intentional freezing of future shipments.

Today's retail refrigeration systems and applications are evident at every turn. Although the preservation of a variety of products remains as the primary function, style and convenience built into refrigerated fixtures play a significant role in marketing as well. Quiet, properly illuminated, and visually appealing equipment attracts the consumer and increases sales, helping to reinforce the feeling of confidence that the contents are safe and properly preserved for consumption.

This module explains the mechanical operation of refrigeration apparatus as applied in the retail refrigeration industry, including a discussion of specialized refrigerant circuit components. Because it is the goal of many systems to provide sub-freezing environments, approaches to defrosting the system will be examined. An assortment of retail fixtures and systems will be discussed, along with controls specific to their purpose. The troubleshooting and maintenance procedures for two distinctly different pieces of equipment are also presented.

2.0.0 ◆ MECHANICAL REFRIGERATION SYSTEMS

Many types and styles of mechanical refrigeration systems are used to cool or freeze foods and other perishables in retail stores. In most cases, consumable products reach the retail level already cooled or frozen. The retail fixture then is used to maintain those conditions while providing easy access to products for the consumer. The operation of the systems used for these applications is basically the same as that of comfort cooling systems, with the addition of components and controls necessary to achieve the desired refrigerated effect in the fixture's environment. *Figure 1* shows a basic air-cooled refrigeration system flow diagram.

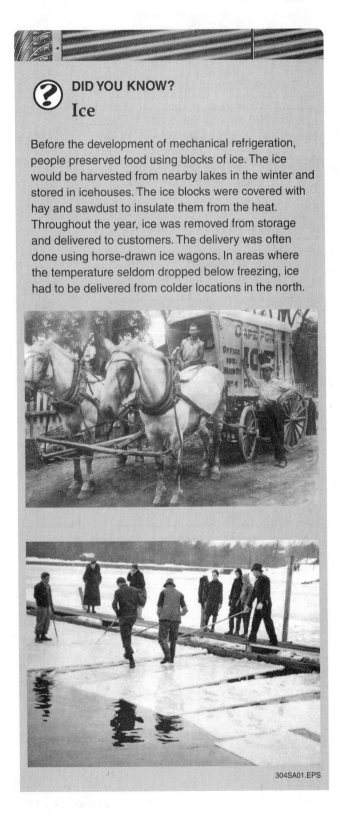

? DID YOU KNOW?

Ice

Before the development of mechanical refrigeration, people preserved food using blocks of ice. The ice would be harvested from nearby lakes in the winter and stored in icehouses. The ice blocks were covered with hay and sawdust to insulate them from the heat. Throughout the year, ice was removed from storage and delivered to customers. The delivery was often done using horse-drawn ice wagons. In areas where the temperature seldom dropped below freezing, ice had to be delivered from colder locations in the north.

304SA01.EPS

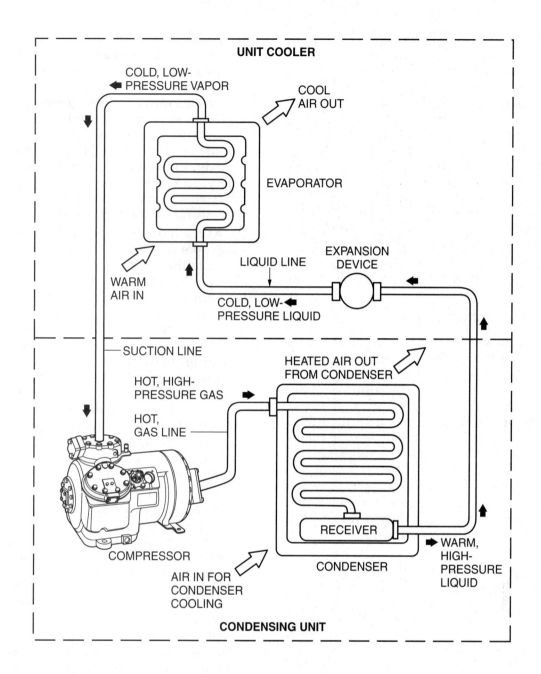

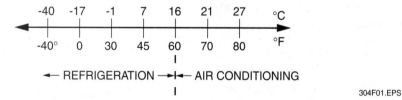

Figure 1 ◆ Basic refrigeration system.

The primary refrigerant circuit components are the compressor, evaporator, condenser, and expansion (or metering) device. A number of other factors are different, however, when the circuit is applied to refrigeration. The installed location of the primary components and interconnecting lines, evaporator temperatures, the condensing environment, and the types of refrigerants employed are all significant variables. The operation of the basic refrigeration system and its components were described in detail in *HVAC Level One*. A review of that material may assist in understanding the differences in the operating cycle when applied to lower temperature applications.

Refrigeration applications are generally classified based on the required space temperature for a given application. It is important to remember these classifications, because some components, such as compressors and expansion devices, are classified for duty based on these temperature ranges:

- *Above +60°F* – Air conditioning/ high-temperature refrigeration
- *–30°F to +60°F* – Medium-temperature refrigeration
- *–40°F to –30°F* – Low-temperature refrigeration

2.1.0 Refrigeration Cycle

The basic cycle in refrigeration applications is similar to that for air conditioning. However, many inexperienced technicians are apprehensive about servicing refrigeration systems because they have spent most of their time working with typical comfort cooling applications using a 40°F evaporator temperature as an established standard. In refrigeration, significantly different evaporator temperatures must be reached in order to achieve the desired product storage temperature. In reality, while simple changes in a few components yield far different results, the theory of operation remains the same.

2.1.1 Medium-Temperature Refrigeration Cycle

Refer to *Figure 2* for an example of a typical medium-temperature refrigeration cycle using HFC-134a as the refrigerant.

Assume that the system is for a retail fixture inside a conditioned space, such as a convenience store. The air temperature in the vicinity of the fixture and unit condenser is 75°F, and the desired box temperature is 38°F. In actual practice, these conditions can vary dramatically, often through the course of a single day. The following numbers correspond to the numbered points shown in *Figure 2*. Note that this example is theoretical and does not account for normal pressure drops through piping or components in a normally operating system.

1. When the refrigerant leaves the expansion device, it is a mixture of liquid and vapor. The vapor is the result of some liquid flashing at the expansion device, absorbing heat in order to cool the remaining liquid to the desired evaporator temperature. A 75 percent liquid to 25 percent vapor volume can be considered reasonably normal. To achieve the desired box temperature of 38°F, you need to create a saturated refrigerant temperature in the evaporator somewhat lower than this. For this example, use an evaporator temperature of 24°F, at an HFC-134a pressure of 21.4 psig. This creates a temperature difference (ΔT) of 14°F between the refrigerant-saturated suction temperature in the evaporator coil and the desired box air temperature.

2. As the refrigerant mixture flows through the evaporator, it absorbs heat from the warmer air in the box and continues to change phase from liquid to vapor. The saturation temperature of the refrigerant remains unchanged until the **phase change** has been completed. The mixture of liquid and vapor will remain at 24°F until all liquid has returned to the vapor phase. To ensure that the maximum amount of evaporator area is providing usable refrigeration, the point where the phase change is complete is ideally located in the last 3 to 7 percent of the coil. Once the phase change is complete, the vapor can begin to increase in temperature and superheat can be measured.

3. Refrigerant exiting the evaporator coil and entering the suction line should now be in a single state—vapor. The completion of the phase change process can be proven by the existence of measurable superheat. Temperature and pressure measurements taken at this point should reveal that the refrigerant has increased in temperature above that shown for saturation on the pressure-temperature (P-T) chart. This provides evidence that the phase change is complete and ensures that liquid refrigerant is not on its way to the compressor, where serious damage could result. In this example, the refrigerant has gained 8°F of superheat, for a suction line temperature of 32°F. It is important to note that the suction line will often form frost, which should not be considered an indication of the presence of liquid refrigerant.

4. The vapor returning to the compressor picked up only a very small amount of heat (2°F) as it traveled through the suction line. In actual practice, this value can vary dramatically, depending on the length of the suction line, quality of insulation, and the ambient temperature in the area of travel. In a small reach-in retail case, the suction line

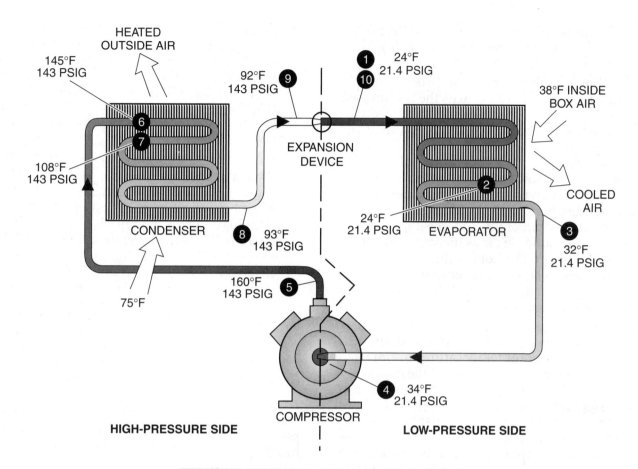

TEMPERATURE/PRESSURE CHART

°F	HFC 134a	°F	HFC 134a
–5	4.1	85	94.9
0	6.5	90	103.9
5	9.1	95	113.5
10	12.0	100	123.6
15	15.0	105	134.3
20	18.4	110	145.6
25	22.1	115	157.6
30	26.1	120	170.3
35	30.4	125	183.6
40	35.0	130	197.6
45	40.0	135	212.4
50	45.3	140	227.9
55	51.1	145	244.3
60	57.3	150	261.4
65	63.9	155	279.5
70	70.9	160	298.4
75	78.4	165	318.3
80	86.4		

304F02.EPS

Figure 2 ◆ Typical medium-temperature refrigeration cycle for HFC-134a.

may be as short as a few inches, while other applications with remote condensing units may require much longer lines.

5. Following the compression process, highly superheated vapor now exits the compressor. In this example, the discharge vapor temperature has reached 160°F at a pressure of 143 psig as the direct result of compression. A quick check of the P-T chart reveals that the superheat is now at 52°F. All of this superheat must be removed from the vapor before condensation back to the liquid state can begin. A small amount of this superheat is often lost to the surrounding air as the vapor travels to the condenser. The hot gas line is rarely insulated, and a significant temperature difference between the line and the surrounding air is most common.

6. As the refrigerant vapor enters the condenser coil, superheat is quickly removed initially. As the refrigerant temperature continues to fall toward the temperature of the condenser air, the T between the condenser air and refrigerant is reduced and the process slows somewhat. All superheat must be removed and the vapor temperature must fall to the saturation temperature for HFC-134a at the given pressure before the condensing process can begin. For this example, the appropriate condensing temperature would be 108°F.

7. When the vapor has cooled to 108°F in the condenser, condensation begins and the temperature will remain unchanged throughout the process. As refrigerant progresses through the condenser, the ratio of liquid to vapor constantly changes until only liquid remains.

8. Once the phase change is complete, the refrigerant liquid can begin to experience measurable reductions in temperature and be subcooled. In this example, the refrigerant temperature has fallen to 93°F at the exit of the condenser. Because the saturation temperature for HFC-134a is 108°F at the pressure of 143 psig, the subcooling is calculated to be 15°F. Remember that without measurable subcooling as an indicator, the condensation process has not been completed and some amount of vapor remains in the liquid. This causes a serious loss of refrigerating capacity in the system, because any vapor that travels through the expansion device and into the evaporator is virtually useless in absorbing any significant amount of heat.

9. As is the case with other refrigerant lines connecting primary components, the heat loss or gain in the liquid line can be dramatic or insignificant. In this example, the liquid line is very short and has experienced a temperature reduction of only 1°F. If the liquid line travels through an area of great temperature difference, a far larger increase or decrease in temperature may be experienced.

10. At this point, the liquid refrigerant has passed through the expansion device, experiencing a pressure drop of over 121 psig in the process. Although the refrigerant was subcooled and entirely in its liquid state as it entered the expansion device, some of the refrigerant instantaneously flashed into the vapor state, absorbing heat and cooling the remaining liquid to the proper saturation temperature for the new pressure condition of 21.4 psig. The cycle is complete and ready to begin again.

2.1.2 Low-Temperature Refrigeration Cycle

Although there is little difference in the cycle used to produce freezing temperatures, an example using HFC-404A refrigerant is provided in a briefer format. Remember that HFC-404A, although classified as a near-azeotropic blend, has a very small glide. This glide can be considered insignificant in normal servicing of the refrigeration circuit.

Figure 3 represents a typical low-temperature refrigeration circuit, such as that used in a reach-in freezer. For this example, again assume that the room/condenser entering air temperature is 75°F. However, the desired box temperature will now be +10°F.

1. To achieve the desired box temperature of +10°F, you need to create a significantly lower saturated refrigerant temperature in the evaporator. For this example, use an evaporator temperature of –5°F, at an HFC-404A pressure of 28.9 psig. This creates a temperature difference (ΔT) of 15°F between the refrigerant in the evaporator coil and the desired box air temperature.

2. The mixture of liquid and vapor in the evaporator will remain at –5°F until all liquid has returned to the vapor phase. To ensure that the maximum amount of evaporator area is providing usable refrigeration, the point where the phase change is complete is ideally located in the last 3 to 7 percent of the coil.

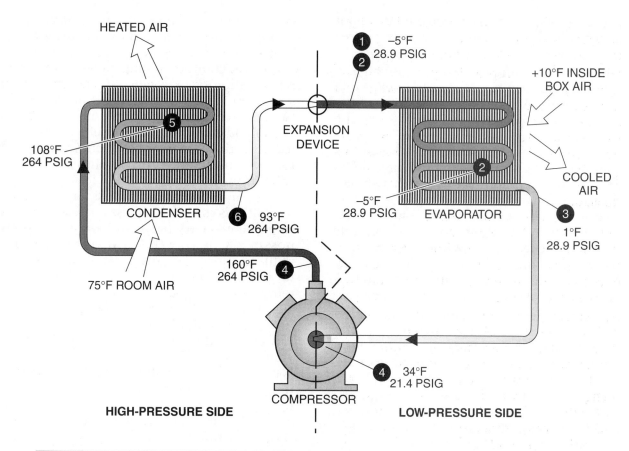

°F	R-404A	°F	R-404A	°F	R-404A	°F	R-404A
−10	24.6	24	62.0	58	121.7	92	210.7
−8	26.3	26	64.8	60	126.0	94	217.0
−6	28.0	28	67.7	62	130.5	96	223.4
−4	29.8	30	70.7	64	135.5	98	230.0
−2	31.7	32	73.8	66	139.7	100	236.8
0	33.7	34	76.9	68	144.4	102	243.6
2	35.7	36	80.2	70	149.3	104	250.6
4	37.7	38	83.5	72	154.3	106	257.8
6	39.8	40	86.9	74	159.4	108	265.1
8	42.0	42	90.4	76	164.6	110	272.5
10	44.3	44	94.0	78	169.9	112	280.1
12	46.6	46	97.6	80	175.4	114	287.9
14	49.0	48	101.4	82	181.0	116	295.8
16	51.5	50	105.3	84	186.7	118	303.8
18	54.0	52	109.2	86	192.5	120	312.1
20	56.6	54	113.3	88	198.4		
22	59.3	56	117.4	90	204.5		

TEMPERATURE/PRESSURE CHART

304F03.EPS

Figure 3 ◆ Typical low-temperature refrigeration cycle for HFC-404A.

Once the phase change is complete, the vapor can begin to increase in temperature and superheat can be measured.

3. Refrigerant exiting the evaporator coil and entering the suction line should now be in a single state—vapor. The completion of the phase change process can be proven by the existence of measurable superheat. Temperature and pressure measurements taken at this point should reveal that the refrigerant has increased in temperature above that shown for saturation on the P-T chart, providing evidence that the phase change is complete and ensuring that liquid refrigerant is not on its way to the compressor, where serious damage could result. In this example, the refrigerant vapor has increased in temperature 6°F (superheat) following the phase change.

4. Following the compression process, highly superheated vapor exits the compressor. In this example, the discharge vapor temperature has reached 160°F at a pressure of 264 psig as the direct result of compression.

5. All superheat must be removed and the vapor temperature must fall to the saturation temperature for HFC-404A at the given pressure before the condensing process can begin. For this example, the appropriate condensing temperature would be 108°F. Once the vapor has cooled to 108°F in the condenser, the temperature will remain unchanged throughout the process of condensation. As refrigerant progresses through the condenser, the ratio of liquid to vapor constantly changes until only liquid remains.

6. In this example, the refrigerant temperature has fallen to 93°F at the exit of the condenser. Because the saturation temperature for HFC-404A is 108°F at a pressure of 264 psig, the subcooling is calculated to be 15°F.

7. At this point, the liquid refrigerant has passed through the expansion device, experiencing a pressure drop of roughly 235 psig in the process. The cycle is complete and ready to begin again.

As you can see, the refrigeration cycle in general remains the same. The differing temperatures and pressures found in various systems are simply the result of different refrigerants and metering device characteristics, but the process remains unchanged. Armed with the correct information and a firm understanding of the cycle, a technician should have no trouble understanding system operating characteristics and identifying when the refrigeration cycle is not operating as designed.

One important note for refrigeration applications compared to comfort cooling regards superheat values. Although typical desired superheat values for comfort cooling may be 10°F to 20°F, refrigeration systems generally operate at lower superheat values. As saturated suction temperatures are reduced, compression ratios become higher and compressor energy efficiency is significantly reduced. Energy efficiency ratios (EERs) for medium-temperature applications are typically around 7.0 to 8.0, while low-temperature systems may operate with EERs as low as 4.0. As a result, it is essential in these applications to gain as much capacity per watt from the system as possible. One way this is achieved is through reduced superheats, exposing as much evaporator internal volume as possible to boiling liquid refrigerant. The phase change process is where all significant heat exchange takes place. Once the liquid has boiled into vapor, little heat is exchanged as it superheats. Reducing superheat values to 6°F to 12°F, measured at the evaporator, is considered acceptable for high- and medium-temperature applications. Superheat values for low-temperature applications may range from 3°F to 8°F to maximize the refrigerating effect in the evaporator. Great care must be taken to ensure adequate compressor protection from liquid floodback that exists when low superheat values are chosen, but there is much to gain.

2.2.0 Primary Refrigeration Cycle Components

Many retail refrigeration applications use air-cooled condensing units (*Figure 4*), a package assembled on a single chassis that contains the compressor, condenser and condenser fan(s), and a variety of possible controls and/or compressor start components. These units can be used in self-contained refrigeration fixtures, or they can be remotely installed and piped to the fixture. Self-contained fixtures used indoors reject the heat from the box, as well as the heat of compression, to the space itself. This additional heat can be a significant comfort problem if not taken into consideration when loads for the comfort cooling system are generated. Although certainly more costly to install, condensing units installed outdoors have the advantage of rejecting their heat outside of the conditioned space. In most cases, remote installation is reserved for larger applications such as walk-in coolers or freezers, as well as larger ice machines. You will also encounter installations where a single condensing unit is used to serve multiple evaporators or fixtures.

2.2.1 Compressors

For retail refrigeration use, hermetic and semi-hermetic reciprocating compressors (*Figure 5*) have long dominated the market. However, scroll compressors (*Figure 6*) are also making their way onto the scene, with design changes that improve their suitability for reduced temperature application.

Early attempts to use scroll compressors in refrigeration applications resulted in overheated and failed compressors due to the high compression ratios and the very low density of the returning refrigerant vapor. However, several manufacturers have now successfully developed scroll compressors for the refrigeration market. For low-temperature applications, many scroll compressors are fitted with a liquid or vapor injection device, designed to provide for improved internal cooling and reduced refrigerant discharge temperatures. Many scroll compressors used in refrigeration are not equipped with internal pressure relief valves. Therefore, providing reliable high-pressure safety controls is a must. Discharge line thermostats will also be incorporated into the controls to further prevent compressor damage or failure from overheating. They are generally set to open at 250°F to 260°F.

Systems based on hermetic reciprocating compressors can be fitted with additional refrigerant flow control components for capacity control. Semi-hermetic compressors with multiple cylinders can be used with integral capacity control to relieve compression in one or more cylinders. Scroll compressors are now available with internal capacity controls that modulate compressor capacity in small increments between 10 and 100 percent. This reduces energy consumption and provides extremely stable temperature control.

2.2.2 Condensers

Most retail refrigeration systems use air-cooled condensers, due to the ease of installation, mobility, and simplicity of the design. The condenser

304F04.EPS

Figure 4 ◆ Refrigeration condensing unit.

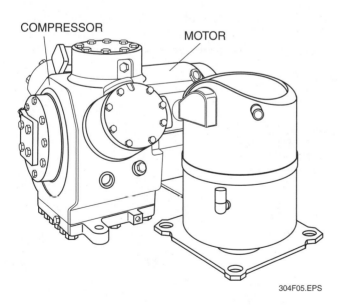

COMPRESSOR MOTOR

304F05.EPS

Figure 5 ◆ Semi-hermetic and hermetic compressors.

304F06.EPS

Figure 6 ◆ Digital scroll compressor.

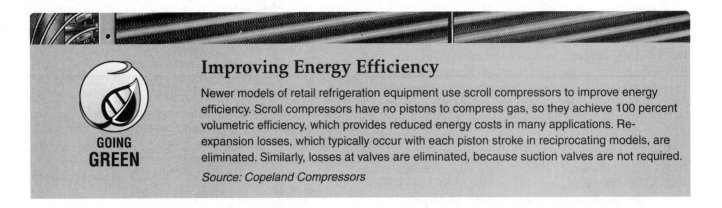

can be mounted on the condensing unit, within inches of the compressor, or remotely installed in other applications. Aluminum fin/copper tube coils are typical (*Figure 7*).

Proper condensing pressures and temperatures must be maintained for the refrigerant circuit to operate effectively. As condensing temperatures rise, capacity and efficiency are reduced. Some retail systems operate in a conditioned environment, but many do not. This must be considered in final equipment selection. Even when placed in a conditioned space, such as a floral showroom or convenience store, it should not be assumed that the condensers will remain clean. Indeed, when used near cooking apparatus, air-cooled condensers can foul rapidly due to airborne grease depositing on the coil fins. Regular condenser cleaning must be an integral part of retail refrigeration system maintenance.

Reduced condensing temperatures can provide increases in energy efficiency. However, low condensing pressures can result in a loss of performance due to insufficient refrigerant flow through the expansion device. This often occurs in winter when condensers or condensing units are exposed to outdoor conditions. Either air-side or refrigerant-side controls can be employed to maintain minimum required condensing pressures. On the air side, the condenser fan(s) can be cycled on and off based on pressure, or the condenser fan speed can be modulated for a more stable result. On the refrigerant side, controls installed in the refrigeration circuit can be used to control the active surface of the condenser by flooding a portion of its internal volume with liquid refrigerant.

2.2.3 Evaporators

Many types and styles of evaporators are necessary in retail refrigeration systems. Although most are very typical air-cooling models, constructed much like comfort cooling coils, evaporators used for liquid cooling and ice making can be unique in their construction and design.

Forced-draft, finned-tube evaporators are most commonly used in smaller retail refrigeration fixtures and systems. For larger applications, such as walk-in freezers or coolers, a unit such as the one in *Figure 8* is used. The evaporator coil is generally a packaged assembly consisting of the coil, evaporator fan, metering device and defrost heaters (when needed). They are also often referred to as **unit coolers**, or unitary coolers.

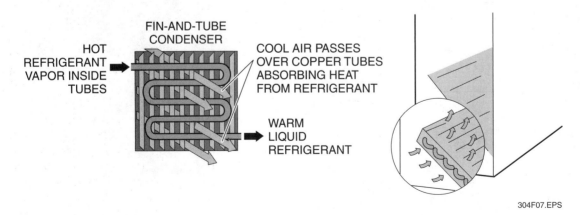

FIN-AND-TUBE CONDENSER

HOT REFRIGERANT VAPOR INSIDE TUBES

COOL AIR PASSES OVER COPPER TUBES ABSORBING HEAT FROM REFRIGERANT

WARM LIQUID REFRIGERANT

304F07.EPS

Figure 7 ◆ Air-cooled fin-and-tube condenser.

Figure 8 ◆ Walk-in refrigeration unit cooler.

304F08.EPS

Unlike most comfort cooling applications, which require blower wheels to overcome external resistance to airflow in an air distribution system, these units generally employ free-blow (no ductwork) propeller-type fans made of plastic or aluminum.

As is the case with comfort cooling, condensation of airborne moisture occurs on the evaporator coil surface when coil temperatures are below the dew point of the enclosure. The condensation must be collected and drained outside the enclosure. Because the volume of condensate collected is relatively small compared to that of comfort cooling, and to further enhance the portability of small, refrigerated fixtures, the condensate is often collected in a small heated pan installed in the fixture cabinet. The electric heating element, embedded in the pan, re-evaporates the condensate to the space. For coils that operate below freezing, the condensate collection pan on the coil assembly generally incorporates an electric heater to prevent freezing and blockage of condensate flow.

Some units may also use the condensate to their advantage. As you may have learned previously from working with window-mounted air conditioners, the condensate can be collected in a pan under the condenser coil. Using a condenser fan blade with a **slinger ring** attached to the tips of the blade, water droplets can be picked up and thrown onto the condenser coil surface to enhance the condensing process. The moisture is evaporated back to the surrounding environment in the process.

Depending on the intended use of a refrigerated fixture, the humidity level in the refrigerated space may be a crucial design and selection consideration. When storing fresh produce,

flowers, and similar products, it is essential that humidity levels remain high to prevent dehydration of the product. In other cases, especially where the stored product is sealed, low humidity levels may be desirable to prevent fog from forming. Proper selection of the evaporator is essential to maintain the desired condition. To minimize dehydration of the product and maintain higher humidity levels, evaporator coils are selected with an extremely low temperature difference between the desired storage temperature and the refrigerant (4°F to 8°F, for example). These coil assemblies generally have large surface areas with high airflow rates to provide the required refrigeration effect with such a small temperature difference. Conversely, when low humidity levels are desirable, the evaporator coil is often smaller and operates with low air flow values. For these applications, the selected air-to-refrigerant temperature difference may be as high as 20°F to 30°F.

Unitary evaporator coils that operate at 30°F and below generally form frost and will require some automated means of frost removal. Defrost strategies will be examined in a later section.

Evaporator coils used for many commercial cubed-ice machines are a variant of simple plate evaporators. A length of tubing is firmly attached to a plate on one side, while the opposite side is used to produce the desired cube size and shape. *Figure 9* shows one such evaporator assembly, where the water flows over the vertical plates to form individual cubes. In other units, the evaporator assembly is a single-sided vertical grid cell plate where water cascades down the face of the grid, building ice cubes in the cell. Most manufacturers offer a variety of cube shapes and sizes to satisfy the preferences of the end user.

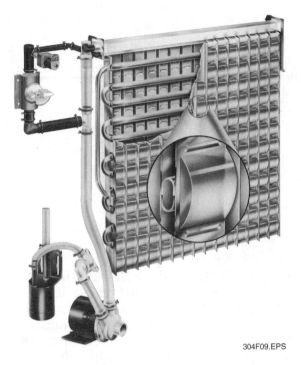

Figure 9 ◆ Commercial cubed-ice machine evaporator.

2.2.4 Expansion (Metering) Devices

The expansion device controls the flow of refrigerant into the evaporator, performing the following two primary functions:

- It provides the essential pressure drop to lower the boiling point of the liquid to the desired temperature.
- It meters the flow of liquid refrigerant (with some flash gas due to the extreme pressure drop) into the evaporator at a rate matching the pace of refrigerant boiling into a vapor inside. In most cases, only refrigerant in its vapor form should leave the evaporator, while still ensuring that the vast majority of the coil internal surfaces are in contact with boiling liquid for maximum capacity.

Although fixed metering devices, such as the capillary tube, remain somewhat popular due to their simplicity, they are not as effective as automatic or modulating metering devices. Capillary tubes and fixed-orifice devices with a replaceable piston (*Figure 10*) feed more or less refrigerant based on the liquid pressure applied on the inlet side. Higher condensing pressures therefore result in higher flow rates, while lower condensing pressures result in reduced flow rates. However, the condensing pressure is not entirely a result of the load on the evaporator itself, and only under a specific set of conditions will the refrigerant flow rate precisely match the evaporator's load. In many cases, the flow rate will be too low, resulting in higher superheat values and lost capacity. The reliability and simplicity of fixed-orifice devices cannot be denied. However, they are not the most effective means of controlling flow because they cannot adjust to changing load conditions. In addition, a precise refrigerant charge is generally required.

A thermostatic expansion valve (*Figure 11*), or TEV, controls the amount of refrigerant flow based on the superheat sensed at the evaporator outlet, modulating to provide a flow rate matching the imposed evaporator load. As the evaporator load changes, the TEV responds by opening or closing slightly. This offers a significant advantage over fixed metering devices and ensures that the maximum amount of evaporator surface is used without overfeeding refrigerant and causing liquid floodback to the compressor.

Retail and commercial refrigeration systems often operate with widely changing load conditions, bringing even greater value to the TEV. At startup, a fixture is generally at room temperature, which could easily be as high as 100°F or more above the desired box temperature. Obviously then, the load imposed on the system at startup is far greater than the load of simply maintaining the desired temperature after pull-

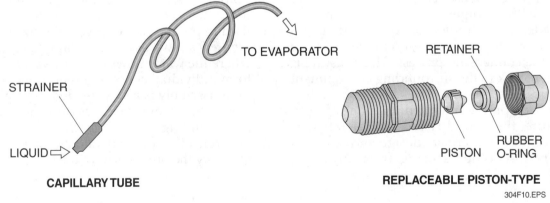

Figure 10 ◆ Fixed metering devices.

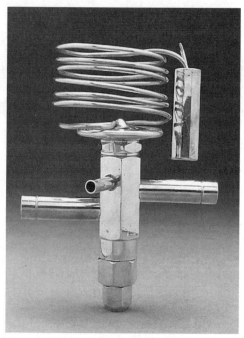

Figure 11 ◆ Thermostatic expansion valve.

down is complete. Freezer evaporators, for example, must be periodically defrosted to remove ice. Cubed-ice machines experience like conditions following each ice harvest, which requires that the evaporator be warmed sufficiently for the cubes to fall free. Following a defrost or harvest period, the evaporator load is significantly higher because the temperature has been raised above freezing. A fixed metering device is unable to compensate for these situations, resulting in a slow pull-down of the coil temperature. With a fixed metering device, the superheat under the above conditions may be as high as 50°F, falling slowly as the load is reduced. The TEV can feed the needed extraordinary amount of refrigerant to the coil for a rapid pull-down during startup or after a defrost/harvest period. One small disadvantage is that TEVs are refrigerant-specific, and the bulb charge must also be properly selected for the type of refrigerant application.

The ideal metering device would be non-refrigerant specific, able to function through a wide range of loads, and be capable of remote adjustment. These issues are all addressed by the electronic expansion valve. Electronic expansion valves, or EEVs, are gaining popularity rapidly in the industry. They offer precise operation and flexibility at extremely low superheats, allowing maximum use of the evaporator's heat transfer surface. Extremely productive advances have been made in the last several years that allow these valves to be very compact, while offering thousands of possible positions, also called steps. This term is derived from the basis for operation, the stepper motor. While older EEVs were equipped with motors offering a few hundred steps or less, over 6,000 steps are now possible. Each valve stroke or movement can be smaller than 0.000008", providing incredibly precise refrigerant metering. The valve positioning is controlled by inputs from one or more temperature sensors and software-based calculations. Most valves can change positions at a rate of 200 steps/second.

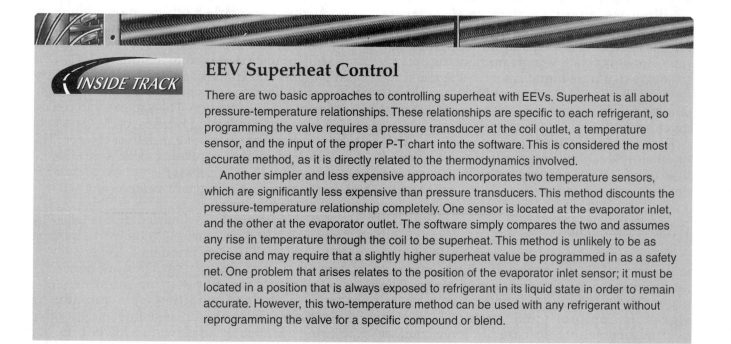

INSIDE TRACK

EEV Superheat Control

There are two basic approaches to controlling superheat with EEVs. Superheat is all about pressure-temperature relationships. These relationships are specific to each refrigerant, so programming the valve requires a pressure transducer at the coil outlet, a temperature sensor, and the input of the proper P-T chart into the software. This is considered the most accurate method, as it is directly related to the thermodynamics involved.

Another simpler and less expensive approach incorporates two temperature sensors, which are significantly less expensive than pressure transducers. This method discounts the pressure-temperature relationship completely. One sensor is located at the evaporator inlet, and the other at the evaporator outlet. The software simply compares the two and assumes any rise in temperature through the coil to be superheat. This method is unlikely to be as precise and may require that a slightly higher superheat value be programmed in as a safety net. One problem that arises relates to the position of the evaporator inlet sensor; it must be located in a position that is always exposed to refrigerant in its liquid state in order to remain accurate. However, this two-temperature method can be used with any refrigerant without reprogramming the valve for a specific compound or blend.

EEVs offer the additional advantage of tight closure during the off cycle. Systems with traditional TEVs are often equipped with a solenoid valve in the liquid line, installed upstream of the TEV. The valve is closed during the off cycle to prevent refrigerant migration. Because refrigerant has a tendency to migrate to cooler areas of the system when the system is off, refrigeration applications provide ideal circumstances for this to happen. The EEV can be driven to a tightly closed position as the refrigerant cycle shuts down, eliminating the need for a solenoid valve.

As the demand for energy-efficient systems continues to escalate, the EEV will become more prevalent in the industry and one day could easily become the standard for retail refrigeration and ice-making systems.

2.3.0 Other Refrigeration Circuit Devices and Components

A variety of supporting components are used in refrigeration systems. Although some of the components discussed here may be used in comfort cooling applications, they are more common in refrigeration applications.

2.3.1 Receivers

Receivers are rarely applied in comfort cooling systems today, but provide a valuable point of liquid refrigerant storage in refrigeration units. Due to the wide variety of load conditions and instantaneous changes in the refrigeration load, having an adequate supply of refrigerant to satisfy the needs of the evaporator under all load conditions is important. Many small receivers incorporate a back-seating shutoff valve on the outlet. Receivers are also equipped with a dip tube, drawing liquid refrigerant from the bottom to ensure that pure liquid is sent to the metering device. The liquid refrigerant level inside will change based on the system flow rate at any given time.

Receivers are used only on systems with a TXV or other modulating metering device. They are of little or no value in systems with a fixed metering device.

2.3.2 Accumulators

Because refrigeration systems tend toward lower superheat values, the prospect of liquid refrigerant leaving the evaporator becomes higher. The accumulator (*Figure 12*) is basically a trap designed to capture liquid refrigerant in the system suction line. Also used in some comfort

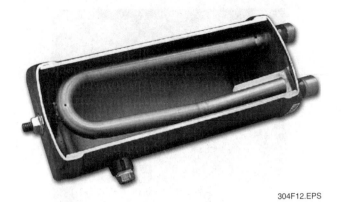

304F12.EPS

Figure 12 ◆ Suction accumulator.

cooling applications and heat pumps, the accumulator incorporates a U-shaped dip tube to separate liquid from vapor. Refrigerant vapor is drawn out one end of the U-tube by the compressor, while the other end terminates near the top of the vessel as well. This ensures that only vapor is drawn in. The U-tube also provides one other valuable feature provided by a small orifice at the lowest point in the tube. This orifice allows small amounts of circulating oil and liquid refrigerant to return to the compressor without the risk of compressor damage.

2.3.3 Crankcase Pressure Regulators

Crankcase pressure regulating (CPR) valves are installed in the suction line just upstream of the compressor, as shown in *Figure 13*. The valve controls the maximum pressure being fed to the compressor to prevent overloading of the compressor motor.

As discussed earlier, refrigeration equipment must often be started up when the entire fixture is at room or ambient temperature even though the fixture may be designed to maintain –10°F. This represents an extraordinary load on the compressor, well outside of its normal design operating range for the application. This same extraordinary load can occur following a defrost cycle, especially in systems that use a hot-gas defrost scheme. The crankcase pressure-regulating valve is installed to ensure that the compressor is not overloaded at these times, and refrigerant vapor is fed to it in a low enough volume.

The valve setpoint is adjusted by changing the internal spring pressure, with the valve moving towards its closed position on a rise of suction pressure above its setting. They are generally quite simple to set by starting the system in a known-overload condition while monitoring compressor current with an ammeter. The valve spring can then be adjusted to a lower pressure

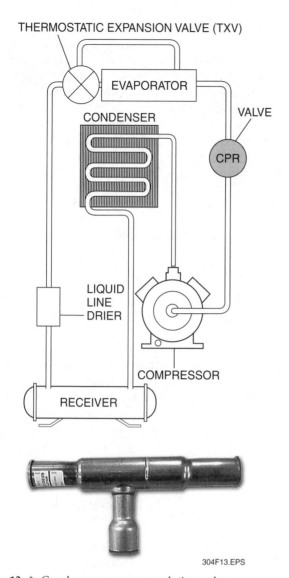

Figure 13 ◆ Crankcase pressure regulating valve.

setting while watching the actual current draw. Once the current falls to, or drops very slightly below, the rated load amperage for the compressor, the valve is properly set.

2.3.4 Evaporator Pressure Regulating Valves

Some advanced and larger retail refrigeration systems may use a single compressor to maintain the temperature in multiple compartments or fixtures, each with its own desired temperature setting. The evaporator pressure regulating (EPR) valve allows the evaporators serving the higher temperature areas to maintain proper refrigerant pressures for their application and, in turn, the saturated suction temperature of the refrigerant. One example of such an installation is shown in *Figure 14*.

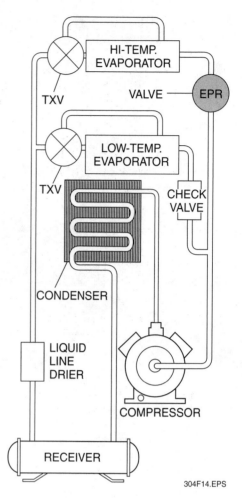

Figure 14 ◆ Evaporator pressure regulator.

The pressure in the common suction line must be sufficiently low to properly refrigerate the coldest fixture or compartment. The EPR, one mounted on each of the higher temperature evaporators, responds to the inlet pressure of the evaporator and opens wider, allowing more vapor to escape, when the pressure rises above the setpoint. On a fall in evaporator pressure, the valve will move toward its closed position. The evaporator will then maintain the proper pressure corresponding to the designated saturated suction temperature.

2.3.5 Refrigerant-Side Head Pressure Control

Refrigeration units installed in locations subject to low ambient temperatures generally require a method of maintaining head pressure. Low ambient air entering the condenser causes a reduced condensing pressure. If the pressure is allowed to fall too low, the pressure drop through the metering device may not be sufficient to ensure adequate refrigerant is fed to the evaporator.

The most accurate and stable means of ensuring that an adequate head pressure is maintained is by using a head pressure control valve (*Figure 15*). Although condenser fan cycling controls are often used, condensing pressures rise and fall through the range of the control and cause unstable operation of the metering device and its flow rate due to the constantly changing condensing pressure. The refrigerant-side head pressure control maintains condensing pressures in a far more stable manner.

304F15.EPS

Figure 15 ◆ Refrigerant-side head pressure control valve.

Figure 16 demonstrates the typical installation of the valve. As the condenser air inlet temperatures fall, there will be a corresponding decrease in the condensing pressure. Once the pressure falls to the setpoint of the valve, the valve routes compressor discharge vapors through port B of the valve and into the liquid line to the receiver through port R. This causes a rise in pressure at the condenser outlet. This rise impedes the flow of refrigerant liquid out of the condenser. Condensed liquid begins to back up in the condenser, effectively reducing the available condensing surface.

For this valve to operate effectively, the system must include a receiver. Although it may seem odd at first thought, systems using this device require a larger volume of refrigerant during low ambient conditions than is needed during periods of high load and high condensing pressures. This additional refrigerant, held in reserve in the receiver, is needed to partially flood the condenser coil with refrigerant as the valve modulates to maintain the appropriate pressure. It should also be noted that the valve displayed here is non-adjustable. This style is refrigerant-specific and is factory preset at an appropriate

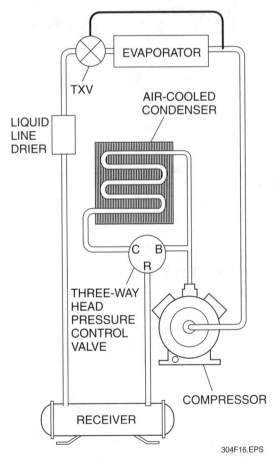

304F16.EPS

Figure 16 ◆ Head pressure control valve installation position.

value for the application and refrigerant used. It has a sensing dome and therefore must be exposed to the ambient temperature. Other models and styles are available that can be adjusted to the end-user's specifications.

3.0.0 ◆ DEFROST SYSTEMS

As long as there is a supply of moisture inside the refrigerated space, it will condense on the surface of the evaporator, as it does in comfort cooling applications. However, when the coil surface is below freezing, frost and ice will build, impeding and eventually blocking airflow through the coil. The coil must be defrosted for it to resume proper operation.

In retail refrigeration applications, these are four primary defrost methods:

- Off-cycle defrost
- Timed defrost
- Electric defrost
- Hot-gas defrost

Regardless of the defrost approach used, there are several common factors regarding frost accumulation on the evaporator coils. Fixtures that were selected and designed to operate at high air-to-refrigerant temperature differences (such as those that are not required to maintain higher humidity levels) will generally build frost more rapidly as their coil temperature will likely be well below the dew point for the box. This encourages greater moisture condensation on the coil. Fixtures that have low air-to-refrigerant temperature differences will not accumulate condensate and frost as quickly.

Another factor in frost accumulation is related to the usage of the fixture and the infiltration of outside air. Most retail refrigeration fixtures are expected to experience high customer usage. In hot, humid weather, a fixture that is opened many times each day for product access will constantly be exposed to hot, moist air. Not only does this significantly increase the volume of airborne moisture in the enclosure, the infiltration of hot air replacing refrigerated air significantly increases the required refrigeration operating time as well. Conditions such as this can overwhelm and interfere with fixture performance, in spite of defrost cycles. Consider the potential difference in performance of an identical reach-in refrigerated fixture installed in two different locations, one in a small, remotely located store with little consumer traffic in Arizona, and the other in a consistently busy, 24-hour convenience store in coastal Florida.

It is important to note that all heat added to the refrigerated enclosure during the defrost cycle, regardless of the source, must again be removed once the process is complete. This fact has a significant impact on the refrigeration load. In addition, the time a system is projected to spend in its defrost mode must be subtracted from the available time for refrigeration. For example, if the total refrigeration load for a fixture were to be calculated as 48,000 Btus over a 24-hour period (2,000 Btuh), but the unit is anticipated to operate in defrost for 2 hours per day, then the refrigeration equipment must be able generate this capacity in 22 hours instead of 24 hours. This would require 2,182 Btuh of capacity instead of 2,000.

3.1.0 Off-Cycle Defrost

Off-cycle defrost is the simplest and most passive of the defrost approaches. In fixtures that maintain temperatures at or above 36°F, the coil is simply allowed to defrost during the normal off cycle. Fixtures in this category will usually operate at a coil/saturated suction temperature of 16°F to 31°F, depending on the desired humidity levels for the stored product and box design. At these temperatures, the coil will certainly build frost, but should defrost naturally once the

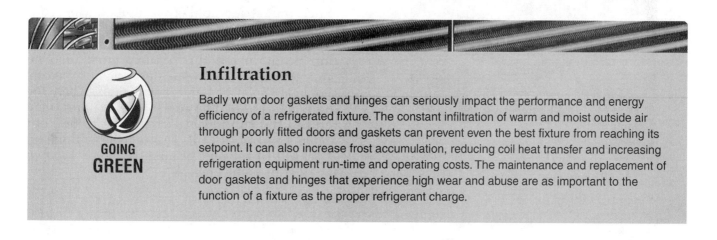

GOING GREEN

Infiltration

Badly worn door gaskets and hinges can seriously impact the performance and energy efficiency of a refrigerated fixture. The constant infiltration of warm and moist outside air through poorly fitted doors and gaskets can prevent even the best fixture from reaching its setpoint. It can also increase frost accumulation, reducing coil heat transfer and increasing refrigeration equipment run-time and operating costs. The maintenance and replacement of door gaskets and hinges that experience high wear and abuse are as important to the function of a fixture as the proper refrigerant charge.

refrigeration apparatus has cycled off, because the fixture temperature is above freezing. The evaporator fan continues to run during the off cycle.

One assist to this natural defrost approach is to ensure that the fixture temperature is not set lower than needed. Maintaining the box even a few degrees colder than necessary can seriously impede or defeat the natural defrost process, resulting in frozen coils and rising box temperatures. Once this occurs, the fixture will require shut down for an extended period to rid the coil of accumulated ice.

3.2.0 Timed Defrost

Systems that operate at slightly lower temperatures than those using simple off-cycle defrost will require a longer period to clear accumulated ice than a normal off cycle provides. Fixtures that operate at 32°F to 36°F often benefit from this approach. Typically, a 24-hour timer (*Figure 17*) with normally closed contacts, placed in series with the compressor operating controls, is set for a reasonable period of time, yet significantly longer than the normal off cycle. The evaporator fan continues to run during the off cycle. The fixture temperature may rise a few degrees above normal during the extended period, but will quickly recover. Like all defrost scenarios, setting the timer for a defrost period that coincides with

the fixture's lightest period of usage (late at night, for example) helps to prevent the fixture temperature from rising beyond an acceptable level. Several cycles per day may be required.

3.3.0 Electric Defrost

Electric defrost systems are prevalent for medium- and low-temperature fixtures that operate at temperatures too cold for off-cycle or timed defrost to be effective, and for freezers in the retail refrigeration class. Electric defrost components are relatively inexpensive and reliable, and are simple to troubleshoot and repair when necessary.

Figure 18 shows a typical arrangement for resistive defrost heaters installed, in this case, on both the face and back of an evaporator coil. For proper operation, the heater must be in good thermal contact with the coil fins to effectively distribute heat and minimize the required defrost period. A timer stops the refrigeration equipment and the evaporator fans during the defrost period, and energizes the electric heater. In many cases, the shape and size of the defrost heater are unique to a given evaporator coil and manufacturer. This requires access to original parts when replacement becomes necessary. Some coil designs require that the heater be placed deep inside the coil assembly, making repair and replacement more challenging, but resulting in a highly effective and rapid defrost cycle.

You may also find some units equipped with radiant electric heaters (*Figure 19*). The resistance heater wire is spirally wound and enclosed inside a quartz tube, with the wires passed through the ends of the sealed tube cap. Using infrared radiant heat, defrost can be accomplished without the heater being in contact with the coil itself. These heaters are capable of reaching a high temperature very quickly and are corrosion resistant. The quartz enclosure can be a bit fragile, however.

When electric defrost is required, the condensate removed from the coil must also be maintained above freezing until it has completely exited the refrigerated environment. Most condensate collection pans, either as part of a unitary evaporator coil assembly or as a separate collection point inside the fixture, must also be equipped with electric heaters such as the one shown in *Figure 20*. Some installations, where the drain line is routed through a freezing area, may require the condensate drain line to be kept above freezing as well. This is usually done by attaching electric pipe heating cable (heat tape) to the exterior of the line.

304F17.EPS

Figure 17 ◆ A 24-hour timer.

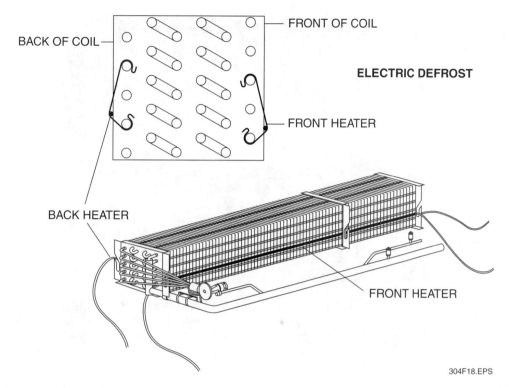

ELECTRIC DEFROST

BACK OF COIL

FRONT OF COIL

FRONT HEATER

BACK HEATER

FRONT HEATER

304F18.EPS

Figure 18 ◆ Evaporator electric heat arrangement.

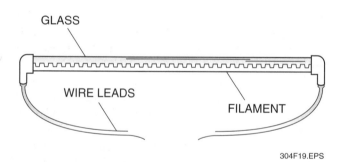

GLASS

WIRE LEADS

FILAMENT

304F19.EPS

Figure 19 ◆ Infrared quartz heater assembly.

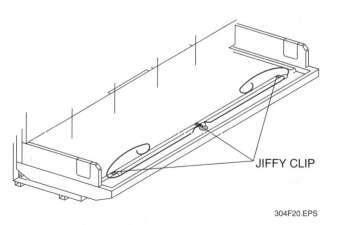

JIFFY CLIP

304F20.EPS

Figure 20 ◆ Condensate pan heater.

Electric defrost systems are generally controlled by a more sophisticated timer than a simple 24-hour model. Electromechanical timers have a dial assembly such as the one shown in *Figure 21*, allowing the timer to be configured for the time, duration, and number of defrost periods during a 24-hour period. The inner defrost dial sets the duration of the defrost period, and defrost is terminated after the period has elapsed. The timer contacts then return to their normal position and the refrigeration cycle resumes.

Electric defrost is often terminated by a temperature sensor or line-mounted thermostat. By monitoring the temperature of the coil, or the refrigerant suction line adjacent to the coil, the thermostat provides an indication that the coil has reached the appropriate temperature and defrost can be terminated before the timer has reached the allowed maximum time. The proper setpoint for temperature termination may differ among products, as it is based on the manufacturer's chosen location and laboratory testing. Line-mounted thermostats are generally non-adjustable, so it is important that the proper device be chosen for replacement.

The final choice of defrost control and termination is often an option provided to the end user, either as a standard feature or at an added cost. Defrost termination can therefore be time-only, temperature-only, or a combination of the two.

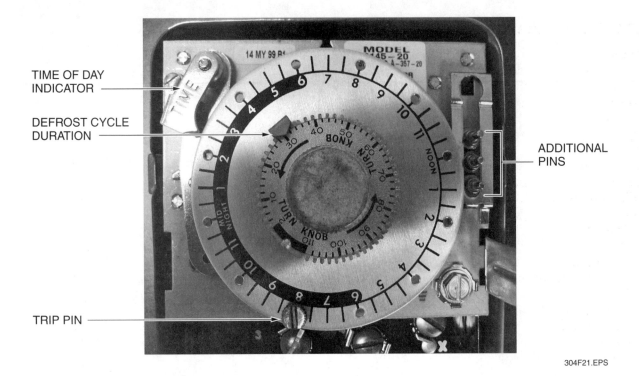

TIME OF DAY
INDICATOR

DEFROST CYCLE
DURATION

ADDITIONAL
PINS

TRIP PIN

304F21.EPS

Figure 21 ◆ Typical defrost timer.

3.4.0 Hot-Gas Defrost

Although far more rare than the other defrost approaches in the retail refrigeration sector, hot-gas defrost is widely considered to be the fastest and most efficient approach. Due to the additional refrigerant circuit components required, its use generally represents a higher equipment cost, impacting its popularity.

Hot-gas defrost sends discharge gas from the compressor directly to the evaporator, bypassing the metering device through the use of several solenoid valves. The coil temperature is allowed to rise to a predetermined pressure and the refrigerant vapor rapidly condenses in the cold evaporator as the process begins. This latent heat of condensation is the primary factor in removing sufficient heat to bring the coil surface above freezing. Care must be taken in these systems to avoid compressor overload and the potential for liquid refrigerant to return to the compressor.

4.0.0 ◆ RETAIL REFRIGERATION EQUIPMENT AND FIXTURES

Retail refrigeration equipment and systems generally include systems that are relatively mobile and can be installed in a variety of locations with a minimal amount of external utilities. Most retail systems, for our purposes, would be considered packaged or unitary in construction. Although the refrigerated cases of supermarket systems can be considered under the "retail" heading, they represent some of the most complex systems in the industry and are rarely self-contained. As a result, they will be examined more thoroughly in another module.

One important point to make regarding many packaged retail refrigeration units is their environment. Many manufacturers are quite clear in their literature that indoor units are designed for indoor spaces maintained at 75°F and 55 percent relative humidity. Although most units are placed in areas that closely conform to these requirements, many are not. Technicians and end-users should not expect proper and reliable operation from units placed in areas that are excessively hot and humid.

In this section, we will review the following types of equipment and their most important features:

• Ice merchandisers
• Reach-in coolers and freezers
• Merchandising walk-in systems
• Commercial ice makers

4.1.0 Ice Merchandisers

Ice merchandisers for bagged retail ice represent a relatively simple design and function. Models are designed for either indoor or outdoor use in sizes ranging from roughly 25 cubic feet to over 100 cubic feet of internal volume. Indoor models (*Figure 22*) are generally equipped with a glass door for a more pleasing appearance, and are often illuminated. Outdoor units (*Figure 23*) have metal doors with padlock provisions and are built more durably to withstand the elements and potential physical abuse. Both styles are usually controlled by a simple thermostat at box temperatures of 20°F to 25°F.

Ice merchandisers are generally offered with two major options—cold-wall construction, which offers no automatic defrost function, and unitary evaporator assemblies, which incorporate electric defrost. In cold-wall models, the walls of the cabinet are used as plate evaporators, with the evaporator tubing bonded to the metal sheets in a serpentine fashion. No evaporator fan is used and defrost is accomplished when the merchandiser is empty by shutting down the unit and allowing it to warm to ambient temperature. Those that use unitary evaporator coil assemblies are equipped with a defrost timer and use electric defrost elements on the evaporator coil.

4.2.0 Reach-In Coolers and Freezers

Reach-in coolers and freezers are available in an array of styles and types to suit almost any storage and merchandising need. In spite of the large variety available, their refrigeration cycles and operating sequences are very much alike. The refrigerated enclosure is insulated with urethane or polystyrene materials, either in board-form or sprayed.

Figure 24 represents a typical two-door reach-in unit. Some typical features and options for these units include the following:

- Interior illumination
- Self-closing doors that can be locked open for stocking
- A variety of shelving options using adjustable shelves and racks for good product display
- Shelf moldings to allow for price tags under products
- Digital thermometers
- A unitary condensing unit that can be removed for service or replacement when necessary
- Adjustable legs to level the unit

304F22.EPS

Figure 22 ◆ Indoor ice merchandiser.

304F23.EPS

Figure 23 ◆ Outdoor ice merchandiser.

304F24.EPS

Figure 24 ◆ Two-door upright reach-in merchandiser.

Figure 25 shows a typical under-counter unit used for the storage of food or beverages. Because units of this type are not generally directly accessible by consumers, colorful presentation and styling of the fixture itself is far less important, giving way to more durable and maintainable design concerns. Typical options and features for this style include the following:

- Stainless steel cabinet construction
- A unitary condensing unit that can be removed for service or replacement when necessary
- Casters for ease of movement, or adjustable legs
- Epoxy-coated evaporator coils to protect the aluminum from natural product vapors that can attack the coil and cause substantial corrosion and deterioration over time
- Exterior thermometers

The condensing units in packaged systems are mounted either above or below the refrigerated case, depending on manufacturer preference and fixture style. Medium-temperature units may incorporate either off-cycle or timed defrost, with

304F25.EPS

Figure 25 ◆ Under-counter reach-in unit.

electric defrost available as a factory-installed option. Low-temperature freezer units are equipped with electric defrost almost exclusively. Units using HCFC-22, HFC-134a, or HFC-404A as a refrigerant dominate the market at this time. The metering device is usually a capillary tube or TEV.

4.3.0 Merchandising Walk-In Systems

Many of today's convenience stores feature merchandising walk-in cooler and freezers (*Figure 26*). Essentially, these units are simply walk-in coolers or freezers with one or more walls replaced by doors, allowing the consumer direct access to a wide variety of products. Specialized racks placed directly behind the doors allow for rapid stocking from within the walk-in unit itself. Products are received and stored inside the same enclosure, never leaving the refrigerated area unless in the consumer's hands. Well-stocked racks at the door opening help to block the consumer's less-attractive view of perishables in storage awaiting use. The merchandising rack system usually incorporates special product feeding, allowing a new product to fall or slide into place as one is removed.

Walk-in coolers generally use unit coolers and remote condensing units, either located outdoors or on top of the refrigerated enclosure inside the retail space. Because the refrigeration capacity need is obviously larger than that of smaller self-contained fixtures, semi-hermetic compressors are often used. TEVs are most often used for the metering device.

Figure 26 ◆ Walk-in merchandiser.

4.4.0 Commercial Ice Makers

Packaged ice-making equipment can be broken down between the two primary types: cubed-ice machines and flaked-ice machines. Either type is commonly used in restaurants, convenience stores, hotels, and other point-of-use locations. Typically, the ice is formed on a surface at roughly 10°F.

The two types of ice machines have some common features. They are designed to operate automatically, control their own water supply, freeze water into the desired shape and size, collect the ice in a storage bin, and shut down once the bin is full. In most cases, the bin is simply an insulated enclosure with a durable plastic or vinyl liner, refrigerated only by the ice it contains. A drain is provided to allow melted ice water to drain away. Smaller ice makers may also be placed on top of a small bin. Such units are equipped with the necessary mechanical drive to dispense small amounts of ice on demand. They units are often coupled with beverage dispensing systems accessible to the consumer in restaurants and convenience stores.

Most ice machines feature air-cooled condensers, although water-cooled models are readily available. Self-contained units can usually be placed in ambient temperatures between 40°F and 115°F, although the environment can significantly impact the speed at which ice is made.

The temperature of the supply water can also impact production speed because warmer water requires more sensible cooling before the latent process of ice building can begin.

More care must be taken in the location of air-cooled models to prevent recirculation of condenser air due to nearby obstructions. Larger ice makers use remote condensing units to move the rejected heat out of the conditioned space. For obvious reasons, most ice makers are not designed to operate in environments that approach or drop below freezing temperatures. If they are to be shut down during the winter but left in an area prone to freezing, the water supply should be turned off, and components such as the water pump and sump should be drained to prevent damage.

4.4.1 Cubed-Ice Machines

Cubed-ice machines (*Figure 27*) generally range in capacity from 65 pounds to 2,000 pounds of ice per 24-hour period. Storage bins mated with ice machines commonly store between 120 pounds and 1,500 pounds of ice. The selection of cuber and bin combination is based on appropriate dimensions and the ice usage requirements. Applications where large quantities of ice are used randomly, versus those expecting steady consumption throughout the day, would likely result in different cuber/bin combinations.

Figure 27 ◆ Cubed-ice machine with bin.

Cubed-ice machines have four basic modes of operation: fill, freeze, harvest, and drain-down. In the fill mode, new water is supplied for the next batch of cubes. In the freeze mode, water applied to the evaporator freezes. When the ice has reached a predetermined size, the evaporator is defrosted to loosen the ice from the evaporator quickly, melting only a small portion of the product in the process. The harvest mode is not unlike a typical defrost mode in other refrigeration applications. Drain-down, or purge, rids the machine of minerals left behind when the water freezes.

Some units harvest the ice in a single solid sheet, which falls from the evaporator onto an electrical wire grid. The grid is heated and the wires melt through the ice in knife-like fashion, separating the ice sheet into cubes of various sizes, such as 1" × 1". *Figure 28* provides an example of this design. Once the cubes are cut by the wire grid, they fall into the storage bin. The other popular design (*Figure 29*) either sprays water into cube-shaped chambers permanently attached to a horizontal plate evaporator, or water is pumped to the top of a vertically positioned evaporator of the same design and allowed to cascade down by gravity. Once the harvest mode begins,

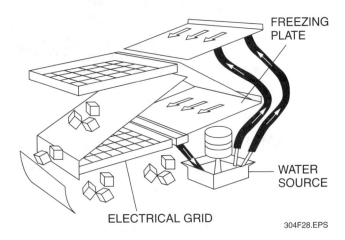

Figure 28 ◆ Cubed-ice maker with an electrical cutting grid.

hot gas refrigerant or electric heat is directed to the plate evaporator, quickly loosening the cubes, which fall individually into the bin. With either style, the ice cube size is one option available to the buyer at the point of purchase.

An ice thickness sensor is typically used on a cubed-ice evaporator. It is set at a precise distance from the evaporator. As ice builds on the plate, the cascading water will eventually make contact with the sensor. Once it has been wetted continuously for 10 seconds, indicating the ice has reached proper thickness, the harvest mode is initiated.

Cubed-ice makers generally have a rather complex sequence of operation to ensure that a consistent ice product is provided. Shutdown of the process is done by either a temperature sensor strategically located near the top of the bin or by using a photoelectric eye that is blocked by the presence of ice, indicating the bin is full.

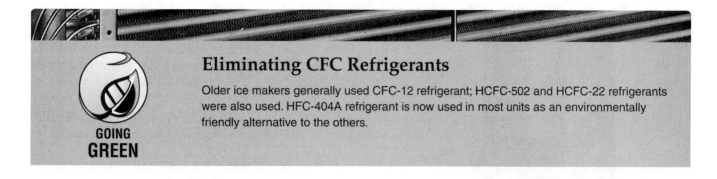

Eliminating CFC Refrigerants

Older ice makers generally used CFC-12 refrigerant; HCFC-502 and HCFC-22 refrigerants were also used. HFC-404A refrigerant is now used in most units as an environmentally friendly alternative to the others.

GOING GREEN

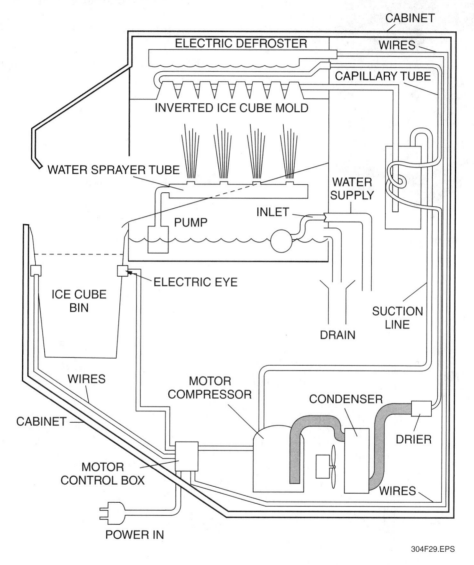

Figure 29 ◆ Cubed-ice maker with cube-shaped chambers on the evaporator plate.

304F29.EPS

4.4.2 Flaked-Ice Machines

Ice flakers (*Figure 30*) operate in a continuous ice-making mode. Although they may be similar in external appearance to cube machines, they are quite different in their operation. Once the unit is started and the evaporator assembly reaches the proper temperature for freezing, it continues to operate until the bin is full.

The freezing assembly for flaked-ice makers, as shown in *Figure 31*, consists of an insulated vertical cylinder. The outer walls of the cylinder form the evaporator, which is filled with boiling, low temperature refrigerant. Ice is formed on the inner walls of the cylinder, which is kept bathed with a consistent supply of water. An auger, driven by a relatively slow gear motor drive,

Figure 30 ◆ Flaked-ice machine with bin.

turns constantly, scraping the ice from the walls of the cylinder. Much like the action of a screw, the flaked ice is pushed out of the top of the cylinder by the auger where it can fall by gravity into the bin.

As is the case with cubed-ice machines, flaked-ice units are now generally designed around HFC-404A refrigerant.

5.0.0 ◆ COMMON REFRIGERATION SYSTEM CONTROLS

Although many of the controls used in retail refrigeration systems are within the same family as those used in comfort cooling applications, they are often applied in different ways and may be quite different in construction. Some of the typical controls to be presented here include thermostats, pressure controls, time delay relays, and solenoid valves.

5.1.0 Thermostats

Thermostats for refrigeration applications provide the same basic function as thermostats used in comfort cooling applications; that is, they

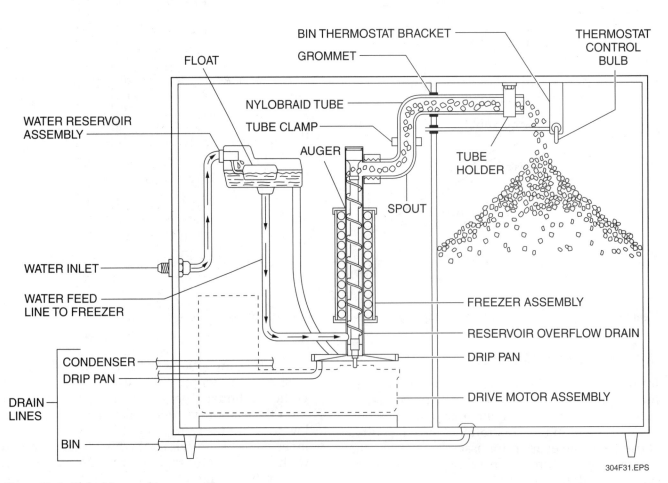

Figure 31 ◆ Flaked-ice machine operation.

sense temperature at a given point and initiate an action based on that temperature. An assortment of styles is available. Some of them are designed specifically for a given refrigeration system manufacturer and end product.

Most refrigeration applications call for the use of remote-bulb thermostats to allow for adjustment outside the refrigerated space. The basic operation of the remote-bulb thermostat is demonstrated in *Figure 32*. Even those that are located in the refrigerated space usually incorporate the same design, using a very short capillary tube rolled up behind the switching assembly.

Electronic temperature controls are also becoming very popular in the industry. They offer reliable operation and a very wide range of operation, allowing identical controls to be installed on high-, medium-, or low-temperature systems. Installing a small temperature sensor and routing flexible wiring back to the control is also much simpler than routing an 8-foot copper or aluminum capillary tube through complex cabinet components. A typical thermostat features a setpoint range of –30°F to 130°F and an adjustable differential of 1°F to 30°F. The range can be modified by using the different available temperature sensors. It also provides an LED indicator of status, and normally open/normally closed operation is field selectable.

The following are a few of the applications for thermostats in refrigeration:

- Control of the refrigeration circuit and compressor to maintain desired fixture temperature
- Provide a low temperature safety control for the refrigeration circuit

- Delay the restart of evaporator fans following a defrost cycle. This allows the refrigeration equipment to operate long enough to re-freeze any remaining moisture on the evaporator coil before the fans start, preventing water from being blown off of the coil and onto product.
- Cycling of one or more condenser fans based on ambient temperature
- Termination of the defrost cycle based on evaporator coil or suction line temperature
- Shutdown of an ice-making process based on bin temperature

Thermostats for these applications are available with a wide variety of ranges and differentials. They can be adjustable or have a fixed setting chosen for specific applications. Single-stage thermostats provide an action based on a single temperature setting, while multi-stage thermostats respond to two or more temperature points. Those used as safety controls generally must be reset manually. Once the setpoint is reached, the contacts open and the process shuts down until the user or technician determines the cause and resets the control to resume operation.

5.2.0 Pressure Controls

Pressure controls, sometimes called pressurestats, are also applied to refrigeration systems in many different ways.

The internal construction of pressure switches varies widely, as shown in *Figure 33*. Many gauges and some pressure switches rely on the operating principle of the Bourdon tube, where pressure applied to a small curved tube causes it

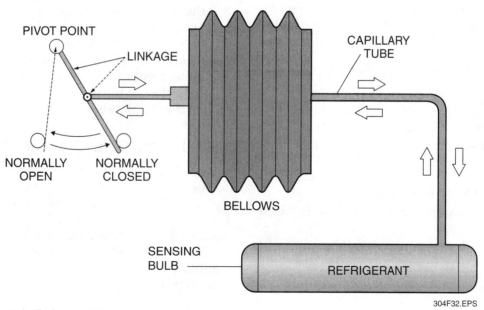

Figure 32 ◆ Remote-bulb thermostat.

304F32.EPS

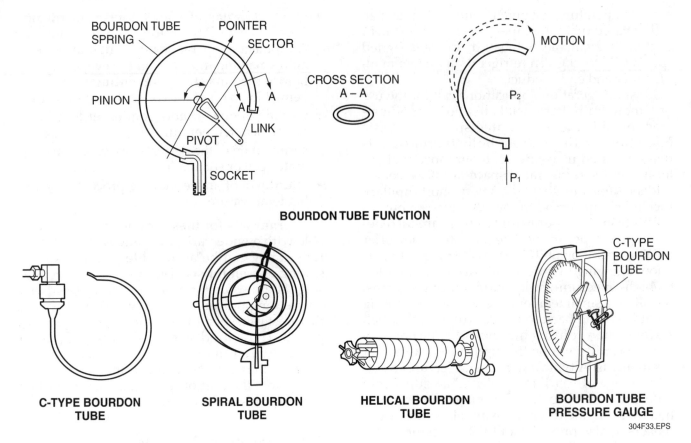

Figure 33 ◆ Bourdon tube function.

to attempt to straighten. Others incorporate several different diaphragm designs to create internal movement of the switch.

Like thermostats, pressure switches perform a variety of different functions and are available in many different operating ranges to suit the application. Adjustable switches with a wide operating range are generally larger than fixed devices. They can use a single high- or low-pressure input, or accept two separate pressure inputs and provide an action based on the differential between the two.

The following are some of the functions that pressure switches provide:

• Control of the refrigeration equipment based on the pressure sensed in the suction line. Since suction pressure is related to saturated suction temperature, pressure switches can provide start-stop control of refrigeration circuits (primarily those using fixed metering devices).

• Control of one or more condenser fans based on condensing pressure.

• Shutdown of the refrigeration equipment based on excessively low pressures caused by problems such as refrigerant loss or evaporator fan failure.

• Shutdown of the refrigeration equipment due to a high-pressure condition caused by condenser fan failure, excessive refrigerant charge, or other problems.

• Pump-down control for systems that are equipped with a liquid line solenoid valve controlled by a thermostat. Once the valve has closed, the compressor continues to run until the majority of refrigerant has been pumped out of the evaporator and suction pressure falls to the setting of the pressure switch.

• Proof of lubricating oil pressure on semi-hermetic compressors by monitoring the differential between the suction pressure and oil pump discharge pressure.

5.3.0 Time Delay Relays

Time delay relays are used to delay a switching action or ensure that a sequence of actions takes place at the proper time. Many are very compact, such as the one shown in *Figure 34*, but still offer adjustability across a range of time periods.

Most time delay relays used in refrigeration applications can be specified as one of the following types:

- *Delay-on-make* – Once energized, the contacts remain in their existing position until the specified time period elapses, then change position. These relays are often used to delay a particular step in a sequence of operation or to prove that a specific condition has existed for a predetermined period before allowing the system to respond.

- *Delay-on-break* – Once de-energized, the relay contacts immediately change position and do not return to the opposite position until the time period has elapsed. Delay-on-break timers are commonly used as compressor short-cycle protectors by opening their contacts when de-energized, preventing the circuit from being completed again until the time period (usually two to five minutes) has elapsed.

- *Bypass duty* – One frequent application for the bypass timer is to bypass a low-pressure switch for a short period of time during a compressor startup. Due to the migration of refrigerant to the condenser during the off cycle, or on systems that pump refrigerant out of the evaporator prior to shutdown, the evaporator may have little or no liquid refrigerant present during startup. As a result, the suction pressure may fall abnormally low for the first minute of operation. This actuates the low-pressure safety switch, causing the compressor to cycle off. Several start cycles like this will often repeat before the pressure stays sufficiently high to allow continued operation. The normally closed switch contacts of the bypass timer (*Figure 35*) are placed in parallel with the low-pressure switch, providing an electrical path around the switch until the selected time period has elapsed. The

timer contacts then open, allowing the low-pressure switch to regain control once the system is in stable operation.

More sophisticated timers can provide multiple timing events or a sequence of timed events repeatedly for special applications, such as ice makers. Timed functions can also be built into printed circuit boards and microprocessor controls.

5.4.0 Solenoid Valves

Solenoid valves (*Figure 36*) are used to stop or allow the flow of either liquids or gases in refrigeration systems. This particular valve is fitted with extended fittings, allowing the valve to be brazed into the line without disassembly and using minimal heat protection on the valve body.

Solenoid valves are selected based on the needed flow rate and the acceptable pressure drop allowed. This often results in a selection that also fits the system line size, but line size is not

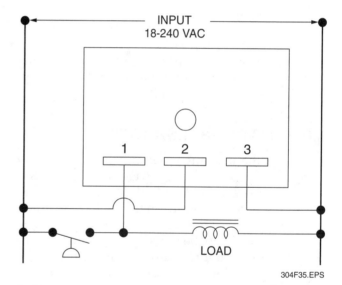

304F35.EPS

Figure 35 ◆ Bypass timer wiring.

304F34.EPS

Figure 34 ◆ Time delay relay.

304F36.EPS

Figure 36 ◆ Solenoid valve.

the primary selection factor. Solenoid valves can be provided with or without a manual lifting stem, allowing the technician to manually open the valve temporarily when a coil failure occurs or the valve sticks closed. The cap protecting the manual lifting stem is pictured on the bottom of the valve in *Figure 37*. Although most are normally closed and are energized to open, they can also be acquired as normally open, and energized for closure. The electrical coil is available in a variety of common voltages to fit the application and they are easily replaced.

The following are some of the applications for solenoid valves:

- Solenoid valves provide flow control of liquid refrigerant to the metering device, controlled by the fixture thermostat. This prevents refrigerant migration during the off cycle and enables a pump-down circuit on systems so equipped.
- On systems with a single compressor that provide cooling using multiple evaporators, solenoid valves may be used in the liquid line to each evaporator, providing independent temperature control.
- Solenoid valves control liquid refrigerant flow to injection-cooling TEVs.
- Solenoid valves control the flow of hot gas refrigerant for defrost, capacity control, or low-ambient condenser circuiting.

6.0.0 ◆ TROUBLESHOOTING

Although the working environment is often very different, troubleshooting refrigeration systems is similar to troubleshooting traditional comfort systems. Rather than working in the backyards, attics, or crawl spaces of homes or businesses, however, you will more likely be working in restaurants in a busy and often more hazardous environment. The repair and troubleshooting of some systems may also require you to work directly under the watchful and curious eye of customers in a retail store or fast-food restaurant. Other than that, all basic troubleshooting skills previously presented for basic controls and wiring schematics continue to apply here.

Troubleshooting and maintenance procedures for two distinctly different retail refrigerant systems are examined: a typical reach-in freezer and a cubed-ice machine. It is important to note, however, that systems can differ dramatically. Therefore, the sequence of operation and service information for a given system should always be consulted before beginning the troubleshooting or repair process.

304F37.EPS

Figure 37 ◆ Solenoid valve with manual lift option.

6.1.0 Troubleshooting and Maintenance of a Reach-In Freezer

In this section, troubleshooting and maintenance procedures are given for a typical reach-in freezer (*Figure 38*) with an electric defrost approach. Of course, a lack of simple maintenance is one of the primary causes of performance problems. Although the periodic maintenance activities required for a reach-in freezer are few, they are critical to proper operation. The most important of these activities is to inspect and maintain both the evaporator and condenser coils. Depending on the environment in which the unit is located, the condenser coil can become fouled very quickly, causing elevated condensing pressures. In the early stages, this may only elevate the operating cost without seriously impacting performance. But as condenser coil fouling worsens, performance will begin to drop. Evaporator coils generally do not foul as quickly, but are subject to becoming clogged and have the potential for rapid corrosion and deterioration from the storage of certain foods. Aluminum fins can become coated with aluminum oxide, which will negatively impact heat transfer.

Generally, with refrigeration systems of this size that hold relatively small refrigerant charges, the operating pressures are not checked during every maintenance visit. The small loss of refrigerant that occurs each time, if done repeatedly, will cause performance problems. Refrigerant operating pressures should not be checked unless you suspect problems. Most motors on units of this size do not require lubrication, but they should be checked to ensure this is the case.

Figure 38 ◆ Reach-in freezer.

Before troubleshooting, review the manufacturer's data and normal operating conditions. For example, suppose the features of a typical freezer include the following:

- The unit power switch is located on the electrical box behind the lower louvered access panel. The condensing unit is also located in this area.
- It is equipped with an electronic temperature control that has no OFF position. The electro-mechanical defrost timer and temperature control are located inside the same enclosure where the power switch is mounted.
- The defrost timer has been factory-preset for two defrost cycles per day at 6:00 AM and 10:00 PM, with a 40-minute fail-safe defrost termination period. The defrost cycle is normally terminated based on a defrost thermostat located on the evaporator suction line. The three-wire thermostat will terminate defrost when a line temperature of 58°F is reached. As refrigeration resumes, it will not allow the evaporator fans to operate until the line temperature has dropped below 32°F. This prevents condensate on the coil from being blown onto the product. Should

the defrost thermostat fail to terminate defrost, the timer will terminate the cycle after 40 minutes.

- The electric defrost heaters are firmly clamped to the fins of the evaporator coil using spring clamps.
- Fluorescent lamps provide interior lighting and can be switched off with a switch located inside the cabinet above the door.
- A fan switch is located at the top of the doors. The evaporator fans stop when the door is opened to prevent cold air from being blown out of the box and into the face of the customer.
- Very low-wattage electric heaters are installed in the structure framework and in the perimeter of the doors to prevent freezing. The heaters are not energized until the box has reached operating temperature. The thermostat controlling their operation is also referred to as the alarm thermostat, as it can be wired to generate a visual or audible alarm when the temperature rises above 18°F. It switches to its normal position, energizing the door heaters, when the box temperature has reached 0°F.
- The refrigeration circuit is equipped with a TEV metering device and a refrigerant receiver is installed in the liquid line.
- A semi-hermetic compressor is used, and the system operates with HFC-404A as a refrigerant. To avoid overloading the compressor, a crankcase pressure regulating valve has been installed and factory set at 10 psig.
- The manufacturer indicates the standard ambient temperature should be 75°F, and lists the expected head pressure as 230 to 240 psig. The expected suction pressure is 5 to 6 psig with the box temperature stable at –10°F. The design air-to-refrigerant temperature difference is 15°F, resulting in an expected evaporator air discharge temperature of –25°F.

These features provide you with all the pertinent information required to recognize proper operation or to recognize if a deficiency exists. Before beginning any troubleshooting procedure, it is essential that you understand the system sequence of operation and the indicators of proper operation. Only then will you be able to determine when system performance is not within expectations. The diagnosis for a given problem can be determined by using the skills you have learned previously, along with the wiring diagram in *Figure 39*, and the troubleshooting chart shown in *Figure 40*.

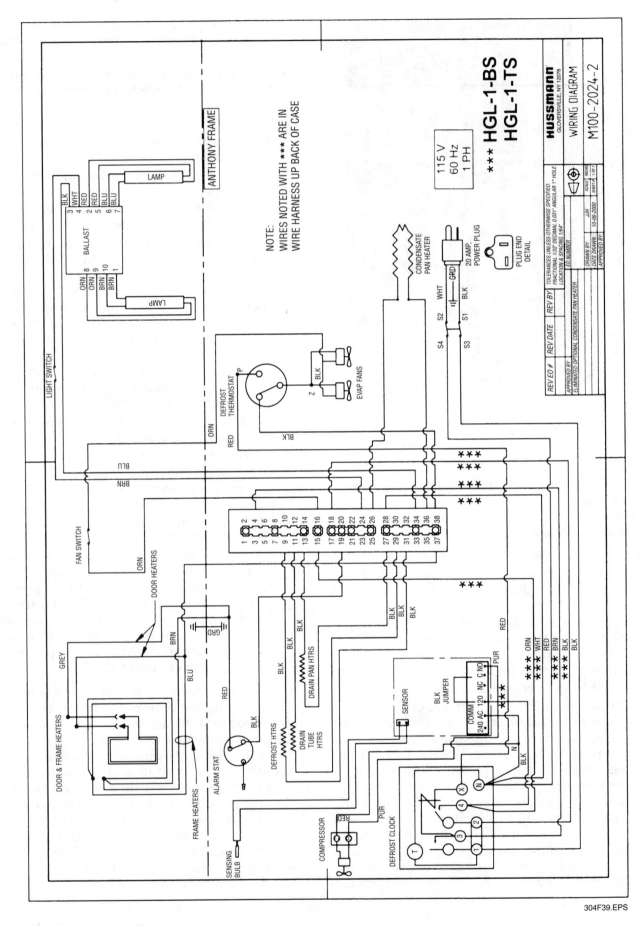

Figure 39 ◆ Freezer wiring diagram.

TROUBLE	PROBABLE CAUSE	SOLUTION
Compressor runs continuously, product too warm	1. Short of refrigerant 2. Inefficient compressor 3. Dirty condenser	1. Leak check, change drier, evacuate and recharge 2. Replace 3. Clean
High head pressure	1. Cabinet location too warm 2. Restricted condenser air flow 3. Defective condenser fan motor 4. Air or non-condensable gases in system	1. Relocate cabinet 2. Clean condenser to remove air flow restriction 3. Replace 4. Leak check, change drier evacuate and recharge
Warm storage temperatures	1. Temperature control not set properly 2. Short of refrigerant 3. Cabinet location too warm 4. Too much refrigerant 5. Low voltage. Compressor cycling on overload 6. Condenser dirty	1. Reset control 2. Leak check, change drier, evacuate and recharge 3. Relocate 4. Change drier, evacuate, and recharge 5. Check power 6. Clean
Compressor runs continuously, product too cold	1. Defective control 2. Control feeder tube not in positive contract 3. Short on refrigerant	1. Replace 2. Assure proper contract 3. Leak check, change drier, evacuate and recharge
Compressor will not start, no noise	1. Blown fuse or breaker 2. Defective or broken wiring 3. Defective overload 4. Defective temperature control 5. Power disconnected	1. Replace fuse or reset breaker 2. Repair or replace 3. Replace 4. Replace 5. Check service cord or wiring connections
Compressor will not start, cuts out on overload	1. Low voltage 2. Defective compressor 3. Defective relay 4. Restriction or moisture 5. Inadequate air over condenser 6. Defective condenser fan motor 7. CRO not set properly	1. Contact electrician 2. Replace 3. Replace 4. Leak check, replace drier, evacuate and recharge 5. Clean condenser 6. Replace 7. Reset to 10 psi
Icing condition in drain pan	1. Low voltage 2. Cabinet not level 3. Defective drain tube heater 4. Defective drain pan heater	1. Check voltage at compressor 2. Check front to rear leveling, adjust legs accordingly 3. Replace 4. Replace

304F40A.EPS

Figure 40 ◆ Troubleshooting chart (sheet 1 of 2).

TROUBLESHOOTING LIGHTING SYSTEM	
Lights won't start	1. Check light switch 2. Check continuity to ballast 3. Check to see if bulbs are inserted properly into sockets 4. Check voltage
Lights flicker	1. Allow lamps to warm up 2. Check lamp sleeve for cracks 3. Check sockets for moisture and proper contact 4. Bulb replacement may be necessary 5. Check voltage 6. New bulbs tend to flicker until used
Ballast hums	1. Check voltage 2. Replace ballast

304F40B.EPS

Figure 40 ◆ Troubleshooting chart (sheet 2 of 2).

Although not every potential situation is covered by this troubleshooting chart, it will help you quickly diagnose the majority of problems you are likely to encounter. Your ability to locate and repair unusual problems will increase with hands-on experience.

6.2.0 Troubleshooting and Maintenance of Cubed-Ice Machines

Most cubed-ice machines have a far more complex sequence of operation than the reach-in freezer. In addition, troubleshooting and properly testing a cubed-ice machine can be quite time consuming because the time period to complete a full cycle can be relatively long. Once any adjustments or changes are made, the unit should be operated through several complete cycles to ensure that it is operating as expected.

A typical air-cooled cubed-ice machine has been selected as a basis to review troubleshooting and maintenance procedures.

WARNING!

It is important to remember that ice machines produce a consumable product, requiring that the unit be properly maintained and kept sanitary. Failure to properly maintain and clean these units could lead to widespread illness from contamination.

INSIDE TRACK

Silver and Biological Films

Silver is being used in products at an increasing rate to assist in the prevention of biological growth. Even clothing items can now be impregnated with silver for this purpose. Many ice machine manufacturers are adding elemental silver in the form of silver ions to plastic components such as ice bin walls to delay biological growth. One such product is known as AlphaSan®. Such additives, compounded directly into the part during manufacturing, release silver ions slowly to the surface of the plastic. Not only do they reduce or prevent biological growth, but the additives also help avoid discoloration of the part over time. They do not impact the taste and flavor of the ice, but the presence of these products should not be cause to ignore proper periodic cleaning.

6.2.1 Cleaning and Maintenance

A properly executed maintenance inspection and cleaning process for an ice machine can be a significant benefit to the technician as well as to the equipment. The time spent allows the technician to become familiar with the location of all components and observe its operation without the pressure of trying to isolate a problem. The primary maintenance activities for cubed-ice machines are outlined in the list that follows. Please note that the process presented here contains generalized instructions. Cleaning or maintenance should not be performed on a unit without access to system-specific literature and guidance from the manufacturer.

Interior cleaning and sanitizing – This maintenance should be conducted at least every six months for most applications. If six-month intervals appear too long, other actions may be indicated such as the installation of improved water filtration or relocation of the unit to a conditioned area (from an excessively hot area) to help impede bacterial growth.

Ice machine cleaners are used to assist in the removal of lime scale and other types of mineral deposits left behind on surfaces from the water supply. Ice machine sanitizers disinfect and remove algae and slime. Always ensure that cleaning and sanitizing agents chosen are consistent with the manufacturer's specifications and are used in their proper concentrations.

For this process, the ice machine should be shut down following a harvest operation so that the evaporator is clear of ice. All ice should be removed from the bin. Thorough cleaning requires significant disassembly of the unit and a considerable amount of time to complete, so it is important to schedule these activities with the owner to ensure that an adequate supply of ice will be on hand when needed.

Ice machine cleaners should be added and circulated throughout the water distribution system using the water pump. Most manufacturers provide a cleaning control switch position for this operation. The cleaner should remain in circulation for at least 30 minutes unless specific instructions direct otherwise. Then, disassemble and remove all components possible that are in contact with water—water curtains, distribution tubes, water sumps, water level probe, and pump assembly. Clean all these components manually using fresh solution, allowing them to soak 15 to 20 minutes for heavily concentrated mineral deposits. Use a soft bristle brush or sponge for cleaning. With the components removed, use fresh cleaner solution to wipe down all accessible surfaces of both the internal ice machine and the bin. Then, rinse or wipe down again with clean water.

Prepare sanitizer solutions and again clean removed components and wipe down all accessible internal surfaces of the unit and bin. Reassemble the unit and add sanitizing solution to the water sump. Circulate the sanitizing solution through the water distribution system for approximately 30 minutes. Disassemble the unit a second time and wipe down removed components and accessible areas of the bin and unit. Rinsing with clear water is generally not needed.

Reassemble the unit, restore power and water, and return the unit to service. Observe the unit's operation through several ice cycles to ensure that all components have been properly reinstalled and performance is as expected.

Ice machine cabinet and bin exterior – The external surfaces should also be cleaned for an attractive appearance. Users cannot see your work behind the panels, but they will be repulsed by a poorly maintained exterior.

Air-cooled condenser coils – As is true for all condensers, they must be kept clean using a soft brush and vacuum. Other coil cleaning solutions may be needed if removal of grease or other deposits is required. Some ice machines are equipped with washable filters to help protect the coil; they can be washed using a mild soap and water solution. Straighten bent fins with a fin comb.

Water filters and conditioning systems – Many systems are equipped with water filters to remove odor, taste, and particulates from the water supply. The filter elements should be replaced during the cleaning/sanitizing process, or when fouled. Specialized water treatment systems are often installed to help prevent the formation of biological growth by adding controlled amounts of chlorine dioxide to the water. These products do not prevent mineral deposits, nor should they be considered a replacement for proper manual cleaning and sanitizing. They may, however, reduce the amount of time required for a cleaning operation.

6.2.2 Operation and Troubleshooting

Before beginning a troubleshooting process, ensure that the condenser coil and other heat transfer surfaces are clean. Fouled surfaces and clogged coils will cause abnormalities in operation and refrigerant pressures. Gauges should be connected and employed only when thought to be necessary to the process. The proper refrigerant charge is vital to their proper operation.

The electrical system of the ice machine (*Figure 41*) includes circuits for the compressor, water pump, condenser fan(s), water and refrigerant solenoid valves, control transformer, control relays, and microprocessor (if used). Basic electrical troubleshooting principles apply to these items when a deficiency exists.

It is essential that the precise timing and sequence of operation are known and understood. Although units may appear to be identical, the precise sequence of events may differ. Always ensure that the proper literature is on hand for the equipment being serviced. The following sequence of operation applies to a very specific line and size of cubed-ice machine, using the wiring diagram shown in *Figure 41*. Also refer to the energized parts chart shown in *Figure 42* for details associated with the operational sequence. A watch is very helpful to time the individual events.

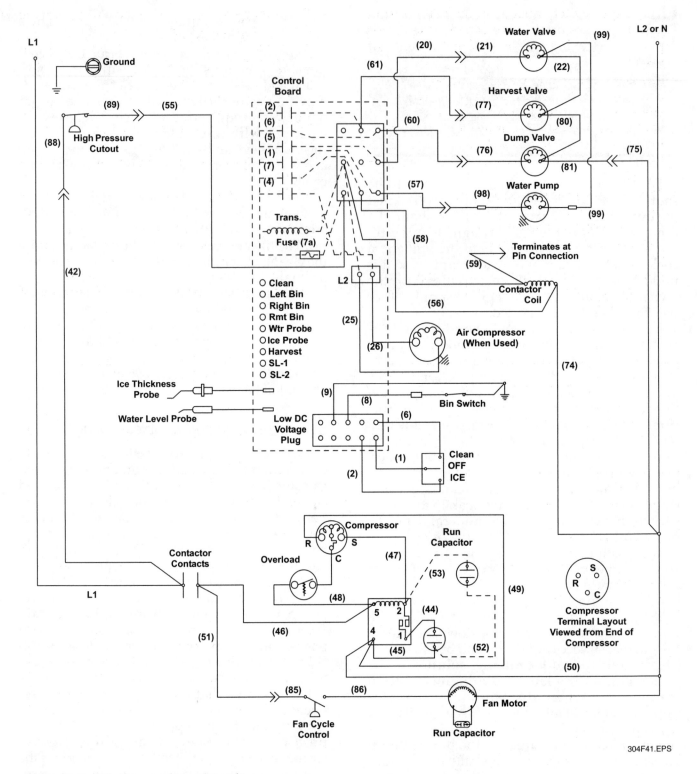

Figure 41 ◆ Cubed-ice machine wiring diagram.

- The toggle switch must be in the ICE position and the water curtain covering the evaporator plate must be in position before the unit will start.
- The water pump and water dump solenoid are energized for 45 seconds before the compressor starts in order to purge the unit of old water. The harvest valve is also energized during the water purge and will remain energized for an additional 5 seconds after the purge process stops.
- The compressor is started after the water purge process and remains in operation throughout the ice-making and harvest cycles. The water fill valve is also energized along with the compressor. The condenser fan motor circuit receives power, but is controlled by a fan cycling control that may cause it to cycle on and off.

- The compressor runs for 30 seconds before the water pump starts in order to chill the evaporator plate. The water fill valve remains on until the probe has indicated that a sufficient water level exists in the sump.
- The water pump starts after the 30-second period. Water is evenly distributed across the top of the plate and trickles down into the individual cube cells. Unfrozen water collects in the evaporator trough. Some water remains behind in the form of ice in the cells, continuing to build in thickness as the cycle continues. The water fill valve may cycle open one more time to replenish the water level.
- As ice builds, water flow is pushed out slightly away from the evaporator plate. Eventually, the flowing water makes contact with the ice thickness probe. After roughly 10 seconds of

Ice Making Sequence of Operation	Control Board Relays							Contactor	Length of Time
	1 Water Pump	2 Harvest Valve (Left) HRP Valve	3 Harvest Valve (Right) (When Used)	4 Air Comp. (When Used)	5 Water Inlet Valve	6 Water Dump Valve	7 Contactor Coil Liquid Line Solenoid	7A Compressor and 7B Compressor Fan Motor	
Initial Start-Up 1. Water Purge	On	On	On	35 sec. off 10 sec. on	Off	On	Off	Off	45 seconds
2. Refrigeration System Start-up	Off	On	On	On	On	Off	On	On	5 seconds
Freeze Sequence 3. Prechill	Off	Off	Off	Off	May cycle on/off during prechill	Off	On	On	Initial start-up is 60 seconds 30 seconds thereafter
4. Freeze	On	Off	Off	Off	Cycles off then on one more time	Off	On	On	Until 10 seconds water contact w/ice thickness probe
Harvest Sequence 5. Water Purge	On	On	On	On after 35 sec.	30 sec. off 15 sec. on	On	On	On	Factory set at 45 seconds
6. Harvest	Off	On	On	On	Off	Off	On	On	Bin switch activation
7. Automatic Shut-off	Off	Off	Off	Off	Off	Off	Off	Off	Until bin switch re-closes and 3 minute delay

304F42.EPS

Figure 42 ◆ Energized parts chart.

constant contact, proving that ice has formed, the harvest cycle can begin. The unit allows at least 6 minutes to elapse in the ice-making cycle before initiating a harvest cycle.

- The harvest valve is opened, allowing hot discharge refrigerant vapor to circulate through the evaporator. At the same time, the water pump continues to run as it did during the ice-making cycle, but the water dump valve is energized for 45 seconds and the water is purged from the sump. The water fill valve is energized during the last 15 seconds of this period to help flush out the existing water. After the 45-second period, the water pump, fill valve, and dump valve are all de-energized.

- The harvest valve remains open, warming the evaporator plate and breaking the bond between the ice and the plate. The cubes, as a single sheet with a thin bridge between the individual cubes, slide out and fall into the storage bin. The cubes should break apart as the result of the fall. The water curtain swings out as the sheet falls, actuating the bin switch in the process. If the curtain and, consequently, the bin switch remain open for more than 30 seconds, it is a signal that the ice has not fallen completely out, and the unit stops, assuming that the storage bin is now full. When the bin is not full, the actuation of the bin switch is momentary and is a signal to the unit to return to the ice-making cycle.

- If the bin switch is open for more than 30 seconds and the unit shuts down, it must remain off for at least 3 minutes before it will restart. Once a sufficient amount of ice is removed from the bin to allow the curtain to swing closed, ice making can again begin.

- The maximum amount of time allowed to complete the ice-making cycle is 60 minutes, after which the unit will automatically initiate a harvest cycle.

- The maximum amount of time allowed for a harvest is 3½ minutes, after which the unit will return to the ice-making cycle.

Over a period of time, water vapors will collect and freeze on the back of the evaporator plate as well. To rid the unit of this accumulation, the microprocessor will initiate a warm-water rinse after 200 freeze/harvest cycles have been completed, in the following sequence:

- Both the CLEAN and HARVEST LEDs light, indicating that the unit has entered a warm-water rinse cycle.

- The compressor and harvest valve are energized.

- The water pump is energized.
- The water inlet valve is energized until it makes contact with the water level probe.
- The water is warmed by this process for 5 minutes, then the compressor and harvest valve are de-energized.
- The water pump remains on for an additional 5 minutes, then is de-energized.
- The process can be terminated by moving the control switch to OFF, then back to ICE.

Most ice machines are precisely operated at the factory and adjusted as necessary. The following tests and adjustments should be conducted at initial startup, after cleaning and sanitizing, and after a prolonged shutdown.

- The water level sensor is designed to maintain proper water level in the sump and is not adjustable. If the water level appears incorrect, check the probe for physical damage and clean it if necessary.

- The ice thickness probe is factory set to maintain an ice bridge thickness of ⅛" (*Figure 43*). This is sufficient to keep the ice cubes together as a sheet until the mass falls free of the evaporator, yet thin enough to break apart easily when it falls to the bin. If the ice bridge appears to be too thick or too thin, set the ice thickness probe distance using the adjusting screw. Begin with a ¼" gap between the thickness probe and the evaporator as a starting point.

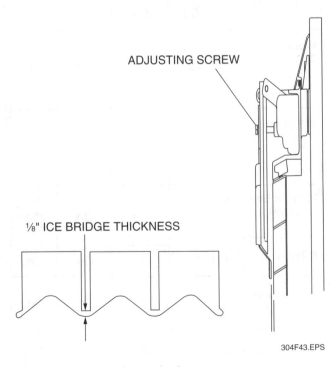

304F43.EPS

Figure 43 ◆ Ice thickness check.

This particular unit is controlled primarily by the microprocessor, which can handle the large number of timed events precisely. Microprocessors are highly reliable, but it is usually impractical or impossible to troubleshoot its individual components. Before condemning the microprocessor, which often happens when technicians do not fully understand its operation, make sure that all external controls and devices are working properly to the best of your ability. Here are some other troubleshooting hints:

• Always listen carefully to the customer's complaint. Even though they likely do not understand the details of the operating sequence, valuable information can be gained from their knowledge of the unit's performance. In addition, technicians who develop good listening skills are often the first to advance.

• Pay close attention to the unit when it is operating properly. Using sight, sound, and touch (carefully), you can often detect abnormalities before ever employing service tools and instruments if you are familiar with normal unit operation.

• Develop a logical approach to troubleshooting all systems. Patience and logical thinking are some of the most important tools at your disposal, especially when working with ice makers and other systems that employ a complex sequence of operation. Each step depends on the proper execution of previous steps. The technician can easily be fooled into an incorrect diagnosis without patience and the use of available data to confirm suspicions. One of the best ways to evaluate a problem is by using the proper sequence of operation and a watch to time each step in the ice-making or harvest cycle carefully, looking for events that do not occur when expected. Using the LED lights on the microprocessor, you can generally determine what it is attempting to do, then see if the controls external to the board itself respond properly. When they do not, you can use common troubleshooting skills to test the affected components. If the microprocessor does not appear to be signaling for the correct step at the appropriate time, test for outputs to external devices at the board itself.

As noted previously, refrigeration gauges should only be applied when absolutely necessary. When adjustments to the refrigerant charge are indicated, or when refrigerant cycle components must be replaced, use proper refrigeration practices and precise measurement of the additional refrigerant to ensure accuracy. Always charge liquid when charging with blended refrigerants.

1. A refrigerated fixture that is designed to operate at an evaporator temperature below –30°F would be classified as a _____ application.

   a. high-temperature
   b. medium-temperature
   c. food processing
   d. low-temperature

2. In the condenser, the refrigerant liquid and vapor mixture will _____ until all of the vapor has condensed to a liquid.

   a. begin to vaporize
   b. remain at the same temperature
   c. freeze
   d. decrease in pressure

3. The technician can determine if all the refrigerant in a condenser has changed to its liquid state by _____.

   a. the presence of measurable superheat
   b. an increase in pressure
   c. a decrease in pressure
   d. the presence of measurable subcooling

4. All superheat must be removed from a refrigerant before the process of _____ can begin.

   a. subcooling
   b. vaporization
   c. condensation
   d. oil separation

5. One way to help compensate for the increased energy consumption in retail refrigeration systems, as compared to comfort cooling systems, is to _____.

   a. operate at lower superheat values
   b. decrease the amount of subcooling
   c. minimize the refrigerant charge
   d. increase the voltage

6. Condensate from a refrigerated fixture can be _____.

   a. picked up and thrown on to the condenser coil for re-evaporation
   b. re-circulated
   c. placed back into the refrigerated area
   d. used to cool the compressor

7. For perishable products requiring storage at high humidity levels, a unit cooler should be selected to operate with a(n) _____.

   a. high refrigerant-to-air temperature difference
   b. electronic expansion valve only
   c. very low airflow
   d. low refrigerant-to-air temperature difference

8. The refrigerant feed rate of the _____ is controlled only by the refrigerant pressure applied.

   a. electronic expansion valve
   b. capillary tube
   c. receiver
   d. injector

9. The primary function of the suction accumulator is to _____.

   a. store enough refrigerant for system operation at heavy load
   b. separate oil from refrigerant before it enters the condenser
   c. capture liquid refrigerant to prevent compressor damage
   d. prevent too much refrigerant vapor to reach the compressor after defrost

10. Allowing the head pressure to drop very low due to low ambient conditions can cause _____.

    a. insufficient flow through the metering device
    b. the compressor to overheat
    c. damage to the air-cooled condenser
    d. excessively low superheat

11. Defrost cycles should be programmed _____.

    a. for the lightest expected periods of usage
    b. at least 6 times per day
    c. for periods of heavy fixture usage
    d. for longer periods in extremely dry environments

12. To melt accumulated frost, _____ do *not* require physical contact with the evaporator.
    a. resistive electric heaters
    b. hot gas defrost coils
    c. infrared electric heaters
    d. three-phase resistive electric heaters

13. Ice merchandisers using cold-wall construction are defrosted _____.
    a. using a hot gas defrost cycle
    b. manually with all ice removed
    c. electrically
    d. several times per day

14. The refrigerants that dominate the new equipment reach-in refrigeration market for medium- and low-temperature units are _____.
    a. HCFC-22 and CFC-12
    b. HCFC-22, HFC-134a, and HFC-404A
    c. HFC-134a and HCFC-22
    d. HCFC-22, HFC-134a, and CFC-12

15. Ice thickness sensors on cubed-ice machines function _____.
    a. instantaneously on contact with the ice surface
    b. based on water temperature
    c. by monitoring refrigerant operating pressures
    d. by continuous contact with water for a specified period of time

16. Electronic temperature controls generally _____.
    a. have a very narrow operating range
    b. are applied to low-temperature systems only
    c. use a copper capillary tube filled with refrigerant to sense temperature
    d. are much easier to install than standard remote bulb thermostats

17. Bypass timers are often used to bypass the _____ during startup.
    a. fixture thermostat
    b. contactor
    c. low-pressure switch
    d. defrost timer

Use the wiring diagram in *Figure 1* to answer *Question 18.*

18. Once the fan switch has been closed, what other device controls power to the evaporator fans?
    a. Defrost thermostat
    b. Defrost timer motor
    c. Alarm thermostat
    d. Defrost heaters

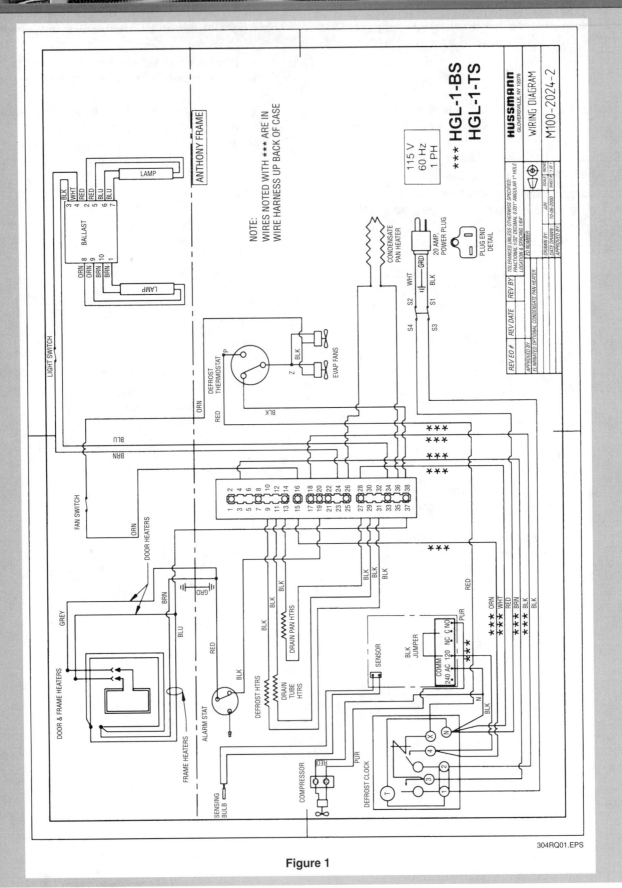

Figure 1

304RQ01.EPS

Use the energized parts chart in *Figure 2* to answer *Question 19*.

19. The ice machine compressor is de-energized during which operating event(s)?
    a. The harvest water purge cycle.
    b. The initial startup water purge cycle and the automatic shut-off cycle.
    c. The freeze pre-chill cycle.
    d. The initial startup water purge cycle and the harvest water purge cycle.

20. When charging a refrigeration unit with a blended refrigerant, you should _____.
    a. release the residual charge to the atmosphere
    b. charge in the liquid state
    c. charge in the vapor state
    d. use any available blended refrigerant

Ice Making Sequence of Operation	Control Board Relays							Contactor	Length of Time
	1 Water Pump	2 Harvest Valve (Left) HRP Valve	3 Harvest Valve (Right) (When Used)	4 Air Comp. (When Used)	5 Water Inlet Valve	6 Water Dump Valve	7 Contactor Coil Liquid Line Solenoid	7A Compressor and 7B Compressor Fan Motor	
Initial Start-Up 1. Water Purge	On	On	On	35 sec. off 10 sec. on	Off	On	Off	Off	45 seconds
2. Refrigeration System Start-up	Off	On	On	On	On	Off	On	On	5 seconds
Freeze Sequence 3. Prechill	Off	Off	Off	Off	May cycle on/off during prechill	Off	On	On	Initial start-up is 60 seconds 30 seconds thereafter
4. Freeze	On	Off	Off	Off	Cycles off then on one more time	Off	On	On	Until 10 seconds water contact w/ice thickness probe
Harvest Sequence 5. Water Purge	On	On	On	On after 35 sec.	30 sec. off 15 sec. on	On	On	On	Factory set at 45 seconds
6. Harvest	Off	On	On	On	Off	Off	On	On	Bin switch activation
7. Automatic Shut-off	Off	Off	Off	Off	Off	Off	Off	Off	Until bin switch re-closes and 3 minute delay

304RQ02.EPS

Figure 2

Summary

Retail refrigeration equipment provides an essential service — the preservation of consumable and perishable goods. This segment of the HVAC&R trade includes a variety of equipment and systems designed for the specific needs of the products to be maintained. Although there are similarities in the refrigeration circuit when compared to comfort cooling systems, a much wider variety of refrigerants and operating parameters are found in the retail refrigeration classification.

Retail refrigeration equipment is often placed in direct view of the consumer, providing direct access to food products. As a result, it is essential that the equipment be aesthetically pleasing as well as reliable.

Systems that operate under the classification of high-temperature refrigeration are often very simple in design and require limited trouble-shooting skills. However, creating storage temperatures below freezing requires equipment capable of eliminating frost build-up through a variety of approaches. Components that enable equipment operation in severe environments also increase the overall complexity for the service technician.

Commercial ice-making systems generally represent the most complex equipment in the segment, with many individual and well-timed steps required for the equipment to operate properly and produce ice in the specified volume. Trouble-shooting these systems requires a great deal of patience and knowledge on the part of the technician. Acquiring and reviewing detailed literature specific to the equipment being serviced will be critical for the technician to provide the high quality repair and maintenance services demanded by equipment owners.

Notes

Trade Terms
Introduced in This Module

Phase change: The conversion of refrigerant from one state to another, such as from liquid to vapor.

Slinger ring: A ring attached to the outer edge of the fan that picks up water as the fan rotates and slings it onto the condenser coil.

Unit cooler: A packaged unit containing evaporator coil, metering device, and evaporator fan(s) that serves as the evaporator section for a refrigeration appliance.

Additional Resources and References

Additional Resources

This module is intended to be a thorough resource for task training. The following reference work is suggested for further study. This is optional material for continued education rather than for task training.

Refrigeration and Air Conditioning, An Introduction to HVAC/R, Latest Edition. Larry Jeffus. Upper Saddle River, NJ: Prentice Hall.

Figure Credits

Cape Pond Ice Company, 304SA01

Emerson Climate Technologies, 304F04, 304F06, 304F11, 304F12, 304SA02

Krack Corporation, 304F08

Hoshizaki America, Inc., 304F09

Topaz Publications, Inc., 304F13, 304F21, 304F26

Courtesy Sporlan Division – Parker Hannifin, 304F15, 304F36, 304F37

Intermatic, Inc. 304F17

Hussman Corporation, 304F18, 304F20, 304F39, 304F40, 304RQ01

Master-Bilt Products, 304F22, 304F23, 304F24, 304F38,

True Manufacturing, 304F25

Manitowoc Ice, 304F27, 304F30, 304F41, 304F42, 304F43, 304RQ02

ICM Controls, 304F34, 304F35

NCCER CURRICULA — USER UPDATE

NCCER makes every effort to keep its textbooks up-to-date and free of technical errors. We appreciate your help in this process. If you find an error, a typographical mistake, or an inaccuracy in NCCER's curricula, please fill out this form (or a photocopy), or complete the online form at **www.nccer.org/olf**. Be sure to include the exact module ID number, page number, a detailed description, and your recommended correction. Your input will be brought to the attention of the Authoring Team. Thank you for your assistance.

Instructors – If you have an idea for improving this textbook, or have found that additional materials were necessary to teach this module effectively, please let us know so that we may present your suggestions to the Authoring Team.

NCCER Product Development and Revision

13614 Progress Blvd., Alachua, FL 32615

Email: curriculum@nccer.org
Online: www.nccer.org/olf

❑ Trainee Guide   ❑ AIG   ❑ Exam   ❑ PowerPoints   Other _____

Craft / Level: _____   Copyright Date: _____

Module ID Number / Title: _____

Section Number(s): _____

Description: _____

_____

_____

_____

_____

Recommended Correction: _____

_____

_____

_____

_____

Your Name: _____

Address: _____

_____

Email: _____   Phone: _____

03305-08

Commercial
Hydronic Systems

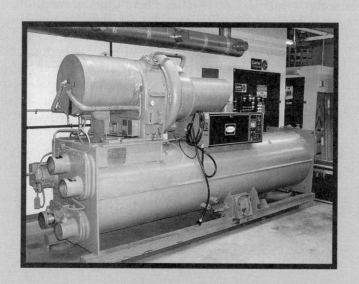

03305-08
Commercial Hydronic Systems

Topics to be presented in this module include:

1.0.0   Introduction 5.2

2.0.0   Water Concept Review 5.2

3.0.0   Commercial Hot-Water Heating System Components 5.4

4.0.0   Chilled-Water Cooling Systems 5.24

5.0.0   Chilled-Water System Components 5.26

6.0.0   Dual-Temperature Water Systems 5.36

7.0.0   Commercial Water Piping Systems 5.36

8.0.0   Water System Balancing 5.41

Overview

In an earlier module, you learned about hydronic systems that use boilers and piping to provide heating for residential applications. In commercial applications, water can be used as the heat exchange medium for both heating and cooling. Heating is still provided by boilers, but they are much larger than those used in residential systems. These systems may also use components and accessories not commonly found on residential hydronic systems. In most hydronic cooling systems, refrigerant is used to chill the water, which is then used as a heat transfer medium. The main component of such a system is called a chiller. Chillers are available for applications ranging from 5 tons to about 2,500 tons. Most chillers use the mechanical refrigeration cycle to chill the water, but other types, called absorption liquid chillers, use a chemical process. Because of their complex nature, the maintenance of commercial hydronic systems requires specialized training. One of the main servicing requirements is maintaining the proper chemical balance in the water in order to avoid biological growth that can be harmful to building occupants, as well as to the system itself.

Objectives

When you have completed this module, you will be able to do the following:

1. Explain the terms and concepts used when working with hot-water heating and chilled-water cooling systems.
2. Identify the major components of hot-water heating, chilled-water cooling, and dual-temperature water systems.
3. Explain the purpose of each component of hot-water heating, chilled-water cooling, and dual-temperature water systems.
4. Demonstrate the safety precautions used when working with hot-water/chilled-water systems.
5. Explain the differences between reciprocating, rotary screw, scroll, and centrifugal chillers.
6. Identify the common piping configurations used with hot-water heating and chilled-water cooling systems.
7. Explain the principles involved, and describe the procedures used, in balancing hydronic systems.
8. Select, calibrate, and properly use the tools and instruments needed to balance hydronic systems.
9. Read the pressure across a water system circulating pump.

Trade Terms

Boiler horsepower (Bohp)
Cavitation
Firetube boiler
Flash economizer
Flow meter
Head pressure
Hydronic system
Latent heat of vaporization
Lithium bromide
Pressure drop
Secondary coolant
Static pressure
Superheated steam
Thermal economizer
Venturi tube
Watertube boiler

Required Trainee Materials

1. Pencil and paper
2. Appropriate personal protective equipment

Prerequisites

Before you begin this module, it is recommended that you successfully complete *Core Curriculum*; *HVAC Level One*; *HVAC Level Two*; and *HVAC Level Three*, Modules 03301-08 through 03304-08.

This course map shows all of the modules in the third level of the *HVAC* curriculum. The suggested training order begins at the bottom and proceeds up. Skill levels increase as you advance on the course map. The local Training Program Sponsor may adjust the training order.

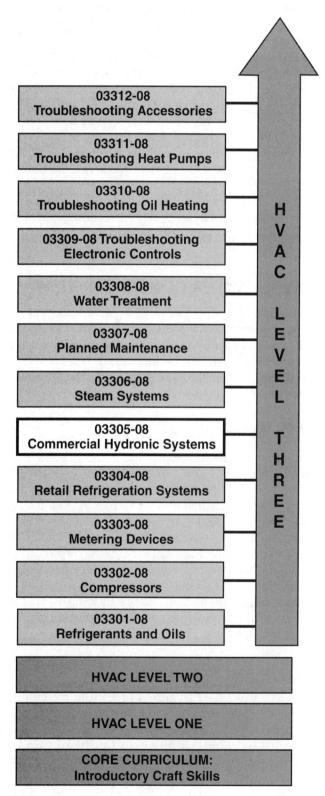

03312-08
Troubleshooting Accessories

03311-08
Troubleshooting Heat Pumps

03310-08
Troubleshooting Oil Heating

03309-08 Troubleshooting
Electronic Controls

03308-08
Water Treatment

03307-08
Planned Maintenance

03306-08
Steam Systems

03305-08
Commercial Hydronic Systems

03304-08
Retail Refrigeration Systems

03303-08
Metering Devices

03302-08
Compressors

03301-08
Refrigerants and Oils

HVAC LEVEL TWO

HVAC LEVEL ONE

CORE CURRICULUM:
Introductory Craft Skills

H V A C   L E V E L   T H R E E

305CMAP.EPS

1.0.0 ◆ INTRODUCTION

As you learned in a previous module, hydronic heating and cooling systems use piping systems to transport heated and cooled fluids for comfort and process purposes. The term *hydronic* refers to all types of water systems, generally excluding domestic potable water systems. In this module, the application of **hydronic systems** in the commercial environment will be examined in detail.

Hydronic systems provide several advantages over forced-air systems in the commercial environment. Large volumes of water can be transported through building structures more economically than air. Piping systems also occupy less space than ducted air distribution systems. In general, the greater the distance that the heating or cooling medium must be transported, the more economy hydronic systems offer.

Subjects covered in this module include commercial and industrial boilers, mechanical water chilling equipment, and chilled and hot water piping systems.

2.0.0 ◆ WATER CONCEPT REVIEW

Before addressing more complex subjects, a review of basic water terminology and concepts will prove helpful. Many problems with hydronic system operation can be identified or resolved if the technician has a thorough understanding of these basic concepts.

2.1.0 Water

Water exists in the three primary states of solid, liquid, and gas. The state of water at any given moment depends upon the surrounding conditions of temperature and pressure. At sea level, water boils at 212°F and freezes at 32°F. As water increases in temperature, its weight is reduced as individual molecules spread farther apart. Between the temperatures at which water freezes and boils, the weight change per cubic foot is just over 2.5 pounds. Due to this change in weight, water can circulate in a piping system without mechanical assistance through the effects of gravity.

2.2.0 Water Pressure

Figure 1 provides the general formula for calculating the pressure exerted by water in pounds per square foot at various depths, as well as the formula to calculate the same information in pounds per square inch. You can use the formula

shown, or simply divide your resulting answer in pounds per square foot units by 144 to obtain pounds per square inch. One cubic foot of water at sea level weighs 62.4 pounds, which provides the basis for the calculations in *Figure 1*.

2.3.0 Pressure Drop

Defined as the difference in pressure between two points, **pressure drop** is an important factor in all hydronic systems. Because of the length and volume of flow required in many commercial applications, as opposed to smaller residential systems, pressure drop in piping must be fully understood and given careful consideration. Where there is no flow or water movement, no pressure drop is encountered. When flow is present, friction is encountered and increases in direct proportion to the velocity of flow. The change in pressure drop resulting from a change in flow in gallons per minute (gpm) can be calculated as follows:

(Final gpm ÷ initial gpm)² × initial pressure drop
= final pressure drop

Hydronic system pumps must provide sufficient power to overcome the effects of pressure drop encountered in the piping system and all components. Because piping systems differ, the pressure drop for piping must be calculated for each system. The manufacturer's specifications for other hydronic system components, such as coils and heat exchangers, provide pressure drop data for various flow rates. This information may be expressed in a variety of units, so the conversions below may be helpful:

- 1 pound per square inch = 2.31 feet of water
- 1 foot of water = 0.433 pounds per square inch
- 1 foot of water = 12,000 milli-inches
- 1 inch of water = 1,000 milli-inches

2.4.0 Head Pressure

Head pressure is another measure of pressure, expressed in feet of water. A head pressure value is also often referred to as *feet of head*. This term is most often used to describe the capacity of a circulating pump, indicating the height of a column of water that the pump can lift, without consideration of pressure drop from friction losses.

Since a column of water 1 foot in height produces a pressure of 0.0433 psi, simple calculations prove that a column of water 2.31 feet in height results in a pressure of 1 psi. *Figure 2* provides examples of these simple formulas.

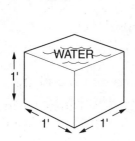
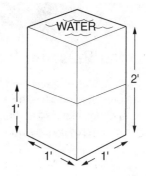

FORMULA FOR COMPUTING PRESSURE IN POUNDS PER SQUARE FOOT (psf)
WHERE: P = PRESSURE PER SQUARE FOOT   D = DEPTH OF WATER IN FEET

$$P = 62.4 \times D$$
$$P = 62.4 \times 1$$
$$P = 62.4 \text{ psf}$$

$$P = 62.4 \times D$$
$$P = 62.4 \times 2$$
$$P = 124.8 \text{ psf}$$

FORMULA FOR COMPUTING PRESSURE IN POUNDS PER SQUARE INCH (psi)
WHERE: P = PRESSURE PER SQUARE INCH   D = DEPTH OF WATER IN FEET

$$P = \frac{62.4}{144} \times D$$
$$P = \frac{62.4}{144} \times 1$$
$$P = 0.433 \times 1$$
$$P = 0.433 \text{ psi}$$

$$P = \frac{62.4}{144} \times D$$
$$P = \frac{62.4}{144} \times 2$$
$$P = 0.433 \times 2$$
$$P = 0.866 \text{ psi}$$

305F01.EPS

Figure 1 ◆ Water pressure concepts.

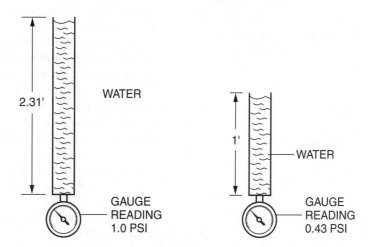

FORMULA FOR CONVERTING HEAD PRESSURE OF WATER IN FEET TO PRESSURE IN PSI
WHERE: P = PRESSURE (PSI)   H = HEAD PRESSURE (FEET OF WATER)

$$P = H \times 0.43$$

FORMULA FOR CONVERTING PRESSURE IN PSI TO HEAD PRESSURE OF WATER IN FEET
WHERE: H = HEAD PRESSURE (FEET OF WATER)   P = PRESSURE (PSI)

$$H = 2.31 \times P$$

305F02.EPS

Figure 2 ◆ Relationship of head pressure in feet to pounds per square inch.

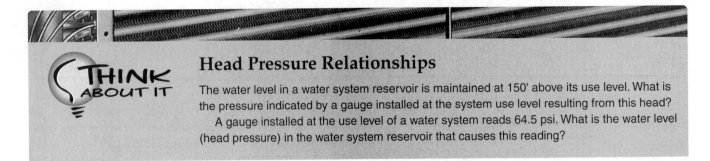

THINK ABOUT IT

Head Pressure Relationships

The water level in a water system reservoir is maintained at 150' above its use level. What is the pressure indicated by a gauge installed at the system use level resulting from this head?

A gauge installed at the use level of a water system reads 64.5 psi. What is the water level (head pressure) in the water system reservoir that causes this reading?

2.5.0 Static Pressure

Static pressure is created solely by the weight of the water in a hydronic system. Since 1 foot of water creates a pressure of 0.0433 psi, static pressure in a system is equal to 0.0433 psi for each foot of height above the system gauge. The measured static pressure will differ from the lowest point of the system to the highest. At the highest point in the system, the static pressure will be 0 psi.

3.0.0 ◆ COMMERCIAL HOT-WATER HEATING SYSTEM COMPONENTS

A number of hydronic devices and components were examined in a previous module, but they apply to commercial systems as well. Those that are applied commercially in their same configuration and style will be briefly reviewed, while components of the same family that differ in commercial applications will be presented in greater detail. The primary commercial hot-water system components are:

- Hot-water boilers
- Operating and safety controls associated with commercial boilers
- Expansion and compression tanks
- Air control devices
- Circulating pumps
- Valves
- Terminal heating devices used in comfort heating applications

3.1.0 Commercial Hot-Water Boilers

A boiler heats water using gaseous fuels, oil fuels, solid fuels, or electricity as the heat source. Geothermal heat pumps are sometimes used to produce hot water for comfort heating. Some boilers can be fired with more than one fuel. This is done by burner conversion or by using dual-fuel burn-

ers. Gaseous fuels include natural gas, manufactured gas, and liquefied petroleum (propane and butane). Oil fuels include both lightweight and heavy oils. Solid fuels include coal and wood. Commercial and industrial markets are dominated by fuel oil and natural gas, due to the availability of these fuels and their high heat content. Natural gas, although not as economically dominant as it once was due to rising costs, is particularly favored in large applications. It is delivered to a given facility via pipeline, eliminating dependence upon ground transportation and requiring no on-site storage. Although rare, large electric boilers are also used in some applications. They are often used by electric power utilities in company-owned facilities as a means of proving their commitment to the electric power they provide.

Combustion in a boiler occurs by combining the fuel with oxygen and igniting the mixture. The methods used for combustion in gas-fired and oil-fired boilers are basically the same as those used with furnaces. The ignition of the gas or oil can be achieved by the use of a standing gas pilot, electric spark-type intermittent ignitor, or hot surface ignitor. The presence of a pilot or flame is proven by a protective device before the main gas or oil valve is allowed to open. Most boiler combustion systems have a purge and prepurge sequence to make sure that there is not a combustible mixture in the boiler that might cause an explosion during startup. The methods used to vent boilers are similar to those used with furnaces.

The construction and operation of hot-water and steam boilers are similar, with two exceptions. The operating and safety controls used with hot-water boilers are different than those used with steam boilers. Also, hot-water boilers are entirely filled with water, while steam boilers are not. Low-temperature boilers are the most widely used type of boiler. Low-pressure hot-water boilers can be built to have working pressures of up to 160 psi. Normally, they are designed for a 30 psi maximum working pressure, but are frequently operated below that pressure level, with 12 to 15 psig being common. Low-pressure hot-water boilers are limited to a maximum operating temperature of 250°F. Above this temperature, even water under low pressure will begin to boil and begin to change to steam, creating a dangerous condition. Medium- and high-pressure hot-water boilers are built to operate at pressures above 160 psig and temperatures well above 250°F. In extremely large systems, such high pressures and temperatures allow the water to be circulated great distances (from a central heating plant to other buildings, for example) without losing all heating capacity before arriving at the intended destination. Such systems can also be extremely dangerous.

Larger hydronic heating systems often use high-temperature, high-pressure water, especially in central plant applications. Water at such extreme temperatures, in excess of 450°F in some cases, and under high pressure, instantly flashes to steam when even a small fracture or tiny leak develops in piping or components. Quite often, evaporation of the **superheated steam** occurs so rapidly, no steam plume is seen. This is due to its sudden exposure to atmospheric pressures, where water boils at 212°F.

WARNING!
Serious physical harm can result by simply walking near or moving a hand across the path of the escaping steam. Use extreme caution when servicing or in the presence of such systems.

Most boilers used in larger commercial applications are made of cast iron or steel. Small commercial facilities are more likely to use boilers of copper-finned tube construction. This design continues to increase in popularity due to heat transfer efficiency, reduced installation cost, and application flexibility.

3.1.1 Copper-Finned Tube Boilers

Boilers using the copper-finned tube design offer some distinct advantages in commercial applications. Because traditional steel and cast-iron units generally have a significant volume of hot water stored on board, they are subject to greater heat losses when heating needs are low, and generally require more space for installation and service. With the copper-finned tube design, heat transfer is greatly enhanced, providing for rapid recovery and eliminating the need for hot water to be

stored within the boiler itself. These boilers are gas-fired, generally using natural gas, or propane where natural gas is not readily available.

Atmospheric styles, such as the one shown in *Figure 3*, often offer modulating or staged heating capacity to further enhance their operating efficiency. When compared to traditional cast iron or steel alternatives, they are also very lightweight, simplifying both initial installation and replacement. One other unique feature these boilers offer is outdoor installation. The unit pictured in *Figure 3* does not have a top connection for flue venting and can be installed outdoors without any additional protection. Boilers installed outdoors are obviously subject to damage from freezing in the event of failure of the burner or pumping system. For outdoor installations, ethylene or propylene glycol is added to protect the water circuit from freezing and the resulting damage to the boiler or piping. The addition of glycol impacts the heat transfer and viscosity of the water, and both issues must be considered in the pump selection process. Generally, system flow rates will need to be increased when glycol is used in order to achieve the desired overall heat transfer effect. Galvanized piping is not recommended when glycol is used, as the zinc in the coating may react with the glycol to form a type of sludge in the piping system.

CAUTION
Proper water flow in gpm through a copper finned-tube boiler is critical to the life of the boiler. Consult the manufacturer's instructions to determine the correct size pipe and pump.

Other copper-finned boiler models feature induced-draft combustion, allowing for additional installation flexibility and higher levels of efficiency. *Figure 4* provides an example of an induced-draft model. The induced-draft design allows the combustion chamber to be sealed, and air for the combustion process can be ducted in from outdoors.

Although very high heating capacities are available in copper-finned tube units, multiple units are often selected and piped in parallel to further enhance efficiency by matching capacity to the heating load. Using multiple boilers to achieve the desired maximum capacity also provides redundancy when one or more units are in need of repair or maintenance.

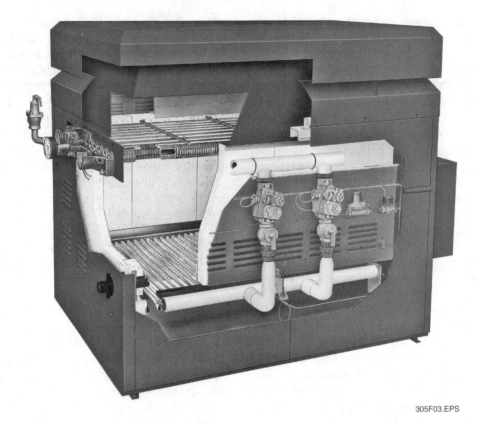

305F03.EPS

Figure 3 ◆ Commercial copper-finned tube boiler cutaway.

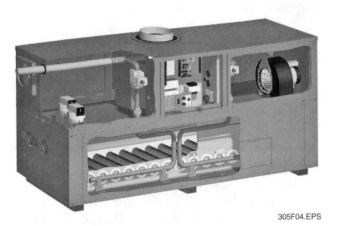

305F04.EPS

Figure 4 ◆ Boiler with sealed combustion chamber.

3.1.2 Cast-Iron Boilers

Cast-iron boilers are formed by assembling individual cast-iron heat exchanger sections together (*Figure 5*). Each section is basically a separate boiler. The number of sections used determines the size of the boiler and its energy rating. Cast-iron boiler capacities range from those required for small residences up to large commercial systems of 13,000 MBh (one MBh is equal to 1,000 Btuh). In a cast-iron boiler, the water circulates inside the cast sections with the flue gases on the

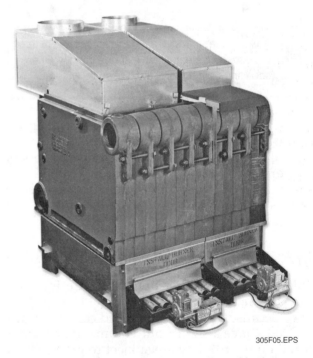

305F05.EPS

Figure 5 ◆ Cast-iron sections in a packaged boiler.

outside of the sections (*Figure 6*). The cast sections are usually mounted vertically, but they can be mounted horizontally. In both arrangements, the heating surface is large relative to the volume of water. This allows the water to heat up quickly.

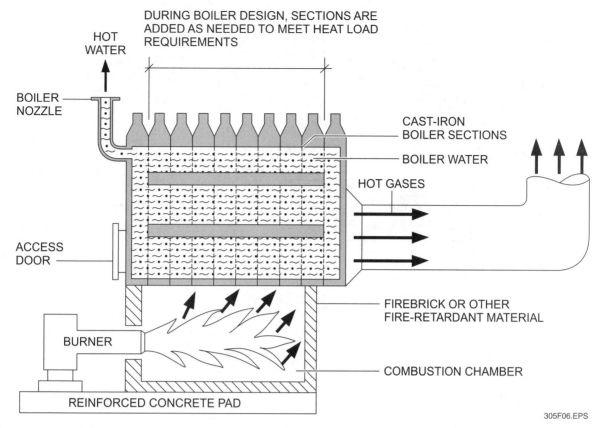

Figure 6 ◆ Cast-iron packaged boiler operation.

The boiler depicted in *Figure 6* uses modular construction. This allows different combinations of boiler sections, bases, and flue collectors to be assembled to match heating requirements. Cast-iron boilers used for residential and smaller commercial jobs usually are supplied completely assembled (packaged). A packaged boiler is one that includes the burner, boiler, controls, and auxiliary equipment. Larger boilers can be packaged units, or they can be assembled on the job site.

3.1.3 Steel Firetube and Watertube Boilers

Steel boilers are fabricated into one assembly of a given size and rating. The heat exchanger surface is usually formed by vertical, horizontal, or slanted tubes (*Figure 7*). Two types of steel boilers are the **firetube boiler** and **the watertube boiler**. In the firetube boiler, the flue gases are contained inside the tubes, with the heated water on the outside. In the watertube boiler, the heated water is contained inside the tubes, with the flue gases

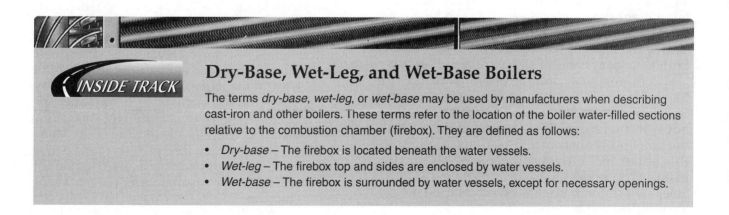

Dry-Base, Wet-Leg, and Wet-Base Boilers

The terms *dry-base*, *wet-leg*, or *wet-base* may be used by manufacturers when describing cast-iron and other boilers. These terms refer to the location of the boiler water-filled sections relative to the combustion chamber (firebox). They are defined as follows:

• *Dry-base* – The firebox is located beneath the water vessels.
• *Wet-leg* – The firebox top and sides are enclosed by water vessels.
• *Wet-base* – The firebox is surrounded by water vessels, except for necessary openings.

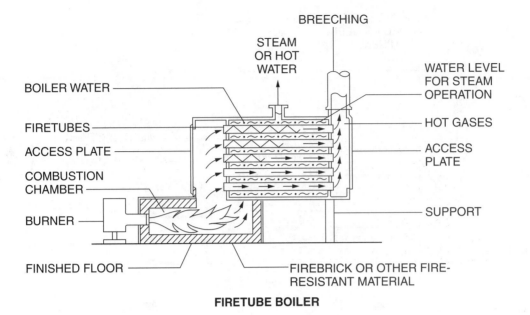

BREECHING

STEAM
OR HOT
WATER

BOILER WATER

FIRETUBES

ACCESS PLATE

COMBUSTION
CHAMBER

BURNER

FINISHED FLOOR

WATER LEVEL
FOR STEAM
OPERATION

HOT GASES

ACCESS
PLATE

SUPPORT

FIREBRICK OR OTHER FIRE-
RESISTANT MATERIAL

FIRETUBE BOILER

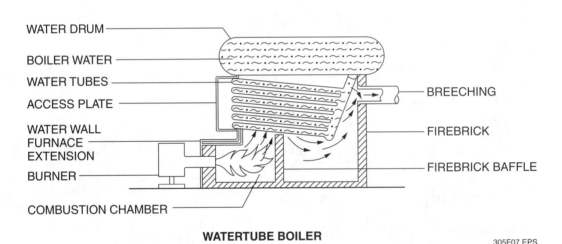

WATER DRUM

BOILER WATER

WATER TUBES

ACCESS PLATE

WATER WALL
FURNACE
EXTENSION

BURNER

COMBUSTION CHAMBER

BREECHING

FIREBRICK

FIREBRICK BAFFLE

WATERTUBE BOILER

305F07.EPS

Figure 7 ◆ Firetube and watertube boilers.

on the outside. Smaller capacity steel boilers are usually vertical firetube units. Medium- and large-capacity steel boilers normally use horizontal or slant-mounted tubes. They can be either firetube or watertube boilers. Steel boilers range in size from small residential units of 150,000 Btuh to large systems of 23,500 MBh and above.

A common steel firetube boiler is the Scotch Marine boiler (*Figure 8*). As with most boilers, Scotch Marine boilers are made to produce either hot water or steam. They are used mainly for heating and industrial applications. As shown, this boiler has a low profile, allowing it to be installed in areas where headroom is limited. Steel boilers, including the Scotch Marine boiler, are usually built as packaged units.

3.1.4 Steel Vertical Tubeless Boilers

Steel vertical tubeless boilers are another type of boiler used to produce hot water or steam (*Figure 9*). In this boiler, the water vessel surrounds the combustion chamber. A top-mounted, fuel-fired burner sends a spinning (cyclonic) flame down the length of the furnace. Flame retarders in the furnace retain the flame for maximum heat transfer. As the flame swirls downward, the water spirals upward in the water vessel. Hot gases from the flame are carried up the outside of the water vessel in a secondary flue passage convection area. Full-length convection fins welded on the outside of the water vessel transfer the remaining heat to the water.

305F08.EPS

Figure 8 ◆ Scotch Marine boiler.

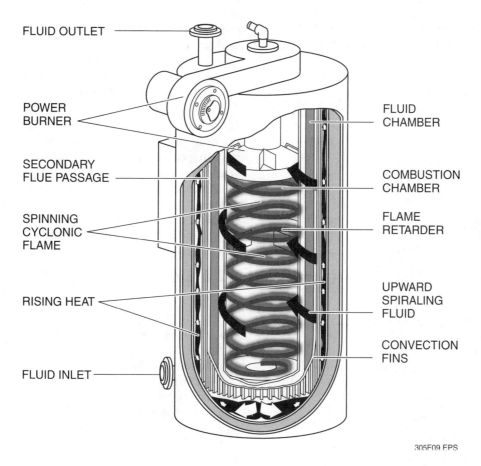

FLUID OUTLET

POWER
BURNER

FLUID
CHAMBER

SECONDARY
FLUE PASSAGE

COMBUSTION
CHAMBER

SPINNING
CYCLONIC
FLAME

FLAME
RETARDER

RISING HEAT

UPWARD
SPIRALING
FLUID

CONVECTION
FINS

FLUID INLET

305F09.EPS

Figure 9 ◆ Vertical tubeless boiler.

3.1.5 Electric Boilers

Electric boilers use immersion resistance heater elements to heat the water in the boiler (*Figure 10*). Heating results from electrical current flowing through the heater elements. These elements are totally immersed in the water within the boiler pressure vessel. The heater elements are uniformly stretched, helical resistance elements made from stainless steel. They are installed in flanges mounted at the top of the boiler assembly. This makes them easily removable for inspection and maintenance.

In smaller boilers, the heater elements are often mounted horizontally. Capacities of resistance-type electric boilers can range from those required for small residences and light commercial use up to a maximum capacity of about two megawatts, or 200 **boiler horsepower (Bohp)**. (One Bohp equals 9.803 kilowatts.) Boiler operation is controlled by an electric/electronic controller. This controller activates and deactivates the individual heater elements in response to load demands. It also allows for incremental loading of the electric service to reduce line fluctuations and power surges during startup.

Another type of electric boiler, called an electrode boiler, relies on the electrical resistance of water for creating heat. In this boiler, an electric current passes through the water between several solid metal electrodes that are immersed in the water. Current flow through the electrical resistance of the water causes the water to be heated. The capacity of this type of boiler is controlled by regulating the amount of water that comes in contact with the electrodes. If there is no contact, there is no current flow, and therefore no heating of the water. Partial contact will cause some heat to be generated. Maximum heat is generated when there is total contact between the electrodes and the water. Hot-water electrode boilers are available with capacities up to 20 megawatts (about 2,000 Bohp).

3.2.0 Boiler Operating/Safety Controls

A variety of different controls are used to control and maintain safe conditions on hot-water boilers. Most of these devices, such as the pressure/temperature gauge, pressure (safety) relief valve, and thermal/electronic probe safety and operating controls, were discussed earlier in your training. If you are unfamiliar with these devices, it is suggested that they be reviewed before proceeding further in this module.

One device not previously presented is the float-operated, low-water cutoff control (*Figure 11*). Float-operated, low-water fuel cutoff controls can be used to protect boilers from damage caused by operation with too low or no water. This control consists of a float that actuates a related electrical switch. The switch contacts are

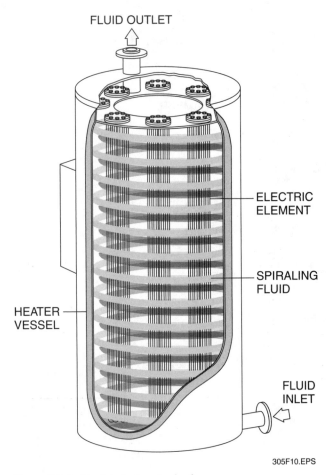

FLUID OUTLET

ELECTRIC ELEMENT

SPIRALING FLUID

HEATER VESSEL

FLUID INLET

305F10.EPS

Figure 10 ◆ Electric hot-water boiler.

connected in the boiler burner control circuit. If the water in the boiler drops below the minimum safe level, the float drops and the switch contacts open, resulting in the fuel supply to the burners being turned off. The operating levels for the control are factory-set and nonadjustable. Some low-water fuel cutoff controls have an opening through which a tool can be inserted and positioned to simulate a low-water condition. Other controls have this feature built in. *Figure 11* shows a typical installation using a float-operated, low-water fuel cutoff control. However, several locations and piping arrangements can be used. Because there is no normal water line to be maintained in a hot-water boiler, any location of the control above the lowest permissible water level is satisfactory.

Copper-finned tube style units transfer heat effectively with no significant volume of water contained on board. Because of this, they are subject to damage very quickly if the water flow is stopped while combustion continues. For this reason, copper-finned tube units should be equipped with water flow switches (*Figure 12*). These simple units have a paddle attached, which is placed into the water piping. They are adjusted so that a flow volume that is insufficient for safe boiler operation will not push the paddle forward, allowing the integral switch to remain open. The switch is usually wired directly in series with the combustion controls to shut the unit down when water flow is too low or non-existent.

Because of the large amount of gas used to fire-off large commercial boilers, these boilers are equipped with special flame safeguard controls.

3.3.0 Expansion and Compression Tanks

Commercial hot water systems also incorporate expansion or compression tanks (*Figure 13*), as is the case with smaller systems. The tanks will often be larger to accommodate the greater volume of water in the system, since the water will require more room for expansion as it is heated.

If a pressurized expansion tank fails to hold its pre-charge of air, you may suspect a leaking air valve or a leaking bladder/diaphragm. Most bladder-type tanks used commercially are larger than those used in smaller applications, and the bladder is accessible for replacement through a large flange, noted in *Figure 13*. With most diaphragm tanks (now used less often than bladder tanks), the diaphragm is not accessible and the tank must be replaced.

305F12.EPS

Figure 12 ◆ Flow switch.

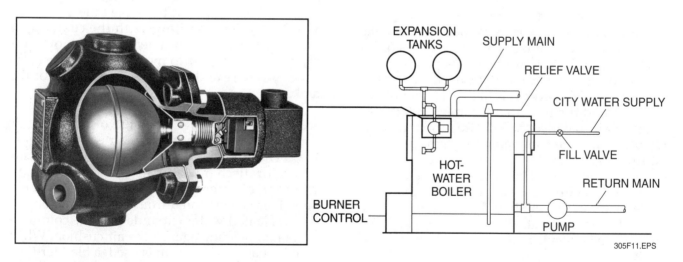

305F11.EPS

Figure 11 ◆ Float-operated, low-water cutoff control.

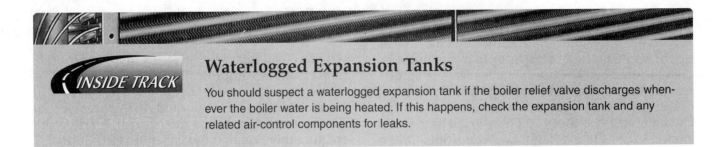

Waterlogged Expansion Tanks

You should suspect a waterlogged expansion tank if the boiler relief valve discharges whenever the boiler water is being heated. If this happens, check the expansion tank and any related air-control components for leaks.

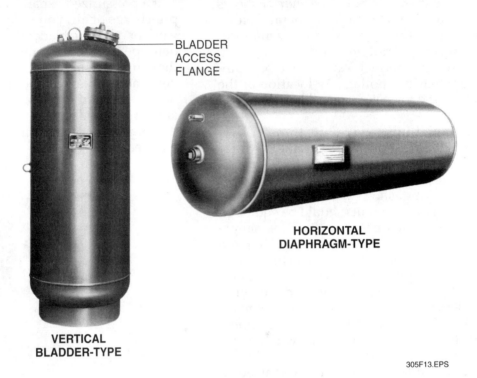

BLADDER
ACCESS
FLANGE

HORIZONTAL
DIAPHRAGM-TYPE

VERTICAL
BLADDER-TYPE

305F13.EPS

Figure 13 ◆ Pressurized expansion tanks.

3.4.0 Air Management

Air management in commercial systems is extremely important to ensure proper water circulation and prevent noise. The same types of air management devices are generally employed in commercial systems as those used in smaller applications. However, due to the greater volume of water and the potential for much larger volumes of air to collect and create problems with circulation, large commercial systems are often equipped with an air separator, such as that shown in *Figure 14*. This type of air separator is also used in chilled-water systems, to be presented later in this module. They are highly efficient at collecting entrained air from water and enable large systems to operate much more effectively.

This particular unit, available in sizes large enough to accommodate 36" pipe, is designed to reduce the water velocity dramatically, to a speed at which the buoyancy of the air bubbles cause them to rise and they are no longer swept along by the water stream. At the same time, it imparts a spin on the water flow, with the centrifugal force created causing heavier water to move outward and lighter air to move inward. The air is collected there and discharged at the top of the tank.

Air management in hydronic systems is accomplished through one of two strategies, air control or air elimination. In air control (*Figure 15*), a compression tank is mounted on the air separator discharge and air is discharged to it. The air and water are in contact inside the tank, but all air is eliminated from the moving loop of water. Heated water expands into the compression tank, pushing against the air cushion. When air elimination (*Figure 16*) is used, a bladder-type tank is used on the suction side of the pump, providing the area for heated water to expand into. All air is vented to atmosphere through a high-capacity air vent on top of the separator.

Figure 14 ◆ Commercial air separator.

3.5.0 Circulating Pumps

A pump is a mechanical device designed to increase the energy of a fluid so that a quantity of the fluid can be transported from one location to another. Pumps are generally driven by electric motors and are used to pump water, stock, coating, additives, oil, and hydraulic fluid.

Air Separation Science

The solubility of air in water at different temperatures and pressures is based on a law of physics known as Henry's law. This law describes how both the pressure and temperature of water impacts solubility. As pressure increases, water is able to keep more air in suspension. Conversely, as pressure decreases, air separates more readily. Higher temperature also decreases water's ability to hold air in suspension. As temperature falls, more air can be held in suspension. The result is that air is separated more readily from water at high temperature and lower pressure. This also means that chilled water systems, which must operate at higher pressures due to the needs of the water distribution system, represent the most challenging air management condition.

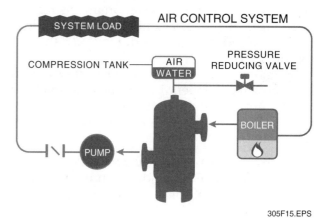

Figure 15 ◆ Air control component arrangement.

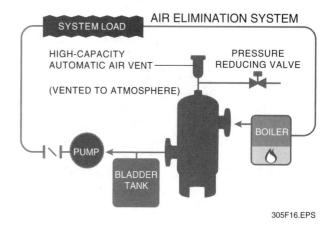

Figure 16 ◆ Air elimination component arrangement.

All pumps operate under the general principle that fluid is drawn in at one end (the suction side) and forced out at the other end (the discharge side) at an increased velocity.

Pumps are classified as either positive-displacement or centrifugal. Positive-displacement pumps are generally used to pump liquids that have very high viscosities; that is, the liquids are thick and flow slowly. Positive-displacement pumps are generally high-pressure pumps. Centrifugal pumps are used to pump liquids that have low viscosities, such as water or thin fluids. Centrifugal pumps are generally used in applications that do not require extreme pressure differentials, and are by far the most common types used in commercial hydronic systems. To accommodate a variety of design situations and installations, they are available in a number of different configurations.

3.5.1 Centrifugal Pump Construction

A centrifugal pump operates by increasing the velocity of a liquid. Fluid entering the pump is rotated by an impeller. This rotation increases centrifugal force within a stationary casing. The force raises the pressure and causes the fluid to be discharged at high speed. *Figure 17* shows a centrifugal pump.

The parts of centrifugal pumps may vary in size and shape, depending on the manufacturer, but will have the same functions. Centrifugal pump parts, as shown in *Figure 18*, include the following primary parts:

*   *Pump casing* – The part surrounding the shaft, bearings, packing gland, and impeller. Pump casings can be of split or solid design.

*   *Suction port* – The place where fluid enters the pump.
*   *Discharge port* – The place where fluid is discharged into the piping system.
*   *Pump shaft* – A bearing-supported part that holds and turns the impeller when the shaft is coupled to a motor.
*   *Bearings* – The parts that support the shaft and impeller in the casing.
*   *Impeller* – A rotating part that increases the speed of the fluid. There are many different types of impellers used for different purposes.

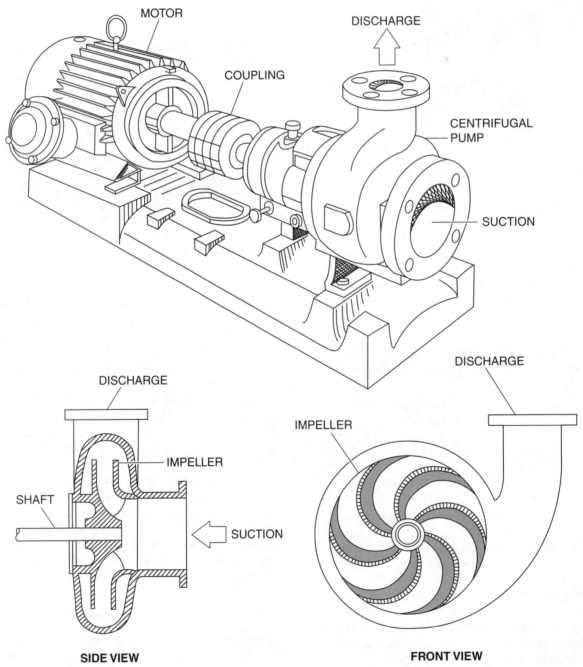

SIDE VIEW

FRONT VIEW

305F17.EPS

Figure 17 ◆ Centrifugal pump.

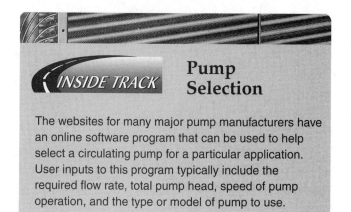

- *Impeller vanes or blades* – The parts of the impeller that direct the flow of fluid within the pump.
- *Impeller shrouds* – The parts that enclose the blades and keep the flow of fluid in the impeller area.
- *Wearing rings* – Replaceable rings used in some pumps to allow some fluid leakage between the impeller and the casing in the suction area. The leakage makes a hydraulic seal and helps the pump operate more efficiently.
- *Packing gland* – Contains an adjustable follower that exerts force upon the packing to control fluid leakage around the shaft.

- *Mechanical seal* – Seals the fluid flow in the pump. In some pumps, they are used instead of packing.

A variety of impeller styles can be chosen, depending upon the application and operating characteristics desired. *Figure 19* provides examples of some different styles. The most common designs used in commercial hydronic comfort systems are the single-suction closed impeller and the double-suction impeller. Some impellers, such as the single-suction closed style, can be precisely machined to a given diameter to provide very specific flow performance in order to meet unique project design specifications.

Cavitation can occur in a water circulating pump when the pressure at the inlet of the impeller falls below the vapor pressure of the water, causing the water to vaporize and form bubbles. These bubbles are carried through the pump impeller inlet to an area of higher pressure where they implode (burst inward) with terrific force that can cause the impeller vane tips or inlet to be damaged due to pitting or erosion. The problem of cavitation can be eliminated by maintaining a minimum suction pressure at the pump inlet to overcome the pump's internal losses. This minimum suction pressure is called the net positive suction head required (NPSHR). Some symptoms that indicate cavitation is occurring in a pump include:

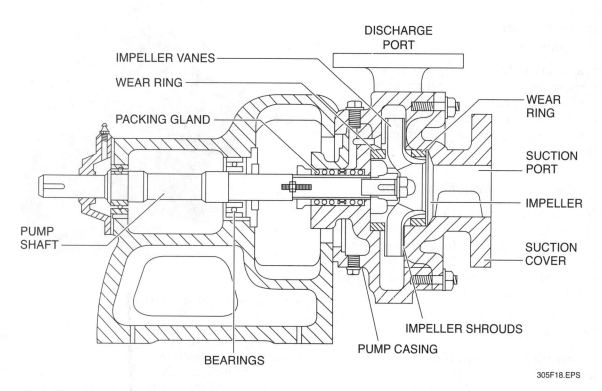

305F18.EPS

Figure 18 ◆ Centrifugal pump components.

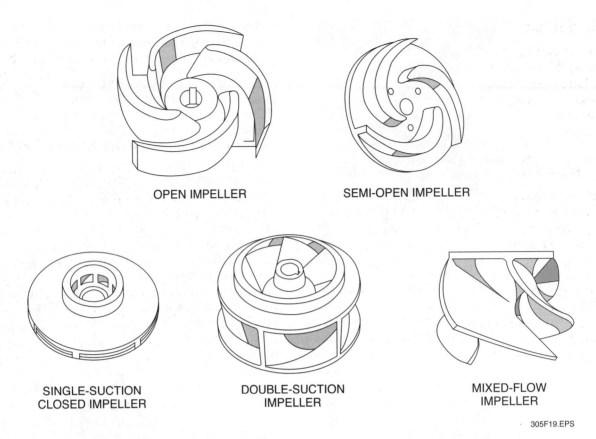

OPEN IMPELLER

SEMI-OPEN IMPELLER

SINGLE-SUCTION
CLOSED IMPELLER

DOUBLE-SUCTION
IMPELLER

MIXED-FLOW
IMPELLER

305F19.EPS

Figure 19 ◆ Impeller types.

- Snapping and crackling noises at the pump inlet
- Pump vibration
- Reduced or no water flow
- A drop in pressure

The type of centrifugal pump used depends upon the requirements of the system. Some systems require that large amounts of fluids be pumped, others require high pressures. Two types of centrifugal pumps are double-suction and multistage.

3.5.2 Double-Suction Centrifugal Pumps

A double-suction centrifugal pump is used to pump large volumes of fluid. Fluid is drawn in through openings on both sides of the impeller and passes out through a single discharge opening. *Figure 20* shows a double-suction centrifugal pump with a horizontally split case.

3.5.3 Centrifugal Pumps

Multistage centrifugal pumps (*Figure 21*) contain two or more stages and can be either single- or double-suction pumps. Fluid is discharged from

305F20.EPS

Figure 20 ◆ Double-suction centrifugal pump.

one stage to the next through passages in the pump casing. Each stage has an impeller, which is used to increase the velocity of the fluid being pumped until the desired pressure is reached.

3.6.0 Valves

Common valves and their uses, as well as some specialty valves, were discussed in earlier modules. Larger commercial systems generally use these same valves, with some additional types of valves needed for proper and flexible system operation.

3.6.1 Multi-Purpose Valves

In larger systems, a multi-purpose valve is commonly installed in the discharge side of the circulating pump (*Figure 22*). This single valve functions as a shutoff valve, a check valve, and a balancing valve. It has a calibrated stem used to return the valve to the set position after the valve is used as a shutoff valve. When in the set position, it acts as a balancing valve. The valve is tapped for the installation of two readout valves that enable it to be used as a **flow meter**. When balancing a system, the pressure drop across the

valve can be measured using a differential pressure gauge connected to these readout valves. The measured pressure drop can then be used with a conversion chart to find the corresponding total flow rate of the water passing through the valve.

3.6.2 Balancing Valves and Flow Meters

Balancing valves and flow meters provide a means of balancing a water system. These devices can be separate assemblies or they can be combined into one balancing valve assembly like the ones shown in *Figure 23*. A balancing valve can function as a precision system balancing valve, an isolation valve, or as a variable-orifice flow meter. As a balancing valve, it is adjusted to a predetermined setting that relates to the degree of closure needed to get the desired water flow rate in its flow path. This setting is usually determined by the system designer, and the installer sets it accordingly. When set to the closed position, the

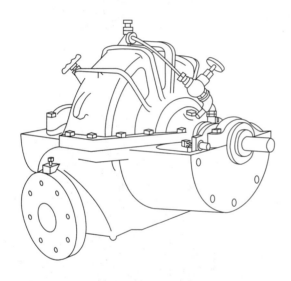

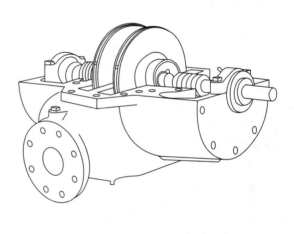

305F21.EPS

Figure 21 ◆ Multistage pump.

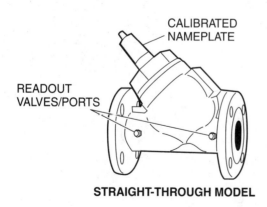

STRAIGHT-THROUGH MODEL

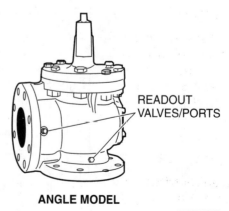

ANGLE MODEL

305F22.EPS

Figure 22 ◆ Multi-purpose valves.

valve is fully closed and serves as an isolation valve. Some balancing valves are equipped with two readout valves that enable them to be used as a flow meter. When a differential pressure gauge is connected to these readout valves, the pressure drop across the balancing valve can be measured. By use of a conversion chart or calculator, the measured pressure drop can then be converted to the corresponding total flow rate of the water

through the valve. Some balancing valves give direct readouts of flow.

Flow measurements made to balance a water system are done by measuring the pressure drops at specific locations throughout the system. **Venturi tubes** and orifice plate devices installed in the system can be used for this purpose (*Figure 24*). They are commonly called flow meters. These devices insert a specific fixed-area restric-

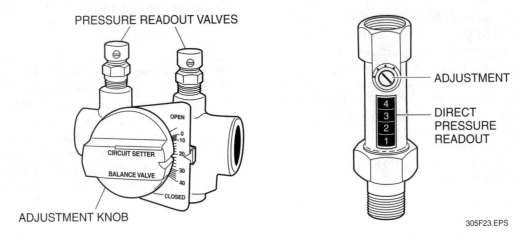

Figure 23 ◆ Combined balancing valve/flow meters.

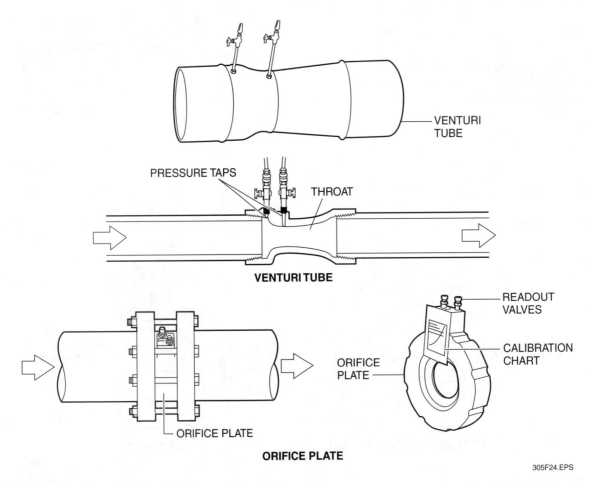

Figure 24 ◆ Venturi tube and orifice plate flow metering device.

tion in the path of water flow, thereby creating a pressure drop. The flow rate through the device can be determined by accurate measurement of this pressure drop. Measurement of the pressure drop is typically made using a manometer or differential pressure gauge (readout meter). For proper operation, the manufacturer's installation instructions must be followed.

Because of its streamlined shape, a venturi tube is more accurate than an orifice plate. Water passing through the reduced area of the venturi throat increases in velocity, creating a pressure differential between the inlet and throat areas. After passing through the throat, the flow area is gradually increased, which decreases the velocity and allows pressure recovery. The orifice plate may be installed between a single pair of flanges that have pressure differential taps built in by the manufacturer.

3.6.3 Butterfly Valves

A butterfly valve (*Figure 25*) has a round disc that fits tightly in its mating seat and rotates 90 degrees in one direction to open and allow fluid to pass through the valve. The butterfly valve can be operated quickly by turning the hand wheel or hand lever one-quarter of a turn, or 90 degrees, to open or close the valve. These valves can be used completely open, completely closed, or partially open for non-critical throttling operations. Butterfly valves are typically used in low- to medium-pressure and low- to medium-temperature applications. They generally weigh less than other types of valves because of their narrow body design. When the butterfly valve is equipped with a hand lever, the position of the lever indicates whether the valve is open, closed, or partially open. When the lever is perpendicular to the flow line through the valve, the valve is open. Butterfly valves that are 12 inches in diameter and larger are usually equipped with a hand wheel or gear-operated actuator because of the large amount of fluid flowing through the valve and the great amount of pressure pushing against the seat when the valve is fully closed.

Butterfly valves have an arrow stamped on the side, indicating the direction of flow through the valve. They must be installed in the proper flow direction, or the seat will not seal and the valve will leak.

There are three common types of butterfly valves in use:

- Wafer-type
- Wafer lug
- Two-flange

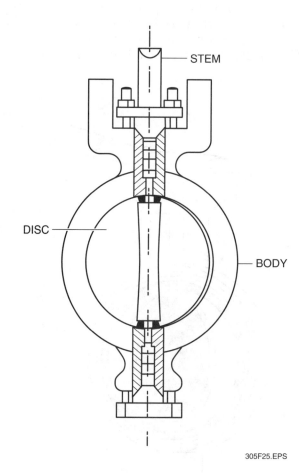

Figure 25 ◆ Butterfly valve.

305F25.EPS

The wafer-type butterfly valve (*Figure 26*) is designed for quick installation between two flanges. No gasket is needed between the valve and the flanges because the valve seat is lapped over the edge of the body to make contact with the valve faces. Bolt holes are provided in larger wafer valves to help line up the valve with the flanges. During installation of the valves, the valve must be in the open position prior to tightening the flange mating hardware in order to prevent damage to the valve seat.

Wafer lug valves (*Figure 27*) are the same as wafer valves except that they have bolt lugs completely around the valve body. Like the wafer valve, no gasket is needed between the valve and the flanges because the valve seat is lapped over the edges of the body to make contact with the valve faces. The lugs are normally drilled to match ANSI 150-pound steel drilling templates. The lugs on some wafer lug valves are drilled and tapped so that when the valve is closed, downstream piping can be dismantled for cleaning or maintenance while the upstream piping is left intact. When tapped wafer lug valves are used for pipe end applications, only one pipe flange is necessary. The body of a two-flange butterfly valve is made with a flange on each end (*Figure 28*). The valve seat is not lapped over the

flange ends of the valve, so gaskets are required between the flanged body and the mating flanges. The two-flange valve is made with either flat-faced flanges or raised-face flanges, and the mating flanges must match the valve flanges.

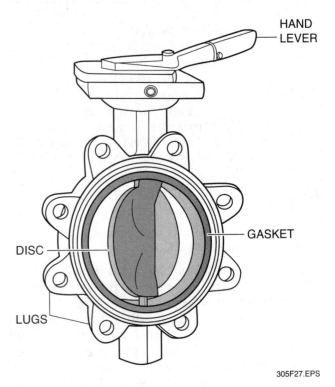

Figure 27 ◆ Wafer lug butterfly valve.

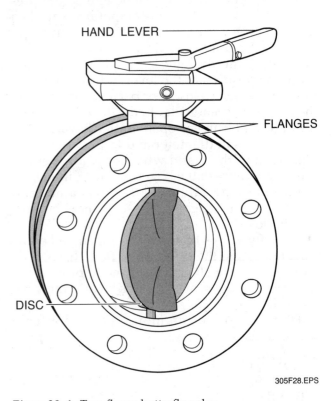

Figure 26 ◆ Wafer butterfly valve.

Figure 28 ◆ Two-flange butterfly valve.

3.6.4 Two-Way and Three-Way Valves

Two-way and three-way valves are used to control the flow and/or temperature of hot water, chilled water, or both. Their operation is normally controlled by a signal applied to the valve's actuator. Actuators used with these valves may be thermostatic, pneumatic, electric, or electronically controlled. In two-way valves, the water enters the input port and exits the output port, as shown in *Figure 29(A)*. Flow at the output can be at full volume or at a reduced volume, depending on the valve's position.

The three-way valve is often used as a mixing (blending) or diverting valve. As a mixing valve, it has two inputs and one output. It mixes the two water streams into one, based on the position of the valve's plug in relation to its upper and lower valve seats. It is commonly used to mix or blend hot water and chilled water inputs so that the single water stream leaving the valve has a controlled temperature.

Another common use is to select either the hot-water or the chilled-water input for application to a multi-purpose coil, as shown in *Figure 29(B)*.

When the three-way valve is used as a diverting valve, it has one input and two outputs. In this application, it splits the single input water stream into two smaller output streams to achieve temperature control. Typically, the flows of the two output water streams are routed in the following manner: one travels through a heat-transfer coil, and the other travels through a bypass pipe around the transfer coil, as shown in *Figure 29(C)*.

3.6.5 Pressure Independent Characterized Control Valves™

A relatively new valve used in commercial hydronic systems offers precise flow control to the load regardless of pressure variations experienced in the hydronic system. All hydronic systems experience operating periods when very few or no loads are active. Conversely, all potential hydronic loads may be active at other times. Balancing the hydronic system to ensure that proper flow rates are achieved under all conditions can be quite challenging. Improper balancing, especially during periods when loads are

(A) TWO-WAY VALVE CONTROLLING THE FLOW OF CHILLED WATER THROUGH A COOLING COIL

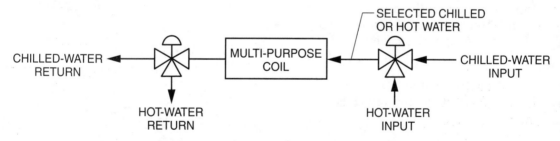

(B) THREE-WAY VALVES USED TO SELECT CHILLED OR HOT WATER FOR APPLICATION THROUGH A MULTI-PURPOSE COIL

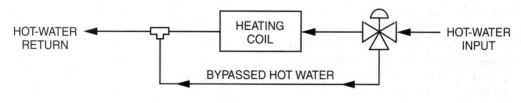

(C) THREE-WAY VALVE USED AS A DIVERTING VALVE

305F29.EPS

Figure 29 ◆ Two-way and three-way valves.

high, can result in a lack of sufficient flow to some loads.

A Belimo Pressure Independent Characterized Control Valve™ *(Figure 30)* helps to eliminate this problem by combining flow balancing with modulating of open-closed flow control. Because it is a two-way valve (as opposed to a three-way valve), it is primarily designed for use with hydronic systems using variable-speed pumps. Once the desired flow rate is set and the valve opens, the flow rate is monitored internally and the valve position is adjusted as necessary by the modulating actuator to meet the setpoint. This type of valve simplifies the valve selection process. It also reduces system problems due to incorrect flow rates and simplifies system balancing and commissioning.

3.7.0 Heating System Terminals

Hot-water heating terminals transfer the heat carried by the system hot water to the conditioned space. This can be done in many ways and using many kinds of devices. In combined heating and cooling systems (dual-temperature systems), different kinds of terminals are used than those found in heating-only systems. The focus of this section is on terminals used in heating-only systems. The terminals used in dual-temperature systems are covered later in this module. In heating-only systems, the transfer of the heat from the terminals results from a combination of radiation to the space and convection to the air in the space. Normally, these terminals are installed at the points of greatest heat loss, such as under windows, along exposed walls, and near door openings. The following types of terminals are commonly used in heating-only systems:

• Convectors
• Baseboard and finned-tube units
• Unit heaters, unit ventilators, and fan coils
• Heating coils

3.7.1 Convectors

A convector *(Figure 31)* is a heating device that depends mainly on natural conductive heat transfer. The heating element is a finned-tube coil or coils, mounted in an enclosure designed to increase the conductive flow. The enclosure can have many shapes. The room air enters the enclosure below the heating element, is heated in passing through the element, and leaves the enclosure through the grill at the top. Convectors are usually mounted at or near the floor on an outside wall of the room.

3.7.2 Baseboard and Finned-Tube Units

Baseboard units are mounted on the wall in place of the usual baseboard. They can be either a finned-tube system, similar to a convector but much smaller, or a cast-iron section with convective heat channels. Heat transfer takes place by convection. Baseboard radiation is usually installed in a continuous run along the outside walls of a room. *Figure 32* shows baseboard and finned-tube units.

Self-contained hydronic baseboard heaters made in various lengths can also be used. These heaters have a thermostat-controlled, electric heating element. When heated by this element, water contained in a closed-loop finned-tube system circulates in a continuous cycle. The control thermostat may be built into the baseboard unit, or it can be wall-mounted. This type of unit can

305F30.EPS

Figure 30 ◆ Belimo Pressure Independent Characterized Control Valve™.

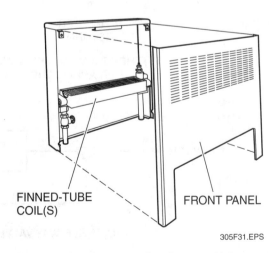

FINNED-TUBE COIL(S)   FRONT PANEL

305F31.EPS

Figure 31 ◆ Convector-type heating terminal.

also be floor-mounted, and is normally used in commercial and special residential applications.

Finned-tube units are room heaters similar to baseboard units. They are composed of larger tubing or pipe, typically 1¼" to 2", with fins bonded to the pipe. Fins can be 3½" to 4½" square. The finned-tube assembly can be housed in a variety of enclosures depending on where it is installed. Finned-tube units are used mostly for perimeter heating, especially in large glassed-in areas.

3.7.3 Unit Heaters and Unit Ventilators

A unit heater consists of heating elements and a circulating fan in an enclosure. The fan blows room air across the heating elements and heat is transferred to the conditioned space. The heating elements can be electric, gas combustion, or hydronic. For gas combustion or hydronic unit heaters, the heating element can be either copper or steel. Unit heaters have a relatively large heating capacity for their size and can project heated air over a considerable distance. They are used mainly in commercial systems, such as those in garages, factories, and warehouses.

Both horizontal and vertical units are used. Horizontal units (*Figure 33*) blow heated air in a horizontal direction. They are used in areas with low to medium ceiling heights. Vertical units blow air in a vertical (downward) direction and are normally used in areas with high ceilings. They are also used when the unit must be installed in an out-of-the-way location due to floor or space limitations. Both types of unit heaters usually have an adjustable diffuser used to vary their air discharge

pattern. Unit ventilators are used for both heating and cooling. They are covered later in this module.

3.7.4 Heating Coils

Heating coils (*Figure 34*) are used mainly in commercial and industrial heating systems. The heat transfer results from natural radiation and convection. These units consist of coils of copper or aluminum tubing through which the heated water flows. The tubing or pipes may be staggered or placed in line with the airflow. To aid in the transfer of heat, uniformly spaced thin metal plates (fins) or a spiral wound metal ribbon is usually attached along the tubing.

305F33.EPS

Figure 33 ◆ Horizontal unit heater.

305F34.EPS

Figure 34 ◆ Hot water coil.

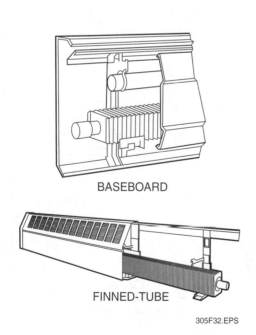

BASEBOARD

FINNED-TUBE

305F32.EPS

Figure 32 ◆ Baseboard and finned-type heating terminals.

4.0.0 ◆ CHILLED-WATER COOLING SYSTEMS

The use of a chilled-water system eliminates the need to circulate refrigerant through refrigeration system components installed in a building to provide comfort cooling. Instead, a device called a chiller (*Figure 35*) uses refrigerant to cool the water. The chilled water is circulated throughout the building and back to the chiller where the process is repeated. This allows the use of basic water piping through the building, rather than refrigerant piping, which is more expensive and represents a unique challenge in large buildings due to pressure drop and oil return issues. Once the chilled water is delivered to the building, it may be circulated through a large cooling coil in a central air handler, or it can be delivered to the conditioned space and circulated through a small cooling coil to cool the room.

The focus of this module is on chilled-water systems that operate above 40°F and are used for comfort cooling. Other types of liquid chilling systems are also used to provide comfort air conditioning. Instead of water, these systems cool brine or other liquid refrigerant, and typically operate at temperatures below 40°F. Chilled-water systems are used independently to cool buildings that have either a separate heating system or no heating system. They can also be used as part of a dual-temperature system. Chilled-water cooling as part of a dual-temperature system is covered later in this module.

Figure 36 shows a simplified diagram of a basic chilled-water system. It consists of a chiller unit (chiller), cooling terminal, chilled-water pump, and the related chilled-water piping. The chiller

unit shown in the figure is a mechanical refrigeration system. Some chilled-water systems use absorption-type chillers to produce cooling.

The chiller shown operates the same as the cooling systems discussed earlier. Its operation is based on the refrigeration cycle and mechanical compression of a refrigerant gas. The chiller evaporator (cooler) is a water-to-refrigerant heat exchanger. It has both the system chilled water and chiller system refrigerant flowing through it. In the evaporator, the chilled water is cooled by the refrigerant liquid evaporating at a lower temperature. This causes the refrigerant to boil or vaporize. For comfort cooling applications, the chilled water temperature supplied to the building is generally around 45°F, with the returning water temperature expected to be around 10°F warmer (around 55°F). The chilled water produced is described as a **secondary coolant**, which is any liquid that is cooled by a system refrigerant and is then used to transmit heat from the conditioned space without a change in state.

The refrigerant vapor from the chiller evaporator is drawn into the chiller compressor. There,

305F35.EPS

Figure 35 ◆ Packaged air-cooled chiller.

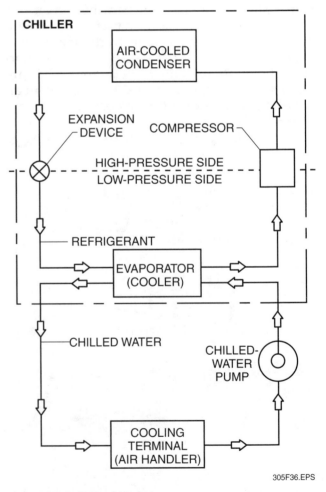

305F36.EPS

Figure 36 ◆ Basic chilled-water system.

the low-temperature, low-pressure refrigerant vapor is converted into a high-temperature, high-pressure vapor so that it can be condensed into a liquid in the condenser. The system shown in the example has an air-cooled condenser.

In some systems, the condenser may be a separate unit and not part of the chiller. The air-cooled condenser in this type of chiller has a series of tubing coils through which the refrigerant flows. As cooler outside air is forced across the condenser tubing, the hot refrigerant vapor condenses into a liquid, giving up its heat to the outside air. The high-temperature, high-pressure liquid refrigerant then flows to the evaporator through the expansion device.

The expansion device regulates the flow of refrigerant into the evaporator and decreases its pressure and temperature. This converts the high-temperature, high-pressure liquid refrigerant into the low-temperature, low-pressure refrigerant needed to absorb heat in the evaporator.

The chilled water cooled in the chiller evaporator is circulated through the chilled-water system piping by the chilled-water pump. When there is a call for cooling, the water is continuously recirculated through the chiller evaporator and cooling terminal.

Circulation of the chilled water through the cooling terminal cools the conditioned space. In a typical system, the terminal is an air handler containing cooling coils and a fan. Cooling of the room air takes place by blowing the warm room air at the input to the terminal over the terminal cooling coils in which the chilled water is flowing. This transfers the heat from the warmer room air through the coil tubing to the chilled water.

A chilled-water system similar to the one previously described is shown in *Figure 37*. Note that the chiller unit used in this system is a packaged air-cooled reciprocating liquid chiller.

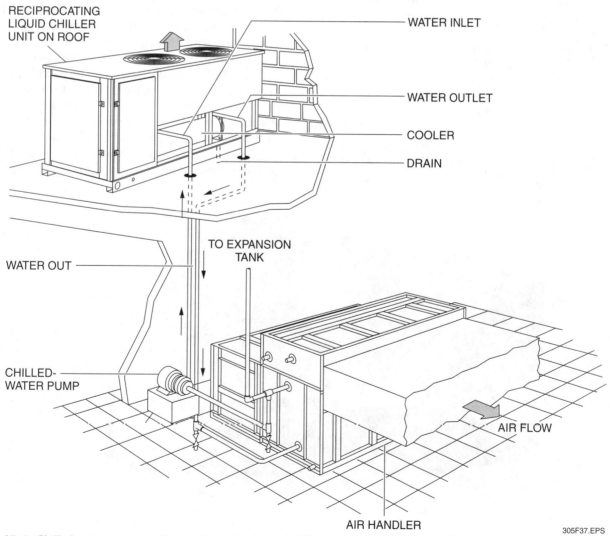

Figure 37 ◆ Chilled-water system with a packaged air-cooled chiller.

305F37.EPS

5.0.0 ◆ CHILLED-WATER SYSTEM COMPONENTS

Chilled-water systems can be configured in a variety of ways. The components of a chilled-water cooling system can include the following:

- Water chillers
- Chiller operating/safety controls
- Cooling towers and evaporative condensers
- Circulating pumps
- Terminals
- Air-cooled condenser

5.1.0 Water Chillers

Water chillers fall into two classes: mechanical and nonmechanical. Mechanical chillers use compressors to provide cooling. Compressor operation lowers the pressure and temperature of the refrigerant in the chiller evaporator, allowing the refrigerant to boil and absorb heat from the chilled water. Also, the pressure and temperature of the refrigerant in the chiller condenser is raised, allowing the refrigerant to condense and give up its heat to either air or water. The compressor also maintains a pressure difference between the high-pressure and low-pressure sides of the chiller system. This pressure difference causes the refrigerant to flow throughout the chiller system.

Nonmechanical chillers do not have a compressor. Instead, they use a chemical cycle to provide cooling. This type of chiller is called an absorption liquid chiller.

Three common types of mechanical chillers are reciprocating or scroll compressor liquid chillers, centrifugal liquid chillers, and screw liquid chillers.

Mechanical water chillers can be assembled from components in the field or they may be assembled at the factory into a packaged unit. This section will focus on packaged chillers. For our discussion, a packaged mechanical chiller is one that includes the compressor, condenser, evaporator, expansion device, related controls, and auxiliary equipment. Also, almost all absorption chillers are packaged units. The basic refrigeration components used in mechanical chillers operate the same as similar components used with the cooling system studied earlier in this curriculum.

5.1.1 Reciprocating and Scroll Liquid Chillers

Reciprocating and scroll liquid chillers (*Figure 38*) use one or more compressors to provide the necessary cooling capacity. When multiple compressors are used, the compressors are generally staged by the temperature controls. As load increases and the water temperature begins to drift farther from the setpoint, more compressors are started. In some cases, two compressors may be connected on a single refrigeration circuit. Although hermetic reciprocating and scroll compressors dominate the small- to medium-sized chiller market today, models using one or more semi-hermetic compressors remain in use and are

305F38.EPS

Figure 38 ◆ Packaged air-cooled scroll chiller.

still available from some manufacturers. The chiller condenser can be air-cooled, water-cooled, or evaporative. Water-cooled condensers may be either tube-in-tube, shell-and-tube, or shell-and-coil.

Tube-in-tube evaporators are sometimes used in smaller units. The expansion device controls refrigerant flow from the condenser to the evaporator to maintain enough suction superheat to prevent liquid refrigerant from flooding the compressor. HCFC-22 refrigerant is used in most installed reciprocating water chillers, but models using non-chlorine refrigerants are being phased in. Reciprocating and scroll liquid chillers have cooling capacity sizes ranging from about 5 tons to 225 tons. Capacity control is normally accomplished using a combination of cylinder unloading, on-off cycling of the compressor(s), or compressor speed control.

5.1.2 Centrifugal Liquid Chillers

The centrifugal chiller (*Figure 39*) uses a centrifugal compressor consisting of one or more stages. Both open and hermetic compressors can be used. Open compressors can be driven by gas or steam turbines or engines, or electric motors. Centrifugal compressors use a high-speed impeller with many blades that rotate in a spiral-shaped housing. The impeller is driven at high speed inside the compressor housing. Refrigerant vapor is fed into the housing at the center of the impeller. The impeller throws the incoming vapor outward in a circular path from between the blades and into the compressor housing. This action, called centrifugal force, creates pressure on the high-velocity gas and forces it out the discharge port.

The condenser in a centrifugal liquid chiller is usually water-cooled by a cooling tower, with the refrigerant condensing on the outside of the tubes. Air-cooled condensers can also be used, but their use greatly increases the unit's power consumption. The evaporator (cooler) can be a direct-expansion type in smaller units, but is usually a flooded type in the larger units. HCFC-123 and

305F39.EPS

Figure 39 ◆ Centrifugal chiller.

HFC-134*a* are common refrigerants used in centrifugal liquid chillers. A variety of low pressure CFC refrigerants were used in the past, but most have either been replaced or modified for use with new, environmentally-friendly refrigerants. Although relatively rare, in light commercial applications some chillers using CFC refrigerants do remain in service. As all refrigerants that bear chlorine will eventually be phased out, those based on HCFC-123 will eventually be replaced or modified.

Centrifugal liquid chillers have cooling capacity sizes ranging from about 200 tons to 8,000 tons. Capacity control is normally accomplished using adjustable compressor inlet guide vanes or compressor speed control.

Subcooling or intercooling devices are often used in a centrifugal chiller to conserve energy. Subcooling is a process by which the condensed refrigerant is subcooled in either a subcooling section of the water-cooled condenser or in a separate heat exchanger. This is done by bringing the warm condensed refrigerant into contact with the coldest (inlet) water tubes in the condenser or in a heat exchanger. This causes the condensed refrigerant to subcool to a temperature lower than its

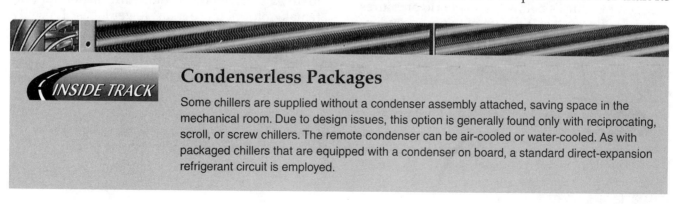

INSIDE TRACK

Condenserless Packages

Some chillers are supplied without a condenser assembly attached, saving space in the mechanical room. Due to design issues, this option is generally found only with reciprocating, scroll, or screw chillers. The remote condenser can be air-cooled or water-cooled. As with packaged chillers that are equipped with a condenser on board, a standard direct-expansion refrigerant circuit is employed.

Central Plant Facilities

Central plant facilities that provide hot water and chilled water to large service areas have long been used for college campuses, airports, high-technology manufacturing centers, and some municipalities. Central plant technology is experiencing renewed interest from energy-conscious system designers. Designers of new buildings and retrofitted high-density complexes are leaning more toward this technology. Many of these facilities combine cogeneration capability that uses the heat byproducts of stand-alone electricity generation to drive the heating water system or power chillers. Typical cooling tonnages for these facilities begin in the 3,000-ton range and are seen consistently in the 8,000- to 10,000-ton range.

condensing temperature. This increases the refrigeration cycle efficiency and lowers power consumption. Devices that accomplish subcooling in this way are referred to as **thermal economizers**.

Intercooling is a process in which the condensed refrigerant leaving the condenser is passed through a flash (surge) chamber before reaching the evaporator. Pressure in this chamber is between the condenser and evaporator pressures. Because its pressure is lower than the condenser pressure, some of the liquid refrigerant flashes into gas, cooling the remaining liquid. The flash gas, having absorbed heat, is drawn off and returned directly to the inlet of the compressor second stage. There, it is mixed with discharge gas that is already compressed by the first-stage impeller. Since the flash gas has to pass through only one stage of compression to reach condenser pressure, there is a savings in power. The liquid refrigerant remaining in the flash chamber is metered through a valve and flows into the evaporator. Because pressure in the evaporator is lower than the flash chamber pressure, some of the liquid flashes and cools the remainder of the refrigerant to evaporator temperature. Devices that accomplish intercooling using a flash chamber or similar device are referred to as **flash economizers**.

Purge units are another type of device used on centrifugal chillers that operate with low-pressure refrigerants, such as HCFC-123. Use of these refrigerants causes below-atmospheric pressures to exist in the system. Because of this, noncondensibles such as air and water tend to leak into and accumulate in the system. A purge unit must be used to remove the noncondensible gases and water vapor from the system. Today's purge units are generally small refrigeration units mounted on the chiller assembly. They draw in refrigerant vapors from the chiller, then chill the vapor below the point of condensation for the refrigerant itself. Any vapor that remains is considered a noncondensible gas other than refrigerant and is released to the atmosphere. Any moisture is captured by filter-driers. The condensed liquid refrigerant is then returned to the chiller. Older purge systems used with CFC refrigerants were quite inefficient and tended to release a significant amount of refrigerant along with the air. Because the refrigerants were relatively inexpensive, this issue presented little concern and the chiller was simply recharged periodically. To comply with new EPA guidelines, purge units have been redesigned to operate much more effectively.

5.1.3 Screw Liquid Chillers

Screw liquid chillers (*Figure 40*) use an oil-injected hermetic or open screw compressor. Screw compressors use a matched set of screw-shaped rotors, one male and one female, enclosed within a cylinder. The male rotor is driven by the compressor motor, and in turn drives the female rotor. Normally, the driven male rotor turns faster than the female rotor because it has fewer lobes. Typically,

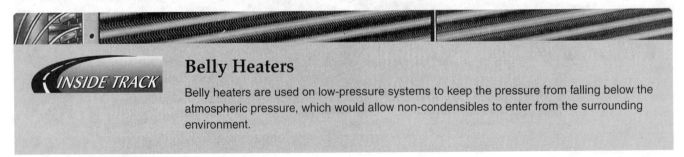

Belly Heaters

Belly heaters are used on low-pressure systems to keep the pressure from falling below the atmospheric pressure, which would allow non-condensibles to enter from the surrounding environment.

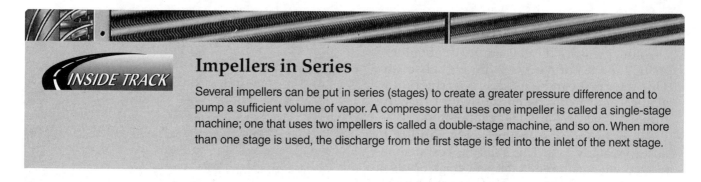

the male has four lobes, the female six. As these rotors turn, they mesh with each other and compress the gas between them. The screw threads form the boundaries separating several compression chambers. The gas entering the compressor is moved through a series of progressively smaller compression stages until it exits at the compressor discharge in its fully compressed state.

Screw liquid chillers can have a water-cooled, shell-and-tube condenser, with the refrigerant condensing on the outside of the tubes. Air-cooled condensers are also very popular. When the chiller is used with a remotely located air-cooled or evaporative condenser, a liquid receiver may replace the condenser on the package assembly. The evaporator (cooler) can be either a direct-expansion or flooded type. When the direct-expansion evaporator is used, the refrigerant usually flows through the tubes. With a flooded evaporator, the refrigerant is normally outside the tubes. Because the screw compressor is oil-injected, an oil separator is also part of the package. A flash economizer, similar to the one described for the centrifugal liquid chiller, can be used to improve chiller efficiency. In the screw compressor, the flash gas produced in the economizer is injected into the compressor at an intermediate point of the compression cycle.

Most packaged screw liquid chillers use HCFC-22, HFC-134a, HFC-407C, or HFC-410A refrigerant. Screw liquid chillers have cooling capacity sizes ranging from about 40 tons to 850 tons. Capacity control is normally accomplished using a hydraulically-operated slide valve.

5.1.4 Absorption Liquid Chillers

Absorption liquid chillers use a chemical cycle to provide cooling. In the absorption cycle, evaporation and condensation of the refrigerant occur at different pressure levels. The cycles differ in that the absorption cycle uses a heat-operated generator to produce the pressure difference, while the mechanical refrigeration cycle uses a compressor. Both cycles require energy for operation, heat in the absorption cycle, and mechanical energy in the compression cycle.

The refrigerant (water) and absorption solution (**lithium bromide**) are the working fluids of the absorption cycle. Lithium bromide is a salt solution commonly used in air conditioning absorption chillers. The absorption cycle (lithium bromide cycle) is based on two principles:

305F40.EPS

Figure 40 ◆ Air-cooled screw liquid chiller.

- Water can be made to boil in the temperature range of 40°F to 45°F by maintaining a low absolute pressure (high vacuum) of about 0.25 to 0.30 inches of mercury (in. Hg) absolute. This means it can be used as a refrigerant to absorb heat from the chilled water flowing through tubes in the chiller evaporator.

- Lithium bromide solution has the ability to absorb water vapor. This ability is best demonstrated when common table salt clogs up a salt shaker in humid weather. This happens because the salt has absorbed moisture from the air.

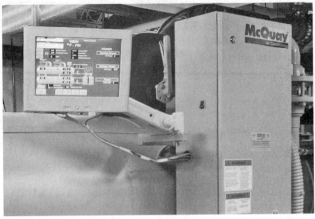

305F41.EPS

Figure 41 ◆ Chiller operator interface.

WARNING!

Lithium bromide is potentially very harmful. It reacts with skin moisture to cause burns, damages the respiratory tract and lungs if inhaled, and is corrosive. Contact with the eyes can cause irritation, blurred vision, and possibly serious eye damage. When handling lithium bromide, always use appropriate personal protection equipment as recommended in the MSDS.

Disposal of lithium bromide is controlled by local, state, and federal agencies. Check local, state, and Federal EPA regulations and comply with all disposal requirements.

Absorption chillers are available in capacities ranging from 50 tons to as high as 1,800 tons. As a rule, they are used in applications where a significant amount of steam is readily available, often where refuse-burning facilities are nearby or on board sea-going vessels with large steam boilers in operation. They are relatively rare in commercial applications, and the operating cycle is quite complex and detailed. For this reason, an in-depth study of their operation is not included here.

5.2.0 Chiller Operating/Safety Controls

Almost all modern chillers have one or more control or interface panels used by the operator to control or monitor the operating status of the chiller (*Figure 41*). Operation of the chiller from these panels is under the control of a microprocessor programmed for the task. Status inputs are supplied to the microprocessor by remote sensors so that the processor can provide the needed control. The sensors may be thermistors for measuring temperature, transducers for measuring pressure, and contacts that supply or interrupt voltage. Typically, chiller performance status is displayed on liquid crystal display (LCD) indicators. Status messages can be generated whether or not the chiller is running. Operation of the chiller from the panel(s) is usually done by entering commands using a computer-type keyboard in response to menu-driven displays. When troubleshooting, these same panels can be used to receive diagnostic messages or run tests. Common chiller operations performed either automatically or manually by a typical microprocessor control center are:

- Startup and shutdown of the chiller
- Adjustment of chiller compressor capacity control to maintain the proper leaving chilled-water temperature
- Control of system operation per the programmed occupied/unoccupied schedules
- Registering cooler, condenser, and lubricating system pressures and temperatures
- Generating alarm and alert messages for abnormal conditions and de-energizing the chiller when system safety dictates
- Recording the total machine operating hours

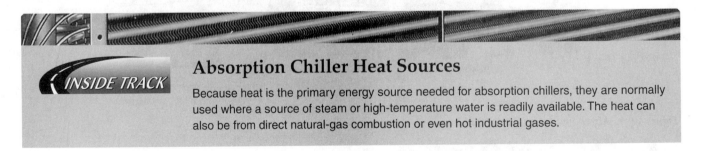

Absorption Chiller Heat Sources

INSIDE TRACK

Because heat is the primary energy source needed for absorption chillers, they are normally used where a source of steam or high-temperature water is readily available. The heat can also be from direct natural-gas combustion or even hot industrial gases.

Because chilled water systems represent a significant investment, use large amounts of energy, and are essential to the comfort of large numbers of occupants, they are often integrated into computer-controlled building automation systems. Through the use of these systems, even the smallest details of the chiller's present operation and operating history are readily available to building managers on site and other personnel in remote locations. When problems occur, service personnel can review detailed system information and the nature of the condition before arriving on site.

5.3.0 Cooling Towers/Evaporative Condensers

Condensers are used to remove heat from the chiller refrigerant system. They take in high-pressure, high-temperature refrigerant gas from the compressor and change it into high-temperature, high-pressure liquid refrigerant. They do this by removing the superheat and **latent heat of vaporization** from the refrigerant by transferring this heat to air, water, or both.

For the condenser to operate properly, the condensing medium of air or water must always be at a lower temperature than the refrigerant it is condensing. Condensers used with chillers can be air-cooled, water-cooled, or evaporative. An evaporative condenser is one that combines both air and water cooling. Water-cooled condensers are normally used in conjunction with a cooling tower. Air-cooled and water-cooled condensers used with chillers are the same kind, and operate in the same way as air-cooled and water-cooled

WARNING!

All cooling towers provide the opportunity for water stagnation in or around the cooling tower. This stagnaton can be from the mist plume, the catch basin of the tower, or puddling on the surrounding roof or ground-mounted mechanical area. If any of these areas of water collection are within 10 feet of a fresh air intake to the building HVAC systems or equipment, they can promote bacterial contamination such as legionella. Although this is a portion of the system design and may be an ongoing issue with older buildings, there should be an awareness of the potential health risk. The building owner should be notified if such conditions exist.

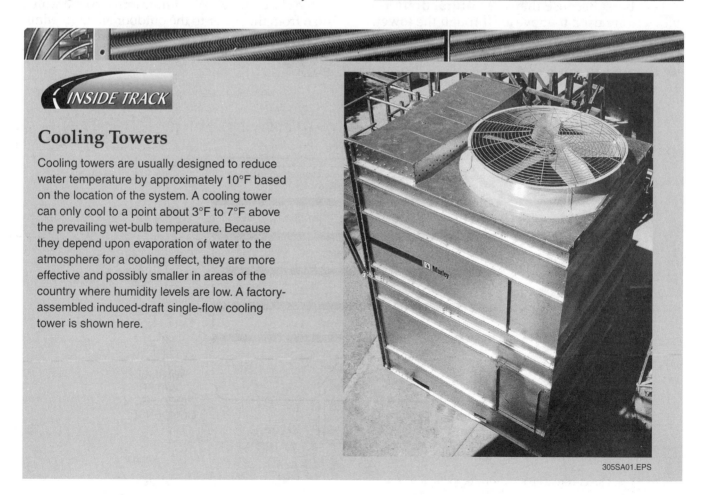

INSIDE TRACK

Cooling Towers

Cooling towers are usually designed to reduce water temperature by approximately 10°F based on the location of the system. A cooling tower can only cool to a point about 3°F to 7°F above the prevailing wet-bulb temperature. Because they depend upon evaporation of water to the atmosphere for a cooling effect, they are more effective and possibly smaller in areas of the country where humidity levels are low. A factory-assembled induced-draft single-flow cooling tower is shown here.

305SA01.EPS

condensers used with other cooling systems. This section will focus on the cooling tower used with a water-cooled condenser and the evaporative condenser.

5.3.1 Cooling Towers

The water portion of a chiller's water-cooled condenser is connected by piping to a cooling tower that is usually located outdoors. In the cooling tower, the heat absorbed by the water in the condenser is rejected from the system into the atmosphere by evaporation. The cooled water is then returned from the cooling tower to the condenser for reuse.

Cooling towers pass the outdoor air over the warm water to remove the system heat from the water. There are many kinds of cooling towers used with water-cooled condensers. One type is a direct-contact, natural-draft tower (*Figure 42*). It is called a direct-contact tower because the heated water comes in direct contact with the air. These towers are mounted outdoors, usually on the roof, to make use of natural air currents and heights for the net positive suction head on the pump. They are made of a metal frame covering several layers or tiers of fiberglass-reinforced plastic decks or splash bars. Because they use natural draft, no blowers are used to move air through the tower. Water is piped up from the condenser located in the building below to the top of the decks and is discharged in sprays over the decks. Spaces between the boards in the decks permit the water to drip or run from deck to deck, while being spread out and exposed to breezes that enter the tower from the open sides. The cooled water is collected in a sump at the bottom of the tower, and is pumped back to the condenser for reuse. Because cooling towers work partially on evaporation, any water lost in the evaporation process must be replaced in order to maintain correct system operation. This is done by a float-controlled makeup water system (*Figure 43*) that senses the water level in the sump and adds water as needed.

A variation of the natural-draft tower is an indirect-contact tower, sometimes called a closed-circuit cooler (*Figure 44*). The indirect-contact tower has two separate water circuits. One circuit consists of cooling tubes through which the water from the chiller condenser is flowing. This circuit is closed to the air, thereby protecting the chiller condenser from being contaminated by airborne dirt and impurities. The second water circuit is open to the air and consists of a pump that delivers water from a sump at the bottom of the tower to the top of the tower. There it is discharged in sprays over the tower decking in the same way as in the direct-contact tower. In this tower, the heat in the closed tubes is first transferred to the water, then from the water to the outdoor air. Drift eliminators prevent moisture discharge from the cooling tower, reducing the amount of makeup water needed.

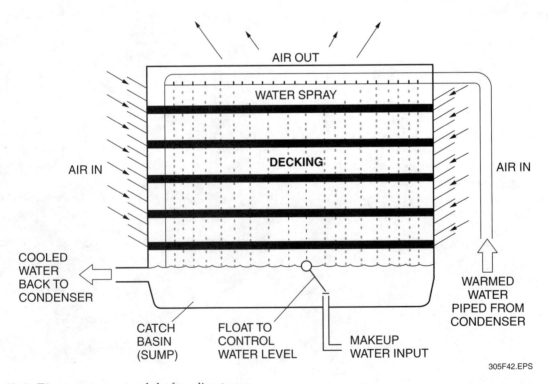

305F42.EPS

Figure 42 ◆ Direct-contact natural-draft cooling tower.

Centrifugal and other types of fans may be used to increase the speed of airflow in a cooling tower. When fans are used, the tower is called either a forced-draft tower or an induced-draft tower.

A forced-draft tower is one in which the fans push the air so that it flows up and over the wet decking surfaces. In an induced-draft tower, the fans pull the air rather than push it, so that it flows up and over the decking. Because they use fans, forced-draft and induced-draft towers tend to be small in comparison to natural-draft cooling towers. *Figure 45* shows an example of an indirect-contact, induced-draft cooling tower.

5.3.2 Evaporative Condensers

Like the water-cooled condenser and water tower combination, evaporative condensers first transfer heat to the water and then from the water to the outdoor air. The evaporative condenser combines the functions of a water-cooled condenser and cooling tower in one enclosure or package (*Figure 46*). The condenser water evaporates directly off the tubes of the condenser with each pound of water evaporated, removing about 1,000 Btus from the refrigerant flowing through the tubes. Air enters the bottom of the unit and flows by induced convection upward over the refrigerant-filled condensing coil. At the same time, water is being sprayed over the condensing coil. Both mediums absorb heat from the refrigerant in the coil. Liquid eliminators located above

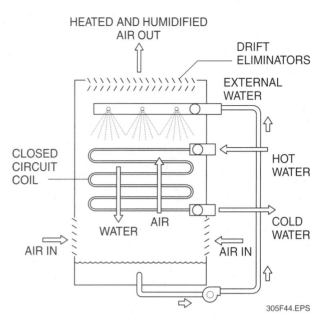

Figure 44 ◆ Indirect-contact natural-draft cooling tower.

the water spray remove water from the rising air, which is then pushed out the top of the unit by one or more fans. Cooled by both air and water, the refrigerant in the coil condenses into a subcooled liquid at the output of the coil.

Relocating the refrigerant condenser to the base of the evaporative condenser generally requires a considerable amount of additional refrigerant piping. Because many water-cooled systems that use an evaporative condenser are quite large, this fac-

305F43.EPS

Figure 43 ◆ Cooling tower float-controlled makeup water system.

Cooling Tower Water Treatment

Proper water treatment is essential to prevent and control biological growth and mineral deposits in cooling towers. Without proper treatment, a wide variety of biological growth will occur, sometimes with hazardous results to humans in the vicinity of the evaporating moisture. These same growths also cause a loss of capacity by impeding the heat transfer process.

Mineral deposits also impede heat transfer; as the water is evaporated, the minerals contained in the water supply remain behind. Although chemical treatment can assist in reducing these deposits, most cooling towers require bleed-off, or a consistent minimal draining of the tower water, to dilute and flush away minerals in circulation. The chemical treatment approach should be designed with consideration for local conditions, the anticipated or identified types of biological growth, and the surrounding environment.

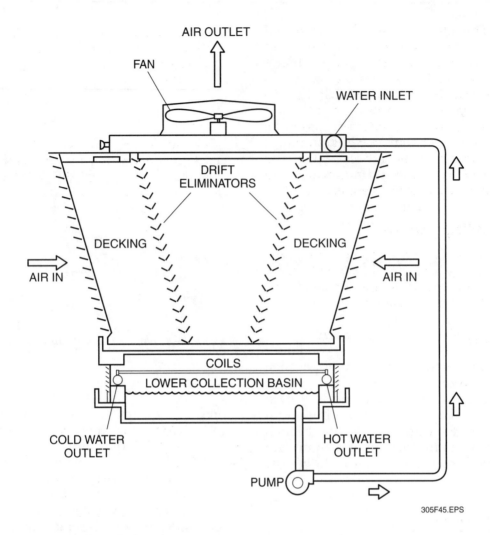

Figure 45 ◆ Indirect-contact induced-draft cooling tower.

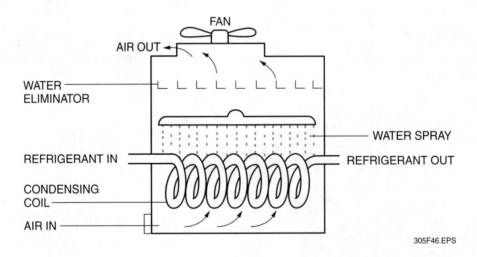

Figure 46 ◆ Evaporative condenser.

tor also significantly increases the refrigerant charge required and exposes the circuit to a greater potential for damage and leakage. Refrigerant piping is usually more expensive, and its installation more complex than water piping, so cooling towers are far more popular for most applications. Evaporative condensers are more likely to be found in refrigeration and food processing applications than in comfort cooling, and are often used in large refrigeration systems using anhydrous ammonia as the refrigerant.

5.4.0 Circulating Pumps

The pumps used to circulate chilled water, look and operate the same as the pumps used to circulate hot water in heating systems. These pumps were described earlier in this module.

5.5.0 Terminals

Terminals commonly used in chilled-water systems can consist of a fan or fans, cooling and/or heating coils, and air filters. Some units also include a damper-controlled mixing box with outside air and return air connections. Three common types of terminals are fan coil units or air handling units, unit ventilators, and air-water induction units.

5.5.1 Fan Coil Units

Fan coil units (*Figure 47*) are terminals that can be used for chilled-water cooling only, or for both cooling and heating. A fan coil unit includes a fan, finned-tube heating/chilled water coils, air filters, and sometimes an outside air connection with a manual damper. Control is provided using a multi-speed blower or control valves. The blower fan causes the room air to recirculate continuously from the conditioned space through the coil that contains the chilled water. This causes the heat in the room air to be transferred to the chilled water flowing through the coil. The air filter prevents clogging of the coil with dirt or lint carried by the recirculated air. Ventilation boxes with manual dampers can be connected to a fan coil unit. Fan coil units range in size from 100 to 2,000 cfm.

The term *fan coil unit* generally refers to units that produce up to 2,000 cfm of airflow. Above that size, they are commonly referred to as *air handling units* for clarity. However, there is little difference in the two beyond their physical size and capacity. Commercial air handling units are usually equipped with belt-driven blowers and do not use multi-speed blowers as a result. Depending upon the overall system design, they typically

are either simple single-speed units or are equipped with variable-speed drive components to modulate the blower motor speed based on the demand for cooling.

5.5.2 Unit Ventilators

Air conditioning unit ventilators (*Figure 48*) are used for both heating and cooling. They are similar to fan units, except they are larger and have

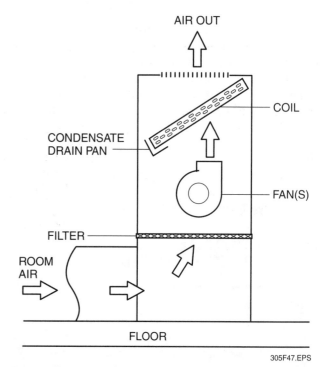

Figure 47 ◆ Fan coil unit.

Figure 48 ◆ Unit ventilator.

access to outdoor air via damper-controlled mixing boxes. The unit may be floor- or ceiling-mounted and uses return and outside air as required by the space. When temperature conditions permit, this allows outside air to be used for ventilation and cooling.

Unit ventilators are used mainly in areas where heavy occupancy requires a high rate of ventilation.

5.5.3 Air-Water Induction Units

Air-water induction units are similar to unit ventilators. Instead of a blower in each cabinet, circulation is provided by a central primary air system that handles part of the load. High-pressure air (primary air) from the central system flows through nozzles arranged to induce the flow of room air (secondary air) through the unit's coil. The room air is either cooled or heated at the coil, depending on the season. Mixed primary and room air are then discharged from the unit. Induction units are encountered in older systems. *Figure 49* shows a typical air-water induction unit.

6.0.0 ◆ DUAL-TEMPERATURE WATER SYSTEMS

CAUTION

Because of the common piping and valve arrangements involved, there is an inherent risk of damage to boilers and chillers used with some dual-temperature water systems. Never open a valve that will permit hot water to enter a low-pressure chiller, as this will cause the maximum chiller pressure of 15 pounds to be exceeded. This will cause the chiller safety relief valve disc to rupture, and the chiller will lose all its refrigerant to the atmosphere. Also, never open a valve that will permit the flow of cold water into a boiler. The application of low-temperature water to a boiler will shock it, possibly resulting in the cracking of the heat exchangers.

A dual-temperature system is a water system that circulates hot and chilled water to heat or cool with common piping and heat transfer terminals. It operates within the pressure and temperature limits of low-temperature water systems with usual winter supply temperatures of 100°F to 150°F and summer supply water temperatures of 40°F to 55°F.

Almost all of the components used in a dual-temperature system are the same as those used in the individual heating or chilled-water systems. The main differences are in the system piping and the valves or other system controls used to

select heating, cooling, or both. These differences, along with piping systems, are covered in the next section.

7.0.0 ◆ COMMERCIAL WATER PIPING SYSTEMS

Water piping systems provide for the routing and distribution of hot water, chilled water, or both. The three general classifications of piping systems are two-pipe systems, three-pipe systems, and four-pipe systems. However, combinations of all three types can exist in any particular water system.

7.1.0 Two-Pipe Systems

Two-pipe systems have terminals connected across separate supply and return main piping. Two-pipe systems are classified as either direct-return or reverse-return systems.

7.1.1 Two-Pipe, Direct-Return Systems

In the direct-return system (*Figure 50*), the supply water and return water flow in opposite directions. Also, the return water from each terminal takes the shortest path back to the boiler. This means the water flowing through the terminal nearest to the boiler is the first back to the boiler, because it has the shortest run. The water flowing to the terminal farthest away from the boiler has the longest run and is the last to return. This arrangement results in an unequal length of travel for water flow through the terminals. This causes an imbalance in the system, with the result being poor heat distribution. Circuit balancing valves are used with direct-return systems to aid in balancing the system. Although circuit balancing valves will likely be found in all commercial heating/cooling piping systems, improper adjustment of one or more valves in the direct-return system will likely be far more obvious in terminal performance and comfort.

7.1.2 Two-Pipe, Reverse-Return Systems

In the reverse-return system (*Figure 51*), the supply water and return water flow in the same direction through parallel pipe runs. In this system, the supply water going to the first terminal is the last to be returned to the boiler. The supply water going to the last terminal is the first to be returned to the boiler. This arrangement equalizes the distance of the water flow in the system, thus providing a balanced system. Because the reverse-return system is more easily balanced, it is used more often than the direct-return system.

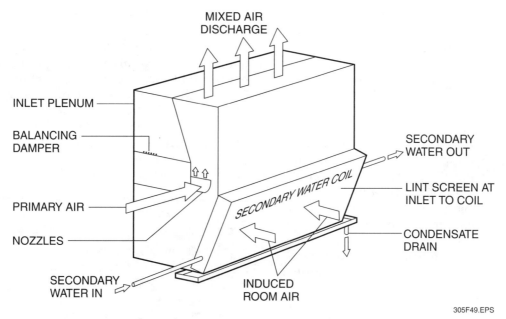

MIXED AIR
DISCHARGE

INLET PLENUM

BALANCING
DAMPER

PRIMARY AIR

NOZZLES

SECONDARY
WATER IN

SECONDARY
WATER OUT

LINT SCREEN AT
INLET TO COIL

CONDENSATE
DRAIN

INDUCED
ROOM AIR

SECONDARY WATER COIL

305F49.EPS

Figure 49 ◆ Air-water induction unit.

7.1.3 Two-Pipe, Dual-Temperature Systems

A two-pipe system is often used in dual-temperature systems to distribute hot water for heating, or chilled water for cooling, to common air handling terminals. In these systems, both the boiler and chiller are connected through valves or other controls to the common two-pipe network. During cold weather, hot water is circulated to all terminals in the circuit or zone. During hot weather, chilled water is circulated to all the terminals. Changeover from hot to chilled water can be for the entire circuit, or for a single zone in zoned systems. Depending on the complexity of the system, changeover can be initiated manually or automatically.

Although the installation of two-pipe systems is a less expensive approach than four-pipe systems, it is important to note that the changeover from one mode to the other can be quite problematic. Hot water introduced to a chiller may cause over-pressurization and opening of the refrigerant pressure relief device. In addition, the chiller compressor may attempt to operate in an overloaded condition. Cold water introduced to some

boilers may cause cracking of cast-iron sections or steel welds. Chillers and boilers can sustain serious damage from the introduction of water outside of their normal temperature range. For that reason, there are often periods of time where the changeover cannot be made, especially during the change of seasons. For example, a commercial building may require heating in the morning and cooling by mid-day, as a result of solar gain, indoor heat sources, and the occupants themselves. If the heating water temperature in the loop remains above a safe temperature for introduction to the chiller, then the changeover must be postponed until the water in the piping cools naturally. Systems are rarely equipped with any

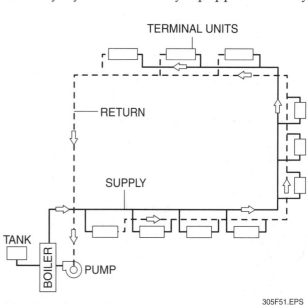

TERMINAL UNITS

RETURN

SUPPLY

TANK

BOILER

PUMP

305F51.EPS

Figure 51 ◆ Two-pipe, reverse-return system.

TANK

BOILER

SUPPLY TERMINAL

PUMP

RETURN

305F50.EPS

Figure 50 ◆ Two-pipe, direct-return system.

alternate means of dissipating this heat, and discomfort often occurs while the changeover process awaits a natural temperature change.

Some two-pipe, dual-temperature systems are used with air handling units that allow up to 100 percent outdoor air to enter the building. In these systems, cooling is accomplished during the summer with chilled water. During the spring and fall seasons, outdoor air may be introduced while hot water is supplied to the system. This method allows for simultaneous heating or cooling as long as the outdoor temperature is below the chilled-water cooling level temperature.

Two-pipe, dual-temperature systems are often zoned so that hot water can be supplied to terminals in one zone, while chilled water is supplied to terminals in another zone. The more zones used, the greater the system's ability to meet changing loads. Changeover and flow of hot or chilled water in a zoned system can be accomplished in many ways. The most common way is to use two-way or three-way zone valves and/or zone pumps. *Figure 52* shows a typical two-zone system in which the selection of hot or chilled water in each zone depends on the positions of the three-way zone valves. The selected hot or chilled water is circulated by the hot-water or chilled-water pump.

7.2.0 Three-Pipe Systems

The three-pipe system is a dual-temperature piping system. It uses two supply pipes and one return pipe. One supply pipe carries hot water and the other carries chilled water. *Figure 53* shows a simplified three-pipe system. In this system, a three-way valve at the input to each terminal is used to select either hot or chilled water for input to the terminal coil. Hot water is circulated only during the heating cycle and chilled water only during the cooling cycle. The hot water and chilled water are not mixed in the coil. The selected hot or chilled water is circulated by the hot-water or chilled-water pump.

WARNING!

Three-pipe systems, in addition to being very inefficient, also provide the greatest threat to major central plant components. The major risk occurs when a rapid weather change causes a temperature swing that can send the unit into the cooling mode right after being in heating, and vice versa. The potential exists for dangerously hot water to be introduced to the chiller from the system common return leg. This can easily burst the chiller refrigerant rupture disc. Additionally, the introduction of chilled water to the boiler can easily shock the boiler and potentially damage the boiler pressure vessel. Typically, controls are in place to help prevent these occurrences, but in the event of a control failure, the risk is high. It is for this reason, as well as poor energy performance, that these systems should not be installed in new or retrofitted systems.

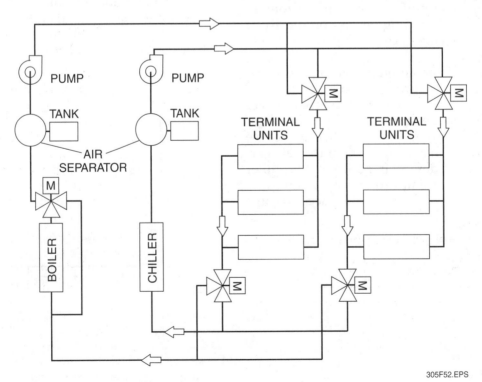

305F52.EPS

Figure 52 ◆ Two-pipe, dual-temperature water system.

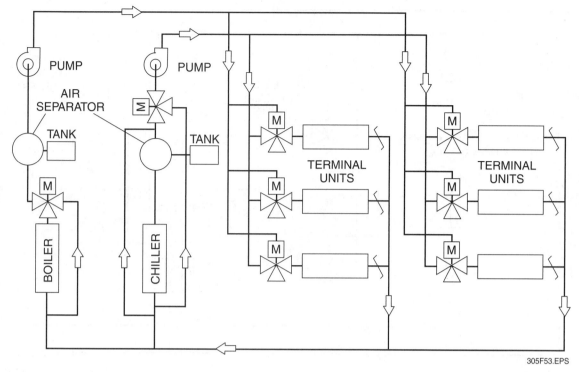

Figure 53 ◆ Three-pipe water system.

7.3.0 Four-Pipe Systems

The four-pipe system is a dual-temperature piping system. Because of its design simplicity, room control, and operational economy, it is commonly used in large commercial applications. It is constructed of two, two-pipe circuits that use separate supply and return lines to carry hot and chilled water to the same terminals. One circuit carries the hot water, while the other carries the chilled water. *Figure 54* shows a simplified four-pipe system. No water is mixed in the four-pipe system because the cold water returns to the chiller and the hot water to the boiler.

In this system, three-way valves are used at the input and output of each single-coil terminal. These valves are used to select either hot or chilled water for input to and output from the terminal coil. The system also uses double coil-terminals. These terminals have separate coils for heating and cooling. This allows heating and cooling to occur at the same time, thus providing for humidification and reheating of the circulated air. Two-way valves are used at the input to each of the double-coil terminals. These valves control the flow of hot and chilled water to the heating and cooling coils in the terminal.

The hot- and chilled-water systems are separate, so pressure regulation is less critical in the four-pipe system than in other piping systems. Also, much higher hot water supply temperatures

are possible, especially when separate heating coils are used in the terminals. This allows a reduction in the size of the piping in that circuit. Another advantage of the four-pipe system is that other types of heating terminals, such as baseboard and finned-tube units, can be connected to the system in spaces that do not need cooling.

7.4.0 Primary-Secondary Water Systems

Primary-secondary hot-water or chilled-water systems are used in applications that have peak and varying loads. They supply variable-flow, constant-temperature chilled/hot water using methods that provide for energy conservation. *Figure 55* shows a simplified diagram for a primary-secondary chilled-water system. For a hot-water system, the chillers would be replaced by boilers. As shown, a primary chiller loop for this system consists of two 500-ton chillers and one 250-ton chiller, each with a related circulating pump. Each of the pumps is sized to handle only the low head and steady gpm for the related chiller. The building load is divided into three secondary loops, each with a circulating pump, terminal devices, and control valves. Again, each of the pumps in the secondary loops is sized to handle the flow and pressure requirements for the related loop. In this system, the primary loop

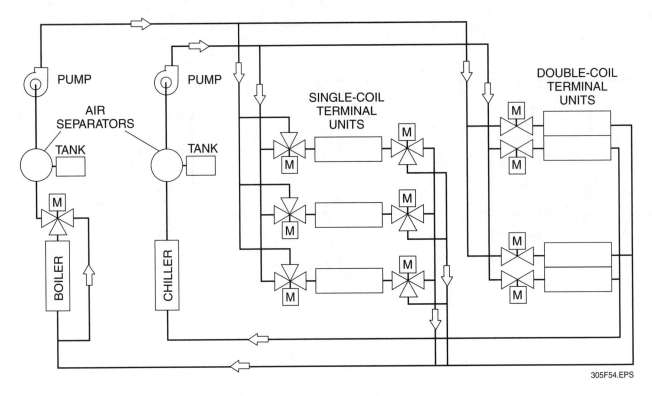

Figure 54 ◆ Four-pipe water system with one-coil and two-coil terminals.

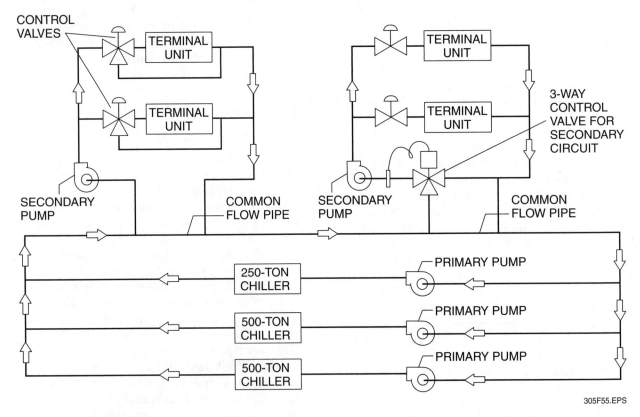

Figure 55 ◆ Simplified primary-secondary chilled-water system.

Typical System Parameters

The following system parameters are typical for chilled-water systems using R-22 refrigerant chillers. All values are based on 67°F wet-bulb air entering the evaporator coil.

Air-Cooled Chiller Operation

Outside ambient air temperatures (°F)	75	85	95
System head pressure (psig)	213	246	280
System suction pressure (psig)	63	66	69

Water-Cooled Chiller Operation

Outside dry-bulb air temperatures (°F)	75	85	95
Outside wet-bulb air temperatures (°F)	65	73	80
Tower basin temperature (°F)	73	81	88
System head pressure (psig)	170	198	213
System suction pressure (psig)	63	66	69

Chilled-Water Temperatures

Supply water leaving chiller: 42°F to 48°F, typically 45°F
Temperature differential (return water – leaving water): 5°F to 10°F, typically 8°F
Freezestat setting (in leaving water): 35°F to 38°F
Leaving chilled-water control thermostat setting: 44°F to 48°F
Return chilled-water control thermostat setting: 42°F to 45°F

Condenser Water Temperatures

Temperature differential (condenser out – condenser in): 5°F to 10°F, typically 7½°F

Note: A range of less than 5°F through the condenser with normal head pressure usually indicates a dirty condenser. A range greater than 10°F usually indicates restricted water flow.

The following system parameters are typical for a hot-water heating system:

- Boiler water temperatures (cold weather) – Leaving water 180°F, return water 160°F
- Boiler water temperatures (mild weather) – Leaving water 160°F, return water 140°F
- Boiler water temperatures (minimum setting) – Leaving water 140°F, return water 120°F

pumps circulate the water through the primary loop; the secondary loop pumps move water through the secondary loops. The transfer of chilled water between the primary loop and each of the secondary loops is made by a short section of pipe common to both the primary and secondary circuits. For this reason, it is called the common pipe.

During periods of peak load, all three chillers in the primary loop operate to produce 45°F water for application to the secondary loops. In each of the secondary loops, the pump is turned on and calling for maximum cooling (maximum flow). During periods of minimum load, only the 250-ton chiller and its pump need to operate in order to supply 45°F water to the secondary loads. With variable-speed pumps used in the secondary loops, each of the secondary pumps may be operating differently to satisfy the individual loop flow requirements. For cooling loads, anywhere between the maximum and minimum

loads, different combinations of chillers and pumps are brought on line in the primary and secondary loops.

8.0.0 ◆ WATER SYSTEM BALANCING

The basic concepts regarding water balance, as well as the calculation of needed flow rates and pumping requirements, were presented in the *HVAC Level Two* module, *Introduction to Hydronic Systems*. Here, we will discuss the process of performing a hydronic system balance procedure.

Testing, balancing, and adjusting hydronic heating or cooling systems can best be accomplished by following a systematic procedure. The entire process should be carefully planned and performed. Scheduling all activities, including organization, procurement of test instruments and equipment, and execution of the balancing procedure, is a must. Coordinate air balancing with water balancing, because many systems function differently on a seasonal basis and temperature performance is a significant factor when testing and balancing.

Before attempting to balance a water system, the boiler and/or chiller and related components should have been made ready for operation in accordance with the manufacturer's instructions. Normally, the system startup procedures include making sure that the chiller is correctly charged with refrigerant and that the boiler and chiller water systems are correctly filled with appropriately treated water. They also include making sure that all chiller/boiler related safety and operating controls have been tested, adjusted, and set for proper operation.

8.1.0 Pre-Balance Checks

Before beginning the balancing procedure for a selected water system, complete the following pre-balancing procedure:

Step 1 Obtain all technical data, including original plans and specifications or as-built plans pertaining to the hydronic equipment. This includes:

- Pump curves and performance data
- Pressure, flow-temperature characteristics, and other performance data for boilers, chillers, heat exchangers, and coils
- Motor nameplate data and ratings
- Performance characteristics of balancing and flow-meter devices

Step 2 Locate and study the following diagrams:

- Temperature control and control valve characteristics
- Special piping circuits

Step 3 Locate and record the ratings and settings of pressure-reducing valves, pressure-relief valves, and temperature-relief valves.

Step 4 Locate the operating and maintenance instructions for important equipment.

Step 5 Prepare a schematic layout of each system, if necessary. Identify connection points of primary and secondary piping circuits, manual valves, automatic valves, flow meters, air vents, coils, and expansion tanks. Note any changes in the design layout that were made during construction. Number all water-flow balance points and metering points for rapid identification and reporting.

Step 6 Study the water system piping arrangement to determine the steps and sequence required for balancing and the instrumentation needed for the task. Note system features that would contribute to an unbalanced condition, such as dirty fins on finned-tube terminals. Determine which circuits must be balanced by temperature rather than fluid flow due to the absence of flow-regulating devices. Note the location of strainers and other piping configurations that may contribute to the collection of dirt and sludge.

Step 7 Obtain all necessary instruments and calibration data. If more than one instrument of a similar type is used, calibrate each one precisely.

Step 8 Make a complete visual inspection of the system and make sure that all air distribution equipment and controls are functioning properly. Do the same for hydronic equipment and controls.

8.2.0 Water System Balancing Procedure

Proper water balancing is essential. A general procedure that can be used for testing, balancing, and adjusting a water system is as follows:

Step 1 Make sure that all preliminary visual checks and service and maintenance tasks have been performed.

 WARNING!
Watch out for rotating, pressurized, or hot components. Follow all safety instructions labeled on the equipment and given in the manufacturer's service manual.

Step 2 Check all electrical wiring in power and control circuits.

Step 3 Check that all manual valves are open or properly preset.

Step 4 Verify that all automatic valves are in the fully open position.

Step 5 Look for broken thermometers and gauges on the equipment.

Step 6 Read and record the system pressure at the pump with the pump de-energized.

Step 7 Operate the system, bleed the air from the piping, and allow flow conditions to become stabilized.

Step 8 Record the operating voltage and amperage draw of all electrical motors.

Step 9 Record the pump speed (rpm).

Step 10 Slowly close the balancing cock in the pump discharge line with the pump running and record the shutoff discharge and suction pressures at the pump gauge connections.

Step 11 Use the shutoff head to determine the pump operating curve (impeller size) and compare it with the design curve data. Check the pump differential at the pump. Use one gauge to read the differential pressure. If two gauges are used, the gauge readings must be corrected to the centerline elevation of the pump. If there is a discrepancy in the pump output as compared to the design specifications, either correct the velocity head or report the information to the proper authority.

Step 12 Record the pump discharge pressure by opening the discharge balancing cock slowly to the fully open position. Record the suction pressure and total head in feet at the same time.

Step 13 Using the total head and the pump curve established in Step 11, read the water flow. If the total head is higher than design specifications, the water flow will be less than design. If the total head is less than design, water flow will be more than the selected rating. If the water flow is more than the selected rating, the pump discharge pressure should be increased by partially closing the balancing cock until water flow is approximately 110 percent of design specifications.

Step 14 Start any secondary pumps and readjust the balancing cock in the primary circuit pump discharge if necessary. Record the new adjusted readings.

Step 15 Take a complete set of pressure-drop readings through all equipment such as chillers, boilers, hot-water coils, and chilled-water coils before adjusting any balancing cocks at the equipment. Compare the readings with design specifica-

tions. Determine which units are above and which units are below design requirements.

Step 16 Make a preliminary adjustment in the balancing device on all the units with high water flow. Adjust them to approximately 10 percent above the design flow rate.

Step 17 Perform another complete set of pressure and amperage draw readings in the pump system. If the pump total flow has fallen below design, open the balancing cock and the pump discharge to raise the pump output, bringing the flow rate within 105 percent to 110 percent of design.

Step 18 Make a second adjustment on the balancing devices on all units that are more than 10 percent above design output. Set them down to increase flow through those units that are under design flow.

Step 19 Keep repeating the procedure (adjusting unit flow devices then pump output) until all units are within ±10 percent of the design water flow.

Step 20 Make a final check of the water flow rates at the pump and equipment and of the pump motor amperage draw and record the findings.

Step 21 After all the balancing is completed, mark all balancing cocks, gauges, and thermometers at the final settings or range of operation for future reference.

Step 22 Check all water flow safety shutdown controls for proper action.

Step 23 Prepare and submit all necessary reports.

These steps are the basic procedures that can be used for balancing regular water (hydronic) systems. However, special types of hydronic distribution systems, such as condenser water and cooling tower systems, steam and hot-water boilers, and heat exchangers will require additional procedures by certified personnel.

1. The difference in weight of a cubic foot of water as its temperature is raised from its freezing point to its boiling point _____.
   a. is roughly 2.5 pounds
   b. is roughly 14.7 pounds
   c. is roughly 62.4 pounds
   d. cannot be determined

2. Assume a hydronic piping system was originally designed for a flow rate of 50 gpm with an initial pressure drop of 6 psi. With an increase in flow rate to 100 gpm, the new pressure drop would be _____ psi.
   a. 2
   b. 12
   c. 24
   d. 72

3. Low-pressure hot-water boilers are limited to a maximum operating temperature of _____°F.
   a. 180
   b. 212
   c. 250
   d. 500

4. One characteristic the copper-finned tube atmospheric boiler offers compared to other commercial types is _____.
   a. the availability of a greater total heating capacity
   b. an extended life
   c. that they cannot be installed outdoors
   d. that they are very lightweight

5. A boiler design that allows the heating capacity to be increased by adding more sections is called a _____ boiler.
   a. firetube
   b. watertube
   c. cast-iron
   d. vertical tubeless

6. In steel firetube boilers, _____.
   a. flue gases are contained in the tubes, with heated water on the outside
   b. heated water circulates through the tubes, with the flue gases on the outside
   c. cast-iron sections contain the flue gases
   d. the water and the flue gases both travel through separate tubes

7. Which type of boiler generally uses a water flow switch?
   a. Steel firetube
   b. Steel watertube
   c. Copper-finned tube
   d. Cast-iron

8. Commercial air separators allow air to separate from the water stream by significantly reducing the water _____.
   a. pressure
   b. velocity
   c. temperature
   d. flow volume

9. In some centrifugal pumps, the packing gland _____.
   a. reduces pump shaft friction
   b. attaches the impeller to the shaft
   c. stores lubricating oil
   d. exerts force on the packing to control fluid leakage around the shaft

10. Multi-purpose valves are commonly installed _____.
    a. on the inlet side of the a pump
    b. on the discharge side of a pump
    c. at the inlet of each terminal device
    d. on the boiler makeup water line

11. The chilled water produced by a chiller is considered to be _____.
    a. a secondary coolant
    b. the primary coolant
    c. a brine solution
    d. more difficult to pump

12. The purpose of a purge unit on a low-pressure chiller is to _____.
    a. provide additional subcooling of the refrigerant
    b. automatically maintain the proper refrigerant charge at all times
    c. remove noncondensibles from the refrigerant circuit
    d. remove the remaining air from the chilled water

13. Evaporative condensers differ from cooling towers in that _____.

    a. they do not use water for cooling
    b. cooling towers do not transfer heat through evaporation
    c. the refrigerant condenser itself is contained inside the evaporative condenser
    d. they cannot be used with most refrigerants

14. A fan coil unit can typically produce up to _____ cfm of airflow.

    a. 100
    b. 2,000
    c. 5,000
    d. 10,000

15. The minimum number of pumps needed in a primary-secondary water piping system is _____.

    a. 1
    b. 2
    c. 3
    d. 4

Summary

Residential and commercial water systems (hydronic systems) use pipes to transport heated or cooled water from its source to where it is needed. The term *hydronic* refers to all water systems. An air conditioning system that circulates chilled water to cooling coils is known as a chilled-water system.

Hot-water heating systems are used mainly to provide comfort heating. They can be separate systems used to heat buildings that have separate or no cooling systems. They can also be used in dual-temperature systems. Commercial hot water systems are of the forced hot-water type, because mechanical pumping is generally necessary to overcome the friction loss and length of large piping systems. A boiler is a heat exchanger that uses radiant heat and hot flue gases to generate hot water for heating and process loads. Low-pressure, hot-water boilers can have operating pressures up to 160 psi, but their operating temperatures must not exceed 250°F. The main differences between hot-water and steam boilers are in the operating and safety controls and accessories. Also, hot-water boilers are filled to capacity with water, while steam boilers are not.

Chilled-water systems eliminate the need to circulate refrigerant through refrigerant system components installed in a building in order to provide comfort cooling. Instead, a chiller uses refrigerant to cool the water. Then, the chilled water is circulated throughout the building and back to the chiller where the process is repeated.

Chillers are either the mechanical compression type or the absorption type. Mechanical compression chillers may use reciprocal, centrifugal, or screw-type compressors. The compressor produces the pressure differential needed to maintain the refrigeration cycle. Absorption chillers use a heat and chemical process involving water as a refrigerant, and lithium bromide as an absorbent to produce the pressure differential for the refrigeration cycle. Both types of chillers evaporate and condense the refrigerant liquid.

Dual-temperature water systems circulate both hot and chilled water to heat or cool with common piping and heat transfer terminals.

Piping in two-pipe, three-pipe, or four-pipe systems, or a combination of all three configurations, can be used in hydronic systems to provide for the routing and distribution of hot water, cold water, or both.

In hydronic comfort systems, water balancing means the proper delivery of hot water, chilled water, or both, in the correct amounts to each of the areas in a structure being conditioned. The satisfactory distribution of conditioned water depends upon a well-designed piping system and properly chosen water system components. Balancing the water system requires the adjustment of the flow controls so that the right amount of hot or chilled water is circulated in the required spaces at the proper velocity to provide satisfactory heating and cooling.

Notes

Trade Terms Introduced in This Module

Boiler horsepower (Bohp): One Bohp is equal to 9.803 kilowatts. Also, it is the heat necessary to vaporize 34.5 pounds of water per hour at 212°F. This is equal to a heat output of 33,475 Btus per hour or about 140 square feet of steam radiation.

Cavitation: The result of air formed due to a drop in pressure in a pumping system.

Firetube boiler: A boiler in which the flue gases are contained inside the tubes, with the heated water on the outside.

Flash economizer: An energy-conservation process used in chillers that accomplishes inter-cooling of the refrigerant between the condenser and evaporator using a flash chamber or similar device. Flash gas of coolant produced in the flash chamber flows directly to the second stage of the chiller compressor, bypassing the evaporator and the first compressor stage.

Flow meter: A venturi tube or orifice plate measuring device installed in a water or other fluid system for the purpose of making pressure drop and velocity measurements.

Head pressure: A measure of pressure drop, expressed in feet of water or psig. It is normally used to describe the capacity of circulating pumps. It indicates the height of a column of water that can be lifted by the pump, neglecting friction losses in piping.

Hydronic system: A system that uses water or water-based solutions as the medium to transport heat or cold from the point of generation to the point of use.

Latent heat of vaporization: The heat that is gained in changing from a liquid to a gas (water to steam).

Lithium bromide: A salt solution commonly used in air conditioning absorption chillers.

Pressure drop: The difference in pressure between two points. In a water system, it is the result of power being consumed as the water moves through pipes, heating units, and fittings. It is caused by the friction created between the inner walls of the pipe or device and the moving water.

Secondary coolant: Any liquid, such as water, that is cooled by a system refrigerant and then used to absorb heat from the conditioned space without a change in state. Also known as indirect coolant.

Static pressure: In a water system, static pressure is created by the weight of the water in the system. It is referenced to a point such as a boiler gauge. Static pressure is equal to 0.43 pounds per square inch, per foot of water height.

Superheated steam: Steam that has been heated above its saturation temperature.

Thermal economizer: An energy-conservation process used in chillers in which warm condensed refrigerant is brought into contact with the coldest (inlet) water tubes in a condenser or heat exchanger. This causes the condensed refrigerant to subcool to a temperature below the condensing temperature.

Venturi tube: A short tube with flaring ends and a constricted throat. It is used for measuring flow velocity by measurement of the throat pressure, which decreases as the velocity increases.

Watertube boiler: A boiler where the heated water is contained inside the tubes, with the flue gases on the outside.

Additional Resources and References

Additional Resources

This module is intended to be a thorough resource for task training. The following reference works are suggested for further study. These are optional materials for continued education rather than for task training.

Air-Cooled Chillers, TDP-622, Latest Edition. Syracuse, NY: Carrier Corporation.

Refrigeration and Air Conditioning, An Introduction to HVAC/R, Latest Edition. Larry Jeffus. Upper Saddle River, NJ: Prentice Hall.

Water-Cooled Chillers, TDP-623, Latest Edition. Syracuse NY: Carrier Corporation.

Figure Credits

NCCER CURRICULA — USER UPDATE

NCCER makes every effort to keep its textbooks up-to-date and free of technical errors. We appreciate your help in this process. If you find an error, a typographical mistake, or an inaccuracy in NCCER's curricula, please fill out this form (or a photocopy), or complete the online form at **www.nccer.org/olf**. Be sure to include the exact module ID number, page number, a detailed description, and your recommended correction. Your input will be brought to the attention of the Authoring Team. Thank you for your assistance.

Instructors – If you have an idea for improving this textbook, or have found that additional materials were necessary to teach this module effectively, please let us know so that we may present your suggestions to the Authoring Team.

NCCER Product Development and Revision

13614 Progress Blvd., Alachua, FL 32615

Email: curriculum@nccer.org
Online: www.nccer.org/olf

❏ Trainee Guide   ❏ AIG   ❏ Exam   ❏ PowerPoints   Other _____

Craft / Level: _____   Copyright Date: _____

Module ID Number / Title: _____

Section Number(s): _____

Description: _____

_____

_____

_____

_____

Recommended Correction: _____

_____

_____

_____

_____

Your Name: _____

Address: _____

_____

Email: _____   Phone: _____

03306-08

Steam Systems

03306-08
Steam Systems

Topics to be presented in this module include:

1.0.0   Introduction 6.2
2.0.0   Fundamentals and Properties of Water 6.2
3.0.0   Steam Cycle Principles of Operation 6.5
4.0.0   Steam Boilers, Boiler Controls, and Accessories 6.6
5.0.0   Valves 6.13
6.0.0   Heat Exchangers/Converters 6.16
7.0.0   Terminals 6.18
8.0.0   Steam Traps and Strainers 6.18
9.0.0   Troubleshooting Steam Traps 6.25
10.0.0   Condensate Return/Feedwater System Components . . .6.27
11.0.0   Flash Tanks 6.30
12.0.0   Steam System Piping 6.30
13.0.0   Steam and Condensate Pipe Sizing 6.35
14.0.0   Boiler Blowdown and Skimming 6.39
15.0.0   Boiler Water Treatment 6.41

Overview

Steam is the most widely used form of energy for industrial use in the world. Water itself is generally plentiful, as well as inert, making it a desirable compound to work with in many applications. The energy stored within water, once it is has been sufficiently heated to the point of vaporization and beyond, can be five to six times greater than liquid water can generally carry. Coupled with the fact that pumps are not required to move steam from the point of creation to the intended load, it easily becomes the best means of storing and moving large amounts of energy in a variety of applications.

Steam plays an important role in the heating of large industrial and multi-building applications. Anyone working in such environments must be knowledgable about the generation and control of steam.

Objectives

When you have completed this module, you will be able to do the following:

1. Explain the terms and concepts used when working with steam-heating systems.
2. Identify the major components of steam-heating systems.
3. Explain the purpose of each component of steam-heating systems.
4. Describe the basic steam-heating cycle.
5. Safely perform selected operating procedures on low-pressure steam boilers and systems.
6. Install and maintain selected steam traps.
7. Identify the common piping configurations used with steam-heating systems.
8. Identify the types of common piping configurations used with steam-heating systems.
9. Safely perform selected operating procedures on low-pressure steam boilers and systems.
10. Install and maintain selected steam traps.
11. Identify the types of common piping configurations used with steam-heating systems.

Trade Terms

Blowdown
Drip leg
Dry steam
Flash steam
Latent heat
Latent heat of condensation
Latent heat of fusion
Pressure drop
Saturated steam

Sensible heat
Skimming
Specific heat
Subcooling
Superheat
Superheated steam
Trim
Turndown ratio
Water hammer

Required Trainee Materials

1. Pencil and paper
2. Appropriate personal protective equipment

Prerequisites

Before you begin this module, it is recommended that you successfully complete *Core Curriculum*; *HVAC Level One*; *HVAC Level Two*; and *HVAC Level Three*, Modules 03301-08 through 03305-08.

This course map shows all of the modules in the third level of the *HVAC* curriculum. The suggested training order begins at the bottom and proceeds up. Skill levels increase as you advance on the course map. The local Training Program Sponsor may adjust the training order.

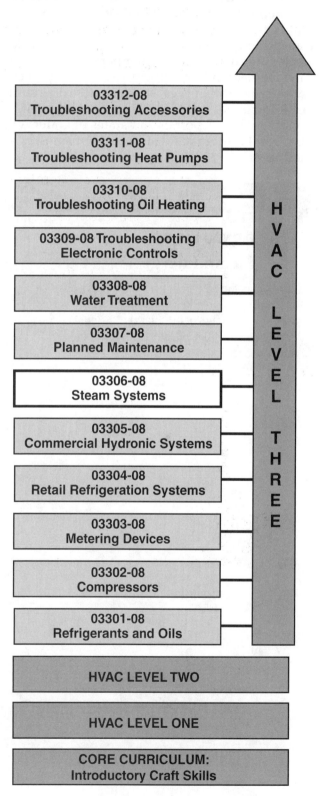

03312-08
Troubleshooting Accessories

03311-08
Troubleshooting Heat Pumps

03310-08
Troubleshooting Oil Heating

03309-08 Troubleshooting
Electronic Controls

03308-08
Water Treatment

03307-08
Planned Maintenance

03306-08
Steam Systems

03305-08
Commercial Hydronic Systems

03304-08
Retail Refrigeration Systems

03303-08
Metering Devices

03302-08
Compressors

03301-08
Refrigerants and Oils

HVAC LEVEL TWO

HVAC LEVEL ONE

CORE CURRICULUM:
Introductory Craft Skills

HVAC LEVEL THREE

306CMAP.EPS

1.0.0 ◆ INTRODUCTION

Steam has a wide variety of uses in the commercial and industrial sectors. With the ability to carry large quantities of stored energy, steam can be applied in comfort heating, process heating, mechanical drive, pressure control, and a variety of other applications. The majority of heat content is stored in its latent form, so steam can be transported efficiently at low cost. It is generally a safer alternative to high-temperature water for carrying heat in large quantities. Although there are certainly differences, steam shares some similarities with hot water systems and equipment. A review of the properties and states of water will help you to understand steam systems and how they operate.

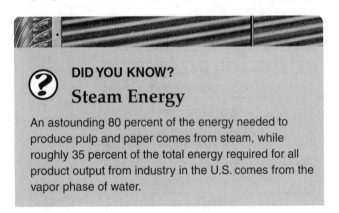

DID YOU KNOW?

Steam Energy

An astounding 80 percent of the energy needed to produce pulp and paper comes from steam, while roughly 35 percent of the total energy required for all product output from industry in the U.S. comes from the vapor phase of water.

2.0.0 ◆ FUNDAMENTALS AND PROPERTIES OF WATER

Depending on its heat content and temperature, water can exist in three states: ice (solid), water (liquid), and water vapor (gas, such as steam).

When water changes from one state to another, some interesting things occur. *Figure 1* shows this graphically for water at sea level. If heat is added to one pound of ice at 0°F, a thermometer will show a rise in temperature until the reading reaches 32°F. This is the point at which the ice starts changing into water. If heat is continually added, the thermometer reading remains fixed at 32°F instead of rising. It continues to read 32°F until the entire pound of ice is melted. The increase in temperature from 0°F to 32°F registered by the thermometer is **sensible heat**. Sensible heat is heat that can be sensed by a thermometer or by touch. The heat that was added to the ice and caused its change in state from a solid to a liquid, but did not register on the thermometer, is **latent heat**. Latent heat is the heat energy absorbed or rejected when a substance is changing state (solid to liquid, liquid to gas, or vice versa) without a change in the measured temperature.

If you continue to add heat to the pound of water, the thermometer will once again show an increase in temperature until the temperature reaches 212°F. At this point, the water starts

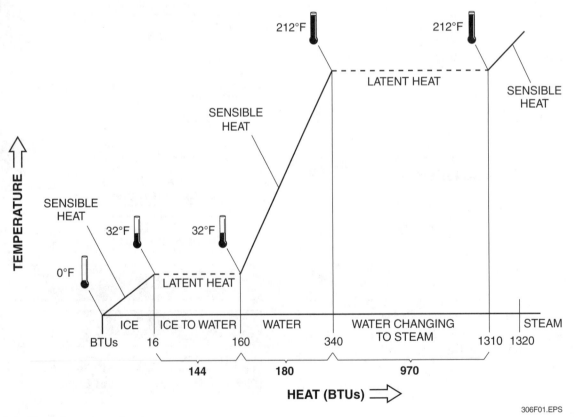

Figure 1 ◆ Changing states of water.

boiling and changes state from water into steam (water vapor). As even more heat is added, the thermometer reading remains at 212°F until all the water has turned into steam. If you continue to add heat, the thermometer will once again register sensible heat.

As shown in *Figure 1*, it takes a great deal more heat to cause a change of state in water than is needed for a degree change in its temperature. It requires 144 Btus of latent heat to melt the pound of ice before the temperature began to rise. This is 144 times as much heat as is needed to raise the temperature of water one degree (**specific heat**). The change from water to steam requires an even greater amount of latent heat. It takes 970 Btus to change the water to steam. None of the latent heat consumed during the change of state of water registers on the thermometer. Once water reaches the threshold of changing states and begins the change process, the thermometer becomes stable and unchanged until the phase change is complete. This property of latent heat helps to explain

why large amounts of energy can be stored and transported through the use of steam.

Figure 2 shows the common terms used in steam systems when defining the latent heat added or removed in changing water to and from the solid, liquid, or vapor states. It also shows the terms **subcooling** and superheat as they relate to sensible heat changes in water. These terms can be applied to steam in the same manner as they are applied in the refrigeration circuit.

- *Latent heat of fusion* – The heat gained or lost in changing ice to water or water to ice.
- *Latent heat of vaporization* – The heat gained in changing water to steam at a given pressure. The temperature at which this occurs is known as the boiling point or the saturation temperature. After all the water has been vaporized into pure steam at the saturation temperature, the steam is referred to as **dry steam** or **saturated steam**.

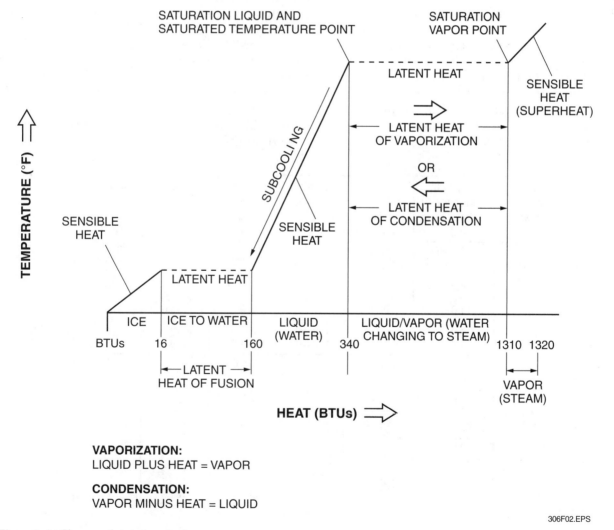

Figure 2 ◆ Change of state terminology.

- *Latent heat of condensation* – The heat given up or removed from steam in changing back to water.

- *Superheat* – The measurable heat added to the steam after all the water has reached its boiling point and completely changed into steam. For example, the saturation temperature of water at atmospheric pressure is 212°F. If the resulting steam is superheated 10°F, the steam temperature is 222°F. Steam that has been heated above the saturation temperature is called **superheated steam**.

- *Subcooling* – The measurable heat that is removed after all the water has condensed from the vapor state to the liquid state. It is the reverse of superheat, and the subcooling process cannot begin until the change of state to liquid is complete. For example, the condensing temperature of water at atmospheric pressure is 212°F. If the water is subcooled 10°F, the temperature is cooled to 202°F.

Figures 1 and *2* show that most of the heat content of steam is stored as latent heat. It is this latent heat (970 Btu/lb at sea level) that makes steam so efficient as a carrier of heat. Large quantities can be transmitted efficiently with little change in temperature. This same amount of latent heat is released from the steam when it is condensed back into water. Steam is pressure- and temperature-dependent. This means that the system can be controlled by varying either the steam pressure or temperature.

2.1.0 Pressure-Temperature Relationships

Steam tables illustrate the pressure-temperature relationships and other properties of saturated steam over a broad range of temperatures. *Figure 3* shows entries from such a table. One entry is for the properties of steam at sea level as just described. The other entries are given for the purpose of providing comparisons.

Columns A through C show the pressure-temperature relationships. Columns A and B show the corresponding gauge and absolute pressures. Absolute pressure is the pressure in pounds per square inch (psia) above a perfect vacuum. Gauge pressure (psig), which starts at zero psig, is pressure in pounds per square inch above atmospheric pressure. Gauge pressure plus 14.7 equals the related absolute pressure.

For each pressure, there is a corresponding temperature. Column C shows that the saturation temperature of steam at sea level (0 psig/14.7 psia) is 212°F. As steam cools down to these conditions, the latent process of condensation begins. If water is heated up to these conditions, the latent process of vaporization is initiated.

Column D shows the heat of saturated liquid. This is the amount of sensible heat required to raise the temperature of a pound of water from 32°F to the boiling point at the pressure and temperature shown. It is expressed in Btu/lb. In some tables, this value is called the enthalpy of the liquid (hf).

	Column A Gauge Pressure	Column B Absolute Pressure PSIA	Column C Steam Temp. °F	Column D Heat of Sat. Liquid btu/lb	Column E Latent Heat btu/lb	Column F Total Heat of Steam btu/lb	Column G Specific Volume cu ft/lb
Inches of Vacuum	29.743	0.08854	32.00	0.00	1075.8	1075.8	3306.00
	29.515	0.2	53.14	21.21	1063.8	1085.0	1526.00
	27.886	1.0	101.74	69.70	1036.3	1106.0	333.60
	19.742	5.0	162.24	130.13	1001.0	1131.1	73.52
	9.562	10.0	193.21	161.17	982.1	1143.3	38.42
	7.536	11.0	197.75	165.73	979.3	1145.0	35.14
	5.490	12.0	201.96	169.96	976.6	1146.6	32.40
	3.454	13.0	205.88	173.91	974.2	1148.1	30.06
	1.418	14.0	209.56	177.61	971.9	1149.5	28.04
PSIG	0.0	14.696	212.00	180.07	970.0	1150.4	26.80
	1.3	16.0	216.32	184.42	967.6	1152.0	24.75
	2.3	17.0	219.44	187.56	965.5	1153.1	23.39
	5.3	20.0	227.96	196.16	960.1	1156.3	20.09
	10.3	25.0	240.07	208.42	952.1	1160.6	16.30
	15.3	30.0	250.33	218.82	945.3	1164.1	13.75
	20.3	35.0	259.28	227.91	939.7	1167.1	11.90
	25.3	40.0	267.25	236.03	933.7	1169.7	10.50
	30.3	45.0	274.44	243.36	928.6	1172.0	9.40
	40.3	55.0	287.07	256.30	919.6	1175.9	7.79
	50.3		297.97	267.50	911.6	1179.1	6.66
	60.3			277.43	904.5	1181.9	5.82
			306.39		897.8		

306F03.EPS

Figure 3 ◆ Example table entries for properties of saturated steam.

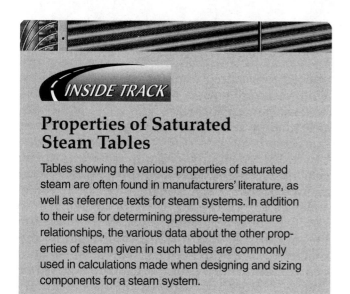

Properties of Saturated Steam Tables

Tables showing the various properties of saturated steam are often found in manufacturers' literature, as well as reference texts for steam systems. In addition to their use for determining pressure-temperature relationships, the various data about the other properties of steam given in such tables are commonly used in calculations made when designing and sizing components for a steam system.

Column E shows the latent heat. This is the amount of heat required to change a pound of boiling water to steam, expressed in Btu/lb. This same amount of heat is released when a pound of steam is condensed back into a pound of water. This heat quantity is different for every pressure-temperature combination. In some tables, this value is called the enthalpy of the evaporation (hfg).

Column F shows the total heat of the steam. This is the sum of the sensible heat and the latent heat given in Columns D and E, expressed in Btu/lb. It is the total heat in the steam above 32°F. In some tables, this value is called the enthalpy of the steam (hg).

Column G shows the specific volume of steam. This is the volume per unit of mass for the steam. It is the space that one pound of steam will occupy at the pressure-temperature shown. It is expressed in cubic feet per pound. Note that, as pressure and the total heat contained within the steam increases on the chart, the volume of the steam decreases.

3.0.0 ◆ STEAM CYCLE PRINCIPLES OF OPERATION

Figure 4 shows the basic components of a simple steam cycle. They include the following:

- Boiler
- Control valve
- Heat exchanger
- Trap
- Deaerator and condensate receiver
- Condensate pump
- Vacuum breaker

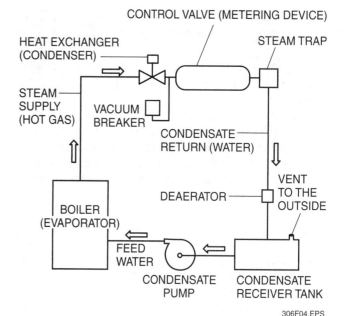

Figure 4 ◆ Basic steam cycle.

The steam cycle begins at the boiler. In the boiler, water is vaporized to make steam. The initial source of energy used for this purpose is typically from the combustion of a fuel such as oil, gas, or coal. Once the water has reached the boiling point and vaporization begins, the pressure begins to rise. Remember that the water temperature required for boiling also increases due to this change in pressure. As shown in *Figure 3*, once the pressure reaches 1.3 psig, the boiling point increases to over 216°F and continues to increase as pressure rises. Steam is found at the highest temperature and pressure in the boiler itself. Steam flows through the supply piping and components as the result of natural movement from an area of higher pressure to an area of lower pressure, much like air through an air distribution system.

The control valve is an automatic, temperature-operated control device. Its purpose is to regulate or meter the flow of steam into the heat exchanger at a rate comparable to the load. This ensures that only the proper amount of steam needed to take full advantage of its stored latent heat is admitted. The steam system is designed so the control valve has the needed residual pressure to assure proper operation of the valve. **Pressure drop** across steam valves must be used to calculate the entering steam pressure to the final user of the steam.

The heat exchanger may be a steam coil in a system airstream, a steam-to-water heat exchanger, or any other type of terminal used with steam systems. Its purpose is to transfer the heat carried in the steam to the conditioned space or other medium. The heat exchanger is like the condenser

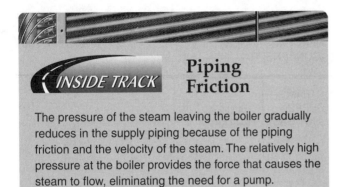

in a refrigeration system, where the steam (gas) is converted back into water (liquid). There, the latent heat and any sensible superheat gained in the boiler by vaporization is rejected to the cooler conditioned space, causing the steam to cool and condense back into water. The condensing process in the heat exchanger occurs at a lower pressure than in the boiler. The amount of pressure depends on the temperature of the medium being heated, the pressure losses in the supply piping, and the pressure drop across the control valve. It is this pressure difference that causes the steam to flow without a pump.

Steam traps are among the most important devices used in the steam cycle. There are a variety of styles and types suited to different applications. Traps also require the most attention and maintenance to ensure their proper operation, because failure of a steam trap can create a number of problems. The trap in the piping at the output of the heat exchanger holds the steam in the heat exchanger until it all condenses. Otherwise, the live steam would pass through the heat exchanger into the return line, wasting the useful heat released as a result of condensation. The trap will allow only liquid condensate and/or noncondensible air, never steam, to flow away from the heat exchanger.

Where elevations permit, steam systems can use a gravity return to route the flow of condensate water directly back to the boiler for re-evaporation into steam. However, most systems use gravity flow to route the condensate into a receiver tank, where it collects and is stored. When the boiler controls call for the water to replenish the amount boiled off, a condensate pump is turned on and produces the necessary pressure differential to pump the condensate (feed water) from the tank to the boiler. This method allows gravity to be used to the extent possible, with the condensate pump doing only the remaining work. Return of the condensate to the boiler for re-evaporation completes the steam cycle.

Steam systems that use a condensate receiver tank and pump in the condensate return line operate at atmospheric pressure, with the return opened to the atmosphere in one or more places. This has an advantage. When steam is applied to heat exchangers that have coils subjected to outside airflow, or cold water such as in domestic hot-water heaters, the pressure in the heating coil can suddenly drop below atmospheric pressure (vacuum range). In extreme cases, this sudden drop in pressure can cause the tubes in the coil to collapse. This can be prevented by allowing air to enter the heat exchanger by way of a vacuum breaker device or through a thermostatic trap that senses the drop in pressure. A small amount of the air that enters the heat exchanger will be absorbed into the condensate, but most of it flows in the return line to the condensate receiver tank. From there it is discharged by vents back into the atmosphere.

4.0.0 ◆ STEAM BOILERS, BOILER CONTROLS, AND ACCESSORIES

Many of the boiler controls and accessories used with steam systems are the same as those used with hot-water boilers and described earlier. Typically, the main difference is that these controls and accessories have different temperature and pressure ratings. For this reason, this section focuses on the controls and accessories that are unique to steam boilers and systems.

4.1.0 Steam Boilers

Steam boilers are divided into three broad categories: those that produce low-pressure steam;

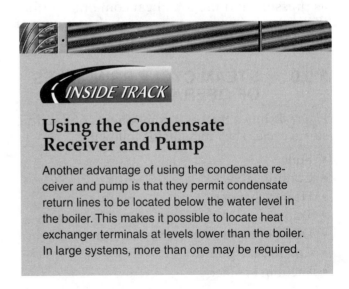

those that produce medium-pressure steam; and those that produce high-pressure steam. Low-pressure steam boilers have working pressures of up to 15 psi (250°F). Medium-pressure boilers operate from 15 psi to 60 psi. High-pressure boilers operate at pressures from 60 psi and up. It is important to note that these ratings indicate the maximum working pressure of the boiler. For safety purposes, the actual operating pressure of the system would be somewhat lower than the maximum. Low-pressure steam boilers (*Figure 5*) are normally used in commercial applications, apartment houses, and single-unit industrial facilities. High-pressure boilers are used mainly to produce process steam for facilities such as college campuses, hospitals, or industrial plants.

With the exceptions that follow, steam boilers operate in the same manner as hot-water boilers:

• Steam boilers are partially filled with water to allow room for steam creation and expansion.

• The pressure-temperature levels at which steam and hot-water boilers operate are different.

• The operating and safety controls are different.

Although larger boilers and higher-pressure boilers are generally designed specifically for steam applications, many low-pressure boilers are identical in construction to those used for hot water applications. For added safety, a number of such boilers are rated at a maximum working pressure of 30 psig for hot water, but have a reduced maximum rating of 15 psig for steam applications. Low-pressure steam boilers often differ from their hot water components only in the **trim**;

that is, the external operating and safety controls or accessories attached to the boiler itself.

Overall, steam boiler designs vary more widely than hot-water designs. Typical designs include the following:

• Cast-iron, often built in matching sections for field assembly

• Steel firetube and watertube boilers

• Steel vertical tubeless boilers

• Electric boilers

Boilers used in residential and light commercial low-pressure applications are generally of cast-iron sectional or one-piece designs. Although many older units were quite inefficient, federal minimum efficiency standards, which took effect in 1992, require boilers to have a minimum annual fuel utilization efficiency (AFUE) rating of 80 percent. Boilers earning the EPA's ENERGY STAR designation must have a minimum AFUE rating of 85 percent. These boilers generally incorporate a number of energy-saving options, such as electronic ignition and sealed combustion systems using outside air for the combustion process. At this writing, the vast majority of high-efficiency steam boilers for residential use are oil-fired, such as the unit pictured in *Figure 6*.

306F05.EPS

Figure 5 ◆ Low-pressure steam boiler.

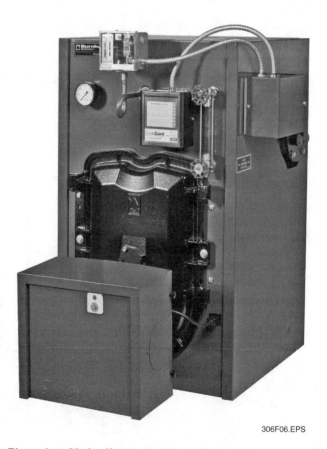

306F06.EPS

Figure 6 ◆ High-efficiency residential steam boiler.

4.2.0 Boiler Operating/Safety Controls

In a well-designed steam system, the boiler is equipped with several operating and safety controls needed to guarantee safe and proper operation of the boiler (*Figure 7*). Operating and safety controls normally used with all steam boilers are:

- Pressure gauge
- Water gauge glass/water column
- Low-water cutoff/water feeder
- High-pressure limit control
- Pressure-relief (safety) valve

4.2.1 Pressure Gauge

The pressure (steam) gauge shows the pressure within the boiler in pounds per square inch gauge (psig). This gauge is installed at the highest point of the boiler's steam space. A siphon assembly is generally installed between the boiler and the gauge. Two different types of siphons are used: pigtail siphons and gauge siphons. Both provide separation for the gauge or other control from the raw steam, which may cause damage to the gauge, as well as erroneous readings. *Figure 8* provides an example of a pigtail siphon. Pigtail

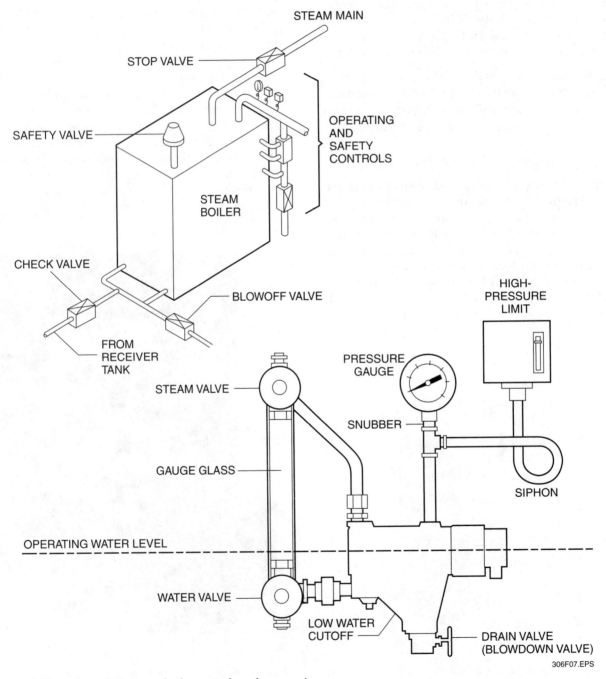

306F07.EPS

Figure 7 ◆ Typical cast-iron steam boiler controls and accessories.

siphons are always installed vertically and are filled with water before installing the gauge.

4.2.2 Water Gauge Glass/Water Column

The water gauge glass is used to check the water or steam level in the boiler (*Figure 9*). The steam level can be checked by opening the steam valve at the top of the water gauge. To check the water level, the water valve at the bottom of the water gauge is opened. Normally, the water gauge glass is connected directly to the boiler. A drain valve (**blowdown** valve) is usually connected to the bottom of the water gauge. When opened, the valve is used to purge the gauge passages of trapped water and any buildup of sludge and sediment.

WARNING!

Blowdown or drain valves opened while a steam boiler is in operation can release water at a temperature above the atmospheric boiling point, causing flash steam. Use extreme caution when using the blowdown valve. The blowdown valve discharge must always be piped to an appropriate drain, and in a manner that will prevent hot water or flash steam from contacting personnel.

NOTE

The terms *bleed-off* and *blowdown* are often used interchangeably, but they apply to different systems. Bleed-off is generally used to define the process of bleeding water from a cooling tower or other hydronic loop to reduce the concentration of dissolved solids in the remaining water. Blowdown is associated with the process of removing dissolved solids and/or sludge from the surface or bottom of a steam boiler. Use of the proper term will ensure clear communication.

306F08.EPS

Figure 8 ◆ Pigtail siphon.

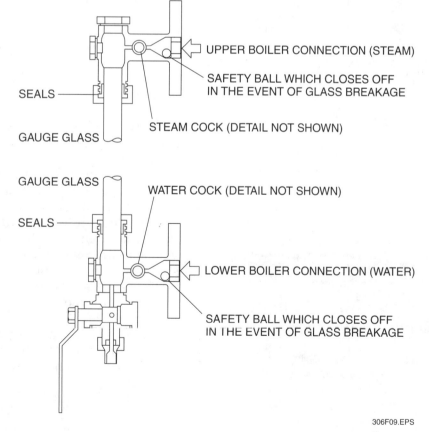

UPPER BOILER CONNECTION (STEAM)

SAFETY BALL WHICH CLOSES OFF IN THE EVENT OF GLASS BREAKAGE

SEALS

STEAM COCK (DETAIL NOT SHOWN)

GAUGE GLASS

GAUGE GLASS

WATER COCK (DETAIL NOT SHOWN)

SEALS

LOWER BOILER CONNECTION (WATER)

SAFETY BALL WHICH CLOSES OFF IN THE EVENT OF GLASS BREAKAGE

306F09.EPS

Figure 9 ◆ Water gauge glass.

On larger boilers, the water gauge may be attached to a water column (*Figure 10*). It diverts water or steam from the boiler for application to the water gauge. This helps to stabilize water levels for more accurate water gauge readings.

Three globe valves on the water column provide a backup method for checking the boiler steam and water levels if the water gauge is damaged. These globe valves are often called gauge cocks or try cocks. Opening the top, middle, or bottom valve allows either steam, a mixture of steam and water, or water to escape. A blowdown valve connected to the bottom of the water column is used to clear the column of trapped water and any buildup of sludge and sediment.

In a steaming boiler, the gauge glass will generally indicate a lower water level than is accurate, as shown in *Figure 11*. This is because the water in the gauge glass is not subject to currents or agitation, it is cooler than the water in the boiler, and it does not contain steam bubbles.

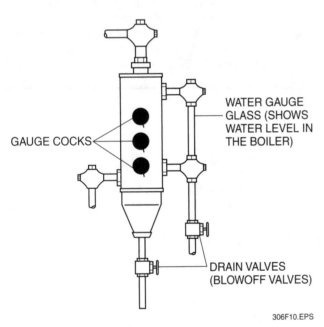

GAUGE COCKS

WATER GAUGE GLASS (SHOWS WATER LEVEL IN THE BOILER)

DRAIN VALVES (BLOWOFF VALVES)

306F10.EPS

Figure 10 ◆ Water column.

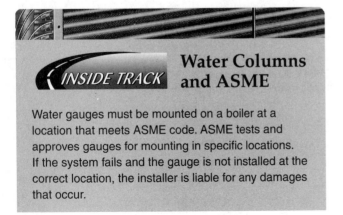

INSIDE TRACK **Water Columns and ASME**

Water gauges must be mounted on a boiler at a location that meets ASME code. ASME tests and approves gauges for mounting in specific locations. If the system fails and the gauge is not installed at the correct location, the installer is liable for any damages that occur.

4.2.3 Low- and High-Water Cutoff and Water Feeder Controls

Low-water cutoff controls provide valuable protection for steam boilers by interrupting power to the heat source when the water falls below the minimum safe level. Water feeders are used to control the flow of makeup water to the boiler. In systems that effectively collect and return steam condensate to the boiler for reuse, the water feeder uses the condensate reservoir and associated pumps as the primary water source. Fresh water, generally from a municipal water supply, becomes the backup source when condensate levels are insufficient. These boilers are generally classified as steam heating boilers. However, steam boilers are also used in many applications where little or no condensate is collected for reuse. In these cases, the water feeder may be quite active, using only the fresh water supply to make up for the lost steam. Boilers used in this manner are classified as steam-process boilers.

Typically, low-water cutoff control for steam boilers is done using the same methods as those for hot-water boilers, employing float-operated controls (*Figure 12*). However, steam systems and boiler operation are adversely affected by a high water level as well. As a result, both low- and high-water cutoff controls are often used. Some steam boilers may be equipped with electronic probe-type controls (*Figure 13*) to provide these functions, much like hot-water boilers.

A marker on the float bowl of the float-operated control used with steam boilers shows the cutoff level for the device. The control is normally installed so that the marker on the float bowl is about ½ inch higher than the lowest visible point in the water gauge glass. For some float-operated controls, the operating level is established automatically by fittings on the device that permit its installation right in the gauge glass tappings of the boiler. If a low-water condition occurs, float-actuated switch contacts in the cutoff control open the electrical circuit to the boiler burner, preventing the burner from firing. Note that water will still be visible in the gauge glass. However, the boiler is shut down and cannot be operated until makeup water is manually added to the boiler. If this situation persists, it can cause a loss of heating until the water in the boiler is returned to a safe level. This problem is usually eliminated by installing a water feeder control between the boiler's supply water line input and its condensate return main line. This control functions to automatically maintain a constant water level in the boiler.

In some boiler installations, the water feeder control is a separate unit (*Figure 14*). There are many kinds of feeder controls. Typically, they use

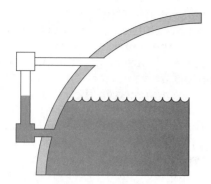

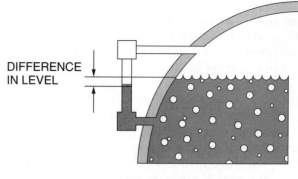

BOILER OFF
NO STEAM BUBBLES, AND THE LEVEL GAUGE GLASS SHOWS THE TRUE WATER LEVEL IN THE SHELL

DIFFERENCE IN LEVEL

BOILER AT HIGH LOAD
MANY STEAM BUBBLES AND A LOWER INDICATED LEVEL IN THE GAUGE GLASS

306F11.EPS

Figure 11 ◆ Water levels.

306F12.EPS

Figure 12 ◆ Low-water cutoff for steam.

306F14.EPS

Figure 14 ◆ Electric water feeder.

the float closes the switch contacts, completing the electrical circuit to the water feeder. This energizes the water feeder solenoid-operated valve, allowing makeup water to flow through the valve and into the boiler. When the boiler water level is restored, the float level actuates the switch to open the electrical circuit to the water feeder. This shuts off the supply of makeup water to the boiler.

In other boiler installations, the water feeder control is combined with the low-water cutoff control into one assembly. These controls are called feeder-cutoff combination controls (*Figure 15*). Usually, the water feeder part of the control is mechanically actuated by the float. It is inactive as long as the condensate water returning to the boiler keeps the water in the boiler at the normal operating level. If for some reason the boiler water drops below the normal operating level, the first thing the feeder cutoff combination control does is to add the amount of water required to keep the boiler in operation. It mechanically opens its feed valve, adds the small amount of

306F13.EPS

Figure 13 ◆ Electronic low-water cutoff.

a solenoid-operated valve controlled by the related low-water cutoff control. The low-water cutoffs have a second switch that is actuated by the cutoff float. This switch is used to control the related solenoid-operated water valve (electric water feeder). This second switch operates at a higher water level than the cutoff level switch. If the boiler water drops to this first operating level,

makeup water necessary, and when the float reaches the normal water level, it closes again. A marker on the float bowl casting shows the feeder closing level. Feeders are normally installed so that the marker on the float bowl is about 2 to 2½ inches below the normal water line.

If the feeder part of a combined feeder cutoff control cannot keep the water in the boiler at a safe level, the low-water fuel cutoff portion of the control will activate. When the water in the boiler drops below the minimum safe level, the float drops and opens the switch contacts, resulting in the fuel supply to the burners being turned off.

The selection of low-water cutoff and water feeder devices generally depends on design system pressure and desired mounting style. It is recommended that steam boilers over 400,000 Btuh input and/or operating above 15 psig be fitted with two low water cut-off devices for redundancy. Water feeder and low water cutoff controls provide valuable protection and are essential for proper system operation. They should be disassembled, cleaned, calibrated, and reinstalled annually, at a minimum.

4.2.4 High-Pressure Limit and Other Controls

The high-pressure limit control (*Figure 16*), also known as a pressuretrol, is activated by a rise in boiler pressure. The high-pressure setpoint should always be lower than the setting of the pressure relief valve, and a pressuretrol style that requires a manual reset should always be used for high-limit duty. Although a few models use snap switches, most use a mercury bulb much like older wall thermostats. Control circuit power is routed through the mercury bulb. Should the boiler pressure rise to the point where it exceeds

the pressuretrol setpoint, the switch contacts open and break the burner control circuit. This shuts down the boiler.

Pressuretrols are also commonly used to provide on-off and sequential control of boilers. They can be adjusted in the field so that the boilers all start at high-fire (unison), or in sequence where the boilers start at some firing rate other than high-fire. The control is activated when the steam pressure drops below a preset level. This causes automatic feeding of an air/fuel mixture to the furnace, ignition of the burners, and opening of the main air damper(s). As the boiler steam pressure rises, it crosses fixed control points. At each control point, a predetermined mixture of fuel and airflow is selected and applied to the burners. When the boiler pressure reaches the operational pressure limit, the control shuts off the boiler.

4.2.5 Pressure (Safety) Relief Valve

A pressure (safety) relief valve is used to protect the boiler and the system from steam pressures exceeding the pressure rating of the boiler or the system (*Figure 17*). It does not operate unless an overpressure condition exists. The low-pressure steam boiler is built for a maximum working pressure of 15 psi. The safety relief valve is set to open at the same pressure and discharge the steam into the atmosphere. More than one pressure relief valve must be used on boilers having over 500 square feet of heating surface. This is a precaution to guard against the failure of a single valve. Safety relief valves should always be installed as directed by the manufacturer. They

306F15.EPS

Figure 15 ◆ Combined feeder cutoff control.

306F16.EPS

Figure 16 ◆ High-pressure limit control.

are usually installed in an upright position at the top of the boiler and at the highest point on the steam side of the boiler.

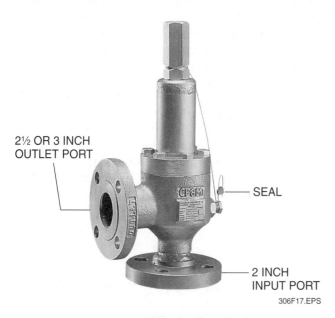

2½ OR 3 INCH OUTLET PORT

SEAL

2 INCH INPUT PORT

306F17.EPS

Figure 17 ◆ Pressure (safety) relief valve.

Safety relief valves should be checked per the manufacturer's instructions intermittently during operation and after long periods of inactivity.

5.0.0 ◆ VALVES

Many valves used to regulate and direct the flow of steam within a steam piping system are the same as those used in hot-water piping systems. These valves include gate valves, ball valves, globe valves, angle valves, and check valves. The focus of this section is on pressure-reducing valves and thermostatic valves used in steam systems.

5.1.0 Pressure-Reducing Valves

Pressure-reducing valves are used to reduce a higher steam pressure to a controlled lower pressure. Such a condition often exists when a process steam boiler supplies steam at a higher pressure than can be used in a related secondary steam-heating system. The valve output pressure level must be held constant, regardless of any pressure fluctuations in the input steam supply or variations in the downstream load. There are several kinds of pressure-reducing valves. One simple type of pressure reducing valve (*Figure 18*) controls the output pressure by achieving a balance between an internal spring pressure acting on a diaphragm and the pressure of the delivered steam. The spring pressure is usually adjusted manually to set the desired pressure downstream.

Larger and more complex pressure-reducing valves use external tubes to sense output pressure further downstream. Other valves are positioned using pneumatic or electric actuators for greater accuracy. This allows them to be interfaced with building or process automation systems using a variety of different sensor inputs.

Figure 19 shows a typical single-stage pressure regulator. Notice that the pressure regulator in

GOING GREEN

Controlling Mercury Hazards

Mercury is a major environmental pollutant that is known to cause cancer. Every effort is being made to remove mercury from the environment and to regulate its disposal. The pressure switches in older boiler systems contain a large amount of mercury, and are therefore a target of this effort. As mercury switches reach the end of their useful lives, or as boilers are overhauled, the mercury switches are being replaced with other devices such as snap-action bimetal and electric switches. Disposal of switches and other devices containing mercury is carefully regulated. Most states have enacted laws governing the storage and disposal of mercury, so it is important to know the applicable regulations in your state.

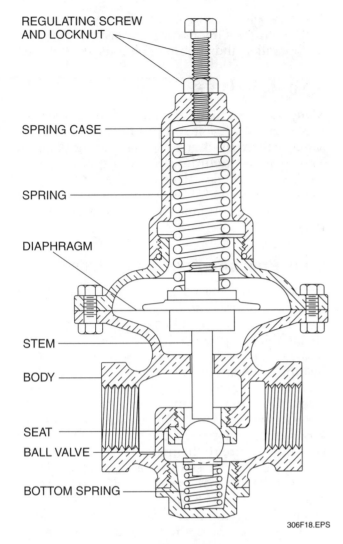

REGULATING SCREW
AND LOCKNUT

SPRING CASE

SPRING

DIAPHRAGM

STEM

BODY

SEAT

BALL VALVE

BOTTOM SPRING

306F18.EPS

Figure 18 ◆ Pressure-reducing valve.

the figure can be bypassed by opening a globe valve. This allows steam to be supplied to the downstream load should the pressure regulator valve malfunction. Also, the gate valve installed on each side of the pressure-relief valve allows the pressure-reducing valve to be isolated for repair without shutting down the system.

Suitable drains and steam traps must be used to prevent accumulation of condensate on the input side of the pressure-reducing valve. The assembly, which consists of the pressure regulator and related valves and traps, is often called a pressure-reducing station.

Pressure reduction can be accomplished with one or more valves (stages). Multi-stage pressure reduction is commonly used where an intermediate pressure is required or where the total pressure reduction is large. A large pressure reduction, also referred to as the **turndown ratio**, is one in which the ratio of downstream pressure to upstream pressure is greater than 50 percent. *Figure 20* demonstrates the installation of pressure-reducing valves in series.

When there is a wide variation in the steam load, such as can occur between summer and winter load conditions, pressure-reducing valves are often installed in parallel. *Figure 21* shows a typical pressure-reducing station with valves installed in parallel. Normally, these valves have different operating pressures that provide for sequenced valve operation. A smaller regulator valve sized to handle about one-third of the maximum load is connected in parallel with a larger valve sized to handle the remaining load. The smaller regulator

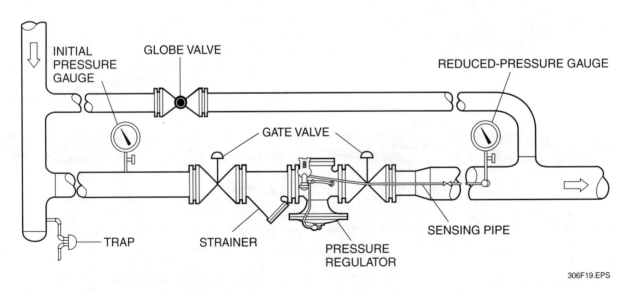

INITIAL
PRESSURE
GAUGE

GLOBE VALVE

REDUCED-PRESSURE GAUGE

GATE VALVE

SENSING PIPE

TRAP

STRAINER

PRESSURE
REGULATOR

306F19.EPS

Figure 19 ◆ Single-stage pressure-reducing station.

valve is set for a slightly higher pressure than the larger one. During times when the load is low, the smaller regulator handles the load. As the load increases, the reduced pressure causes the larger regulator to operate and provide any additional required flow.

There are several common factors that should be considered for a successful pressure-reducing valve installation:

- Follow the manufacturer's guidelines for proper sizing. Oversized valves tend to exceed the desired pressure output and experience greater wear, especially at light loads. Undersized valves are unable to pass a sufficient amount of steam to satisfy downstream needs.

- Pressure-reducing valves operate properly in one direction only. Make sure the valve is properly oriented in the line. Most have an

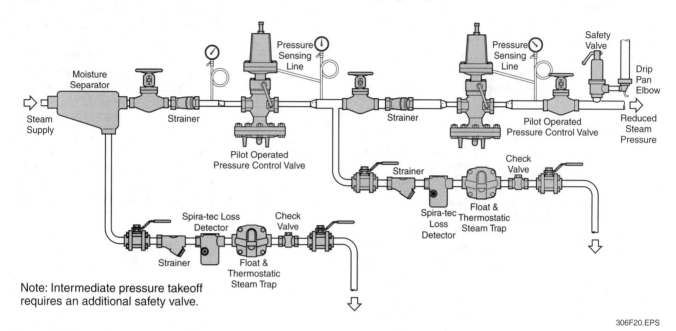

Note: Intermediate pressure takeoff requires an additional safety valve.

306F20.EPS

Figure 20 ◆ Use of pressure-reducing valves connected in series.

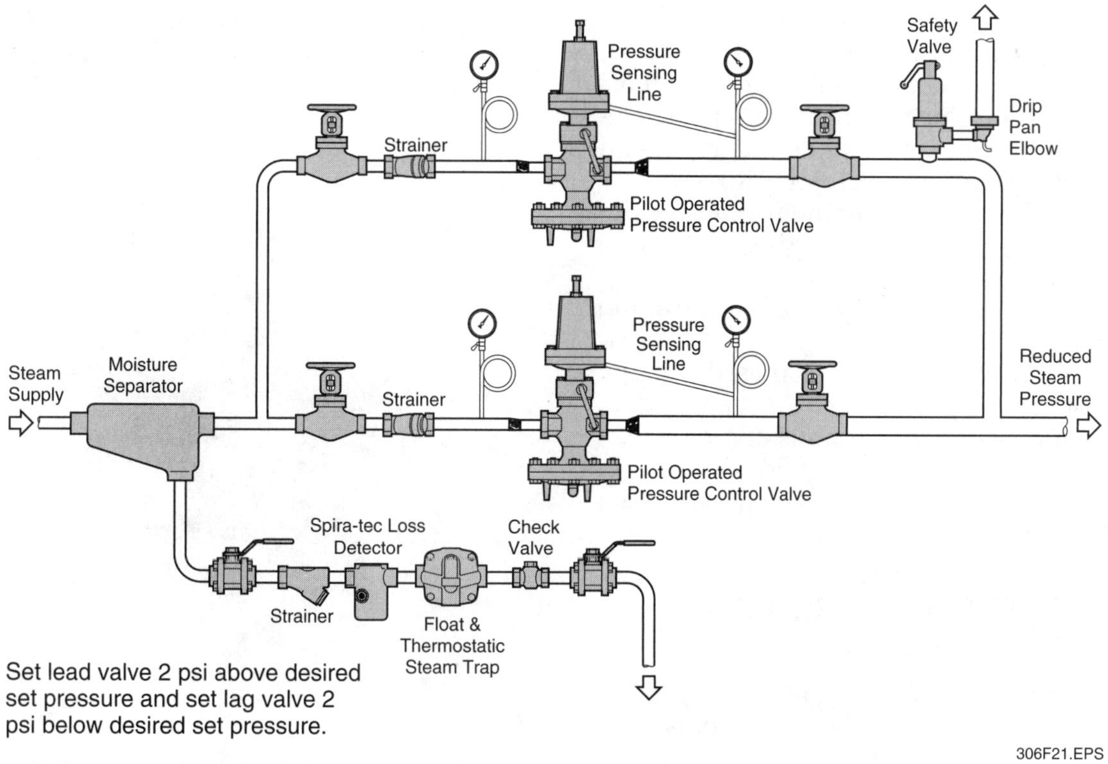

Set lead valve 2 psi above desired set pressure and set lag valve 2 psi below desired set pressure.

306F21.EPS

Figure 21 ◆ Use of pressure-reducing valves connected in parallel.

arrow cast into the valve body to indicate the proper flow direction. Pressure-reducing valves also must be installed horizontally, and should always have a steam trap installed immediately upstream.

- Pressure gauges should be installed near both the inlet and outlet. Valves that use a sensing tube for remote pressure input should also have a gauge installed at the same location as the sensing port.

- Dirt and other foreign matter are the most likely causes of valve failure. The use of strainers upstream of the valve is recommended. Strainers should be removed and cleaned soon after a new installation has been placed in operation.

5.2.0 Thermostatic Valves

Thermostatic valves control the flow of steam based on temperature inputs. They can be applied many different ways. Some possible applications include comfort heating radiation, steam water heaters, steam-jacketed vessels, various types of component molds, and tank heating coils. In the simplest applications, for example, a thermostatic valve can be used to control the volume of steam flow to a radiator or other type of steam terminal (*Figure 22*) to provide zone temperature control. A dial provides for operator setpoint adjustment. Much like pressure-reducing valves, many thermostatic valves are self-contained, with a liquid-filled temperature-sensing element attached directly to the valve such as the one in *Figure 23*. A number of thermostatic elements are available, with selection dependent upon the desired temperature range. As temperature at the sensing element increases, the liquid expands and applies an opening force against the closing force of the valve spring. Setpoint adjustments change the spring tension, thereby changing the force required to open or close the valve.

Thermostatic valves share most of the same common installation considerations with pressure-reducing valves. Although a pressure gauge upstream of the valve is generally recommended, downstream pressure may not need to be monitored. Instead, thermometers should be placed in the controlled medium to provide a quick visual indication of the temperature.

6.0.0 ◆ HEAT EXCHANGERS/ CONVERTERS

A heat exchanger is used to transfer heat from one medium to another. A heat exchanger is used in steam systems mainly to generate hot water for hot-water heating systems, as a steam-to-

water heat exchanger. *Figure 24* shows a horizontal shell-and-tube heat exchanger (converter) used in this manner. It consists of tubes encased in a shell. The tubes provide a closed water circuit that allows redirection of the water through the converter more than once. The hot water is circulated through the exchanger tubes, while the steam circulates in the shell that surrounds the tubes. There, the heat is transferred from the steam to the water. The result is that hot water at

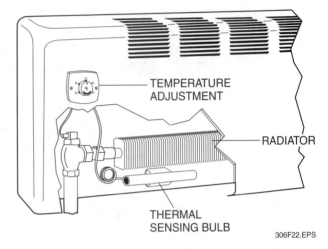

Figure 22 ◆ Typical individual thermostatic control.

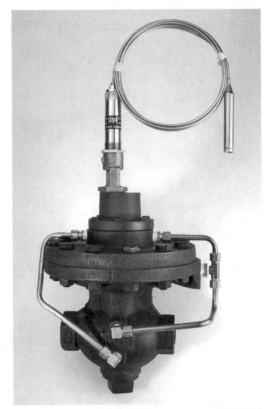

Figure 23 ◆ Temperature-sensing element attached to a valve.

the output of the heat exchanger is available to heat all or part of a building.

Figure 25 shows a typical piping diagram for a heat exchanger used as a hot-water converter in a hot-water heating system. The modulating steam control valve shown is generally thermostatic, regulating the flow of steam based on the leaving water temperature.

A steam-to-water heat exchanger can also be used to heat cold water to produce hot water at

306F24.EPS

Figure 24 ◆ Horizontal shell and tube heat exchanger.

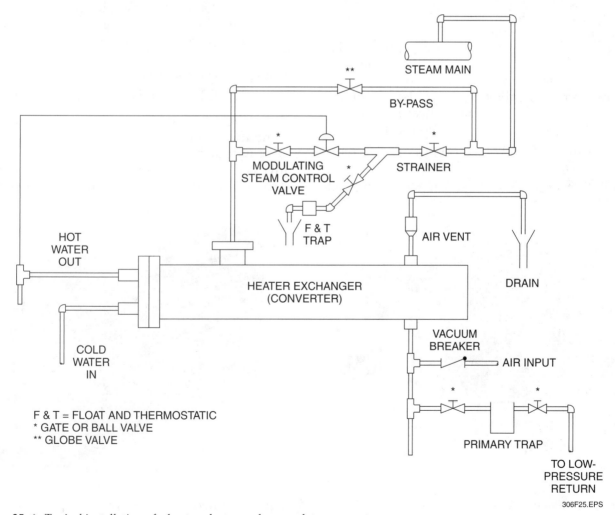

306F25.EPS

Figure 25 ◆ Typical installation of a heat exchanger when used as a converter.

the output of the heat exchanger for domestic use. When used in this way, the heat exchanger is sometimes called an instantaneous heater.

For added safety against contamination, most heat exchangers used to heat potable water are of double-walled construction; that is, two walls separate the steam from the potable water. This is specifically required if conditioning chemicals have been added to the steam system.

7.0.0 ◆ TERMINALS

Many kinds of heating terminals can be used in steam-heating systems. Natural convection units include radiators, convectors, and baseboard convectors. Forced convection units include unit heaters, unit ventilators, cabinet heaters, induction units, fan coil units, and central air handling units.

With the exceptions covered here, the terminals used in steam-heating systems are built and operate in the same way as the terminals used in hot-water heating and dual-temperature systems. Fan coils, induction units, and unit ventilators can be used for both heating and cooling. Dual-coil units have a steam coil used for heating and a separate coil for cooling.

There are two types of steam heating coils, single-row steam coils (*Figure 26*) and steam distribution coils. It can be difficult to distinguish the two types of coils from the outside. Single-row coils are used for general heating applications where the following conditions exist:

- The coil will not be subjected to freezing temperatures
- The flow of steam will likely not be modulated

- Precise control of the air temperature across the entire face of the coil is not necessary

Steam distribution coils (*Figure 27*) are generally specified where one or more of these issues must be considered. These coils are built with perforated inner tubes that allow the steam to be distributed evenly throughout the coil. They are generally considered freeze-resistant. Many HVAC applications require this type of coil. Systems with outside air intakes that use steam to preheat the air are often exposed to temperatures below freezing. Most HVAC applications also require the flow of steam to be modulated, making the steam distribution coil the best overall choice. Single-row coils are used for general heating where no attempt is made to obtain equal air temperatures over the entire length of the coil. They are also used in airstreams where the temperature does not drop below 32°F. Steam distribution coils are made with perforated inner tubes that are used to distribute the steam along the full length of the inner surface of the tubes. They are typically used where the steam must be throttled to obtain control. They are also used where even temperatures are required along the entire length of the tube, or where freezing temperatures are frequently encountered in the air moving over the coil.

8.0.0 ◆ STEAM TRAPS AND STRAINERS

The steam trap is one of the most important components used in steam systems. It is likely that they will demand a great deal of the total time invested in maintenance and repairs. Steam traps provide the point of separation between live

306F26.EPS

Figure 26 ◆ Single-row steam coil.

306F27.EPS

Figure 27 ◆ Steam distribution coil.

steam and water (condensate) in the system and must provide this service effectively for the system to function properly.

As steam is allowed to cool, it condenses to its liquid form. Most of this condensing process takes place at the load, such as a heat exchanger where steam gives up its stored energy to heat water. However, it can also take place in the distribution and piping system and collect in low spots. The condensate can then be picked up by steam as it passes, propelling it at a high rate of speed into bends and fittings. Known as **water hammer**, this phenomenon can cause serious damage and, at a minimum, create noise. For this reason, steam traps are not only used on the leav-ing side of loads such as heat exchangers and coils, but also at strategic low points throughout the piping system.

For steam to remain efficient, it must be dry and at the proper temperature and pressure. Essentially, a steam trap (*Figure 28*) is a self-actu-ating drain valve that maintains effective heat transfer in the steam system by consistently purg-ing it of collected condensate and air.

Air is always present during system startup and in the boiler feed water. In addition, the feed water may contain carbonates that dissolve and release carbon dioxide gas (CO_2). Air and carbon dioxide can cause problems in a steam line. Air is an excellent insulator and can reduce the effi-

306F28.EPS

Figure 28 ◆ Steam traps.

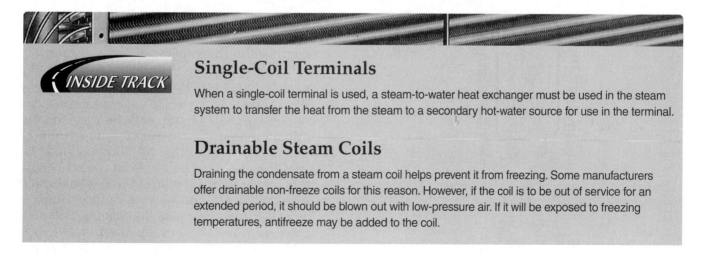

INSIDE TRACK

Single-Coil Terminals

When a single-coil terminal is used, a steam-to-water heat exchanger must be used in the steam system to transfer the heat from the steam to a secondary hot-water source for use in the terminal.

Drainable Steam Coils

Draining the condensate from a steam coil helps prevent it from freezing. Some manufacturers offer drainable non-freeze coils for this reason. However, if the coil is to be out of service for an extended period, it should be blown out with low-pressure air. If it will be exposed to freezing temperatures, antifreeze may be added to the coil.

ciency of the steam at the terminal heat transfer surfaces. Carbon dioxide can dissolve in the condensate and form carbonic acid, which is extremely corrosive. Scale and debris are also transferred with the condensate to areas that are difficult to repair when clogged. Use of properly installed steam traps and strainers throughout the system can prevent these problems.

There are three general categories of steam traps, mechanical, thermostatic, and thermodynamic. Among these many types, there are also various materials used, such as cast iron, stainless steel, forged steel, and cast steel. Steam traps are fitted to the pipe with screwed, flanged, socket-welded, or compression fittings. All types of steam traps automatically open an orifice, drain the condensate, then close before steam is lost.

8.1.0 Mechanical Steam Traps

Mechanical steam traps respond to the difference in density between steam and steam condensate. The trap opens to condensate and closes to steam. The typical mechanical trap is called an inverted bucket trap and has only two moving parts, the valve lever assembly and the bucket (*Figure 29*). The trap is normally installed in the drain line between the steam-heated unit and the condensate return header.

The cycle starts with the bucket down and the valve all the way open. The condensate fills the body and the bucket, causing the bucket to sink

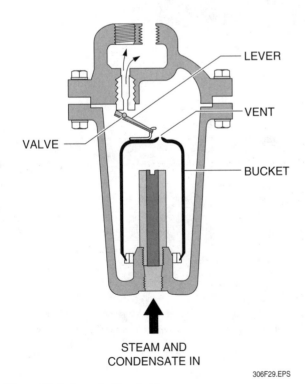

Figure 29 ◆ Inverted bucket trap.

and open the discharge valve. Gases and air flow through first, then steam raises the bucket and closes the valve. The gas slowly bleeds out through a small vent in the top of the bucket. The valve remains open until enough steam has collected to float the bucket again. The inverted bucket trap operates intermittently, responding to the accumulation of condensate by opening, then reclosing to allow steam to pass.

Because the inverted bucket trap operates on a pressure differential, it must be primed before being put into service.

The inverted trap can handle high-pressure steam. It resists damage from pressure surges and water hammer, and tolerates freezing if the body is made of a ductile material. Its disadvantages are its limited air discharge capacity and a tendency to be noisy.

WARNING!

Water hammer can occur when hot steam comes into contact with cooled condensate, builds pressure, and pushes the water through the line at high speeds, slamming into valves and other devices, causing audible noise. In addition, water hammering creates vibrations throughout the system components that can eventually weaken joints and loosen connections, causing them to rupture. Water hammer is most likely to occur in single-pipe systems.

8.2.0 Thermostatic Steam Traps

Thermostatic steam traps respond to temperature changes in the steam line. They open when the cooler condensate is present and close to the higher steam temperatures. The following descriptions, individually illustrated in *Figure 30*, represent four different types of thermostatic steam traps:

- *Liquid expansion thermostatic trap* – Remains closed until the condensate cools below 212°F, opening the thermostat downstream of the valve. This discharges the cool condensate at a constant temperature.

- *Balanced pressure thermostatic trap* – Opens when condensate cools slightly below the steam saturation temperature at any pressure within the trap's range. Controlled by a liquid-filled bellows, the discharge valve closes when hot condensate vaporizes the liquid in the bellows. It opens when the condensate cools enough to lower the pressure inside the element. Balanced pressure traps are smaller than mechanical traps, open wide when cold, readily purge gases, and are unlikely to freeze.

- *Bimetal thermostatic trap* – Has bimetal strips that respond to cool condensate by bending to open the condensate valve. When steam hits the strips, they expand and close the valve. They require considerable cooling to reopen, and the pressure-compensating characteristics vary among the models. The closing force of some designs varies with the steam pressure, approximating the response of a balanced pressure trap. This trap has a large capacity, but responds slowly to process load changes. However, it vents gases well. It is not easily damaged by freezing, water hammer, or corrosion and it can handle the high temperatures of superheated steam.

- *Float and thermostatic (F and T) trap* – Has a float that rises when condensate enters and opens the main valve that lies below the water level. At the same time, another discharge valve at the top of the trap, operated by a thermostatic

bellows, opens to release cooler gases. This trap design vents gases well and responds to wide and sudden pressure changes, discharging condensate continuously. In the failure mode, the main valve is closed but the air vent is usually open. Float thermostatic traps should always be protected from freezing because they usually contain water.

8.3.0 Thermodynamic Steam Traps

Thermodynamic steam traps use the heat energy in hot condensate and steam to control the opening and closing of the trap. They operate well at higher temperatures and can be installed in any position because their function does not depend on gravity.

The thermodynamic disc steam trap is the most widely used. In this unit, the only moving part is the disc (*Figure 31*). Flash steam pressure

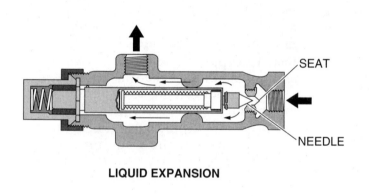

LIQUID EXPANSION

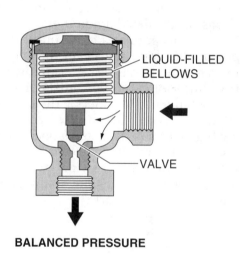

BALANCED PRESSURE

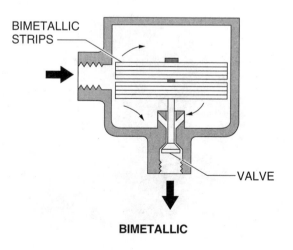

BIMETALLIC

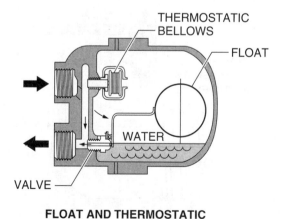

FLOAT AND THERMOSTATIC

306F30.EPS

Figure 30 ◆ Thermostatic steam traps.

from hot condensate keeps the disc closed. **Flash steam** is steam that is formed when hot condensate is released to a lower pressure and flashes back to steam as a result. When the flash steam above the disc is condensed by cooler condensate, the disc is pushed up and remains open until hot condensate flashes again to build up pressure in the cap chamber. At the same time, flashing condensate on the underside of the disc is discharged at a high velocity, lowering the pressure in this area and closing the disc before steam can escape.

The thermodynamic steam trap can withstand freezing, high pressure, superheating, and water hammer without damage. The audible clicking of the disc indicates when the discharge cycle occurs. These traps last longer if they are not oversized.

8.4.0 Fixed-Orifice Steam Traps

As the name implies, the fixed-orifice trap depends on a single opening of a predetermined size to pass a calculated amount of condensate under specific pressure and load conditions. In actual practice, it is rare that steam and condensate flow rates remain precise, rendering the fixed-orifice trap impractical under most conditions. They have a tendency to waterlog (hold back too much condensate) on startup when condensate loads are high, yet pass live steam once the piping system is hot if the load begins to drop. In cases where it can be successfully applied, the fixed-orifice steam trap has the distinct advantage of having no moving parts.

8.5.0 Installing Steam Traps

To mount a steam trap anywhere other than the correct position would render it useless. Two general rules for steam-trap mountings are that they should be lower than any line in the system, and that a strainer has to be upstream of the trap. Isolation drains, couplings, check valves, and **drip leg** placement are also important to each application. Their use depends on the needs and contents of the system. A drip leg is a drain for condensate in a steam line placed at a low point in the line and used with a steam trap.

Each steam trap manufacturer has resource books that detail how to install steam-trap systems. The following is a list of general guidelines for proper steam-trap installation:

- Provide a separate trap for each piece of equipment or apparatus. Short-circuiting may occur if more than one piece of equipment is connected to a single trap.

- When piping from the steam header or main to the load, always tap the steam supply off the top to prevent any collected condensate from draining into the load apparatus.

- Install a supply valve ahead of the steam trap and close to the steam header to allow maintenance or modifications to be performed without shutting down the steam header.

- Install a steam supply valve close to the equipment entrance to allow equipment maintenance work to be performed without shutting down the supply line.

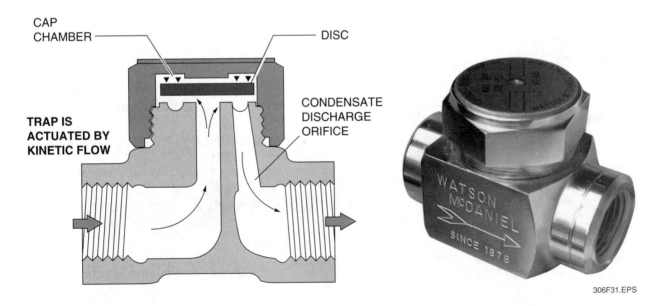

Figure 31 ◆ Thermodynamic steam trap.

306F31.EPS

- Connect the condensate discharge line to the lowest point in the equipment to avoid water pockets and water hammer.
- Install a shutoff valve upstream of the condensate discharge piping to cut off discharge of condensate from equipment and allow service work to be performed.
- Install a strainer and strainer flush valve ahead of each trap to keep rust, dirt, and scale out of working parts and to allow blowdown removal of foreign material from the strainer basket.
- Provide unions on both sides of the trap for its removal or replacement.
- Install a test valve downstream of the trap to allow observation of discharge when testing.
- Install a check valve downstream or upstream of the trap to prevent condensate backflow during shutdown or in the event of unusual conditions.
- Install a downstream shutoff valve to cut off equipment condensate piping from the main condensate system for maintenance work.
- Do not install a bypass unless there is some urgent need for it. Bypasses are an additional expense to install and are frequently left open, resulting in the loss of steam and inefficient equipment operation. When a bypass is necessary, a globe valve should be used to allow manual balancing of condensate flow, as the steam process continues during maintenance of the primary trap.

Figure 32 shows a basic trap installation. A steam trap should have a provision for inspection without interrupting the process flow.

8.6.0 Maintaining Steam Traps

Most steam traps fail in the open position, which is difficult to pinpoint because it does not affect equipment operation. When the trap fails open, the steam flows freely through the system, keeping the heat transfer high but failing to hold back the condensate. Traps that fail in the closed position are easy to identify because the backup of condensate cools the system. Clogged strainers, which protect the trap from debris, also have the same effect.

An effective planned maintenance (PM) program includes scheduled checks of the entire system. Clogged strainers, leaking joints, leaking valve packing, or missing insulation are some of the items to check. When the internal parts of a steam trap wear out, the water seal deteriorates and steam flows through the valve assembly, further worsening the condition. Common causes of steam-trap failure are:

- Scale, rust, or corrosion buildup preventing the valve from closing
- Valve assembly wear
- Defective or damaged valve seat
- Physical damage from severe water hammer
- Foreign material lodged between seat and valve

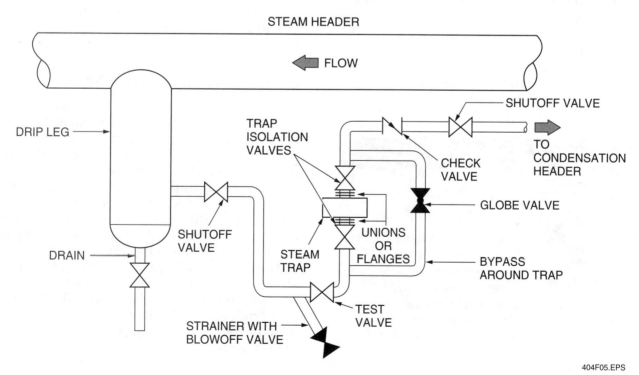

404F05.EPS

Figure 32 ◆ Basic trap installation.

- Blocked, clogged, or damaged strainers
- Increased back pressure

Other failures are specific to the type of steam trap. The two primary indications of failure in a steam trap are allowing live steam to pass freely (will not close), or failure to allow condensate to pass through (will not open). The following sections explain the causes and solutions for failures on each type of steam trap.

8.6.1 Inverted Bucket Trap

If an inverted bucket trap blows steam, check for loss of the water seal. Isolate the trap, wait for the condensate to accumulate, and start the steam flow again. If this solves the problem, try to discover the cause of the water seal loss. It could be caused by superheat, sudden pressure fluctuations, or the trap being installed so that the water seal runs out by gravity.

Another solution is to try installing a check valve ahead of the trap. If the steam blow persists, check for dirt or wear on the valve and linkage. Replace the valve, the seat, and the lever. If this fails, check the bucket. If the bucket or lever is distorted, the trap failure was probably caused by water hammer. Trace the source of the problem and try to eliminate it.

If the trap will not pass condensate, check that the maximum operating pressure marked on the trap is not lower than the actual pressure to which it is subjected. If it is, the valve cannot open. Install a valve and seat assembly with the correct pressure rating. Make sure that it has sufficient capacity to handle the maximum load. Check the internal parts and be sure that the air vent hole in the bucket is not obstructed. This could cause pockets of air to be trapped in the vent. This condition is called air binding.

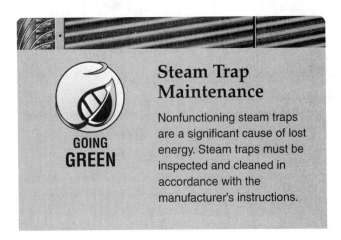

Steam Trap Maintenance

GOING **GREEN**

Nonfunctioning steam traps are a significant cause of lost energy. Steam traps must be inspected and cleaned in accordance with the manufacturer's instructions.

8.6.2 Liquid Expansion Thermostatic Trap

If a liquid expansion thermostatic trap blows steam, check for dirt or wear on the valve and seat. If wear has occurred, replace all of the internal parts. Remember that this type of trap is not self-adjusting to changes in pressure. If it has been set to close at a high pressure, it may not close off at a lower pressure. Try adjusting the trap to a lower setting, making sure that it does not water-log excessively. If it does not appear to react to this change, replace all of the internal parts. If a trap will not pass condensate, check that it has not been adjusted to a setting that is too cold.

8.6.3 Balanced Pressure Thermostatic Trap

If a balanced pressure thermostatic trap blows steam, isolate the trap and allow it to cool before inspecting it for dirt. If the seat is wire-drawn, replace all of the internal parts, including the thermostatic element. The original parts have probably been strained by continuous steam blow.

If the valve and seat seem to be in good order, check the element. You should not be able to compress it when it is cool. Any flabbiness of the element indicates failure. Flattening of the convolutions indicates water hammer damage. If the water hammer cannot be eliminated at its source, a stronger type of trap must be used.

If the trap will not pass condensate, the element may be overextended due to excessive internal pressure, making it impossible for the valve to lift off its seat. An overextended element could be caused by superheat or by someone opening the trap while it is still very hot and before the vapor inside has had time to condense.

8.6.4 Bimetal Thermostatic Trap

If a bimetal thermostatic trap blows steam, check for dirt and wear on the valve. A bimetal trap has limited power to close because of its method of operation, and the valve may be held off its seat by the accumulation of soft deposits.

This type of trap is usually supplied preset. Check that any locking device on the manual adjustment is still secure. If it is not secure, see if the trap responds to adjustment. If cleaning has no effect, replace all of the internal parts.

The valve on a bimetal trap is on the downstream side of the valve orifice, so these traps tend to fail in the open position. Failure to pass cold condensate indicates either gross maladjustment or complete blockage of the valve orifice or built-in strainer.

8.6.5 Float Thermostatic Trap

If a float thermostatic trap blows steam, check the trap for dirt at the main valve and the air vent valve. If the trap has a steam lock release, check to make sure that it is not opened too far. It should be open no more than one quarter of a full turn.

Make sure that the valve mechanism has not been knocked out of line either by rough handling or water hammer, preventing the valve from seating. Check that the ball float is free to fall without fouling the casing, which would cause the mechanism to hang up.

Test the air vent the same way as the element of a balanced pressure trap is tested. If the internal parts of a float trap must be replaced, install a complete set as supplied by the manufacturer.

If the trap will not pass condensate, check that the maximum operating pressure marked on the trap is not lower than the actual pressure to which the trap is subjected. If it is, the valve cannot open. Install a valve and seat assembly with the correct pressure rating. Make sure that this has sufficient capacity to handle the maximum load. If the ball float is leaking or damaged, it is probably caused by water hammer. Also, be sure that the air vent or steam lock release is working correctly.

8.6.6 Thermodynamic Disc Trap

If a thermodynamic disc trap blows steam, the trap will probably give a continuous series of abrupt discharges. Check the trap and strainer for dirt and wipe the disc and seat clean. If this does not correct the problem, it is possible that the seat faces and disc have become worn. The extent of wear is evident by the amount of shiny surface that replaces the normal crosshatching of machining.

If records show that thermodynamic traps on one particular installation suffer repeatedly from rapid wear, the trap may be oversized, associated pipework may be undersized, or there may be excessive back-pressure.

If the trap will not pass condensate, the discharge orifices may be plugged with dirt. This is most likely caused by air binding, particularly if it occurs regularly during startup. Check the air venting arrangements of the equipment. In extreme cases, it may be necessary to install an air vent parallel with the trap, or to use a float trap with a built-in air vent instead of a thermodynamic trap.

8.7.0 Strainers

In all systems, a strainer should be installed ahead of the steam trap. The scale and corrosion in any steam system has to be stopped before it enters the trap, or it will cause clogging and damage. Strainers need to be cleaned and inspected on a regular basis. Strainers are available in many styles and are sometimes incorporated into the steam trap. *Figure 33* shows three types of strainers.

9.0.0 ◆ TROUBLESHOOTING STEAM TRAPS

To diagnose and solve problems with steam traps requires listening to the noises they make and measuring temperature and pressure. The temperature of steam can be measured with hand-held or remote-sensing pyrometers that read in degrees centigrade or Fahrenheit. Steam is produced at 212°F when at atmospheric pressure. When the steam systems are functioning, the temperature will be substantially higher, as it is confined inside the system. The pressure will be higher as well. Pressure readings are in two scales, absolute (psi) and gauge (psig). Absolute pressure is read from zero up, which is to say that normal atmospheric pressure is 14.7 psi at sea level. Gauge pressure is relative to atmospheric pressure at sea level; therefore, zero psig is equal to 14.7 psi absolute. Pressures below zero gauge are expressed in inches of mercury. Steam trap manufacturers have excellent books and web documents that detail how to troubleshoot and repair steam traps.

9.1.0 Diagnostic Methods

Three basic diagnostic methods of reading a steam system are sight, sound, and temperature. The criteria observed include pressure. The other process, less used to date, is the use of a conductance probe, which compares the difference in electrical conductivity between condensate and live steam.

Each of the three main methods has advantages and limitations. Sight is fairly immediate; it is easy to tell whether there is condensate coming from a trap. If the condensate line is open, or if a test valve has been added on the condensate line, the condensate can be vented to the atmosphere. It will be obvious that the trap is removing condensate from the line. If live steam is going into the condensate line, the trap has failed open, and it is necessary to resolve the problem. If no condensate is coming from the line, the trap may

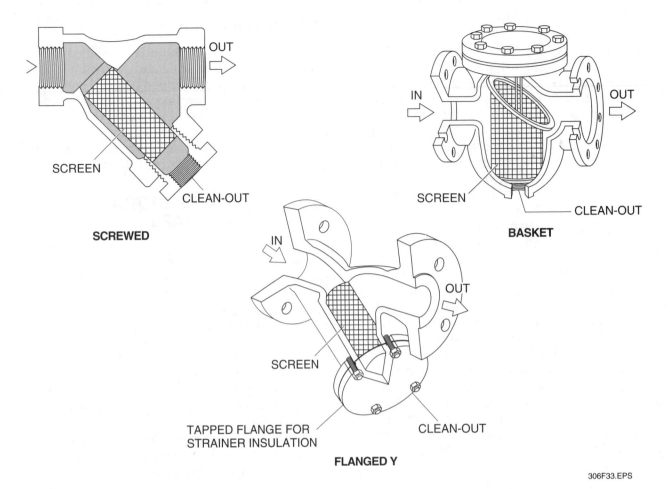

Figure 33 ◆ Strainers.

306F33.EPS

have failed closed. The limitation of sight diagnosis is that the worker must be able to tell the difference between flash steam and live steam, and it is necessary for the worker to check by activating a test valve. It is also possible for the trap to be overloaded or loaded from the return system, producing an apparent open failure.

 WARNING!

In a high-pressure steam system, the condensate will flash back to steam when the pressure drops as the result of exposure to the atmosphere through a leak. This exposure causes a rapid change in volume, which can result in the release of high-pressure steam.

The two most commonly used diagnostic tools are temperature sensing and sound, either at human hearing ranges or ultrasound. Note that temperature and sound together are a much more sensitive test than one or the other alone.

From a diagnostic point of view, there are two general types of traps. Traps either flow or dribble continuously or intermittently (on and off).

Float thermostat (F and T) traps discharge fairly continuously if they are working properly. Inverted bucket, bellows thermostatic, thermodynamic, and most thermostatic traps flow intermittently.

A few common characteristic symptoms of traps when tested might include the following:

• A thermal analysis will show the inlet as a high temperature area for any of the traps. In most functioning traps, the condensate outlet, whether it is a closed return or an open return, will be cooler than the inlet. Remember that the condensate will most frequently be at steam temperature, but it should be lower than the inlet. The live steam outlet should be hotter than the condensate outlet. If this is not the case, the trap is probably malfunctioning.

• Normal failure for an inverted bucket trap is open failure. The trap loses its prime, and steam and condensate blow through steadily. If this happens, the continuous rushing sound of the blowthrough may be accompanied by the bucket banging against the inside of the trap. The temperature of the trap and of the condensate line will rise because it is no longer holding condensate. The sound of the bucket linkage rattling inside the trap may indicate that the trap is beginning to loosen up and needs checking. With ultrasound equipment, the inverted bucket or thermodynamic traps would show a cyclic curve on the screen in normal function.

- Normal failure for a thermodynamic trap is open; that is, passing steam. Usually, the disk clicks shut audibly once for each cycle. When the disk is cycling normally, it will shut about 4 to 10 times a minute. When the trap fails, the disk no longer clicks as it rises and falls, and steam blows through. If the disc produces a continuous, rapid, rattling sound, the disc is worn, and more problems will probably develop. If a continuous hissing sound is produced, the trap is not cycling. Again, the condensate line will heat up due to escaping steam and the absence of cooler condensate.

- Bellows or bimetallic thermostatic traps operate on a difference in temperature between condensate and steam. Both types are intermittent in operation. When there is little condensate, they remain closed most of the time. When there is a lot of condensate, for example at startup, they may run continuously for a long time. In the case of bimetallic traps, misalignment may allow steam leaks. With sound testing equipment, a constant rushing sound in the closed part of the cycle would indicate a leak.

- Float thermostatic traps tend to fail closed. A leaky float will not float on the condensate, or the float may have been crushed by water hammer. In either case, the F and T trap will remain closed and will not cycle. The trap will remain cool, and a temperature sensor will reveal that. With listening equipment, a failed float will be silent in the absence of discharging condensate.

- The thermostatic element in an F and T trap is normally quiet; a rushing sound would indicate that the element had failed open. A rattling or metallic clanging noise might mean that the mechanical linkage had sustained some damage.

The sight method uses a test valve that vents the process steam to the atmosphere for visual inspection. This test is subjective and depends on skill and experience. Most traps are dealing with condensate that is at steam temperatures; that is, somewhere above the boiling point of water. When condensate is released from the pressure of the steam system in the trap or in a condensate release, a certain amount of the condensate will immediately turn into flash steam. The difference is fairly visible once you have seen it a few times;

the live steam comes out at first as a hard, blue-white straight flow, often with a clear bluish area at the outlet. Flash steam usually billows and spreads more quickly, as much of the pressure is dissipated in the initial release. The sound probes are either audible-range or ultrasonic-range probes that pick up noise from the steam system and send an audible signal to headphones. The technician performing this test must also have skill and experience at interpreting the noise as leaks, discharges, or other steam trap problems. *Figure 34* shows a portable ultrasonic tester.

Temperature-sensing tests use pyrometers (*Figure 35*) that are handheld for external readings or use mounted thermowells (*Figure 36*) that read actual steam temperature. A thermowell is a permanently installed well or cavity in a process pipe or tank into which a glass thermometer or thermocouple can be inserted.

10.0.0 ◆ CONDENSATE RETURN/ FEEDWATER SYSTEM COMPONENTS

Condensate is the byproduct of a steam system. It forms in the distribution system because of the heat transfer from the steam to the area or substance being heated. Some condensation is also formed as a result of unavoidable radiation. Once the steam has condensed, giving up its latent heat, the hot condensate must be removed immediately and returned to the boiler. This is because the amount of heat in a pound of condensate is negligible when compared to the heat in a pound of steam.

The components that may be used in a condensate return/feedwater system are a condensate pump and receiver tank, vacuum pump, and de-aerating feedwater heater.

Figure 34 ◆ Ultrasonic tester.

306F34.EPS

10.1.0 Condensate Pump and Receiver Tank

On larger steam-heating systems, a condensate pump and receiver are normally used to accumulate return condensate and pump it back into the boiler. There are many variations and types of equipment that can be used. Some systems may use a condensate receiver equipped with a single pump, while others may use a receiver with two pumps, often referred to as a duplex unit. *Figure 37* shows an example of both. Duplex units are sometimes equipped with an alternator arrangement. In this arrangement, pumps are automatically switched on a timed basis, or when a pump failure occurs, to prevent the system from flooding and shutting down

Figure 38 shows one type of condensate and receiver tank system. The receiver tank is an open reservoir into which the return piping discharges its condensate. This system uses a pump controller installed at the normal boiler water line to control the condensate (boiler feed) pump. Whenever the pump controller senses a drop in the normal boiler water level, it turns on the condensate pump. This causes water from the receiver tank to be pumped into the boiler as needed to replenish the boiler water level. A check valve in the condensate pump discharge line prevents the boiler water from backing up into the receiver when the condensate pump is turned off. A gate valve in the line allows the pump to be serviced without the need to drain the boiler.

The receiver tank is also equipped with a float-actuated makeup water feeder control. If the level of condensate water in the receiver tank drops below the normal operating level, the control float drops and opens its valve to allow makeup water to enter the receiver tank from the city water supply. Additional boiler protection is provided by a combined feeder cutoff control. In the event of a power failure, condensate pump failure, or other problem, this control will first attempt to supply the boiler with sufficient water from the city water supply to prevent it from overheating. If the boiler water drops below a safe level, the control then acts to turn off the boiler burners.

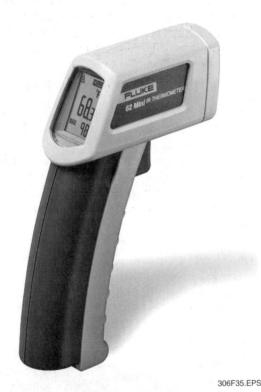

306F35.EPS

Figure 35 ◆ Pyrometer.

306F36.EPS

Figure 36 ◆ Thermowells.

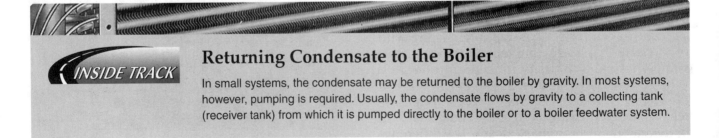

Returning Condensate to the Boiler

In small systems, the condensate may be returned to the boiler by gravity. In most systems, however, pumping is required. Usually, the condensate flows by gravity to a collecting tank (receiver tank) from which it is pumped directly to the boiler or to a boiler feedwater system.

SINGLE

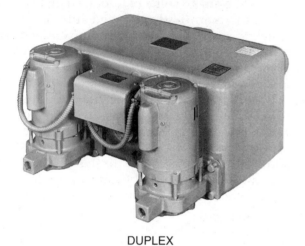

DUPLEX

306F37.EPS

Figure 37 ◆ Condensate receivers.

10.2.0 Vacuum Pumps

Some systems use a vacuum pump in the condensate return line. Vacuum pump operation is similar to condensate pump operation in that the vacuum pump causes condensate from the condensate return line to flow into a receiver tank. Normally, a feed pump is then used to pump the condensate water from the receiver tank into the boiler. The vacuum unit and condensate receiver and pump are often combined into a single unit, such as the unit pictured in *Figure 39*.

At system startup, the condensate vacuum pump produces a vacuum in the system by removing the air and vapor from the system and venting it to the atmosphere. By quickly exhausting the air and condensate from the system, the vacuum pump causes the steam to circulate rapidly, reducing warmup time and providing quieter system operation. In addition to causing condensate flow during system operation, the vacuum pump also turns on whenever needed to maintain the vacuum level in the system.

10.3.0 Deaerating Feedwater Heater

In some larger steam systems, the condensate water pumped out of the receiver tank does not get pumped into the boiler. Instead, it is pumped into a deaerating feedwater heater, also called a deaerator. This device uses system steam to preheat the water before it is supplied to the boiler in order to remove most of the dissolved oxygen.

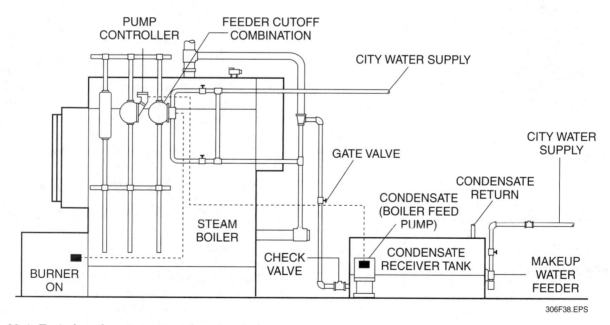

306F38.EPS

Figure 38 ◆ Typical condensate pump and receiver tank.

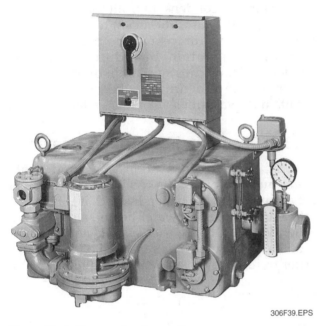

Figure 39 ◆ Vacuum heating units.

306F39.EPS

This is especially important on systems that lose 25 percent or more of the steam condensate from the circuit during normal operation, or are drained and refilled regularly. The reason is that fresh makeup water carries large quantities of dissolved oxygen. Deaeration is also beneficial in systems that operate at pressures above 75 psig.

11.0.0 ◆ FLASH TANKS

When hot condensate under pressure is released to a lower pressure, part of it is re-evaporated and becomes flash steam. This normally happens when hot condensate is discharged into the condensate return line at a pressure lower than its saturation pressure. Some of the condensate flashes into steam and flows along with the liquid condensate through the return lines back to the boiler. This tends to cause an undesirable pressure increase in the condensate return lines. Flash tanks are used in medium-pressure and high-pressure systems to remove flash steam from the condensate lines by venting it either to the atmosphere or into a low-pressure steam main for reuse.

When flash steam is used as an energy source, high-pressure steam returns are usually piped to a flash tank (*Figure 40*). Flash tanks can be mounted vertically or horizontally. However, vertical mounting provides for better separation of steam and water.

When the condensate and any flash steam in the return line reach the tank, the high-pressure condensate is released into the lower pressure of the tank, causing some of it to flash into steam. This flash steam is discharged into the low-pressure steam main. The remaining condensate is pumped back to the boiler or discharged to a waste drain.

For proper flash tank operation, the condensate lines should pitch toward the tank. If more than one condensate line feeds the tank, a check valve should be installed in each line to prevent backflow. The top of the tank should have a thermostatic air vent to vent any air that accumulates in the tank. The bottom of the tank should have an inverted bucket or float and thermostatic-type steam trap.

The demand steam load on the flash tank should always be greater than the amount of flash steam available from the tank. If it is not, the low-pressure system can be over-pressurized. A safety relief valve must be installed at the top of the flash tank to protect the low-pressure line from any over-pressurization. Since the flash steam produced in the flash tank is less than the amount of low-pressure steam that is needed, a makeup valve in the high-pressure steam line is used to supply any additional steam needed to maintain the correct pressure in the low-pressure line. This steam makeup valve is actually a pressure-reducing valve. It is much like those discussed earlier, but has an external pilot line to monitor the low pressure downstream of the valve itself.

12.0.0 ◆ STEAM SYSTEM PIPING

Steam piping systems are classified as either one-pipe or two-pipe systems. The type of system is determined by the piping method used to supply steam to and return condensate from the system terminals. Each of these piping systems is further classified by the method used to return the condensate (gravity or mechanical flow), and the directions of steam and condensate flow in the system.

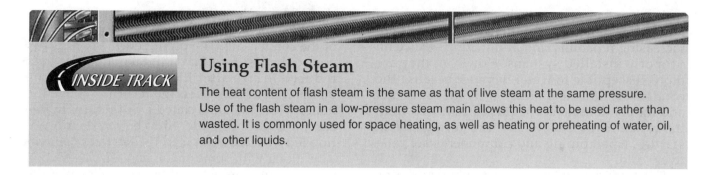

Using Flash Steam

The heat content of flash steam is the same as that of live steam at the same pressure. Use of the flash steam in a low-pressure steam main allows this heat to be used rather than wasted. It is commonly used for space heating, as well as heating or preheating of water, oil, and other liquids.

12.1.0 One-Pipe Systems

In a one-pipe system, steam flows to and condensate is returned from the terminal unit through the same pipe. There is only one connection to each heating terminal unit, which functions both as the supply and the return. Normally, one-pipe systems use a gravity return, but they may use a condensate pump. This is done when there is insufficient height above the boiler water level to develop enough pressure to return the condensate directly to the boiler. One-pipe gravity systems do not use steam traps. Air vents are used at each terminal unit and at the ends of all supply mains to vent air so the system can fill with steam. If a one-pipe system uses a condensate pump, air vents are used at each terminal unit, but steam traps are used at the end of each supply main. One-pipe systems are used mostly in residential and small commercial systems. They are best applied in homes or small buildings with basements, where space and height exist for the needed drainage pitch. Though limited in potential applications,

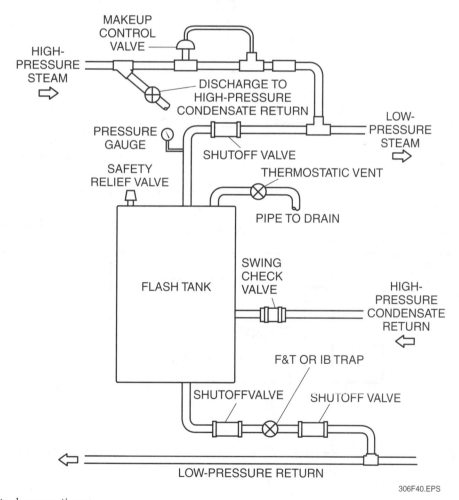

Figure 40 ◆ Flash tank connections.

306F40.EPS

their advantage rests in their simplicity. Very few moving or functional components are required, increasing reliability and minimizing service on properly installed systems. However, they are more susceptible to water hammer because the condensate and steam frequently make contact and the condensate is not quickly removed from the system. This is especially true during system startup, when piping and components are relatively cold.

There are several configurations of one-pipe steam systems. *Figure 41* shows a basic one-pipe up-feed gravity system. The term *up-feed* means that the steam supply main is below the level of the heating terminals that they serve. Steam flow is then fed upward in the riser for input to the terminal units.

As shown in *Figure 41*, the steam supply main rises from the boiler to a high point, then is pitched downward from this point. The pitch should be no less than one inch per 20 feet and should slope in the direction of the gravity flow of the condensate. The piping normally runs full size to the last terminal unit takeoff and is then reduced in size after it drops down below the boiler water line. When the return is above the boiler water line, it is called a dry return. In this case, it is only considered to be dry as it is not flooded or filled with water—it carries condensate, along with some steam and air. The section of return piping installed below the level of the water in the boiler is called the wet return. It is completely filled with water and does not carry steam or air. When the system is in operation, the steam and condensate flow in the same direction in the horizontal mains and in opposite directions in the branches and risers. Because the steam and condensate flow in the same direction in the horizontal mains, this type of system is also called a parallel-flow system.

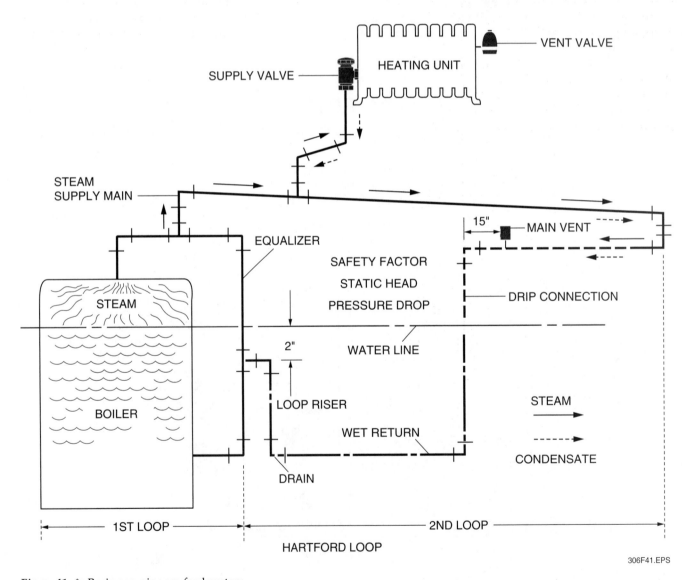

Figure 41 ◆ Basic one-pipe up-feed system.

A Hartford loop is a pressure-balancing loop that introduces full boiler pressure on the return side of the boiler. This pressure prevents reverse circulation or water leaving the boiler by the return line. The Hartford loop is actually two loops; the first is an equalizer line and the second is an extension from the steam main through the return and back to a connection at the equalizer loop.

12.2.0 Two-Pipe Systems

The two-pipe system uses separate pipes to supply the steam and return the condensate from the terminal units.

Thermostatic traps are used at the outlet of each terminal. They keep the steam in each terminal unit until it gives up its latent heat, then the trap opens to pass the condensate and permit more steam to enter the terminal. Float and thermostatic steam traps should be used at the end of each steam main, at the bottom of each riser, and along the horizontal mains. *Figure 42* shows a basic two-pipe gravity system.

Two-pipe systems can have either gravity or mechanical condensate returns. However, most systems use a mechanical return (*Figure 43*) provided either by a condensate pump or a vacuum pump. When steam systems increase in size,

higher steam pressures are needed to obtain steam circulation, and some means other than gravity must be used to return the condensate to the boiler. A condensate pump and receiver tank are commonly used for this purpose. Normally, the condensate from the return piping system flows by gravity into a receiver tank. A condensate pump mounted near the receiver pumps the accumulated condensate into the boiler feedwater return piping. A typical condensate pump and receiver tank used for this purpose were covered earlier in this module. Note that in *Figure 42* air vents have been eliminated from the design. With a mechanical condensate return, air is released from the system through an open vent pipe on the receiver tank. This vent must open at a higher level than the boiler water level.

A vacuum-return system (*Figure 44*) is similar to a condensate return system, except that a vacuum pump is installed to provide a low vacuum in the return line to return the condensate to the boiler. A vacuum-return system is one that operates with a pressure in the condensate-return piping that is below atmospheric pressure. The steam supply lines and terminal units are maintained at a positive pressure. The vacuum pump produces a vacuum in the system return lines by removing the air and vapor and venting it to the atmosphere in a related receiver tank.

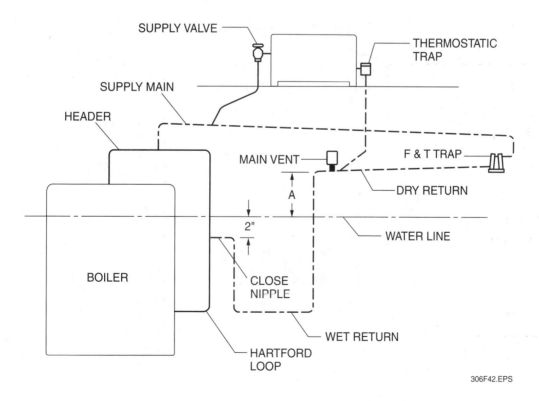

Figure 42 ◆ Two-pipe gravity system.

It is important to note that the vacuum pump does not draw or pump the condensate itself. By quickly exhausting air from the system, the vacuum pump causes the steam to flow more rapidly, reducing warmup times, as well as noise from water hammer. The vacuum pump cycles as needed to maintain a predetermined vacuum level in the return lines. Operating the return line in a vacuum, rather than at or slightly above atmospheric pressure, increases the pressure differential in the overall system. As a result, steam pressure settings at the boiler can sometimes be lowered for significant fuel savings and improved efficiency. Much of this efficiency can easily be lost through air leaks. It is essential that return piping be leak-free to maximize the benefits associated with vacuum return and prevent excessive pump operation.

It is commonly understood that water boils, creating steam, at a lower temperature when exposed to pressures less than atmospheric. For example, at 15 in Hg vacuum, water boils at roughly 180°F. Some vacuum systems, known as variable vacuum systems, rely on this advantage by placing the supply side of the system, as well as the return piping, into a vacuum. This obviously requires greater capacity (expressed in cfm) from the vacuum pump, and air-tight piping systems become even more important. By reducing the boiling point of the water, even greater operating efficiencies can be obtained.

The vacuum pump controller, which determines the vacuum level that the pump must maintain, is generally designed to respond to outdoor temperatures. During mild weather, vacuum levels can be maintained at a lower value (18 to 25 in Hg), resulting in lower temperature steam and reduced fuel usage. As outdoor temperatures fall, creating the need for additional capacity, vacuum levels are maintained at a higher value (2 to 15 in Hg) to increase the temperature of the steam provided to the heating system. The controller is programmed to monitor outdoor temperatures and reset the vacuum level across the range of heating season outdoor temperatures expected for the specific geographic region.

Operation at lower pressures will increase the specific volume of the steam. Because of this increased volume, it takes less steam to fill the system, resulting in more efficient system operation.

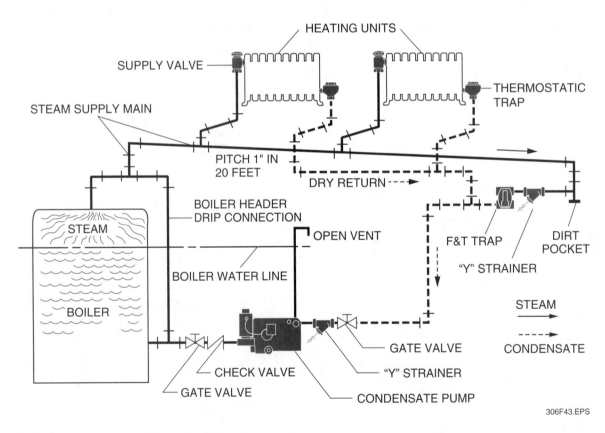

Figure 43 ◆ Two-pipe system with mechanical condensate return.

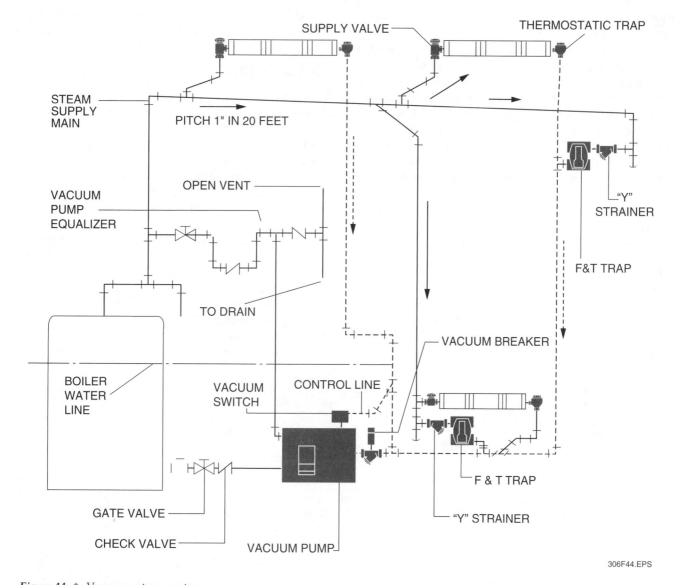

Figure 44 ◆ Vacuum return system.

13.0.0 ◆ STEAM AND CONDENSATE PIPE SIZING

Proper steam and condensate line sizing is essential for any steam system to operate efficiently and provide the needed capacity for each load. To size the piping system, information regarding the volume of steam needed, as well as the steam pressure in the system must be known.

As a general rule, it is best that steam boilers operate at a pressure near their design pressure. In some cases, especially in larger facilities with a significant number of loads, this may require that the steam system pressure be considerably higher than actually required at any one of the operating loads. Operating steam boilers at pressures significantly lower than the boiler is designed for reduces the volume of space in the boiler that nor-

mally is allowed for steam. This places the water level somewhat closer to the steam outlet, and the flow velocity of the steam to the outlet becomes faster due to the smaller area of confinement. When operated in this manner, the likelihood of water droplet carryover is greatly increased.

There are additional advantages to controlling steam pressure higher than is needed downstream. The size of steam distribution mains becomes smaller to carry the same amount of energy, thus saving on installation costs. Due to smaller surface areas with smaller piping, heat loss is also reduced. Steam pressure can then be reduced using pressure-reducing valves near the inlet to each of the individual loads. These valves can be physically smaller, further reducing installation costs.

Appropriate pipe sizing is important to ensure a sufficient steam supply to the loads. Undersized piping causes high pressure drops and greater velocities, increasing noise and erosion of internal piping surfaces. It is also quite common for a new load to be added to a steam system long after the initial installation is complete. A piping system that is already undersized or marginal may prevent even a relatively small load from being added without re-piping, or an insufficient supply of steam to all loads could result. On the other hand, piping that is too large creates unnecessary heat losses and higher installation costs.

Two primary methods are used to size steam piping: sizing by desired velocity or sizing to maintain an acceptable pressure drop. Systems with shorter mains and branch lines are often sized by velocity. Longer piping runs may be sized first by velocity, then re-checked to ensure that acceptable pressure drops have not been exceeded. There a number of factors that impact steam pipe sizing in various parts of the system.

13.1.0 Velocity Sizing

When sizing based on velocity, steam lines should be designed for speeds of 80 to 120 feet per second (fps), or 4,800 to 7,200 feet per minute (fpm). In very limited cases, where noise may not be an issue and pressure drops are not as important, velocities up to 200 fps or 12,000 fpm have been used. Again, these values are not recommended for the reasons stated earlier concerning undersized piping.

Figure 45 provides a steam velocity chart for pipe sizing. As mentioned earlier, the volume of steam to be carried, stated in lbs/hr (pounds per hour) must be known. For new pipe sizing, the system design steam pressure must also be known. Note that this chart is for Schedule 80 pipe. The appropriate chart for the type of pipe being used in the installation must be used.

Capacity of Sch. 80 Pipe in lb/hr steam															
Pressure psi	Velocity ft/sec	1/2"	3/4"	1"	1 1/4"	1 1/2"	2"	2 1/2"	3"	4"	5"	6"	8"	10"	12"
5	50	12	26	45	70	100	190	280	410	760	1250	1770	3100	5000	7100
	80	19	45	75	115	170	300	490	710	1250	1800	2700	5200	7600	11000
	120	29	60	110	175	245	460	700	1000	1800	2900	4000	7500	12000	16500
10	50	15	35	55	88	130	240	365	550	950	1500	2200	3770	6160	8500
	80	24	52	95	150	210	380	600	900	1500	2400	3300	5900	9700	13000
	120	35	72	135	210	330	590	850	1250	2200	3400	4800	9000	14400	20500
20	50	21	47	82	123	185	320	520	740	1340	1980	2900	5300	8000	11500
	80	32	70	120	190	260	520	810	1100	1900	3100	4500	8400	13200	18300
	120	50	105	190	300	440	840	1250	1720	3100	4850	6750	13000	19800	28000
30	50	26	56	100	160	230	420	650	950	1650	2600	3650	6500	10500	14500
	80	42	94	155	250	360	655	950	1460	2700	3900	5600	10700	16500	23500
	120	62	130	240	370	570	990	1550	2100	3950	6100	8700	16000	25000	35000
40	50	32	75	120	190	260	505	790	1100	1900	3100	4200	8200	12800	18000
	80	51	110	195	300	445	840	1250	1800	3120	4900	6800	13400	20300	28300
	120	75	160	290	460	660	1100	1900	2700	4700	7500	11000	19400	30500	42500
60	50	43	95	160	250	360	650	1000	1470	2700	3900	5700	10700	16500	24000
	80	65	140	250	400	600	1000	1650	2400	4400	6500	9400	17500	27200	38500
	120	102	240	410	610	950	1660	2600	3800	6500	10300	14700	26400	41000	58000
80	50	53	120	215	315	460	870	1300	1900	3200	5200	7000	13700	21200	29500
	80	85	190	320	500	730	1300	2100	3000	5000	8400	12200	21000	33800	47500
	120	130	290	500	750	1100	1900	3000	4200	7800	12000	17500	30600	51600	71700
100	50	63	130	240	360	570	980	1550	2100	4000	6100	8800	16300	26500	35500
	80	102	240	400	610	950	1660	2550	3700	6400	10200	14600	26000	41000	57300
	120	150	350	600	900	1370	2400	3700	5000	9100	15000	21600	38000	61500	86300
120	50	74	160	290	440	660	1100	1850	2600	4600	7000	10500	18600	29200	41000
	80	120	270	450	710	1030	1800	2800	4150	7200	11600	16500	29200	48000	73800
	120	175	400	680	1060	1520	2850	4300	6500	10700	17500	26000	44300	70200	97700
150	50	90	208	340	550	820	1380	2230	3220	5500	8800	12900	22000	35600	50000
	80	145	320	570	900	1250	2200	3400	4900	8500	14000	20000	35500	57500	79800
	120	215	450	850	1280	1890	3400	5300	7500	13400	20600	30000	55500	85500	120000
200	50	110	265	450	680	1020	1780	2800	4120	7100	11500	16300	28500	45300	64000
	80	180	410	700	1100	1560	2910	4400	6600	11000	18000	26600	46000	72300	100000
	120	250	600	1100	1630	2400	4350	6800	9400	16900	25900	37000	70600	109000	152000

306F45.EPS

Figure 45 ◆ Steam pipe sizing for steam velocity.

Example 1

A steam system is being designed that must carry a minimum of 1,500 lbs/hr of steam through the steam main. Maximum design velocity must not exceed 100 fps, and the designated steam pressure will be 100 psig. Select an appropriate pipe size for the steam main.

Step 1   Enter the left side of the chart at 100 psig. Since the velocity cannot exceed 100 fps in this example, read across the chart horizontally at the line of values for 80 fps (120 fps is obviously too fast).

Step 2   Continue across the chart until you reach a value equal to, or greater than, 1,500 lbs/hr. The closest value that is not less than our minimum of 1,500 lbs/hr is 1,660 lbs/hr.

Step 3   Read up the column to the top line of steam pipe sizes. Pipe size will be 2 inches.

When troubleshooting steam systems, the chart can also be used to check existing conditions. For example, if the system seems too noisy in operation, check the pipe size against the volume of steam being transported to see if velocities exceed the prescribed maximum of 120 fps for quiet operation. This can happen when new loads are added to an existing piping system, and steam pressures are increased to provide the necessary additional volume of steam through the existing piping.

13.2.0 Steam Line Pressure Drop

Steam lines are generally sized based on velocity first, then checked against the appropriate chart for pressure drop in the system. The allowable pressure drop varies widely, based on the system itself. Systems operating at 100 psig, with one or more loads that require a minimum of 80 psig at the inlet, certainly cannot be sized at a 25 psig allowable loss. Perhaps a more appropriate choice here would be a 5 to 10 psig loss. However, a system operating as low as 5 psig should probably be sized for a pressure drop no greater than 0.3 psig to maintain the needed pressure downstream. As a general rule, the higher the steam pressure, the greater the pressure drop that can likely be tolerated.

Figure 46 is one example of a steam pressure drop chart. Note that this chart is also designated for Schedule 80 pipe. Pressure drops for a given volume of flow will differ in other classes of pipe. Also note that the chart is based on a steam pressure of 100 psig. For pressures other than 100

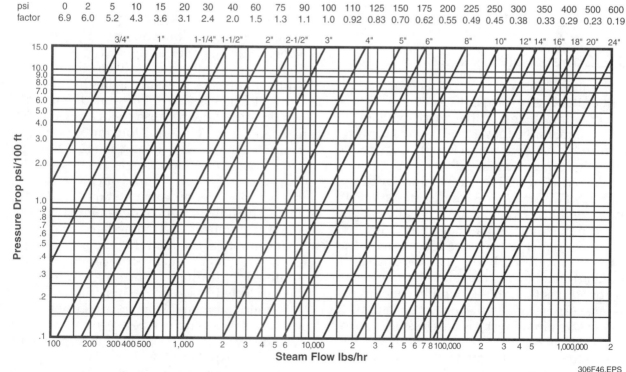

Figure 46 ◆ Steam pipe sizing for pressure drop.

306F46.EPS

psig, a correction factor must be applied to the value read from the chart. The correction factors for various pipe sizes are listed at the top of the chart. For example, if a steam system is to operate at 5 psig, then the value read from the chart at 100 psig must be multiplied by 5.2 to achieve the correct pressure drop value.

The values in the chart provide the pressure drop that exists in 100 feet of pipe. Your final answer must also be adjusted for the actual length of the pipe in your installation.

Example 2

Using the values and resulting pipe size in Example 1, calculate the pressure drop in 75 feet of pipe.

Step 1 Enter the chart at the bottom, locating the vertical line representing 1,500 lbs/hr of steam flow.

Step 2 Follow the line vertically until it intersects with the diagonal line representing 2-inch pipe.

Step 3 Follow the horizontal line left from this point of intersection and read the resulting pressure drop for 100 feet of Schedule 80 pipe. The result would be 2.0 psig of pressure loss per 100 feet of pipe.

Step 4 To adjust for the actual length of pipe (75 feet in our example), multiply 2.0 psig/100 feet by 0.75. The resulting pressure drop will then total 1.5 psig in this particular steam line.

13.3.0 Condensate Line Sizing

The steam condensate can carry as much as 15 percent of the energy originally used in the process of creating steam. With ever-rising energy costs, it is imperative that steam systems capture and return the condensate back to the process to preserve this expended energy.

Condensate line sizing can be more difficult than sizing of the steam lines, due to the variety of differing conditions found in the system. A single set of recommendations does not apply. The lines carrying condensate from the load (such as a steam coil in ductwork for a comfort heating application) and into the trap represent a different set of conditions than those found in condensate lines transporting condensate away from a trap.

Condensate piping systems are classified based on the design pressures. Gravity returns describe those systems where the condensate line is at or very near atmospheric pressure, and the condensate generally returns to a vessel vented to the atmosphere. Low-pressure returns are those operating at a pressure below 15 psig, while medium-pressure returns are those operating between 15 and 100 psig. Systems that operate with the condensate line experiencing pressures above 100 psig are classified as high-pressure returns.

In all cases, the sizing of condensate lines is based on several primary factors:

- *Pressure* – The difference in pressure from one end of the line to the other is certainly a factor. This pressure difference may cause some flash steam to be generated and may help promote flow in the line.

- *Condition of the condensate* – For a proper selection, the designer must know whether the condensate is primarily liquid, or whether a significant amount of flash steam can be expected.

- *Volume* – The anticipated volume of condensate flow is also an important factor. This condition changes dramatically between startup of a system and full, consistent operation. During startup, the temperature difference at the load and throughout the piping system, relative to the temperature of the steam supply, is greater. Far more condensate will be generated during this period in all locations, including steam mains, branch lines, and especially at the system loads.

13.3.1 Drain Lines to Traps

Condensate must be drained away from load apparatus and through the trap to prevent excessive water hammer and potential damage to piping and the internal components of the trap. As a general rule, there is no significant pressure drop in this line to consider, so flash steam resulting from changes in pressure is negligible or nonexistent. There is also no pressure differential that would assist flow because condensate movement is by gravity. The condition and volume of the condensate will, however, change dramatically from startup to operation at full load. It is not uncommon for the volume of condensate to be three times greater during startup than at full load. The flow of noncondensibles (air) through the system must be accommodated as well.

It cannot be assumed that the pipe outlet size of the load or apparatus is the proper size to promote good drainage. Once the steam trap has been selected based on design criteria, however, it is quite common for the trap inlet size to be the correct pipe size to use between the two points. In most cases, sizing the line for a volume twice that of the expected flow during a running load is sufficient to accommodate startup loads without being oversized for normal operating conditions.

13.3.2 Discharge Lines From Traps

Wherever possible, these lines should incorporate gravity as the primary mover. Due to the pressure drop of the steam trap, there is a pressure drop to be considered. Flash steam will result from this pressure difference, and will be more significant at steam pressures above 30 psig. As condensate is discharged from the trap, some small amounts of live steam may also make it past the trap. In the trap discharge line then, liquid and vapor are moving together. This is known as two-phase flow, much the same as the flow of liquid and vapor in sewer systems.

The mixture leaving the trap will exhibit more properties of steam than properties of liquid water, especially at steam pressures above 30 psig. Sizing is therefore evaluated much like steam lines. The same maximum velocities are sometimes used, although lower velocities generally work best to minimize water hammer and the creation of flash steam. Undersized drain lines from the trap, unable to handle the flow as needed, can dramatically slow down or flood the process.

13.3.3 Draining Steam Mains

As a result of heat loss through piping walls, condensate will also gather in steam mains, especially during system startup. Steam mains should be pitched, usually ½ inch per 10 feet of run, to allow the condensate to drain toward a gathering point. Collection legs, also known as drip legs, are installed on the bottom of the main along the horizontal run at 150 to 300-foot intervals. The collection leg should be the same pipe diameter as the main itself. In cases where the pitch of the steam main causes it to be too low, a riser may be needed to route the line up higher. An additional collection leg should be installed at the base of all risers in stem mains. The inlet of a steam trap is then connected to each collection leg to drain the condensate away.

13.3.4 Common Return Lines

In most cases, it is necessary to drain condensate from more than one load into a common condensate return line. The line should never become completely filled, or flooded, with condensate. It should slope toward the vented receiver tank or flash tank, when one is included. In most cases, sizing for liquid-only (no steam, flash or otherwise) condensate drainage should be for velocities at or below 7 fps. The common line is generally sized based on the combined volume of flow used to size the individual trap discharge lines routed to the common line.

14.0.0 ◆ BOILER BLOWDOWN AND SKIMMING

WARNING!

Boilers operate under high pressure and are potentially dangerous. When working with or around boilers, follow all applicable manufacturer's instructions, OSHA regulations, and job site requirements relating to safety and the use of safety equipment.

During operation, makeup water is added to boilers to maintain the proper water level. As a result of adding this water, sediment, dissolved salts, or similar organic matter gradually accumulate in the boiler. Some of these impurities cause foam to form on the surface of the boiler water. This is usually indicated by drops of water appearing with the steam. If all the foreign matter is not purged from the boiler on a regular basis, boiler operation will degrade. The process of purging the boiler is called blowdown. Every boiler is equipped with one or more valves (blowdown valves) for this purpose. Two kinds of blowdown valves may be found on a boiler: surface blowdown (**skimming**) valves and/or bottom blowdown valves.

Surface blowdown valves are used to skim off impurities that cause foaming on the surface of the water inside the boiler. This foaming is caused by high surface tension due to a scum buildup from oil, grease, and/or sediment on the surface of the water. Foaming prevents steam bubbles from breaking through the water surface and hinders steam production. Surface blowdown valves are located on the boiler at the normal operating water level. Not all boilers are equipped with a surface blowdown valve. Use of the surface blowdown valve to skim off the impurities requires some experience. Skimming should always be done according to the instructions given in the boiler manufacturer's service literature.

14.1.0 Skimming Procedure

WARNING!

Never add cold water to a completely empty boiler when the burners are ignited. Doing so will cause serious damage and may create an explosion hazard.

Step 1 If not permanently connected to a drain, run a temporary connection from the boiler's skimming valve to a suitable drain.

Step 2 With the boiler empty and cool, slowly begin to add water. After a quantity of water has entered the boiler, turn on the burners and adjust the firing rate and/or burner controls so that the water being added is kept just below the boiling point. Boiling and turbulence must be avoided.

Step 3 Gradually raise the hot water level in the boiler to the point where the water just flows from the skimming valve, being careful not to raise it above this point.

NOTE

Skimming will not clean the boiler of sediment that has accumulated at the bottom. After skimming, the boiler should be cleaned further by performing the blowdown procedure.

Step 4 Continue to skim the boiler water until there is no trace of impurities. Water may be checked to make sure that it is free from oil by drawing off a sample. If the sample is reasonably free of oil, it will not froth when heated to the boiling point.

14.2.0 Blowdown Procedure

NOTE

Only attempt blowdown at a light load. Ideally, temporarily suspend the boiler's heating process to halt water turbulence and allow the solids to settle out.

Step 1 Check the water level in the boiler.

Step 2 Partially open the bottom blowdown valve. Once the water starts draining, fully open the valve.

Step 3 Remain at the blowdown valve. Monitor the gauge glass during blowdown to make sure that the water level is not lowered to a dangerously low point.

Step 4 When the desired amount of water has been drained from the boiler, close the blowdown valve.

Another method used to purge the boiler of foreign matter is to drain off some of the water from

the bottom of the boiler, then replace it with an equal amount of fresh, clean water. The bottom blowdown valve is used to drain off sediment, scale, and other impurities that settle out of the water as it is heated and accumulate at the lowest point in the boiler. If the boiler is extremely dirty, blowdown may require that the boiler be completely drained and refilled with water one or more times. This process must be continued until the water discharged from the boiler runs clear. On smaller systems, the hot boiler water discharged through the bottom blowdown valve is drained directly into a nearby sewer. In larger systems, it is piped through a line to a tank where it is allowed to cool before going into the sewer. Codes require that waste water from a steam or hot-water boiler not be hotter than 140°F. Boiler blowdown should always be performed as directed in the boiler manufacturer's service instructions.

15.0.0 ◆ BOILER WATER TREATMENT

The water used in a boiler must be kept clean for efficient operation. Never add dirty or rusty water to a boiler. Hard water may eventually interfere with the efficient operation of a boiler. For this reason, hard water should be chemically treated with water softeners before being used. Scale, corrosion, fouling, and foaming can all cause boiler problems. Scale deposits on heating surfaces increase boiler temperatures and lower operating efficiency. Corrosion can damage metal surfaces, resulting in metal fatigue and failure. Fouling clogs nozzles and pipes with solid materials, thereby restricting circulation and reducing the heat transfer efficiency. Foaming results in over-heating and can cause water impurities to be carried along with the steam into the system.

There are various chemical and mechanical methods of water treatment used to help prevent problems caused by water impurities. At a minimum, proper water treatment includes blow-down, pH control, and the addition of chemicals to neutralize other contaminants. The specific methods and devices used for water treatment are covered in *HVAC Level Four*.

Review Questions

1. The amount of latent heat required to vaporize, or evaporate, 5 pounds of water into steam is _____ Btus.
   a. 970
   b. 1,114
   c. 1,294
   d. 4,850

2. The latent heat of condensation is defined as _____.
   a. the heat removed during the change of state from steam to water
   b. the heat gained during the change of state from water to steam
   c. the heat lost in changing water to ice
   d. any heat added after ice changes to water

3. The majority of maintenance attention in a steam system is generally devoted to the _____.
   a. piping system
   b. pressure regulating valves
   c. steam traps
   d. heat exchangers

4. Steam boilers _____.
   a. have exactly the same type of controls as hot water boilers
   b. are entirely filled with water
   c. must operate at extreme pressures only
   d. are partially filled with water

5. A boiler with a maximum operating pressure of 75 psi would be classified as a _____ boiler.
   a. high-pressure
   b. low-pressure
   c. process
   d. medium-pressure

6. A _____ assembly is generally installed between the boiler and an external pressure gauge to protect the gauge from contacting raw steam.
   a. strainer
   b. siphon
   c. heat exchanger
   d. trap

7. Steam-to-water heat exchangers designed for potable hot water use are of _____ construction.
   a. stainless steel
   b. aluminum
   c. hermetically sealed
   d. double-walled

8. Mechanical steam traps take advantage of the difference in density between steam and _____.
   a. air
   b. steam condensate
   c. gases
   d. debris

9. When a bypass line around a trap is installed to facilitate maintenance activities and allow the system to continue operating, a _____ valve should be used to allow manual balancing of flow.

    a. gate
    b. pressure reducing
    c. globe
    d. vacuum breaker

10. Bimetallic steam traps function based on a difference in _____ between condensate and steam.

    a. air volume
    b. temperature
    c. weight
    d. pressure

11. The three basic diagnostic methods for steam systems, especially traps, are _____.

    a. pressure, temperature, and time
    b. temperature, flow, and sound
    c. current, pressure, and sound
    d. sight, sound, and temperature

12. The condensate receiver tank is equipped with a _____ to maintain and ensure a sufficient supply of water for the boiler is available.

    a. flash reservoir
    b. float-actuated makeup water valve
    c. thermodynamic trap
    d. vacuum pump

13. When more than one condensate line is connected to a flash tank, each line should be equipped with its own _____.

    a. check valve
    b. pressure reducing valve
    c. condensate pump
    d. sight glass

14. Collected flash steam is generally discharged into a _____.

    a. floor drain
    b. vacuum pump
    c. boiler
    d. low-pressure main

15. When the condensate return line is above the boiler water line, it is called the _____ line.

    a. steam
    b. Hartford loop
    c. dry return
    d. wet return

16. Most two-pipe steam systems use a _____.

    a. mechanical condensate return
    b. gravity condensate return
    c. fixed-orifice trap
    d. very long startup cycle

17. In a vacuum return system, _____ can cause excessive vacuum pump operation.

    a. too much condensate
    b. low steam pressure
    c. air leaks
    d. heat exchangers

18. Operating steam boilers at pressures significantly lower than their design operating pressure increases the likelihood of _____.

    a. leaks
    b. thermal shock
    c. water droplet carryover
    d. trap damage

19. Steam system piping that is too large causes _____.

    a. excessive pressure drop
    b. some trap failures
    c. high installation costs and greater heat loss
    d. condensate pump overload

20. Steam supply piping for a system designed to operate at 5 psi should generally be designed for a pressure drop no greater than _____ psi.

    a. 0.3
    b. 2
    c. 5
    d. 10

Summary

Steam systems generate and distribute steam used for comfort heating and commercial and industrial processes. Steam-heating systems can be separate systems used only to heat buildings, or they can provide energy in a combined system for both comfort and process use.

Boilers used in steam systems are divided into the two broad categories of low-pressure and high-pressure boilers. Low-pressure steam boilers are normally used in residential and commercial applications for both heating and process use. High-pressure boilers are typically used in larger systems for college campuses, hospitals, and industrial applications.

Steam traps provide a means to separate live steam from condensate, trapping the steam inside the terminal until it has condensed, thereby giving up the vast majority of the energy it contains before being allowed to pass through the trap. Steam traps automatically open an orifice, drain away accumulated condensate, and then re-close once the condensate is removed. They can be classified based on their operating characteristics as mechanical, thermostatic, or thermodynamic. Steam traps that fail can be quite costly in terms of energy loss, so the proper maintenance of steam traps is crucial to control energy costs and ensure good system performance.

Steam piping systems are either one pipe or two pipe in design. The piping approach is determined by the requirements to supply steam to the terminals and by the destination of the condensate. Condensate can be returned by gravity when appropriate, but many systems will require mechanical assistance when the condensate is to be returned to the boiler for re-use.

Notes

Blowdown: The process of purging a boiler of foreign matter.

Drip leg: A drain for condensate in a steam line placed at a low point or change of direction in the line and used with a steam trap.

Dry steam: The steam that exists after all the water has been vaporized into steam at its saturation temperature.

Flash steam: Formed when hot condensate is released to a lower pressure and re-evaporated.

Latent heat: The heat energy absorbed or rejected when a substance is changing state (solid to liquid, liquid to gas, or vice versa) and there is no change in the measured temperature.

Latent heat of condensation: The heat given up or removed from a gas in changing back to a liquid state (steam to water).

Latent heat of fusion: The heat that is gained or lost in changing to or from a solid (ice to water or water to ice).

Pressure drop: The difference in pressure between two points. In a steam system, it is the result of power being consumed as the steam moves through pipes, heating units, and fittings. It is caused by the friction created between the inner walls of the pipe or device and the moving steam.

Saturated steam: The pure or dry steam produced at the temperature that corresponds to the boiling temperature of water at the existing pressure.

Sensible heat: Heat that can be measured by a thermometer or sensed by touch. The energy of molecular motion.

Skimming: The process of removing impurities and foam from the water surface of a steam boiler; also commonly referred to as surface blowdown.

Specific heat: The amount of heat required to raise the temperature of one pound of a substance one degree Fahrenheit. Expressed as Btu/lb/°F. At sea level, water has a specific heat of 1 Btu/lb/°F. At sea level, air has a specific heat of 0.24 Btu/lb/°F.

Subcooling: The reverse of superheat. It is the temperature of a liquid when it has cooled below its condensing temperature.

Superheated steam: Steam which has been heated above its saturation temperature.

Trim: External controls and accessories attached to the boiler itself, such as sight glasses and water feeder controls.

Turndown ratio: In steam systems, the ratio of downstream pressure to upstream pressure, usually applied to steam pressure-reducing valves. More than a 50 percent turndown ratio is generally considered a large pressure reduction.

Water hammer: A condition that occurs when hot steam comes into contact with cooled condensate, builds pressure, and pushes the water through the line at high speeds, slamming into valves and other devices. Water hammer also occurs in domestic water systems. If the cause is not corrected, water hammer can damage the system.

Additional Resources and References

Additional Resources

This module is intended to be a thorough resource for task training. The following reference work is suggested for further study. This is optional material for continued education rather than for task training.

www.spiraxsarco.com

Figure Credits

Carrier Corporation, 306F05

Burnham Hydronics, 306F06

Kele, 306F08

Spirax Sarco Inc., 306F09, 306F11, 306F20, 306F21, 306F45, 306F46
www.spiraxsarco.com

Courtesy ITT, 306F12, 306F13, 306F14, 306F15, 306F18, 306F37, 306F39, 306F41, 306F43, 306F44

Anderson Greenwood, 306F17

Leslie Controls, Inc., 306F23, 306F24

McQuay International, 306F26, 306F27

Armstrong International, Inc., 306F28

Watson McDaniel, 306F31 (photo)

Courtesy of UE Systems, Inc., 306F34

Fluke Corporation, 306F35

Weed Instruments, Co., 306F36

ECR International – Dunkirk Boilers, 306SA01

Courtesy of Honeywell International Inc., 306F16

NCCER CURRICULA — USER UPDATE

NCCER makes every effort to keep its textbooks up-to-date and free of technical errors. We appreciate your help in this process. If you find an error, a typographical mistake, or an inaccuracy in NCCER's curricula, please fill out this form (or a photocopy), or complete the online form at **www.nccer.org/olf**. Be sure to include the exact module ID number, page number, a detailed description, and your recommended correction. Your input will be brought to the attention of the Authoring Team. Thank you for your assistance.

Instructors – If you have an idea for improving this textbook, or have found that additional materials were necessary to teach this module effectively, please let us know so that we may present your suggestions to the Authoring Team.

NCCER Product Development and Revision

13614 Progress Blvd., Alachua, FL 32615

Email: curriculum@nccer.org
Online: www.nccer.org/olf

❏ Trainee Guide   ❏ AIG   ❏ Exam   ❏ PowerPoints   Other _____

Craft / Level: _____ Copyright Date: _____

Module ID Number / Title: _____

Section Number(s): _____

Description: _____

_____

_____

_____

Recommended Correction: _____

_____

_____

_____

Your Name: _____

Address: _____

_____

Email: _____ Phone: _____

03307-08

Planned Maintenance

03307-08
Planned Maintenance

Topics to be presented in this module include:

1.0.0	Introduction	.7.2
2.0.0	Fossil-Fuel Heating Appliances	.7.4
3.0.0	Cooling Units	.7.24
4.0.0	Heat Pumps	.7.31
5.0.0	Accessories	.7.32

Overview

Simple periodic servicing by qualified service technicians, such as replacing air filters and inspecting heat exchangers, allows systems to operate more efficiently and prevents premature equipment failures. In addition, periodic inspection can prevent harmful gases from threatening building occupants. Without scheduled maintenance, HVAC equipment will not perform properly or efficiently over the long term, and the useful life of the equipment will be reduced. In this module, you will learn about planned maintenance for residential and light commercial heating and cooling systems, as well as the accessories used with these systems. Proper tools and service procedures for planned maintenance of these systems and accessories are discussed.

Objectives

When you have completed this module, you will be able to do the following:

1. Describe planned maintenance and service procedures required for selected HVAC equipment and components.
2. Develop a planned maintenance and service checklist for selected HVAC equipment and accessories.
3. Perform identified service and maintenance tasks on selected HVAC equipment, components, and accessories.
4. Identify the tools and materials necessary for performing service and maintenance tasks.
5. State the safety practices associated with the servicing of selected HVAC equipment, components, and accessories.

Trade Terms

Combustion efficiency	Safety drop time
Hantavirus	Static pressure drop
Ionizing wires	Unbalanced flame

Required Trainee Materials

1. Pencil and paper
2. Appropriate personal protective equipment

Prerequisites

Before you begin this module, it is recommended that you successfully complete *Core Curriculum*; *HVAC Level One*; *HVAC Level Two*; and *HVAC Level Three*, Modules 03301-08 through 03306-08.

This course map shows all of the modules in the third level of the *HVAC* curriculum. The suggested training order begins at the bottom and proceeds up. Skill levels increase as you advance on the course map. The local Training Program Sponsor may adjust the training order.

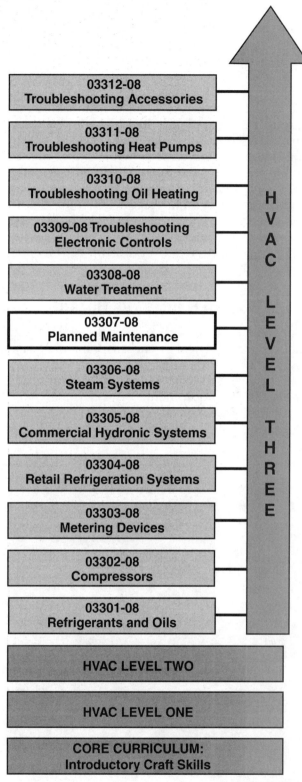

307CMAP.EPS

1.0.0 ◆ INTRODUCTION

Safe and economical operation of HVAC equipment depends on proper installation, service, and maintenance. If the system is deficient in any way, provisions must be made to correct the problem by replacing obsolete parts or servicing the system. The HVAC service technician (*Figure 1*) must be a skilled individual who is able to perform service and maintenance tasks on selected equipment and accessories while exercising caution in regard to personal and equipment safety.

Always include the make, model, and serial number on everything you document for the unit being serviced. This information can be found on the unit nameplate. The serial number is particularly important when obtaining replacement parts for the unit.

1.1.0 Rodent Hantavirus Hazard

Many of the hazards encountered by HVAC technicians are electrical or chemical. In some cases, biological hazards must also be considered. One such hazard is associated with rodents nesting in or near HVAC equipment.

According to the National Center for Infectious Diseases (NCID), a component of the Centers for Disease Control and Prevention (CDC), **hantavirus** pulmonary syndrome (HPS) is a potentially fatal respiratory disease. Caused by hantaviruses carried by rodents, particularly deer mice, HPS can be contracted by humans who come into contact with the rodents' wastes or nesting materials. While HPS is not common in the United States, HVAC technicians must take steps to guard against it.

The most effective way of preventing hantavirus infection is to eliminate contact with rodents. This is especially important when working with equipment mounted on the ground, such as the outdoor sections of a heat pump. While poorly maintained mechanical spaces may be infested, equipment that sits on the ground is much more accessible to mice, rats, and other vermin. These creatures are attracted to the protection afforded and sometimes the warmth generated by the equipment.

You won't normally see rodents, but their nests and droppings will be visible.

WARNING!

Rodent nests and droppings may harbor hantavirus. Wear goggles and heavy latex or rubber gloves when removing these materials. If possible, wear disposable coveralls and rubber boots. If hantavirus has occurred in your area, wear a full-face respirator mask with a high-efficiency particulate air (HEPA) filter. A self-powered half- or full-face respirator or a supplied-air full-face respirator is recommended by NIOSH for maximum protection. Make sure you are properly trained in respirator use. Also, do not use a broom or vacuum cleaner to remove these materials, because this may stir up infected particles. Follow all appropriate health codes.

Use the following procedure to remove rodent wastes and nesting materials:

Step 1 Spray the area to be cleaned with a liquid disinfectant or 10 percent bleach solution.

Step 2 Once the area is wet, use a towel soaked in the cleaning solution to remove the material. If you find any dead rodents, spray them liberally with the cleaning solution before removing them using gloves or tools.

Step 3 Using the cleaning solution, mop and sponge down the cleaned area again.

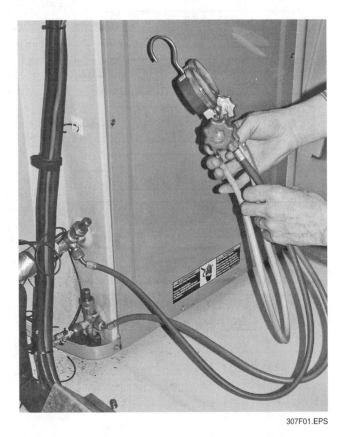

307F01.EPS

Figure 1 ◆ Technician working on HVAC equipment.

Symptoms of HPS

Step 4 Double-bag all used cleaning materials. Dispose of them properly following all appropriate health codes. This normally involves burning or burying the materials. If neither of these options is feasible, contact the state health department for instructions. Make sure to disinfect your gloves before removing them for disposal. Then, wash your hands with an appropriate disinfectant.

For more information on hantavirus and HPS, refer to the Centers for Disease Control website at www.cdc.gov.

1.2.0 Bird and Bat Excrement Hazards

Bird and bat excrement (droppings) can present a hazard to HVAC technicians when servicing attic-mounted and outdoor ground- or roof-mounted units, especially if the units are located under bird roosts at building eaves and in attics where birds or bats are roosting or have roosted. In some rare cases, the equipment itself may contain bird roosts. The following diseases are associated with organisms found in dried bird and bat droppings:

- *Cryptococcosis* – Cryptococcus neoformans (C. neoformans) is commonly found in the dust of pigeon, dove, or bat droppings. Cryptococcosis is a yeast infection contracted by inhaling the mold spores of the organism. The mold cells are 5 to 8 microns in diameter and are easily airborne. An infection may result in cryptococcal meningitis, which is difficult to diagnose and can be fatal if not properly and promptly treated.
- *Histoplasmosis* – Histoplasma capsulatum (H. capsulatum) is a fungus found in soils through-

out the world. It flourishes in soils that have been enriched by bird or bat droppings for three or more years; however, it has been found in droppings not in contact with soil. It is difficult to eliminate from soil even after the nutrient source is removed. Histoplasmosis is contracted by inhaling the spores of the fungus carried by wind and dust. Most infections produce no symptoms, or only a mild flu-like illness. However, untreated chronic or reactivated infections can result in pneumonia, blindness, chronic and progressive lung disease, and even death.

- *Psittacosis (also called ornithosis or parrot fever)* – Psittacosis is caused by inhaling Chlamydia psittaci bacterium that is found in feathers and droppings from any infected birds, including pigeons and doves. Because the organism becomes less infectious with time, active roosts are the greatest concern. The disease results in fever, headaches, and muscle pain, with or without obvious respiratory symptoms. Untreated cases can progress to pneumonia or toxemia and can result in death, especially in older people.
- *Rabies* – Rabies is a viral disease. Contact with the dried blood, urine, or droppings from an infected bat is not a risk factor for contracting rabies once the bats have been excluded from the area. However, dead bats should be picked up and disposed of using a tool or heavy gloves because the rabies virus can remain infectious in a carcass until decomposition is well advanced. If any scratch from a live or dead bat, or a bite from a live bat occurs and the bat cannot be retrieved for testing, prompt rabies vaccinations must begin or death will occur once the symptoms of rabies appear.

If HVAC equipment or the surrounding area is contaminated with bird or bat droppings, the building owner should be promptly notified so that professionals can eliminate the birds or bats from the roost, seal the area of the roost, and clean up the droppings. If the equipment must be serviced, temporary cleaning of the equipment and close surrounding area can be accomplished by using the procedures described in the previous section. NIOSH recommends that gloves, full disposable suits, and supplied-air full-face respirators be used for cleanup of bird and bat droppings. However, individuals using this type of equipment in high ambient temperature and humidity environments require close monitoring for heat exhaustion.

2.0.0 ◆ FOSSIL-FUEL HEATING APPLIANCES

In order to operate safely and efficiently, gas-fired and oil-fired furnaces should be serviced at least once a year by a qualified technician. Safety is an especially important consideration because the burning of fossil fuels produces by-products, such as carbon monoxide, that can be harmful to people as well as the environment.

Carbon monoxide (CO), even in small concentrations, can be dangerous because its effect on the human body is cumulative. It slowly builds up in the bloodstream, replacing oxygen until there is not enough oxygen in the blood to support life. Carbon monoxide is colorless, odorless, and non-irritating, so it can only be detected with a test instrument. For this reason, it is important to check carbon monoxide levels when servicing a furnace. It has become common to install carbon monoxide detectors in buildings.

The appearance of the flame and the presence (or absence) of soot deposits on burners and heat exchangers can help determine something about the **combustion efficiency** of a furnace. To get an accurate picture of combustion efficiency, and to determine the level of toxic gases being released into the building and the outdoors, it is necessary to use more sophisticated measuring equipment. Careful inspection of heating equipment is also helpful in preventing carbon monoxide problems. Cracked heat exchangers, inadequate or obstructed vents, and obstructions around the furnace that prevent adequate airflow are some factors that can cause excessive carbon monoxide levels. As part of every service call, check to make sure that the furnace has sufficient draft to remove flue gases. In natural-draft systems, this can be done with a match or smoke candle. It merely involves checking to be sure there is a positive draft, with no flue gases spilling out of the vent.

In some areas, state or local codes may require testing for carbon monoxide, sulfur dioxide, and oxides of nitrogen (nitric oxide and nitrogen dioxide) that contribute to smog and acid rain. This is primarily applicable to large commercial installations. Instruments are available for measuring these gases.

2.1.0 Gas Heating

Gas heating is the most common type of heating system available. When properly serviced, gas heating equipment (*Figure 2*) is efficient, clean, and operates quietly. To keep gas equipment trouble-free, there are a number of safety and servicing procedures for the HVAC technician to master.

2.1.1 Gas Heating Safety

Safety is the most important part of your job, both for self-protection and for the protection of others. The following safety rules pertain to working with gas heating:

 WARNING!
When performing planned maintenance that requires the dismantling of gas burners or any gas line components for cleaning or repair, or if a misfire occurs, always shut off the gas supply before beginning work. Allow five minutes to elapse so that any unburned gas will dissipate. Liquid petroleum (LP) gas is heavier than air and will not vent upward. If you suspect that a concentration of LP gas is present, *do not* turn on any lights or electrical appliances. These devices may generate a spark and ignite a gas explosion. Open doors and windows to ventilate the area.

- Always conduct a leak test after completing any installation, repair, or service. Never use a flame to check for a gas leak. Use an electronic leak detector or a rich soap-bubble solution applied to tubing connections, pipe joints, and valve gaskets. The bubbles will indicate a leak. Tighten joints and screws or replace gaskets as needed to stop any leaks found.

- Always disconnect the power supply before connecting or disconnecting any wiring (except for millivoltage controls) to prevent the danger of electrical shock, damage to equipment, or danger of explosion. In general, always turn off and lock out the power unless the procedure must be done with power applied.

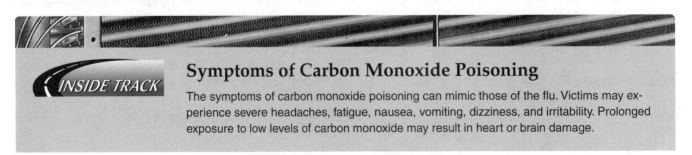

INSIDE TRACK

Symptoms of Carbon Monoxide Poisoning

The symptoms of carbon monoxide poisoning can mimic those of the flu. Victims may experience severe headaches, fatigue, nausea, vomiting, dizziness, and irritability. Prolonged exposure to low levels of carbon monoxide may result in heart or brain damage.

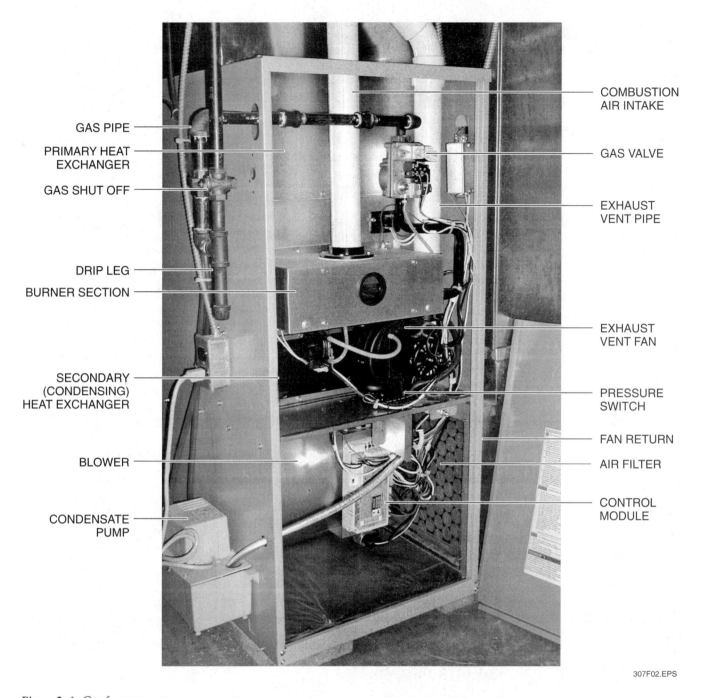

GAS PIPE

PRIMARY HEAT
EXCHANGER

GAS SHUT OFF

DRIP LEG

BURNER SECTION

SECONDARY
(CONDENSING)
HEAT EXCHANGER

BLOWER

CONDENSATE
PUMP

COMBUSTION
AIR INTAKE

GAS VALVE

EXHAUST
VENT PIPE

EXHAUST
VENT FAN

PRESSURE
SWITCH

FAN RETURN

AIR FILTER

CONTROL
MODULE

307F02.EPS

Figure 2 ◆ Gas furnace.

- Never jumper or short the gas valve coil terminals on 24V controls. This could burn out the heat anticipator in the thermostat or short out the 24V transformer.

2.1.2 *Tools, Equipment, and Materials for Servicing Gas Heating Equipment*

Specialized knowledge is required to service gas heating control circuits and devices. You must be able to properly and safely use the following equipment:

- Voltmeter
- Ammeter
- Millivoltmeter
- Milliammeter
- Ohmmeter
- Wire brush
- Soap and water solution
- Vacuum cleaner with HEPA filter
- Respirator
- Carbon dioxide tester (if not part of combustion test equipment)

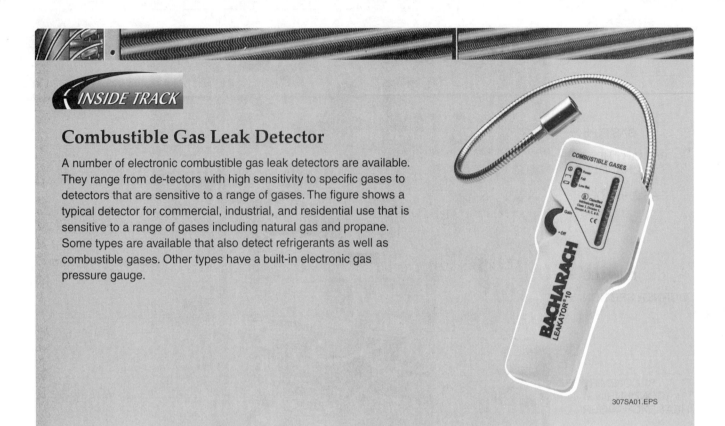
- Lubricating oil
- Mechanic's tool set
- Fiber optic borescope
- Gas burner combustion test equipment
- Carpenter's level
- Carbon monoxide tester
- Oil burner nozzle wrench

2.1.3 Service Procedures for Gas Heating

Before performing any service procedure, check with the customer and listen to any complaints. Make a note of any comment that would indicate a possible source of trouble or equipment malfunction. Then proceed as follows:

- Check the thermostat.

- Move the system switch to the HEAT or AUTO position. Set the heat above room temperature. The heating system should start and the fan should run after a short delay.
- Move the heat lever below room temperature. The heating unit should shut off. The fan should run for a short time and then shut off.

- Remove the thermostat cover. Check the terminal connections for tightness.
- For mechanical thermostats, dust the interior and clean the contacts if it is not a mercury switch type. For battery-operated electronic thermostats, replace the batteries annually.
- Reinstall the thermostat cover.

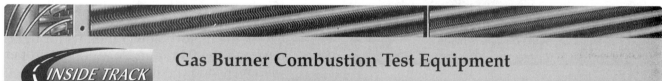

INSIDE TRACK

Gas Burner Combustion Test Equipment

Both mechanical and electronic gas burner combustion test equipment is available. The mechanical equipment shown is available as a tool kit that contains indicators for CO_2, O_2, and CO in flue gas samples, a dial thermometer for flue gas or plenum temperature measurements, a draft gauge, a gas pressure manometer, a gas service calculator, and a fire efficiency slide rule, along with applicable instruction manuals. The slide rule is used to determine combustion efficiency (Annual Fuel Utilization Efficiency or AFUE that is sometimes called EFF) using data from the CO_2/O_2 indicators and the dial thermometer.

With appropriate probes, the electronic gas/oil analyzer directly measures and displays O_2, CO, stack temperature, draft, differential pressure, combustion air temperature, and other optional selected gas emissions. It simultaneously calculates and displays combustion efficiency, excess air (EA), CO_2, NO_X, and O_2 reference values. The analyzer performs combustion calculations for ten fuels including natural gas and propane, as well as #2, #4, and #6 fuel oil.

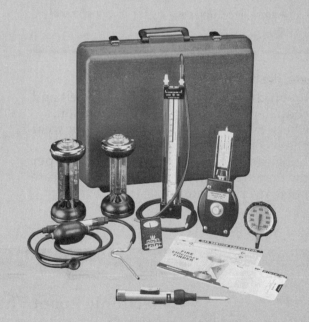

ELECTRONIC GAS/OIL
BURNER COMBUSTION ANALYZER

MECHANICAL GAS BURNER
COMBUSTION TEST EQUIPMENT

307SA02.EPS

NOTE

For some battery-operated programmable electronic thermostats, the programming will be lost when the batteries are replaced. It is good practice to make a record of the programming before replacing the batteries of any battery-operated programmable thermostat.

Do not adjust a room thermostat heat anticipator unless it is absolutely necessary. If the customer has not complained of discomfort, adjusting the control by the book will accomplish nothing. It might even prompt the customer to complain after the change is made.

- Check the blower compartment.
  - Check and record the supply voltage at the unit disconnect.
  - Close the blower compartment door. Check and record the amperage draw of the blower motor and check it against the nameplate. The blower compartment door must be closed to perform this procedure.
  - Turn off the power at the disconnect and check all wiring for loose connections and broken or frayed insulation.
  - Change or clean the filters as applicable.

CAUTION

Wear a proper HEPA respirator when changing/cleaning filters and cleaning the blower compartment.

- Check the blower.
  - If belt-driven, remove the blower belt and check for cracks, wear, and deterioration.
  - Check the motor/blower bearings for wear and record the conditions.
  - Lubricate the motor/blower bearings per the manufacturer's instructions, taking care not to overlubricate.

- Clean (vacuum) the blower wheel and compartment. If the blower wheel is very dirty, remove the motor/blower and clean the wheel with a detergent solution. Dry the blower wheel before reinstalling the motor/blower.
- If belt-driven, check the motor and pulley setscrews for tightness and check the pulleys for proper alignment.
- Check the blower wheel for balance and free rotation.
- If belt-driven, replace the belt and adjust the belt tension, if applicable.
- Check the motor/blower mounts for tightness and the vibration isolators for proper placement and condition.

- Check the gas burner and controls.
  - Check all the electrical wiring for proper connections, tightness, and corrosion. Check the insulation for damage and fraying.

CAUTION

When removing burners, use caution to prevent displacing or damaging any pilot flame assembly, hot surface ignitor, spark ignitor, or flame sensor. These devices may have to be carefully removed first before the burners are removed.

NOTE

With the burners removed, check the interior of the heat exchanger. If heavy carbon deposits are found or the heating unit is over nine years old, the heat exchanger(s) must be cleaned and inspected for corrosion, cracks, or holes as detailed later in this section.

- With the gas and power off, remove and clean the burners and, if present, air shutters inside and out using a wire brush and bottle brush. Reinstall the burners.

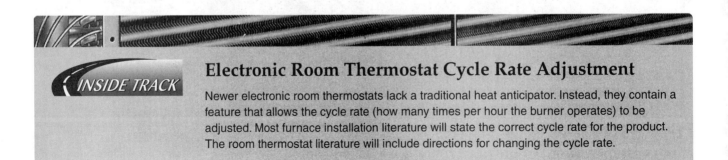

INSIDE TRACK

Electronic Room Thermostat Cycle Rate Adjustment

Newer electronic room thermostats lack a traditional heat anticipator. Instead, they contain a feature that allows the cycle rate (how many times per hour the burner operates) to be adjusted. Most furnace installation literature will state the correct cycle rate for the product. The room thermostat literature will include directions for changing the cycle rate.

- Turn on the power and gas supply. Light the pilot if required; otherwise, initiate a call for heat.
- Check the pilot flame (*Figure 3*). If necessary, change or clean the orifice. Adjust the pilot flame at the gas valve. The flame should cover ⅜" to ½" of the thermocouple tip (*Figure 4*).

- Check for quiet, even burner ignition and operation.
  - Initiate a call for heat and observe the burner flame. It should be a soft blue. If there is yellow in the flame, adjust the primary air supply at the burner if possible.
  - If the flame will not clean up, shut off the main gas supply valve and check the burner orifices for blockage or damage. Replace any blocked or damaged orifices.

If the flame is still too small, check for a clogged pilot orifice. If the orifice is OK, check for a clogged gas filter. Then check for low gas supply pressure and adjust pressure, if necessary.

Check for:
- Clogged orifice filter
- Clogged pilot filter
- Low gas supply pressure

A lazy, yellow flame means lack of air. This problem may come from an overly large orifice, a dirty lint screen, or a dirty primary air opening.

Lack of Air From
- Large orifice
- Dirty lint screen
- Dirty primary air opening

A waving, blue flame means an excessive draft at the pilot location. Install a shield to protect the pilot. A waving blue flame may also indicate a cracked heat exchanger.

Means
- Excessive draft at pilot location
- Recirculating products of combustion

A noisy, lifting flame means high gas pressure.

Relieve this situation by adding a pilot pressure reducer to the pilot gas line. It will help increase thermocouple life and also reduce gas consumption.

Means
- High gas pressure

A hard, sharp flame may mean the orifice is too small. However, this flame is normal for manufactured, butane-air, or propane-air fuels.

Means
- Characteristic of manufactured, butane-air, and propane-air fuels
- Orifice too small

307F03.EPS

Figure 3 ◆ Pilot flame characteristics.

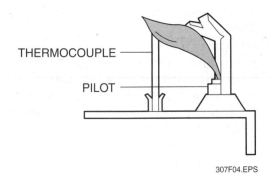

THERMOCOUPLE

PILOT

307F04.EPS

Figure 4 ◆ Flame adjustment.

NOTE

On LP burners, the pressure in the tank changes as the fuel supply tank empties. The pilot and burner flame may be affected; therefore, do not adjust the flame unless the supply tank is at least 50 percent full. A two-stage regulator on the LP tank can reduce or eliminate this problem.

CAUTION

The inside diameter of the orifice is precisely drilled and has a clean, smooth inner surface to provide unimpeded gas flow. Attempting to clean a blockage with a sharp wire or drill bit may scratch the inner surface. This may disrupt the gas flow, causing a variety of burner-related problems. Blow out the orifice with CO_2 or nitrogen.

- Check and record the operating pressure of the gas in the manifold. The recommended pressure for natural gas is 3.0 to 4.0 in. w.c. For LP gas, it is 10.5 to 11.5 in. w.c. Excess gas pressure will overfire the burner, which can damage the heat exchanger. Too little gas pressure will underfire the burner and could cause condensation of water vapor and rusting of the heat exchanger. To check the gas pressure, proceed as follows:
  - Turn off the unit disconnect switch and gas supply.
  - Remove the pipe plug (pressure tap) in the gas manifold or gas valve and insert a fitting or plug valve.
  - Connect a U-tube manometer (*Figure 5*) or gas pressure gauge hose to the plug valve.
  - Turn on the disconnect switch and gas supply. Then turn the thermostat to the highest setting. With the burners operating, open the plug valve, and record the pressure reading.

INSIDE TRACK

Adjustments in Modern Furnaces

Older natural-draft gas furnaces often contain air shutters on the burners to adjust combustion air. Many modern furnaces using induced-combustion technology do not provide for any field adjustment of combustion air. They also use electric hot surface ignitors to fire the burners and electronic flame sensors to shut off the gas control in the event of burner misfire or flameout.

- If the pressure is over or under the recommended setting, adjust the pressure regulator to conform to specifications.
- De-energize the furnace disconnect, remove the plug valve or fitting, and reinsert the pipe plug in the manifold. Check for leaks with soap bubbles.
- Re-energize the furnace disconnect to place the system into service. Check the flame shape and color and make sure the burner ignites properly. Refer to the burner manufacturer's service literature for more specific information.

INSIDE TRACK

Millivolt Systems

Thermocouples are used in standing pilot gas furnaces to provide proof to the gas valve that the pilot flame exists. The thermocouple generates a 30-millivolt signal from the heat of the pilot flame. Gas systems that do not require any external power employ a thermopile instead of a thermocouple. Thermopiles generate 750 millivolts. This voltage is sufficient to power the gas valve, which is the only electrical load in the circuit. Low-resistance thermostats are used to prevent voltage drop in the control circuit. Because the system operates at such a low voltage and current, all electrical connections must be clean, tight, and free of corrosion. This keeps resistance in the circuit to a minimum. Systems operating from a thermopile have no fans or blowers; heating is provided to the space by convection and natural drafts only.

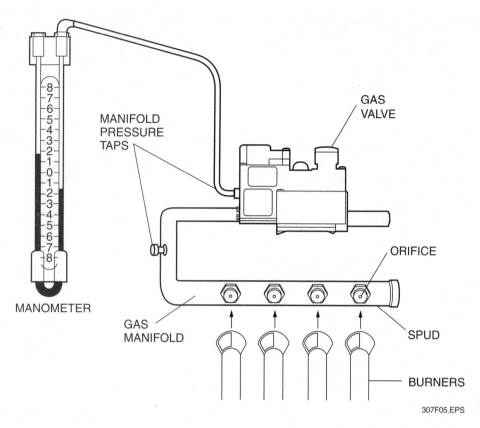

Figure 5 ◆ Gas pressure check.

- If equipped with a pilot flame assembly, check the **safety drop time** of the pilot valve.
  - Turn the pilot flame out and start timing by stop watch or second hand.
  - Listen for a click from the gas valve. When you hear the click, check the elapsed time. The maximum time should not exceed 2½ minutes. The millivoltage (mV) reading of the unheated thermocouple should be less than 4mV, or as specified by the manufacturer.
  - If the gas valve is sticking and does not drop out, replace it.

- Check the automatic vent damper system, if applicable.
  - Adjust the room thermostat to 10°F above the conditioned space. The damper must be fully open before the ignition control attempts to light the pilot burner.
  - When the damper is fully open, the ignition control should light the pilot gas, which in turn should light the burner gas.
  - Test for air spillage at the draft hood (flame or smoke method).

- Turn the thermostat to 10°F below the conditioned space temperature. The pilot and main burner should go out and the damper should close.
  - Return the thermostat to the normal setting as specified by the customer.

- Check the electronic spark ignition control (*Figure 6*) if the unit is so equipped.

Figure 6 ◆ Ignition controls.

Thermal Vent Dampers

Vent dampers are an obsolete technology used on natural draft furnaces. Most vent dampers are electrically operated. However, there are thermally-operated vent dampers. In a thermally-operated vent damper, the damper element is made of a bimetal that warps and opens the damper in the presence of heat. No electrical power is required.

NOTE

Many microprocessor-controlled systems will go through a spark and purge cycle many times over a period of several minutes, before lockout occurs. Check the product literature to see how the particular ignition system is designed.

Many modern, induced-draft furnaces use a hot surface ignitor (HSI) for direct ignition of the main burners. No pilot is used. The hot surface ignition system operating sequence is:

1. The room thermostat calls for heat.
2. The inducer runs and airflow is detected through the vent system.
3. The HSI heats up.
4. The gas valve opens and the HSI lights the gas.
5. A flame sensor detects ignition and the presence of an adequate burner flame within a specified number of seconds. The HSI is then shut off.
6. The gas valve stays energized until the room thermostat stops calling for heat.

Hot surface ignitors are commonly used in conjunction with an electronic furnace control board.

- Check all electrical connections with the power off.
- Set the space thermostat above the room temperature and observe the control for the presence of a spark.
- Using a feeler gauge, check for proper spark gap as stated by the product literature ($\frac{7}{64}$" is common).
- Check for burner ignition. If it does not ignite but the pilot gas does, check the flame sensor.

- Check the safety lockout.
  - Turn off the main gas valve when the furnace is operating on a call for heat.
  - The system should lock out after a specified number of attempts.
- Check the safety limit control.
  - Prevent the blower from operating by removing the belt or disconnecting the common wire terminal.
  - Initiate a call for heat at the thermostat and place a stack thermometer (*Figure 7*) in the heat exchanger plenum.
  - After a few minutes of operation, the limit control should open and shut down the system. The temperature and time period at which this occurs will be specified in the product literature.
  - If the furnace continues to operate after reaching the high-limit temperature, shut it off immediately and replace the safety limit control. Check the operation of the new limit control.

307F07.EPS

Figure 7 ◆ Stack thermometers.

Stack Thermometers

Stack thermometers, like other instruments, are delicate. Treat them carefully when using them. Store them in their boxes when not in use.

Automatic Reset Limit Controls

A limit switch should open quickly if a blower motor fails, and it should reset and resume burner operation after a cooling-off period. Many older, spiral-wound, bimetal limit controls were adjustable, universal devices that could be used in a wide variety of furnaces. With modern furnaces, it may be necessary to obtain an exact replacement limit switch to achieve safe and reliable furnace operation.

- Check the temperature rise. This value is stated as a temperature range and is found on the furnace rating plate. If the owner is satisfied with the comfort conditions in the structure, the blower speed may not need to be changed if it is within the manufacturer's recommendations.
  - Initiate a maximum call for heat at the thermostat. Make sure the blower door is closed on the furnace and the filters are clean.
  - Place one or more thermometers in the supply and return ducts as indicated in *Figure 8*.

NOTE

If two thermometers are used, they must be calibrated to register identical temperatures; otherwise use a single thermometer to conduct each reading separately. Make sure the heating plant operates for at least 5 to 10 minutes before measuring temperatures.

- The thermometer in the supply duct must be located out of the line of sight of the furnace heat exchanger to prevent radiant heat from affecting the thermometer.
- The temperature difference (ΔT) varies from furnace to furnace. The measured temperature rise should be around the center of the range specified by the manufacturer. Refer to the product literature.
- If the temperature differential is less than recommended, reduce the blower speed; if it is above the recommendation, increase the blower speed only after making sure that there are no airflow restrictions and that the burner is not overfiring.
- Operate the system for a few more minutes. Recheck the temperature rise until it becomes stable within the recommended limits.

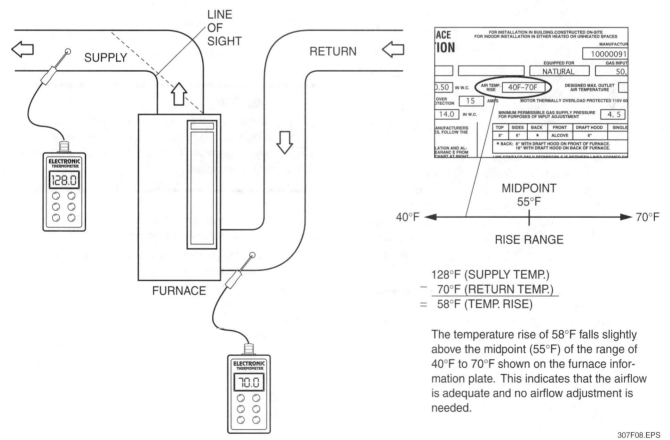

Figure 8 ◆ Temperature rise check.

- Clean and check the heat exchangers.
  - Use a drill with a spring cable and brush to clean out the inside of each heat exchanger.
  - Check the heat exchangers for leaks. A flickering burner flame while the blower is running could indicate a heat exchanger leak. If a leak is suspected because of corrosion or for some other reason, it is a good idea to inspect the heat exchangers thoroughly.

- Determining if a crack or hole exists in a heat exchanger can be done in several ways:

  - Remove access panels and visually inspect the heat exchangers. Shine a flashlight into the opening and see if any light is visible.
  - Remove the burners and inspect the interior of each heat exchanger cell with a fiber-optic borescope. Areas near the burner are susceptible to cracks caused by the expansion and contraction of the metal through repeated heating cycles.
  - Shine a strong light into the bottom of each heat exchanger cell and view the top of the cell through an access opening in the supply air plenum. The presence of light indicates a crack or hole.

There are other detection methods available that require injecting noncorrosive leak-detection chemicals into the burner flame. A special electronic leak detector is used to check for the chemical in the supply air. If the chemical is detected, a leak in the heat exchanger exists.

- On furnaces equipped with induced-draft fans, perform the following procedures:
  - Check the pressure switch by disconnecting the pressure hose while the burners are operating. The burners should shut off immediately.
  - Lubricate the motor if required by the manufacturer.

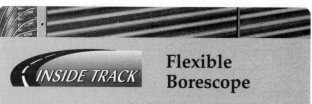

Flexible Borescope

The 48-inch flexible borescope shown uses a fiber-optic cable with a 45-degree side-view mirror in the tip. Light from a bright LED is transmitted through the cable to illuminate the area under inspection and the illuminated image is transmitted back up the cable to the borescope for display at the eyepiece. The borescope has a high-resolution image display in the eyepiece and can be focused with one hand. Other types of borescopes have a selection of tips with different viewing angles, or articulated tips controlled from the borescope, and/or video monitors for displaying images.

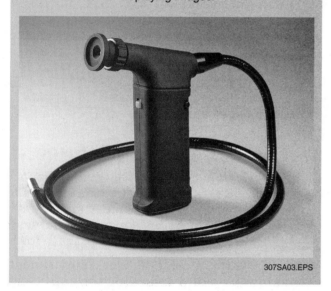

307SA03.EPS

- If it is necessary to remove the inducer motor, replace the gasket when you replace the motor. Scrape away old sealant and use RTV-type high-temperature sealant or other specified sealant to guarantee an airtight seal. It is a good idea to coat the mating surface with a non-stick vegetable oil spray, such as those sold for cooking. This will make it a little easier to take apart the next time.

- On condensing furnaces, perform the following procedures:
  - Check the condensate drain to make sure it is not blocked or leaking. Where required, add an antibacterial agent to the condensate pan.
  - If a fin-type condensing coil is used, clean the coil. Follow the cleaning procedure in the product literature to avoid damaging the fins.
  - Check the vent pipe for restrictions.

- Place the system into operation and observe it through at least one complete heating cycle to make certain that all controls are functioning properly.
- With the heating plant in operation for five or more minutes, conduct a CO check in the plenum using an appropriate CO detection device and probe. CO levels of less than 50 parts per million are acceptable.
- With the heating plant in operation for five or more minutes, conduct a gas burner combustion analysis using mechanical or electronic combustion test equipment in accordance with the equipment manufacturer's instructions. The calculated efficiency should be close to the AFUE rating of the heating plant. If not, troubleshooting is required. When a satisfactory efficiency is achieved, record the results and provide a copy to the customer for future reference.
- Leave a copy of a completed service and maintenance checklist with the owner.
- Clean up your work area and complete the service call.

2.2.0 Oil Heating

An oil-burning furnace (*Figure 9*) uses fuel that contains more carbon than natural or LP gas. Oil is burned most efficiently by atomizing burners that separate the fuel into tiny droplets and then spray them into the combustion chamber.

307F09.EPS

Figure 9 ◆ Oil furnace.

A blower that is driven by the burner motor supplies a turbulent airstream that breaks the oil into a mist and also supplies combustion air to the burner. An electric spark provided by the ignition transformer ignites the fuel/air mixture.

Oil-fired heating systems require more servicing than gas-fired systems because of the need to maintain proper burning efficiency. Inefficient burning will cause soot buildup, which will result in a further loss of efficiency.

2.2.1 Oil Furnace Safety

In addition to the specific safety precautions provided in the product literature, the following general safety precautions should be followed when servicing an oil furnace:

WARNING!

Never fire an oil furnace if oil has seeped into the combustion chamber and formed a puddle. There is a high risk of explosion or fire. Turn off the oil supply, then pump out or soak up the oil before firing the furnace.

Be very careful when starting a furnace that the customer has tried to turn on with the reset button. This procedure pumps oil into the combustion chamber, so a fire or even an explosion can occur when the burner ignites.

The 10,000V high-voltage transformers used on oil burners are potentially lethal if a defective or damaged high-voltage lead to the electrodes is contacted while the spark is active.

- Unless it is necessary to have the power on for a particular procedure, always turn off and lock out the power when working on the equipment. If it is necessary to check voltage or current, be sure to keep one hand outside the unit and carefully avoid touching bare wires and connections.
- Do not use compression fittings on oil piping. Use flare fittings to prevent air and oil leaks in the fuel lines.
- Wear protective respiratory equipment when cleaning the soot out of a furnace.
- Do not try to clean the oil burner nozzle. It should be replaced.

CAUTION

A nozzle tool must be used to remove the nozzle; pliers and wrenches may cause damage.

2.2.2 Tools, Equipment, and Materials for Servicing Oil Furnaces

The following items are required when servicing oil furnaces:

- AC ammeter
- Bucket
- Carbon dioxide tester (if not part of the combustion test equipment)
- Oil burner combustion test equipment
  - Draft gauge, smoke spot tester, CO_2 tester
- Electrode gauge
- Flashlight
- Hand mirror
- Insulated jumper wire
- Multimeter
- Oil pressure gauge kit
- Nozzle wrench
- Stiff brush
- Temperature probes
- Utility knife
- Vacuum cleaner

2.2.3 Service Procedures for Oil Furnaces

Before performing any service or maintenance procedure, check with the customer and listen to any complaints. Make note of any comment that would indicate a possible source of trouble or equipment malfunction. Then proceed as follows:

- Check the thermostat, blower compartment, and blower motor following the applicable portions of the same basic procedure provided earlier for gas heating.
- With the furnace disconnect turned off, check the electrical wiring for loose connections, insulation breakdown, and corrosion or dirt on the terminals.
- Inspect the combustion chamber using a hand mirror and trouble light. Look for carbon buildup, holes in the combustion chamber, and erosion of the oil burner head.
- Check for soot in the heat exchanger.
  - Remove the flue or vent pipe.
  - Use a flashlight or trouble light to check inside the heat exchanger flue for carbon buildup. Do not clean the heat exchanger unless there seems to be an excessive amount of carbon or scale on the heat exchanger surfaces.
  - If the heat exchanger needs cleaning, refer to the manufacturer's service manual for the proper cleaning procedure.

Oil Burner Combustion Test Equipment

Both mechanical and electronic oil burner combustion test equipment is available. The mechanical equipment shown is available as a tool kit that contains an indicator for CO_2 in flue gas samples, a dial thermometer for flue gas or plenum temperature measurements, a draft gauge, a spot smoke detector with a smoke scale, and a fire efficiency slide rule, along with applicable instruction manuals. The slide rule is used to determine combustion efficiency (EFF) using data from the CO_2 indicator and the dial thermometer.

With applicable probes, the electronic gas/oil analyzer shown directly measures and displays O_2, CO, stack temperature, draft, differential pressure, combustion air temperature and other optional selected gas emissions. It simultaneously calculates and displays combustion efficiency, excess air (EA), CO_2, NO_x, and O_2 reference values. The analyzer performs combustion calculations for ten fuels, including natural gas and propane, as well as #2, #4, and #6 fuel oil.

ELECTRONIC GAS/OIL
BURNER COMBUSTION ANALYZER

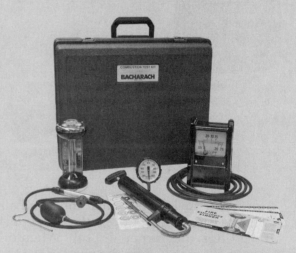

MECHANICAL OIL BURNER COMBUSTION
TEST EQUIPMENT

307SA04.EPS

- Change the oil filter.
  - Check the filter bowl for the name and type of filter element.
  - Place a container below the filter to catch any spilled fuel oil.
  - Close the fuel line valve and unscrew the bowl mounting screw or attaching hardware. Gently pry the bowl loose from the casting, remove the filter, and place it in the waste oil container. Pour the oil from the filter bowl into the container also.
  - Clean the inside of the filter bowl with a clean, soft cloth.
  - Remove the ring gasket at the top of the filter bowl. It will probably be stuck to the casting. Scrape away any filter gasket particles with a wide-blade putty knife, then clean the casting and the top of the filter bowl by wiping with a soft cloth.
  - Replace the filter with one of proper size and construction. Fill the cartridge with fresh oil. Install a new ring gasket, replace the bowl, and firmly tighten the mounting hardware. Be careful not to overtighten to a point at which damage can occur to the filter or oil supply piping.

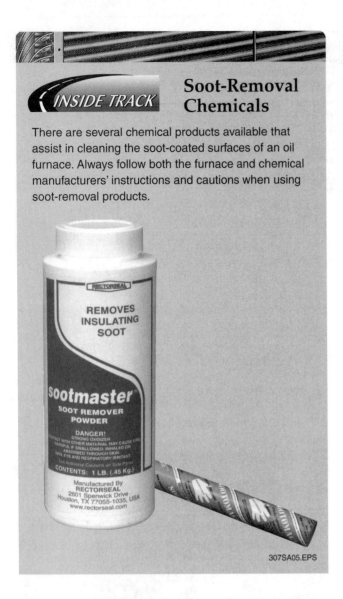
   – Turn on the fuel supply valve and bleed air from the filter bowl through the bleed port at the filter, the burner, or both. When oil starts to discharge from the ports, close the tank valve and then the bleed port(s). Open the oil supply control valve, then check for leaks.

• Check the oil piping.

   – Check all oil supply lines between the supply tank and burner for dents, rust, kinks, loose connections, cracks, or holes. Wetness around fittings, or oil spots on the equipment or the floor, may indicate a leak.

   – Check for proper support of oil piping. Check the routing for areas where potential damage could occur.

• Inspect and service the oil burner (*Figure 10*).

   – Turn the power off.

   – Disconnect all wiring and piping to the gun and blower assembly.

307F10.EPS

Figure 10 ◆ Oil burner.

   – Remove the assembly, being careful not to damage any of the components of the burner or ignition system and controls. Do not disturb the slip ring for the air gate or lose track of the nozzle adjustment or the insertion dimensions.

   – Clean the blower wheel with a stiff bristle brush.

   – Clean the burner tube with a soft cloth or disposable towels.

   – Remove, clean, and inspect the electrodes. Use fine steel wool for cleaning. Do not use sandpaper.

• Replace the nozzle. Nozzles are inexpensive; always replace the nozzle as part of any planned maintenance procedure.

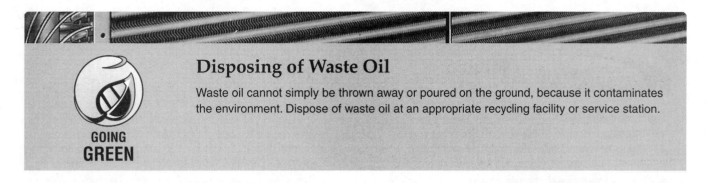

Disposing of Waste Oil

GOING GREEN

Waste oil cannot simply be thrown away or poured on the ground, because it contaminates the environment. Dispose of waste oil at an appropriate recycling facility or service station.

– Determine the gallon per hour (gph) rating and spray angle (marked on the nozzle body) and replace it with a matching nozzle.
– Seat the new nozzle with the nozzle wrench; do not overtighten.

– Reinstall the electrodes and check the position of the electrode tips. Adjust the gap and spacing from the nozzle (*Figure 11*) as specified by the oil burner manufacturer.

CAUTION

Always make sure a replacement nozzle is an exact match for the old nozzle. Replacing a nozzle with one that is not an exact match may cause burner problems.

NOTE

Electrode insulators should be clean and free of cracks and carbon buildup. Clean or replace them as necessary.

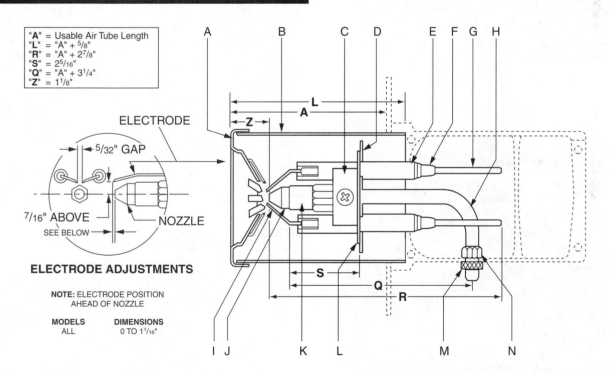

A. BURNER HEAD, SPECIFY TYPE F
B. AIR TUBE
C. ELECTRODE CLAMP, STATIC PLATE, AND NOZZLE LINE SUPPORT ASSEMBLY
D. CENTERING SPIDER
E. PORCELAIN INSULATOR
F. ELECTRODE ROD EXTENSION ADAPTER, AS REQUIRED

G. ELECTRODE ROD EXTENSION, AS REQUIRED
H. NOZZLE LINE AND VENT PLUG
I. ELECTRODE ROD AND TIP
J. NOZZLE
K. NOZZLE ADAPTER – SINGLE
L. STATIC PLATE
M. LOCKNUT BULKHEAD FITTING
N. BULKHEAD FITTING

307F11.EPS

Figure 11 ◆ Typical electrode adjustment.

INSIDE TRACK

Nozzle Wrench

The nozzle wrench can save time and reduce equipment damage. Nozzle removal normally requires two common wrenches. The awkward effort required to handle the two wrenches with the assembly intact can easily cause damage to the porcelain insulators on the electrodes. Keep in mind that the nozzle must be tight to prevent oil leakage during burner operation, so considerable effort may be required when tightening the nozzle.

The nozzle wrench consists of two sockets, with the outer socket larger than the second socket on the inside. The wrench is slipped over the nozzle, where the outer socket seats on the hexagonal section of the oil tube. The inner socket is then pushed over the nozzle, and the nozzle is loosened with minimal effort by rotating the handle attached to the inner socket. This eliminates the need for a second wrench and is far less awkward.

Once the new nozzle is installed using the same wrench, the electrodes can be returned to their proper position, preferably using the electrode setting tool. This is another tool common to all technicians who service oil burners.

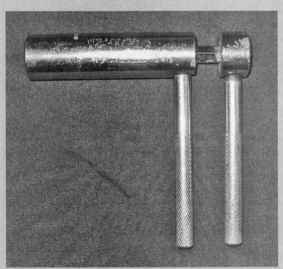

307SA06.EPS

INSIDE TRACK

Electrode Setting Tool

Accurately setting the position of the electrode tips can be very challenging. The angle of the nozzle spray pattern determines how far forward the electrodes are positioned, and this precise measurement must be known to successfully service the burner. Nozzles with narrow angles require that the electrode tips be positioned further forward to ensure that the spark is at the edge of the oil spray. Wider angles require the electrode tips to be moved further back. The distance between the two tips, as well as the vertical distance of the tips above the nozzle, are also crucial to proper ignition.

The simple device pictured eliminates a lot of guesswork and simplifies the process. The electrode setting tool is slipped over the nozzle tip. On the tip of the tool, the various possible spray angles for nozzles are molded into the plastic. Simply select the positioning guide for the angle of the nozzle being adjusted, then lay the electrodes in the appropriate guide. This will properly position the electrodes in all three dimensions. Once the electrode mount clamp is tightened, the tool is slipped off of the end of the nozzle. No additional measuring instruments or nozzle angle charts are required. For technicians servicing oil burners with any regularity, this is an indispensable tool.

307SA07.EPS

- Inspect the cad cell flame detector and clean it, if necessary. Check for proper contact with the mounting bracket and reinstall the unit.
- Replace the gun assembly and reassemble per the heating plant manufacturer's instructions. Check for air (blast) tube length (*Figure 12*) and reinstall.
- Check the burner motor coupling; replace or service as necessary.
- Remove and clean the oil pump screen.
- Perform a combustion efficiency check.

> **NOTE**
> The following procedures are for mechanical oil burner combustion testing equipment.

- Make sure the inspection door and the blower compartment door are closed properly.
- Operate the burner for at least 10 minutes. Confirm that the oil stops pumping when the pump is shut off.
- Verify that the pump operates at 100 psi or per manufacturer's specifications.

- Select a draft gauge and insert it into the test hole (*Figure 13*) in the vent pipe. Plug the test hole when testing is complete.

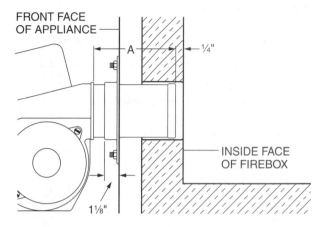

THE AIR TUBE LENGTH (DIMENSION A) IS THE DISTANCE FROM THE FRONT OF THE BURNER HOUSING TO THE DRAIN HOLE IN THE BURNER HEAD

NOTE: ADJUSTABLE FLANGE WIDTH - 1⅛"

307F12.EPS

Figure 12 ◆ Typical air tube length.

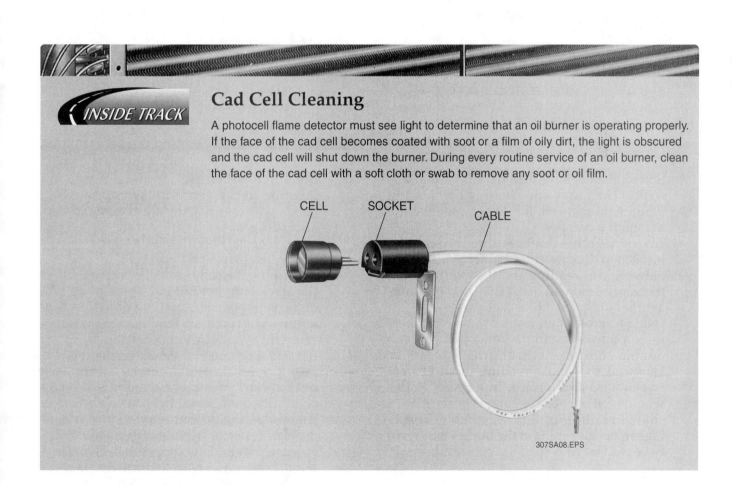

Cad Cell Cleaning

A photocell flame detector must see light to determine that an oil burner is operating properly. If the face of the cad cell becomes coated with soot or a film of oily dirt, the light is obscured and the cad cell will shut down the burner. During every routine service of an oil burner, clean the face of the cad cell with a soft cloth or swab to remove any soot or oil film.

307SA08.EPS

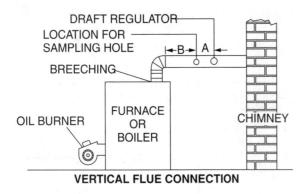

VERTICAL FLUE CONNECTION

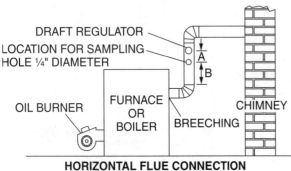

HORIZONTAL FLUE CONNECTION

DESIRABLE LOCATION FOR ¼" FLUE PIPE SAMPLING HOLE FOR TYPICAL CHIMNEY CONNECTIONS:

A. LOCATE HOLE AT LEAST ONE FLUE PIPE DIAMETER FROM THE FURNACE OR BOILER SIDE OF THE DRAFT CONTROL.

B. IDEALLY, HOLE SHOULD BE AT LEAST TWO FLUE PIPE DIAMETERS FROM BREECHING OR BELOW.

C. PLUG THE HOLE WHEN TESTING IS COMPLETE.

307F13.EPS

Figure 13 ◆ Test hole location.

- Record the negative reading. Check that it is within the range specified by the heating plant manufacturer. If it is not, readjust the barometric damper.
- Take a negative draft reading through the inspection door (over the flame). A difference between this reading and the reading from the vent of more than 0.002 in. w.c. indicates a dirty or restricted heat exchanger.
- Using a smoke spot test kit (*Figure 14*), perform a smoke test in the vent pipe. For residential equipment that burns No. 2 oil or kerosene, the smoke reading should be between zero and a trace. In a heavy boiler where regular tube cleaning is performed, a smoke reading of 2 is acceptable. If it is higher, the air setting at the burner may have to be readjusted.

Overfire Draft

Traditionally, the overfire draft should be a negative (–) value. However, some modern oil furnaces may specify a positive (+) overfire draft. Before taking the draft reading, check the oil furnace manufacturer's service literature to determine what the overfire draft value should be.

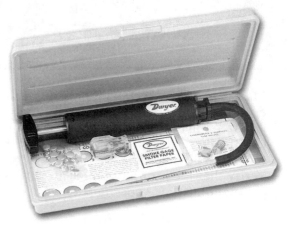

307F14.EPS

Figure 14 ◆ Smoke spot test kit.

NOTE

Run the unit for 10 to 15 minutes before doing the smoke test.

- Perform a carbon dioxide check at the same location in the vent pipe where the draft check was made. Operate the instrument according to the manufacturer's instructions. The carbon dioxide reading should not be less than 8 percent.
- Determine the combustion efficiency. Check the stack temperature in the sampling hole of the vent pipe. Subtract from this value the ambient or room air temperature to determine the net stack temperature. The stack temperature must be recorded after the temperature levels out and does not increase over a short period of time. Determine the combustion efficiency by interpolating from the slide rule accompanying the carbon dioxide tester or by consulting *Table 1*. The efficiency should not be less than 75 percent. If it is less than 75 percent, further inspection or troubleshooting is warranted.

Table 1 Combustion Efficiency (Carbon Dioxide)

					NET STACK TEMPERATURE (°F)								
PERCENT CO$_2$	300°	350°	400°	450°	500°	550°	600°	650°	700°	750°	800°	850°	900°
15	87½	86½	85¼	84¼	84¼	82	81	79¾	78¾	77½	76½	75½	74¼
	87½	86¼	85	84	83	81¾	80¾	79¼	78½	77¼	76	75	73¾
14	87¼	86	84¾	83¾	82¾	81¾	80¼	79	78	76¾	75½	74½	73
	87	85¾	84½	83½	82½	81¼	80	78¾	77½	76¼	75¼	74	72¼
13	87¾	85½	84¼	83¼	82	80¾	79½	78¼	77	75¾	74½	73½	71¾
	86½	85¼	84	83	81½	80¼	79	77¾	76½	75¼	73¾	72¾	71
12	86¼	85	83¾	82¼	81¼	79¾	78½	77¼	75¾	74½	73	71½	70¼
	86	84¾	83½	82	80¾	79¼	78	76	75¼	73¾	72¼	70¾	69½
11	85¾	84½	83	81½	80¼	78¾	77¼	75¼	74½	73	71½	70	68½
	85½	84	82½	81	79½	78	76½	75½	73¾	72	70½	69	67½
10	85	83½	82	80½	78¾	77¼	75¾	74¾	72¾	71	69½	68	66¼
	84½	83	81½	79¾	78	76½	75	73	71¾	70	68¼	66¾	65
9	84	82¼	80¾	79	77¼	75¾	74	72	70¾	68¾	67¼	65¼	63½
	83½	81¾	80	78¼	76½	74¾	73	71	69½	67½	65½	63¾	62
8	83	81	79¼	77½	75½	73¾	71¾	70¾	68	66	64	62	60
	82¼	80¼	78½	76½	74½	72½	70½	68½	66½	64¼	62¼	60	58
7	81½	79½	77¼	75¼	73¼	71	69	67	64¾	62½	60¼	57¾	55½
	80¾	78½	76¼	74	71¾	69½	67¼	65¼	62¾	60¼	57¾	55½	53
6	79¾	77¼	75	72½	70	67¾	65¼	62¼	60¼	57½	55	52½	50
	78½	76	73½	71	68	65½	63	60	57½	54½	51¾	49	46½
5	77¼	74½	71¾	69	65¾	63	60	57	54	51	48	45½	42½
	75½	72½	69½	66¼	63	60	56¾	53¾	50¼	47	43½	40¼	36¾
4	73¼	69¾	66¼	62¾	59¼	55¾	52	48	45	41¼	37½	33¾	30

307T01.EPS

- Check the flame propagation. An **unbalanced flame** (*Figure 15*) indicates a bad nozzle.
- Check the cad cell flame detector. While the burner is operating, disconnect the cad cell leads at the primary control. The burner should stop in less than 90 seconds. If it does not, either the cad cell or the leads are defective and should be replaced.
- Check the safety limit control.
  - Prevent the blower from operating by removing the belt or disconnecting the common wire terminal.
  - Initiate a call for heat at the thermostat and place a stack thermometer in the heat exchanger plenum.
  - The safety limit control should shut down the furnace at 200°F or after 2½ to 3 minutes of operation.
  - If the furnace continues to operate after reaching the high-limit temperature, shut it off immediately and replace the safety limit control.
- Check the temperature rise. If the owner is satisfied with the comfort conditions in the structure, the blower speed may not need to be adjusted if it is within the manufacturer's recommendations.
  - Initiate a call for heat at the thermostat. Make sure all the doors are closed on the furnace and the filters are replaced or clean.
  - Place one or more thermometers in the supply and return ducts.
  - The thermometer in the supply duct must be located out of the line of sight of the furnace heat exchanger to prevent radiant heat from affecting the thermometer.
  - The temperature difference (ΔT) varies from furnace to furnace and is indicated on the furnace nameplate.

307F15.EPS

Figure 15 ◆ Unbalanced flame.

- If the temperature differential is below the specified range, reduce the blower speed. If it is above the range, make sure that there are no airflow restrictions and that the burner is not overfiring, then increase the blower speed.
- Operate the system for a few more minutes and recheck the temperature rise until it becomes stable within the recommended limits.

• Place the system into operation and observe it through at least one complete heating cycle to make certain all controls are functioning properly.

- With the heating plant in operation for five or more minutes, conduct a CO check in the plenum using an appropriate CO detection device and probe. CO levels of less than 50 parts per million are acceptable.

• Check the flue pipe or vent for rust, soot, deterioration, and proper sealing.

• Leave a copy of a complete service and maintenance checklist with the owner.

• Clean up your work area and complete the service call.

3.0.0 ◆ COOLING UNITS

This portion of the module covers a systematic approach to seasonal maintenance checks for residential and small commercial cooling equipment

(*Figure 16*). The previous sections can be used for checking thermostats, blower compartments, and blowers that are usually a part of a cooling system.

3.1.0 Cooling Unit Safety

In addition to the safety precautions specified in the product literature, the following general safety precautions must be followed when servicing cooling units:

• Be sure all electrical power to the equipment is turned off. Open, lock, and tag disconnects. If it is necessary to check voltage or current, be sure to keep one hand outside the unit and carefully avoid touching bare wires and connections.

• Watch out for pressurized, hot, or rotating components.

307F16.EPS

Figure 16 ◆ Air conditioning unit.

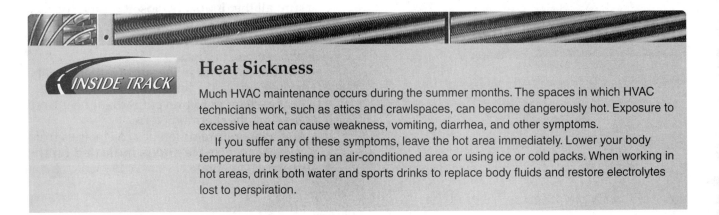

Heat Sickness

Much HVAC maintenance occurs during the summer months. The spaces in which HVAC technicians work, such as attics and crawlspaces, can become dangerously hot. Exposure to excessive heat can cause weakness, vomiting, diarrhea, and other symptoms.

If you suffer any of these symptoms, leave the hot area immediately. Lower your body temperature by resting in an air-conditioned area or using ice or cold packs. When working in hot areas, drink both water and sports drinks to replace body fluids and restore electrolytes lost to perspiration.

WARNING!

If compressor terminals are damaged, touching them while the system is under pressure could cause them to blow out. Serious injury can result. When it is necessary to check voltage at the compressor terminals, do not touch probes to the terminals unless the refrigerant charge has been recovered. Make the measurement at a terminal strip or other location away from the compressor.

3.2.0 Tools, Equipment, and Materials for Servicing Cooling Equipment

Always arrive at the job with the equipment, tools, and material you will need. The following items are required when servicing cooling equipment:

- Ammeter
- Coil cleaner
- Drill
- Gauge manifold
- Inclined manometer
- Multimeter
- Plumber's putty (or permagum)
- Thermometer, temperature probe, or infrared thermometer
- Leak detector

3.3.0 Service Procedures for Cooling Equipment

Cooling equipment, like heating equipment, requires scheduled maintenance. To perform a seasonal or periodic check of a residential or small commercial cooling system, proceed as follows:

- Check the electrical wiring, including the wiring size, routing, terminal connections, and insulation.
- Check the evaporator coil.

  – Remove the evaporator access panel. Check the evaporator fins for accumulated dirt, dust, lint, or scale.
  – Clean the clogged fins with a vacuum cleaner, drapery brush, wire brush, or fin comb. Be careful not to damage the evaporator coil tubing-and-fin connections. On coils that are fouled with dirt or grease, chemical cleaners may be used in accordance with the manufacturer's recommendations.
  – Check the tubing, fin connections, and tubing connections for signs of oil leaks that would indicate refrigerant leakage. If a leak is suspected, check the tubing with a leak detector.

- Check the condensate tray or pan.

NOTE

Some geographic areas require condensate water-conditioning chemicals or special services.

  – Clean the condensate tray of any accumulation of dirt and rust.
  – Blow out the drain hole and make sure it is not plugged. Algae may be removed with suitable chemicals; generally, household bleach is acceptable.
  – Pour some water into the drip tray and check for proper drainage.

- Check the **static pressure drop**.

CAUTION

Make sure that test holes to be drilled in a coil plenum are located so that the drill will not penetrate into the evaporator coils.

NOTE

This procedure assumes you are starting with a clean, dry coil.

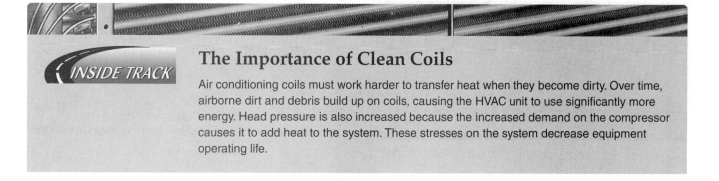

The Importance of Clean Coils

Air conditioning coils must work harder to transfer heat when they become dirty. Over time, airborne dirt and debris build up on coils, causing the HVAC unit to use significantly more energy. Head pressure is also increased because the increased demand on the compressor causes it to add heat to the system. These stresses on the system decrease equipment operating life.

- Drill two ¼" pressure tap holes in the access panel, one on each side of the A-coil (*Figure 17*). Be careful not to drill into or damage the evaporator coil.
- Energize the evaporator blower only; do not energize the heating or cooling unit.
- Obtain an inclined manometer or a draft gauge of correct calibration.
- Level the manometer and check rubber hoses from the gauge for leaks.
- Insert the manometer hose into the holes in the evaporator access panel so that about ¼" extends beyond the inside of the cabinet access panel. Seal around the hose with plumber's putty or permagum. The hose from the lower end of the inclined manometer should go into the bottom hole on the downstream side.
- The pressure differential between the pressure of the air entering and leaving the coil is the pressure drop through the coil.
- Refer to the manufacturer's specifications for correct pressure drop and air volume. See *Table 2* for an example. If the pressure drop is too high, check that the evaporator coil is clean, and clean it if necessary.

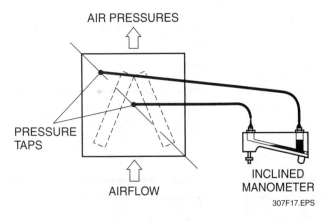

Figure 17 ◆ Static pressure drop check.

Table 2 Pressure Drop (Example Only)

Manometer Reading (Dry Evaporator)			
CFM	cu m/h	in wc	mm water
900	1,529	0.09 – 0.11	2.3 – 2.8
1,000	1,699	0.12 – 0.14	3.0 – 3.6
1,200	2,039	0.16 – 0.18	4.1 – 4.6
1,400	2,379	0.21 – 0.23	5.3 – 5.8
1,500	2,548	0.24 – 0.26	6.1 – 6.6

307T02.EPS

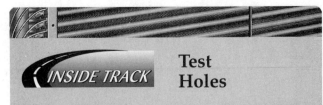

INSIDE TRACK

Test Holes

Some manufacturers offer plugs designed specifically for filling test holes in coil plenums. It is helpful to have some of these plugs on hand when making test holes.

- If the pressure drop is still too high, adjust the blower speed to obtain the proper pressure drop and air volume. Some manufacturers recommend that the volume of air moving across the coil is 350 to 450 cfm per ton of rated cooling capacity.
- Fill the test hole.

NOTE

An alternate method of measuring airflow is the temperature rise method described earlier in this module.

- Check and clean the condenser unit.
  - Turn the unit off at the disconnect.
  - Check that the condenser unit (outdoors) is level.
  - Check the wiring connections, routing, and insulation on the condenser.
  - Remove the access panels. Remove the condenser coil grille and, if necessary, the fan blades.

NOTE

Some condensers contain split coils (one directly behind the other). These coils must be separated for cleaning.

- Flush the coil with water from the inside toward the outside to remove grass, leaves, dirt, and other debris.
- Recheck and clean both sides of the coil. On coils that are fouled with dirt or grease, chemical cleaners may be used in accordance with the manufacturer's recommendations.
- Lubricate the condenser fan motor if required by the manufacturer. Use only the lubricant specified in the product literature.

- Check the refrigerant piping for evidence of leaks and for proper anchoring and routing. Oil around fittings can indicate a refrigerant leak. Look for the following indications of piping problems:
  - A long or uninsulated suction line may develop excessive superheat.
  - A liquid line running through an unconditioned hot or cold space might affect subcooling.
  - Any buried lines might cause refrigerant flood-back.
  - An extremely long liquid line might hold an excessive amount of refrigerant.
- Check the supply voltage.
  - Check the voltage at the disconnect or at the compressor contactor and compare the reading to specifications.
  - Check wire size and connections for any discrepancies.
- Check contactor contacts for pitting.
- Check for loose and discolored terminals and discolored wire insulation.
- Check the refrigerant charge. Consult the equipment manufacturer's service literature for more specific information.
  - Connect the high-pressure side of the gauge manifold set (*Figure 18*) to the service gauge port (valve) on the liquid line. Next, connect the gauge manifold suction line to the suction side of the compressor.

NOTE

For the following checks, make sure the condenser, evaporator, and filter are clean, all fan or blower speeds are correct, and that all vents/registers are completely open to fully load the system.

  - Adjust the thermostat to start the unit and operate it for 10 minutes to normalize conditions.

NOTE

When the compressor starts, observe starting conditions, such as amperage, operating pressures, and temperatures.

CAUTION

The following checks for proper refrigerant charges are for normally operating systems with no known problems. If the pressures and temperatures obtained are significantly outside the system or refrigerant tolerances, further analysis must be accomplished before adding or removing refrigerant.

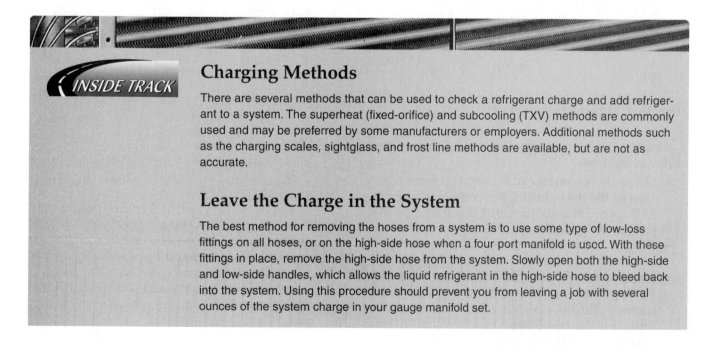

INSIDE TRACK

Charging Methods

There are several methods that can be used to check a refrigerant charge and add refrigerant to a system. The superheat (fixed-orifice) and subcooling (TXV) methods are commonly used and may be preferred by some manufacturers or employers. Additional methods such as the charging scales, sightglass, and frost line methods are available, but are not as accurate.

Leave the Charge in the System

The best method for removing the hoses from a system is to use some type of low-loss fittings on all hoses, or on the high-side hose when a four port manifold is used. With these fittings in place, remove the high-side hose from the system. Slowly open both the high-side and low-side handles, which allows the liquid refrigerant in the high-side hose to bleed back into the system. Using this procedure should prevent you from leaving a job with several ounces of the system charge in your gauge manifold set.

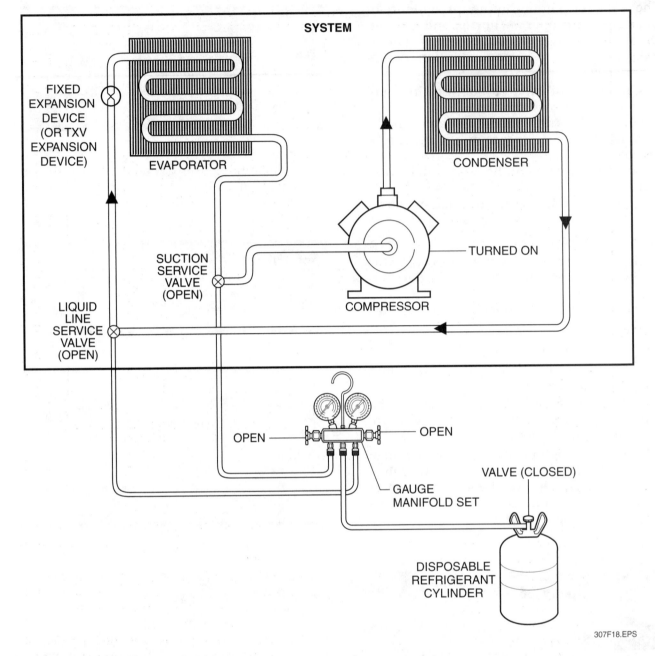

Figure 18 ◆ Refrigerant charge check.

- Charge determination for fixed expansion device systems.
  - Use a thermometer to record the temperature of the air entering the condenser. Read and record the suction and discharge pressure at the gauge manifold.
  - Consult the pressure curve chart on the cabinet and locate the recorded suction pressure in the appropriate column. *Figure 19* is an example of this chart. Follow across the curve to find the correct temperature of air entering the condenser. Mark this point of intersection. Then, read straight down to the discharge pressure. If the discharge pressure registering on the gauge manifold is within 3 psig of the value on the chart, the unit is properly charged.
  - If the pressure reading is more than 3 psig over the proper value on the chart, the system is overcharged and some refrigerant must be recovered and disposed of properly. Close the discharge valve and allow the unit to operate until the pressure stabilizes. Recheck and repeat the procedure if necessary.

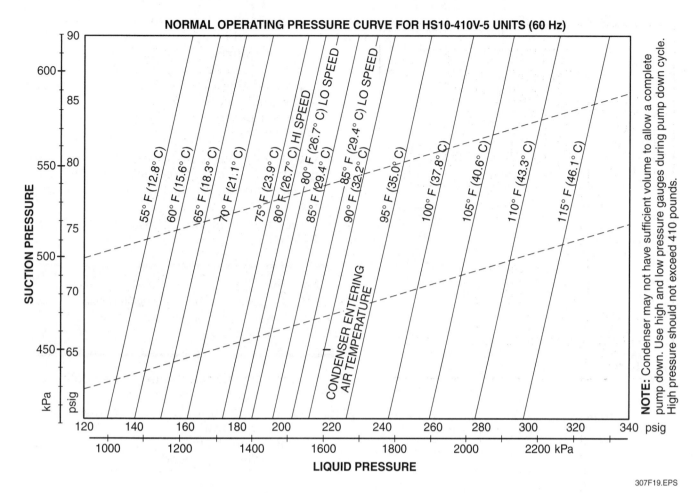

NORMAL OPERATING PRESSURE CURVE FOR HS10-410V-5 UNITS (60 Hz)

307F19.EPS

Figure 19 ◆ Example of a normal operating pressure curve chart.

- If the pressure reading is less than 3 psig below the proper value on the chart, the system is undercharged and must have refrigerant added. Open the valve on the refrigerant drum and the manifold suction valve. Charge a quantity of refrigerant into the suction side of the system and close the manifold valve. Allow the pressure to stabilize and recheck the pressure values. Compare pressure to the normal operating pressure curve chart and continue to add refrigerant until a complete charge is indicated.
- If the system is operating properly, make sure that all service valves are closed and remove the gauge manifold. Remove the suction hose while the unit is operating. Remove the discharge hose while the unit is off.

- Charge determination for TXV systems.
  - Determine the correct subcooling temperature specified by the equipment manufacturer. This can be found on the unit nameplate or in the manufacturer's service literature for the unit being serviced.

- Operate the system until it is stabilized, then measure the system liquid line temperature and pressure.
- Use the measured liquid line pressure and a standard pressure/temperature chart to find the saturated temperature of the refrigerant in the liquid line (*Figure 20*).
- Calculate the subcooling in the liquid line by subtracting the actual temperature measured in the liquid line from the saturated temperature. Compare the manufacturer's value for subcooling and the actual subcooling temperature that was calculated using the measured and saturated liquid line temperatures. If the measured subcooling temperature is within ±3°F of the required temperature, no adjustment in refrigerant charge is necessary.
- If refrigerant has been added or removed, allow the system to operate for five minutes to stabilize and then repeat all of the above until the liquid line temperature falls within the tolerances for the pressure recorded.

°F	R-22	R-134A	R-410A	°F	R-22	R-134A	R-410A
46	77.6	41.1	132.2	100	195.9	124.3	316.4
48	80.7	43.3	137.2	102	201.8	128.5	325.6
50	84.0	45.5	142.2	104	207.7	132.9	334.9
52	87.3	47.7	147.4	106	213.8	137.3	344.4
54	90.8	50.1	152.8	108	220.0	142.8	354.2
56	94.3	52.3	158.2	110	226.4	146.5	364.1
68	97.9	55.0	163.8	112	232.8	151.3	374.2
60	101.6	57.5	169.6	114	239.4	156.1	384.6
62	105.4	60.1	175.4	116	246.1	161.1	395.2
64	109.3	62.7	181.5	118	252.9	166.1	405.9
66	113.2	65.5	187.6	120	259.9	171.3	416.9
68	117.3	68.3	193.9	122	267.0	176.6	428.2
70	121.4	71.2	200.4	124	274.3	182.0	439.6
72	125.7	74.2	207.0	126	281.6	187.5	451.3
74	130.0	77.2	213.7	128	289.1	193.1	463.2
76	134.5	80.3	220.6	130	296.8	198.9	475.4
78	139.0	83.5	227.7	132	304.6	204.7	487.8
80	143.6	86.8	234.9	134	312.5	210.7	500.5
82	148.4	90.2	242.3	136	320.6	216.8	513.4
84	153.2	93.6	249.8	138	328.9	223.0	526.6
86	158.2	97.1	257.5	140	337.3	229.4	540.1
88	163.2	100.7	265.4	142	345.8	235.8	553.9
90	168.4	104.4	273.5	144	354.5	242.4	567.9
92	173.7	108.2	281.7	146	363.3	249.2	582.3
94	179.1	112.1	290.1	148	372.3	256.0	596.9
96	184.6	116.1	298.7	150	381.5	263.0	611.9
98	190.2	120.1	307.5				

SATURATION TEMPERATURE (FROM PT CHART)
— LIQUID LINE TEMPERATURE (MEASURED)

SUBCOOLING VALUE

EXAMPLE:
1. MEASURED LIQUID LINE PRESSURE (R-22)   = 274 PSIG
2. SATURATED TEMPERATURE FROM CHART   = 124°F
3. MEASURED LIQUID LINE TEMPERATURE   = 114°F
4. SUBCOOLING   = 124°F – 114°F = 10°F

307F20.EPS

Figure 20 ◆ Charging for proper cooling.

- Check the amperage draw of the condenser fan motor.
  - Install service access panels so all air flows through the condenser coil.
  - Start the cooling system and allow it to stabilize.
  - After the condenser fan motor has come up to speed, check the amperage draw with a clamp-on ammeter. If the condenser fan motor does not come up to speed, check for defects in the fan or motor.

- Record the amperage draw of the operating motor and compare it with the nameplate specifications. If there is a serious discrepancy, check the available voltage at the power panel, and the size of the wiring and the connections. If all are in order, the condenser fan motor or capacitor may have to be replaced.

- Check the amperage draw of the compressor motor. Follow the same steps outlined for checking the amperage draw of the condenser fan motor.
- Check the temperature differential over the indoor coil.

NOTE

A cooling check on a heat pump should be made only when the outdoor air temperature is 70°F or above.

- Adjust the thermostat to call for cooling and allow it to normalize.
- Insert thermometers in the holes on the upstream and downstream sides of the indoor coil (*Figure 21*). Check the supply air after a 90-degree bend.

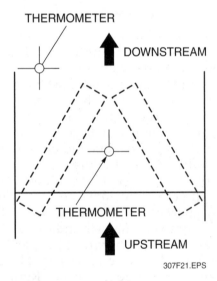

307F21.EPS

Figure 21 ◆ Temperature check.

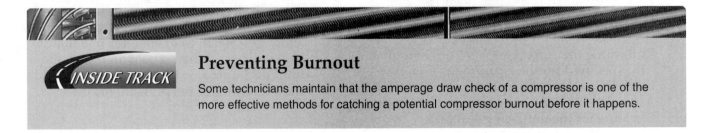

Preventing Burnout

Some technicians maintain that the amperage draw check of a compressor is one of the more effective methods for catching a potential compressor burnout before it happens.

- The difference between the two readings is the temperature differential across the indoor coil in the cooling mode. Generally, it should range between 18°F and 20°F. This may vary, however, due to the temperature and humidity within the conditioned space. Check the manufacturer's instructions.
- If the temperature differential is much less than 18°F, either the blower speed is too high and should be reduced, or there is a problem with the refrigeration system.
- If the temperature differential is much more than 20°F, the blower speed may be too low, there may be a restriction in the air distribution system, or the system may have an incorrect refrigerant charge.

• Place the system into operation and observe it through at least one complete cycle of operation. Make a final walk-around inspection of the complete system.

• Leave a copy of a completed service and maintenance checklist with the owner.

• Clean up your work area and complete the service call.

4.0.0 ◆ HEAT PUMPS

The servicing procedures and safety requirements for heat pumps are essentially the same as those for cooling-only units. A few additional checks are needed because of the reverse-cycle heating and defrost functions. The coil temperature check that follows serves as a basic capacity check, as well as a check of blower cfm. It also applies to cooling-only systems.

• Perform the procedure described earlier for cooling equipment, as appropriate, at the indoor and outdoor units.

• Check the heating mode. A heating check should be made only when the outdoor temperature is below 50°F.
   - Adjust the thermostat to call for heating and set the control above ambient temperature.
   - Operate for a few minutes until the system is normalized.

NOTE

If the heat pump operates in one mode, but not the other, it is possible that the reversing valve or check valve is defective.

- If a reversing valve malfunction is suspected, switch the system back and forth between the heating and cooling modes and observe that the reversing valve solenoid coil activates and de-activates and that the valve moves. If it does not, troubleshoot the solenoid, solenoid circuit, and valve.

NOTE

A differential measure of at least 100 psi is required to reposition the reversing valve.

- If a check valve malfunction is suspected, place the gauge manifold set into position. Operate the equipment in both the heating and the cooling modes. If the equipment operates with normal head and suction pressures in either the heating or cooling mode, but the pressures are abnormal in the remaining mode, the check valve(s) may be faulty.

• Check the defrost control. Depending upon the type of defrost control, refer to the manufacturer's manual to check for proper operation of the unit. Most electronic defrost controls have a feature that allows you to initiate defrost and run it through an accelerated cycle. If the unit you are servicing does not have this bypass feature, you may have to wait for the normal cycle to be completed, perhaps 90 minutes. However, if the outdoor temperature is not low enough (usually below 50°F) for frost to build up on the coil, defrost may not activate unless you bypass the control.

While the unit is in defrost mode, check to make sure that the electric heater is on to provide heat to the conditioned space. This can easily be done by feeling the airflow at a discharge grille.

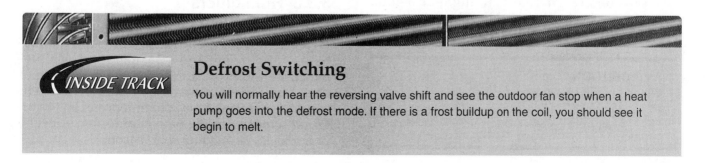

Defrost Switching

You will normally hear the reversing valve shift and see the outdoor fan stop when a heat pump goes into the defrost mode. If there is a frost buildup on the coil, you should see it begin to melt.

- Check the supplemental electric heat.
  - Check all electrical wiring for loose connections, insulation, incorrect routing, and fraying.
  - Check the amperage draw of each heating element.

    Adjust the thermostat about 10°F above the room temperature. If the electric furnace uses a two-stage or an outdoor thermostat, adjust that device to allow all stages of the heating unit to energize.

    Place the jaws of a clamp-on ammeter around one hot wire in the heating element. Position the meter so its dial can be seen. The meter needle should make a 15A to 22A jump (5kW element) as each element comes on.

    The blower should energize prior to, or simultaneously with, the first element. Observe the meter dial for the amperage peaks each time an element energizes. If the elements do not all come on in sequence, there could be a problem with a control device in the element circuit, or the second-stage or outdoor thermostat may be holding some elements out of the circuit. Recheck the setting of the second-stage thermostat.

  - Check the total amperage draw of all the elements.

    Place the clamp-on ammeter jaws around each wire leading from the power source to the elements, and record each wire reading with all the elements energized.

    Add the amperage draw readings for all the wires to obtain the total amperage draw.

    With all elements energized, record the total amperage draw.

    Turn off the unit disconnect and remove the ammeter. Most electric heating elements are rated at 240VAC and the power company will usually deliver 208V to 240V (480V or 575V for commercial installations) to the power panel. If the supply voltage is less than required, the current draw and Btuh output of the element(s) will decrease proportionally. Therefore, it might be a good idea to check the available voltage if there is a complaint of insufficient heating.

CAUTION

It is not advisable to check the limit controls and sequencers at this time because this could damage the elements.

- Check the system outdoor thermostat. Check the cut-in setting and the system's ability to run in emergency heat mode. Record the thermostat setting whether or not changes are made.
- Place the system into operation and observe it through at least one complete cycle of operation. Make a final walk-around inspection of the installation.
- Leave a copy of a completed service and maintenance checklist with the owner.
- Clean up your work area and complete the service call.

5.0.0 ◆ ACCESSORIES

Conventional residential and small commercial HVAC system accessories should be serviced when the regular air conditioning system is checked.

The installation of air cleaners and humidifiers will affect the air volume and performance of the system and may require adjustment of the indoor blower.

5.1.0 Accessory Safety

In addition to the safety precautions specified in the product literature, be sure all electrical power to the equipment is turned off. Open, lock, and tag disconnects. If it is necessary to check voltage or current, be sure to keep one hand outside the unit and carefully avoid touching bare wires and connections. Also, watch out for rotating components.

5.2.0 Tools, Equipment, and Materials for Servicing Accessories

Servicing of accessories generally does not require much in the way of tools, equipment, and materials. The following items are required when servicing accessories:

- Dishwashing detergent
- Large tub
- Multimeter

5.3.0 Humidifiers

Humidifiers, such as the bypass humidifier in *Figure 22*, require periodic cleaning and maintenance for efficient and safe operation. Maintenance should be performed at the beginning of each humidifying season. All humidifiers will accumulate calcium and other deposits on the filtering screen or in the pan. Humidifiers must be periodically cleaned for that reason.

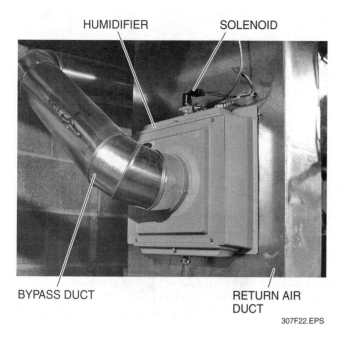

HUMIDIFIER   SOLENOID

BYPASS DUCT   RETURN AIR DUCT

307F22.EPS

Figure 22 ◆ Humidifier installed on a furnace.

Some commercial humidifiers are self-contained units that use service hot water as the heat source for evaporation and also as the source of water.

Room-conditioned air is drawn through the moisture-laden evaporator panel where the evaporation process occurs. The humidified air is then distributed by the fan to the living space.

An example of the maintenance service procedures for a typical bypass or duct-mounted residential humidifier follows:

- Turn off the power and water supplies.
- Remove the water panel evaporator.
  - Remove the evaporator element.
  - Carefully clean the excess mineral deposits from the openings in the water distribution tray. *Do not* scrape the coating off the bottom surface of the tray.
  - Clean the metal water element frame.
  - Assemble the evaporator section with a new water panel. Some panels have a colored spot to indicate which end of the panel should be placed at the top.
  - Make sure the baffles of the distribution tray penetrate into the water panel media.
  - Reposition the assembly into the housing. Make sure the drain spud rests in the opening in the bottom of the housing. Tighten the retaining hardware.

- Service the water solenoid valve on the top of the unit.
  - Remove the cap.
  - Remove the water strainer and flush it with clean water.
  - Replace the strainer on the threaded spud and hand-tighten the cap.
  - Clean the plastic orifice in the copper tube connected to the water solenoid valve using an appropriately sized drill. Be careful not to enlarge the orifice.
  - Flush out the drain line with water under pressure.
  - Lubricate the motor with two drops of SAE 30 nondetergent oil in the oil ports.
- Turn on the power supply to activate the humidifier for normal operation.
- Remind the building occupants that the humidifier should be checked monthly for bacterial growth and scale buildup. Water treatment products should be available from local distributors.

5.4.0 Electronic Air Cleaners

Basic maintenance checks performed periodically will maintain the efficiency and increase the life of electronic air cleaners (*Figure 23*). A lack of maintenance is likely to result in premature failure.

Some of the maintenance checks that should be performed include the following:

- Check to see that power is delivered to and from the main disconnect.
- Check the operation of the sail switch or electrical blower interlock switch. To do this, first energize the furnace blower, then check the electronic air cleaner (EAC) light. If the EAC is not operating and the light is off, the sail switch may be faulty.

 WARNING!
You could be exposed to lethal voltage while performing these tests. Be extremely careful to avoid contact with exposed wiring and connections.

- Check the voltage to the cells.
  - Remove the access door.
  - Check the voltage to the power pack. The reading should indicate a line voltage of 120VAC.

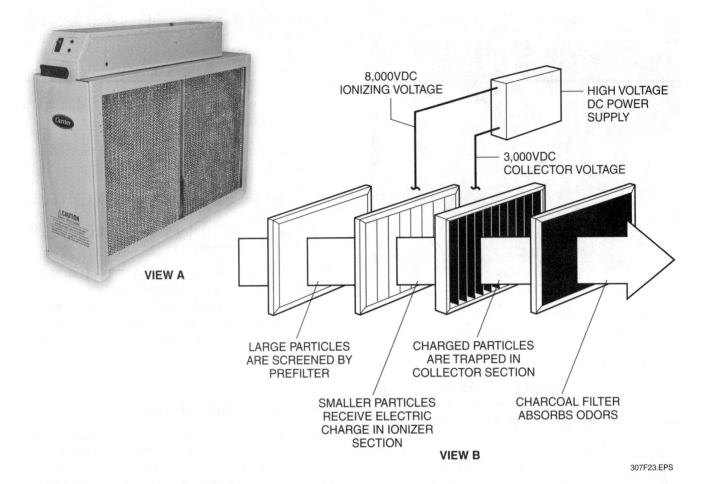

8,000VDC IONIZING VOLTAGE

HIGH VOLTAGE DC POWER SUPPLY

3,000VDC COLLECTOR VOLTAGE

VIEW A

LARGE PARTICLES ARE SCREENED BY PREFILTER

CHARGED PARTICLES ARE TRAPPED IN COLLECTOR SECTION

SMALLER PARTICLES RECEIVE ELECTRIC CHARGE IN IONIZER SECTION

CHARCOAL FILTER ABSORBS ODORS

VIEW B

307F23.EPS

Figure 23 ◆ Electronic air cleaner.

– Check the voltage to the collecting plate section and ionizing plate section. The readings should be specified in the manufacturer's literature.

WARNING!

Wear gloves and a respirator mask when cleaning the charging collector cells. Potentially harmful particles may be released into the air during cleaning.

• Clean the charging collector cells.
  – Turn off the power to the cleaner at the main disconnect.
  – Slide the cells out of the cabinet. Be careful not to damage the cells and also take care to avoid cuts from the sharp edges.
  – Clean the cells using the dishwasher method. Check the manufacturer's recommendations and make sure this method is approved. Also, obtain permission from the customer to use their dishwasher.

  Place the cells on the bottom rack of a dishwasher with the airflow arrows pointing up. Be careful not to damage the **ionizing wires** when handling them.

  Using dishwasher detergent, run the cells through a normal dishwashing cycle.

  After washing, inspect the cells for dirt or residue. Rewash if necessary.

  – Clean the cells using the soak method.

  Find a container large enough to immerse all the cells.

  Fill the container with very hot water and dissolve about ¾ of a cup of dishwashing detergent per cell. Soak for 15 to 20 minutes.

  Rinse with a fine spray and soak once more in clean hot water. Stand the cells upright to drain.

  Reinspect and rewash, if necessary.

- Clean the cells using the chemical spray method.

  Spray on a chemical cleaner, such as Simple Green® or a generic electronic air cleaner solvent. Make sure to apply foam to the complete surface of the cell.

  Let the foam stand for five minutes.

  Use warm water or a fine-pressure spray to remove the foam.

  Inspect the cell. If it is still dirty, spray on more chemical cleaner and let the foamed cell stand for a longer period of time. Rinse the cell and repeat this procedure, as necessary.

> **NOTE**
>
> Cell cleaning is usually performed on a six-week cycle. This may not coincide with normal planned maintenance programs. The equipment owner must be advised of this unless the planned maintenance program takes this schedule into account.

- Check the ionizing wires for damage.
  - Check for damage and replace if they are damaged or broken. Check the cells from the upstream side.
  - Use care to avoid damaging the spring connector when removing broken wires.
  - Replace the fine tungsten wire in the cells with proper replacement materials.

- Check the prefilter.
  - Inspect the prefilter for dirt.
  - If it is dirty, wash it in detergent or vacuum it clean.
- Check the charcoal filter (if so equipped).
  - The activated charcoal filter must be replaced according to the following guidelines:

    A filter used 12 to 24 hours per day must be replaced once every three months.

    A filter used three to six hours per day must be replaced once every six months.

    The activated charcoal filter must be replaced at least once every year, regardless of the amount of use.
- Reassemble the electronic air cleaner. It must be allowed to dry before you proceed.
- Turn the power on at the main disconnect. Energize the blower and allow the cleaner to operate for several minutes.
- Advise the customer of interim maintenance requirements.

5.5.0 Standard Media Filters

Servicing and disposing of standard media filters (*Figure 24*) is one of the key elements of a planned maintenance program. Most systems use standard media filters.

Filters are the first line of defense for system components, but filter maintenance tends to be

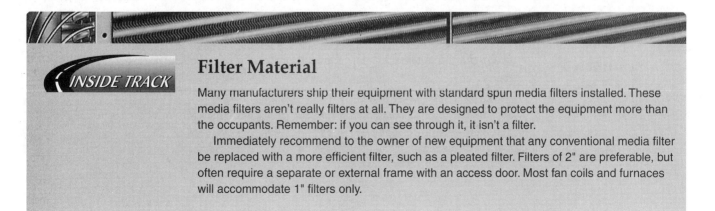

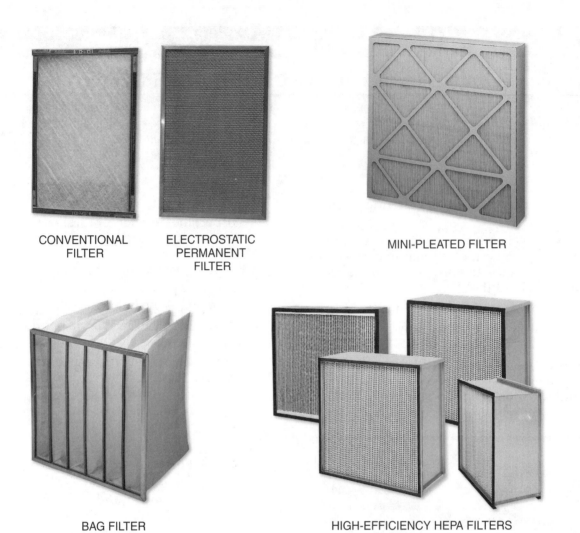

CONVENTIONAL
FILTER

ELECTROSTATIC
PERMANENT
FILTER

MINI-PLEATED FILTER

BAG FILTER

HIGH-EFFICIENCY HEPA FILTERS

307F24.EPS

Figure 24 ◆ Standard media filters.

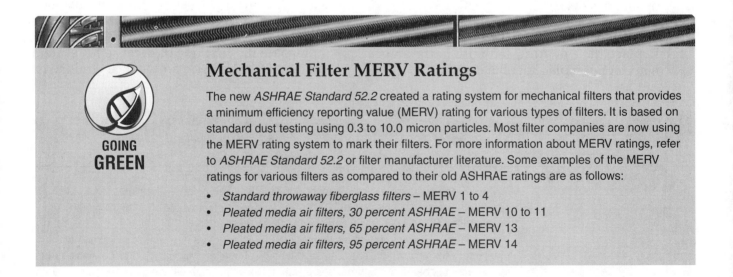

Mechanical Filter MERV Ratings

The new *ASHRAE Standard 52.2* created a rating system for mechanical filters that provides a minimum efficiency reporting value (MERV) rating for various types of filters. It is based on standard dust testing using 0.3 to 10.0 micron particles. Most filter companies are now using the MERV rating system to mark their filters. For more information about MERV ratings, refer to *ASHRAE Standard 52.2* or filter manufacturer literature. Some examples of the MERV ratings for various filters as compared to their old ASHRAE ratings are as follows:

- *Standard throwaway fiberglass filters* – MERV 1 to 4
- *Pleated media air filters, 30 percent ASHRAE* – MERV 10 to 11
- *Pleated media air filters, 65 percent ASHRAE* – MERV 13
- *Pleated media air filters, 95 percent ASHRAE* – MERV 14

GOING GREEN

neglected, and is the root of many system problems.

Change standard media filters as follows:

 WARNING!
Wear gloves and a respirator mask when replacing filters. Potentially harmful particles may be released into the air during cleaning.

 NOTE
Make sure the HVAC system does not have filters in multiple locations that are in series. This can inadvertently create an excessive pressure drop.

• Make sure the fan is off. Remove the filters from the holding frames in the air handling units.

 NOTE
Many units used in residential and light commercial applications have no filters on the units themselves. Filters for these units are located in the return air grilles, in hallways, or in some central location. Take special care when changing these filters to avoid spreading dust and debris in occupied spaces.

• Remove excess dust and lint from the frame with a vacuum cleaner, if necessary.
• Install new filters in the frames. Make sure not to bend or soil the filter media surfaces.
• Bag dirty filter media before disposing of it. Make sure to clean up any dust or debris in the work area.
• Replace and close all access panels. Restart the fan.
• If a record chart is available, log the date of the change on the unit. On the service invoice, record and note the filter sizes used and the units on which they were installed.

5.6.0 Permanent (Washable) Filters

The cleaning and servicing of washable filters (*Figure 25*) should be performed when the pressure drop through the media becomes excessive, as specified by the manufacturer.

 WARNING!
Wear gloves and a respirator mask when cleaning filters. Potentially harmful particles may be released into the air during cleaning.

To clean washable filters, proceed as follows:

• Remove the filter(s) from the holding frames in the air handling units.
• Remove excess dust and lint by tapping the filter, dirty side down, or by vacuuming it.
• Clean the filters by washing them with a stream of water from both the exhaust and intake sides of each filter.

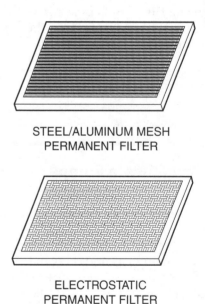

STEEL/ALUMINUM MESH
PERMANENT FILTER

ELECTROSTATIC
PERMANENT FILTER

307F25.EPS

Figure 25 ◆ Washable filters.

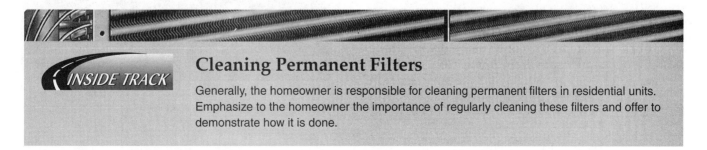

Cleaning Permanent Filters

Generally, the homeowner is responsible for cleaning permanent filters in residential units. Emphasize to the homeowner the importance of regularly cleaning these filters and offer to demonstrate how it is done.

- If the filters are extremely dirty or filled with lint:
  - Fill a container with warm water and mild detergent and swish the filters in the cleaner.
  - Rinse clean and allow to dry before recoating them with filter adhesive.
- Apply the filter adhesive.
  - Dip method

    Dip the filter in adhesive for the maximum amount of adhesive power and optimum filter performance.

    After dipping, drain the filter horizontally to guarantee a more uniform distribution of the adhesive over the face of the filter.
  - Spray method

    Apply the filter adhesive by spraying with a standard garden sprayer or paint spray gun.

    Spray the adhesive on both the intake and exhaust sides of the filter.

    Filters of ½" thickness or less require spraying on only one side.
- Return the filter to its position in the air handling unit. Be careful to note the direction of airflow arrows, if applicable.

CAUTION

Do not reinstall washable filters until they are dry. Extra moisture can encourage microbe growth.

5.7.0 Condensate Pumps

Condensate pumps have long been used to drain indoor evaporator coils when a floor drain is not in close proximity to the system or equipment. With the more recent introduction of high-efficiency condensing gas and oil furnaces, the use of condensate pumps is often required on systems where cooling is not even installed. As a result, service on condensate pumps is becoming much more common.

To service condensate pumps, perform the following procedure:

- Switch off the system power to all components.
- Remove the top and pump from the sump.
- Check and clean the condensate drain. Make sure the condensate drain line is clear of debris and microbial growth. Treat the source drain pan with biocides to eliminate any moss growth in the drain line.

INSIDE TRACK

Filter Selection

Permanent washable filters are not recommended for most commercial applications. Their maintenance schedules and filter performance do not meet the current criteria for indoor air quality.

When a customer is using permanent washable filters, recommend switching to a minimum of 1" (preferably 2") pleated disposable filter media, to be changed at least every 90 days.

Some allergy-reducing electrostatic filters provide adequate performance, but they must be cleaned much more frequently than most planned maintenance schedules allow.

- Clean the condensate pump sump and fill the sump with water.
- Reinstall the condensate pump.
- Reconnect all lines and restore system power. Check that the float switch activates proper pump cycling.

5.8.0 Ultraviolet Lamps

Ultraviolet lamps (*Figure 26*) can be used for microbial control on evaporative coils, drain pans, and duct insulation. Most of these lamps are clearly labeled for safety purposes, because the ultraviolet light from the lamps can damage eyes.

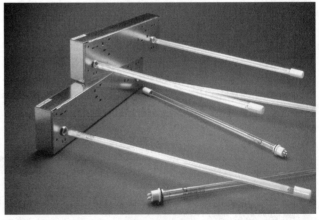

307F26.EPS

Figure 26 ◆ Ultraviolet lamp system.

Perform the following steps to service ultraviolet lamps:

- Check the service records before attempting to service ultraviolet lamps. Most lamps are designed for approximately one year of installed operating life. Change them annually and make sure they appear on the planned maintenance schedule.
- Switch off the system power. Make sure the power to the ultraviolet lamp is disconnected.
- Examine the lamp and the duct chamber where the lamp is located.
- If any dust or debris is present, clean the lamp and the lamp socket.
- Examine and record the condition of the coil and drain pan on the service invoice.
- Close all access panels and restart the system.

5.9.0 Outside Air Dampers

Because of concerns about indoor air quality, the work of technicians who perform regularly scheduled maintenance on HVAC systems is under great scrutiny. Many systems have been installed over the years without proper application of ventilation air to the system. If these conditions are discovered, they must be documented, recorded on the service invoice, and followed up with the owner by both the technician and the technician's supervisor. Methods of correcting any problems should be recommended to the owner.

Perform the following steps for outside air (OSA) systems:

- Check for proper connection and operation of the damper (*Figure 27*) and time clock interlock.
- Cycle the system (and the time clock, where applicable) to make sure the damper closes when the fan is not operating and returns to the appropriate position when the fan is operating. Many OSA dampers are marked at a preset balance point established during original commissioning or startup. If this position is changed for any reason, make sure the service invoice indicates any changes and the reasons for them.
- Return the system to normal operation.

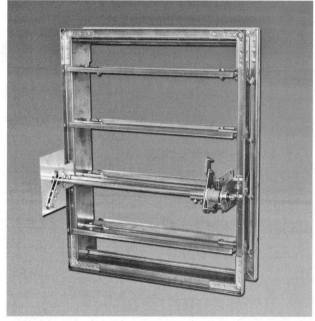

307F27.EPS

Figure 27 ◆ Outside air damper.

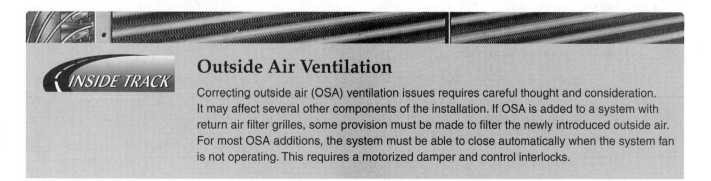

Outside Air Ventilation

Correcting outside air (OSA) ventilation issues requires careful thought and consideration. It may affect several other components of the installation. If OSA is added to a system with return air filter grilles, some provision must be made to filter the newly introduced outside air. For most OSA additions, the system must be able to close automatically when the system fan is not operating. This requires a motorized damper and control interlocks.

1. One indication of inefficient combustion in a gas furnace is the presence of soot.
    a. True
    b. False

2. An effective method of testing a gas valve is to jumper the gas valve coil terminals to see if the gas valve closes.
    a. True
    b. False

3. After repairing a gas furnace, always _____.
    a. clean the heat exchanger
    b. perform a leak check
    c. adjust the thermostat heat anticipator
    d. jumper the gas valve coil terminals to make sure the burners fire

4. All of the following may be used to check a furnace heat exchanger for leaks *except* _____.
    a. visual inspection
    b. an inspection mirror
    c. a salt solution
    d. a flashlight

5. If the gas flame in a natural gas furnace is yellow, _____.
    a. do nothing; yellow is the correct color
    b. add salt to the system
    c. adjust the primary air supply
    d. replace the thermocouple

6. The temperature rise in a natural gas furnace should be _____.
    a. 40°F to 75°F
    b. 60°F
    c. less than 75°F
    d. near the center of the range specified by the manufacturer

7. If there is a problem with the nozzle on an oil burner, _____.
    a. replace the nozzle
    b. clean the nozzle
    c. adjust the nozzle
    d. replace the entire burner assembly

8. In a residential oil furnace burning No. 2 oil, the smoke reading should be _____.
    a. zero
    b. between zero and a trace
    c. between 1 and 2
    d. no greater than 2

9. The net stack temperature in an oil furnace is determined by subtracting the _____ from the measured stack temperature.
    a. oxygen percentage
    b. room air temperature
    c. carbon dioxide percentage
    d. efficiency rating

10. The test hole for checking furnace draft should be drilled _____.
    a. as close as possible to the furnace
    b. at least one flue pipe diameter from the furnace breeching
    c. at least two flue pipe diameters from the furnace breeching
    d. between the draft regulator and the chimney

11. A P-T chart is used for refrigerant charge checks on _____ systems.
    a. dual circuit
    b. fixed expansion device
    c. staged condenser
    d. TXV

12. If the static pressure drop across a clean evaporator coil exceeds the manufacturer's recommendation, the problem should be corrected by _____.
    a. replacing the evaporator
    b. changing the blower speed
    c. adjusting the TXV
    d. cleaning out the condensate drain

13. The primary purpose of the electric heater that comes on during the defrost mode of a heat pump is to melt the frost buildup on the coil.
    a. True
    b. False

14. All of the following preventive maintenance procedures are unique to heat pumps
except the _____.
   a. defrost function check
   b. reversing valve check
   c. compressor current draw check
   d. reverse-cycle heating check

15. If an electronic air cleaner contains a charcoal filter, the filter must be _____.
   a. replaced at least once a year
   b. replaced every six months regardless of
the extent of use
   c. washed at least once a year
   d. turned over at least once a year

Summary

An important duty of HVAC technicians is to service and maintain heating and cooling equipment. Proper and timely maintenance is the key to providing safe, efficient, and effective comfort control. The HVAC technician can face a number of hazards, including toxic gases, explosion, fire, and hantavirus infection. Proper periodic maintenance can protect against these hazards.

The servicing requirements for gas heating, oil heating, electric heating, and cooling equipment and accessories vary from unit to unit. While this module provides general guidelines, the technician must consult the equipment manufacturer's service literature for specific instructions and information.

Service and maintenance tasks must be performed in an orderly and effective manner. The owner should be furnished with a copy of the checklist concerning the condition and operational state of the various components. The owner also should be informed of the need for and value of planned maintenance.

Notes

Combustion efficiency: Producing the most heat with the least amount of energy.

Hantavirus: Any of several viruses that cause hantavirus pulmonary syndrome (HPS), a potentially fatal respiratory disease.

Ionizing wires: The fine wires in the ionizing section of an electronic air cleaner that apply a positive charge to the airborne particles.

Safety drop time: The time it takes for the gas valve to shut off the gas supply after the pilot is disabled.

Static pressure drop: The pressure differential between the air entering the evaporator and the air leaving it.

Unbalanced flame: A flame that is not centered in the combustion chamber opening of an oil furnace. This can be caused by low oil pressure or a defective nozzle.

Additional Resources

This module is intended to be a thorough resource for task training. The following reference work is suggested for further study. This is optional material for continued education rather than for task training.

Refrigeration and Air Conditioning, An Introduction to HVAC/R, Fourth Edition. Larry Jeffus. Air Conditioning and Refrigeration Institute. Prentice Hall.

Figure Credits

Topaz Publications, Inc., 307F01, 307F02, 307F07, 307SA06, 307SA07, 307F22, 307F23 (photo)

Bacharach, 307SA01, 307SA02, 307SA04, 307F07

Courtesy of Honeywell International, Inc., 307F06, 307SA08

Extech Instruments, 307SA03

Photo courtesy of Rheem Manufacturing Company, 307F09

RectorSeal, 307SA05

Carlin Combustion Technology, Inc., 307F10

Dwyer Instruments, 307F14

Courtesy Sporlan Division – Parker Hannifin, 307F20

CLARCOR Air Filtration Products, 307F24

Steril-Aire, Inc., 307F26

Nailor Industries Inc., 307F27

NCCER CURRICULA — USER UPDATE

NCCER makes every effort to keep its textbooks up-to-date and free of technical errors. We appreciate your help in this process. If you find an error, a typographical mistake, or an inaccuracy in NCCER's curricula, please fill out this form (or a photocopy), or complete the online form at **www.nccer.org/olf**. Be sure to include the exact module ID number, page number, a detailed description, and your recommended correction. Your input will be brought to the attention of the Authoring Team. Thank you for your assistance.

Instructors – If you have an idea for improving this textbook, or have found that additional materials were necessary to teach this module effectively, please let us know so that we may present your suggestions to the Authoring Team.

NCCER Product Development and Revision
13614 Progress Blvd., Alachua, FL 32615

Email: curriculum@nccer.org
Online: www.nccer.org/olf

❏ Trainee Guide ❏ AIG ❏ Exam ❏ PowerPoints Other _____

Craft / Level: _____ Copyright Date: _____

Module ID Number / Title: _____

Section Number(s): _____

Description: _____

_____

_____

_____

_____

Recommended Correction: _____

_____

_____

_____

_____

Your Name: _____

Address: _____

_____

Email: _____ Phone: _____

03308-08

Water Treatment

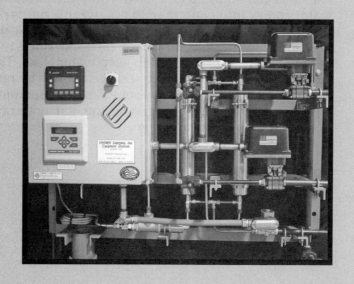

03308-08
Water Treatment

Topics to be presented in this module include:

1.0.0	Introduction	8.2
2.0.0	Water Characteristics and Analysis	8.2
3.0.0	Problems Caused by Using Untreated Water	8.4
4.0.0	Water Treatment in Open Recirculating Water Systems	8.5
5.0.0	Water Treatment in Closed Recirculating Water Systems	8.8
6.0.0	Water Treatment in Steam Boilers and Systems	8.9
7.0.0	Mechanical Water Treatment Equipment	8.11
8.0.0	General Water Treatment Procedures and Guidelines	8.19
9.0.0	Water Treatment Chemical Safety Precautions	8.25

Overview

If untreated water is used in hydronic heating and cooling systems, the systems will not operate efficiently and may eventually fail. Water contains minerals that can build up on system components and cause them to fail. When water comes into contact with air, organic matter, bacteria, and algae can form. Treatment of water with appropriate chemicals is therefore a necessity. In addition to chemical treatment, filters, strainers, water softeners, and other mechanical devices are used to keep water clean. These components require periodic servicing. A service technician may be required to test the water, either by using a test kit or by sending water samples to a designated laboratory. Based on the results of that testing, it may be necessary to add chemicals in order to keep the water clean.

Objectives

When you have completed this module, you will be able to do the following:

1. Explain the reasons why water treatment programs are needed.
2. List symptoms in heating/cooling systems that indicate a water problem exists.
3. Describe the types of problems and related remedies associated with water problems that can occur in the different types of water and steam systems.
4. Recognize and perform general maintenance on selected mechanical types of HVAC equipment that are used to control and/or enhance water quality.
5. Use commercial water test kits to test water quality in selected water/steam systems.
6. Perform an inspection/evaluation of a cooling tower or evaporative condenser to identify potential causes and/or existing conditions that indicate water problems.
7. Clean open recirculating water systems and related cooling towers.
8. Inspect, blowdown, and clean steam boilers.

Trade Terms

Alkalinity
Backwashing
Bleed-off
Colloidal substance
Concentration
Cycles of concentration
Dissolved solids
Electrolysis
Electrolyte
Fouling

Grains per gallon (gpg)
Hardness
Inhibitor
Milligrams per liter (mg/l)
Parts per million (ppm)
pH
Scale
Suspended solids
Total solids

Required Trainee Materials

1. Pencil and paper
2. Appropriate personal protective equipment

Prerequisites

Before you begin this module, it is recommended that you successfully complete *Core Curriculum*; *HVAC Level One*; *HVAC Level Two*; and *HVAC Level Three*, Modules 03301-08 through 03307-08.

This course map shows all of the modules in the third level of the *HVAC* curriculum. The suggested training order begins at the bottom and proceeds up. Skill levels increase as you advance on the course map. The local Training Program Sponsor may adjust the training order.

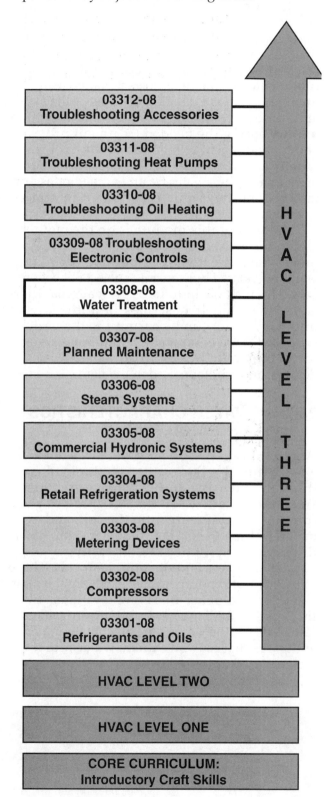

308CMAP.EPS

1.0.0 ◆ INTRODUCTION

Proper treatment of water used in HVAC systems is one of the most important factors contributing to efficient system operation, and is necessary to prevent premature failure of the equipment. It is needed for all steam, hot-water, and chilled-water systems and components, regardless of their use. Depending on the situation, water treatment may be done by chemical treatment, a physical means such as bleed-off filtering, or a combination of the two. Correct water treatment involves making a detailed analysis of the available water, considering its intended use, then implementing the needed treatment program and/or equipment as required. Analysis of water with regard to its specific use and the development of a specific water treatment method is normally done by a qualified water treatment specialist.

The focus of this module is on the methods and devices used to treat water in HVAC systems. Water treatment of potable (drinking) water is a related but separate subject not covered here. This material introduces the main characteristics of water. It describes how the use of untreated water can be harmful to the various types of HVAC equipment, and discusses some of the common water treatment chemicals, devices, and methods used in the field.

2.0.0 ◆ WATER CHARACTERISTICS AND ANALYSIS

Water without dissolved substances does not exist naturally. Pure water is obtained only through extensive and expensive treatment. Water is an effective solvent that dissolves gases from the air, minerals from the soil, and even materials from the piping and other components through which it flows. Water easily dissolves various minerals and organic chemicals. These impurities, and not the water itself, are the main cause of problems in HVAC systems. Even though the quality of a water supply makes it safe to drink, the water can still contain minerals and gases that may be extremely harmful if the water is used untreated in HVAC systems.

The different kinds of dissolved inorganic materials, including gases, and their relative amounts describe the characteristics of a specific water sample. The strength or relative amount of an element present in a water solution is commonly referred to as its **concentration**. The term **parts per million (ppm)** is used to express the level of concentration for a specific material or element that is dissolved in a water sample. For the purpose of comparison, one ounce of a contaminant mixed with 7,500 gallons of water equals a concentration of about one ppm. Another comparison is that one drop of a material or element mixed in 60 quarts of water is about one ppm. Sometimes, an alternate unit of measure called **grains per gallon (gpg)** is used instead of ppm to express a level of concentration. Grains per gallon can be converted to ppm by multiplying the value of gpg by 17. Concentration levels in water are frequently expressed using the metric unit of measure, **milligrams per liter (mg/l)**. One mg/l is equal to one ppm.

The characteristics of water supplied to a given system can vary widely over time. This can be because more than one source is being used or because the characteristics of a single supply source have changed, such as in the case of a lake or river. *Table 1* shows a hypothetical example of the many kinds of substances that can be found in

Table 1 Analysis of an Untreated Water Supply

Substance	Concentration (ppm)
Silica	37
Iron	1
Calcium	62
Magnesium	18
Sodium	44
Bicarbonate	202
Sulfate	135
Chloride	13
Nitrate	2
Dissolved solids	426
Carbonate hardness	165
Noncarbonate hardness	40

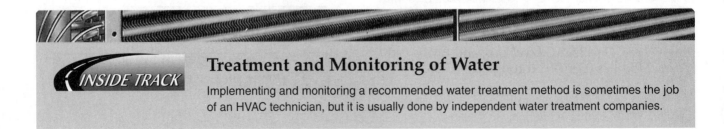

INSIDE TRACK

Treatment and Monitoring of Water

Implementing and monitoring a recommended water treatment method is sometimes the job of an HVAC technician, but it is usually done by independent water treatment companies.

an untreated water supply. Note that only **dissolved solids** are listed in the table. This is because most water analyses omit the dissolved gases.

Gases such as nitrogen have almost no effect on water use. Other gases like oxygen, carbon dioxide, and hydrogen sulfide do affect water systems. The levels of oxygen and hydrogen sulfide must be measured when the water sample is collected. This is because when air comes in contact with water, oxygen easily dissolves into and saturates the water according to its partial pressure in air and the water temperature.

The amount of carbon dioxide in water can be either measured directly or estimated from the water's **pH** and total **alkalinity**. Alkalinity is a measure of water's buffering capacity, which allows it to counteract the amount of pH reduction from the addition of acids. Water with low alkalinity will experience a greater reduction in pH from the addition of acid than water of higher alkalinity. Water above a pH value of 7.0 is considered to be alkaline. Alkalinity is measured using two indicators: the phenolphthalein alkalinity (P-alkalinity) measures the strong alkali present; the M-alkalinity (methyl orange alkalinity) or total alkalinity measures all the alkalinity in the water. The term *pH* is a measure of alkalinity or acidity of water. The pH scale ranges from 0 (extremely acidic) to 14 (extremely alkaline) with a pH of 7 being neutral (*Figure 1*). Specifically, the pH defines the relative concentration of hydrogen ions and hydroxide ions. As pH increases, the concentration of hydroxide ions increases.

The **hardness** of a water sample indicates the amount of dissolved calcium (limestone) salts and/or magnesium in the water. Both calcium and magnesium ions may result from several dissolved chemical compounds. For example, **scale** occurs as a calcium compound such as calcium bicarbonate, calcium chloride, or calcium sulfate.

The amount of calcium and magnesium in water is often reported as if it is all present as calcium carbonate ($CaCO_3$). This allows total hardness to be reported as one number. There are many hardness classification systems. *Table 2* shows the classification of hardness according to the American Society of Agricultural Engineers. As the hardness level increases, the amount of scaling usually increases.

Solid materials can exist in water in the dissolved state and as **suspended solids**. Dissolved solids are not visible and consist mainly of calcium, magnesium, chloride, and sulfate. They may also include nitrates, silicates, and organic material. Dissolved solids can contribute to the corrosion and scale formation in a system. The greater the concentration, the greater the potential

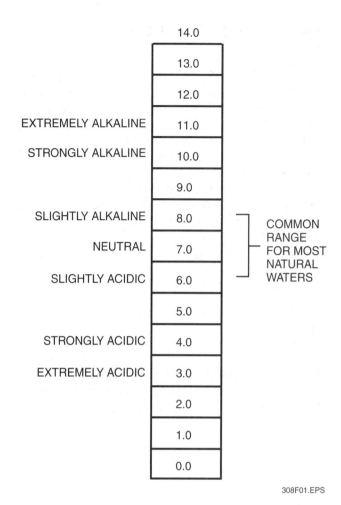

308F01.EPS

Figure 1 ◆ The pH scale.

Table 2 Classification of Hardness

Classification	Total Hardness as Calcium Carbonate (ppm)
Soft	0.0 to 60
Moderately hard	61 to 120
Hard	121 to 189
Very hard	Over 190

for corrosion. Suspended solids may be visible as individual particles or give water a cloudy look. They can include silt, clay, decayed organisms, iron, manganese, sulfur, and microorganisms. Suspended solids can clog treatment devices or shield microorganisms from disinfection. **Total solids** refers to the amount of both dissolved and suspended solids in water.

The term *cycles of concentration* is frequently used when determining the total dissolved solids (TDS) in system water. It indicates when the concentration of an element in the water system has risen above the concentration contained in the makeup water. Cycles of concentration are determined by the ratio of dissolved solids in the existing system water to the quantity of dissolved solids contained in the makeup water. For instance, if the hardness in the system water is determined to be two times as great as the hardness in the makeup water, then the system water is said to have two cycles of hardness. If you know the amount of TDS in the makeup water, it can be multiplied by the cycles to find out how much TDS is being contributed by the makeup water supply. Note that this does not take into account the TDS contributed by water treatment compounds.

Bacteria, algae, fungi, and protozoa can also be present in water systems. Problems resulting from these organisms are usually limited to cooling systems. This is because these organisms tend to die in systems operating at higher temperatures.

3.0.0 ◆ PROBLEMS CAUSED BY USING UNTREATED WATER

Water problems in heating and cooling systems can reduce system efficiency and lead to premature system failure. Some symptoms that can indicate potential water problems include the following:

- A greatly reduced cooling or heating ability of the equipment because insulating deposits have formed on the heat exchanger surfaces
- Reduced water flow due to partial or complete blockage of pipelines, condenser tubes, or other water line orifices
- Excessively rapid wear of moving parts such as pumps, shafts, and seals

The most common problems that untreated water can cause in heating and cooling systems include the following:

- Corrosion
- Scale formation
- Biological growth
- Suspended solid matter

3.1.0 Corrosion

Corrosion is a state of deterioration or the wearing away of metal surfaces within a system. It can occur both internally and externally to system units. External corrosion appears as rust or oxidation. Internal corrosion can be caused by acids, oxygen, and other gases present in the water. It can also be caused by **electrolysis**. Electrolysis is the process of changing the chemical composition of a material by passing electrical current through it. Corrosion occurs in the following forms:

- *Pitting corrosion* – A localized type of corrosion normally caused by the breakdown of the passive film at the metal surface, resulting in a blister or pit tubercle. This blister concentrates ions, which accelerates metal loss and creates the typical pinhole leak.
- *Grooving* – A uniform type of corrosion indicated by narrow grooves. Groove corrosion commonly appears along the edges of riveted joints.
- *Embrittlement (caustic cracking)* – A crystallizing or hardening process that causes fine cracks in metal, especially around riveted joints and holes. This can be a dangerous type of corrosion because it weakens the metal walls and joints in equipment, making them more vulnerable to stress under high operating pressures.
- *Galvanic corrosion* – An accelerated electrochemical corrosion produced when one metal is in contact with another dissimilar metal, and both are in contact with the same **electrolyte** (water). An electrolyte is a substance in which conduction of electricity is accompanied by a chemical action. The corrosion is caused by electrolytic action, which transfers the molecules from one metal to the other dissimilar metal. The rate at which the corrosion occurs is slower when the water has a low mineral content. This is because it has a lower electrical conductivity than water with a high mineral content.
- *Uniform corrosion* – A common corrosion caused by acid.
- *Air corrosion* – The carbon dioxide and oxygen compounds of air released by turbulence in piping systems can corrode pump impellers and other components.

Corrosion control is typically done with simple mechanical filtration and the use of inexpensive chemical **inhibitors** such as sodium silicate or phosphate-silicate mixtures. An inhibitor is a chemical substance that reduces the rate of corrosion, scale formation, or slime production.

3.2.0 Scaling

Scale is a coating of mineral matter that precipitates and settles out on internal surfaces of equip-

INSIDE TRACK

Carbon Dioxide (CO_2) Pitting Corrosion

Left unchecked, air entrainment in water will cause CO_2 pitting corrosion that will eventually cause a failure of the pipe wall.

ment as a result of rising and falling water temperatures. Even a small buildup of scale on a heat exchanger surface will reduce water flow and decrease its ability to transfer heat. Scale particles can also speed up the wear of moving parts.

3.3.0 Biological Growth

Temperatures in a water system ranging between 40°F and 120°F increase the potential for bacterial growth. Most waters contain organisms capable of producing slime when conditions are favorable. These conditions include sufficient nutrients accompanied by proper temperatures and pH conditions. Algae use sunlight to convert bicarbonate or carbon dioxide into the energy needed for growth.

Algae, bacterial slimes, and fungi can interfere with the operation of a cooling system if they clog distribution passages and restrict water flow. A film composed of bacteria, other organisms, and slimes produced by the microorganisms acts as an insulator on heat transfer surfaces. Also, accumulations of organic matter can cause localized corrosion that results in equipment failure. Fungus buildup can also destroy wooden cooling towers because some fungi use wood as a nutrient and thus weaken the tower lumber.

3.4.0 Suspended Solid Matter

Suspended solid matter can include silt, sand, grease, bacteria, algae, oil, and corrosion products. They can accumulate in a system and clog nozzles and pipes, restricting circulation or otherwise reducing heat transfer. Problems resulting from suspended solids are often referred to as **fouling**.

4.0.0 ◆ WATER TREATMENT IN OPEN RECIRCULATING WATER SYSTEMS

Open recirculating systems use cooling towers (*Figure 2*) or evaporative condensers to cool the water leaving the cooling system condenser for reuse. Recirculating and spraying this water in the cooling tower allows it to come in contact with air. As a result, the composition of the water is drastically changed by evaporation, aeration, and other chemical and/or physical processes. Also, acidic gases and other contaminants in the air are absorbed into the water. Open towers are highly susceptible to the growth of algae, bacteria, and other living organisms, especially if located where the water surface is exposed to sunlight. The main reasons for water treatment in an open recirculating system are the prevention of corrosion and scale and the elimination of algae, bacteria, and other living organisms.

4.1.0 Corrosion and Scale Deposits

Cooling tower systems are exposed to many types of corrosion, from simple electrochemical corrosion to pitting caused by deposits or microorganisms. Microbiologically induced corrosion is one of the main types of corrosion in cooling towers. It results from improper control of biocides which can lead to a high algae and/or bacterial count in tower water.

The buildup of solids also contributes to corrosion and scaling in an open recirculating water system. These solids and other impurities are caused by evaporation of the recirculating water in the cooling tower. As the system water recirculates through the tower, some of the water evaporates into the atmosphere. Any dissolved solids that were contained in the evaporated water remain behind in the recirculating system water. These solids consist of calcium and magnesium originally introduced into the system in the makeup water. Calcium carbonate and calcium sulfate are the main ingredients in the scale deposits. Additional solids and other impurities are introduced into the recirculating water each time fresh makeup water is added to the system to compensate for water lost through evaporation, controlled bleed-off, or other methods.

Microbiological Pitting

The pitting damage caused by microbiological activity can be controlled in open cooling systems by chemical treatment and periodic cleaning.

TYPICAL COOLING TOWER

BLEED-OFF

REFRIGERANT

CHILLER

AIR

AIR

WARM WATER

MAKEUP WATER

COOL WATER

CONDENSER

DRAIN

308F02.EPS

Figure 2 ◆ Simplified open recirculating water system.

Chemical treatment of open recirculating systems to control corrosion and scale involves the use of blends of phosphates, phosphonate, molybdate, zinc, silicate, and various polymers. Tolyltriazole or benzotriazole is added to these blends to protect copper and copper alloys from corrosion. Before using any chemical for water treatment, the water chemist or other responsible person must always follow the local, state, and federal environmental guidelines and regulations.

The range of pH in cooling tower water is typically maintained between 7.5 and 8.5. Sulfuric acid is generally used when pH control is needed because of high hardness or alkalinity in the make-up water. The sulfuric acid is used to increase the solubility of calcium, thereby reducing the potential for scaling. If enough acidic gases are absorbed, the water can become very acidic with a correspondingly low pH. This requires that the pH of the recirculating water be carefully regulated by adding caustic or alkali. Because the acid is introduced into the recirculating water from the air and not from the makeup water, the correct amount of caustic or alkali that must be added to maintain

the pH level cannot be calculated and is usually determined by trial and error.

In addition to chemical treatments, corrosion and scaling are controlled by a procedure called **bleed-off**. In an open recirculating system, this procedure involves continuous controlled draining and throwaway (wasting) of a quantity of the existing system water to help maintain impurities within acceptable levels. The bleed rate for most systems should be about 0.5 to 1 gallon per hour per ton of capacity.

4.2.0 Silt Deposits

Large quantities of airborne dirt enter an open system at the cooling tower and settle out as silt deposits. These deposits can cause corrosion, restrict water flow, and promote growth of microorganisms. Silt is commonly controlled by adding polymers that keep the silt in suspension while it flows through the system. This tends to cause the silt to accumulate in the tower basin where it can be manually cleaned during maintenance.

4.3.0 Biological Growths

The water contact with air that occurs in a cooling tower or evaporative condenser allows organic matter, bacteria, algae, and other airborne debris to be introduced into the system recirculating water. Algae are primitive plants that require sunlight for growth. They tend to grow in the perforated head pans of cooling towers and can eventually clog holes or nozzles in the head pan. They can also clog the system pipes and pumps. Slimes are jelly-like materials produced by the growth of bacteria. They tend to cling to many parts of the system. If they are allowed to grow, the flow of water through the nozzles can eventually be restricted. This blockage can cause high head pressure in the condenser, odor in air washers, and increased corrosion of all equipment and components.

Control of algae and slimes can be difficult because they react differently to various chemicals. Also, they can build up a resistance to the chemical being used. Treatment with chemical algaecides or bactericides prevents or slows these growths. Regular shock treatments with chlorine or other chemicals are commonly used to control growth on the system. Typically, two different chemicals are used on an alternating basis to make sure that the algae or microorganisms do not build up a resistance to any one chemical.

Much larger dosages are needed if the material has been allowed to accumulate to any depth. The dosage used depends on the amount of water held in the system. Therefore, the volume of water normally held must be calculated. This can be done by multiplying the tonnage of the cooling tower by 6 and all other open systems by 4. The treatment frequency is usually determined by trial and error and the rate of growth of the algae or slime. Chemical treatment alone does not usually control algae or slime. For this reason, the equipment should also be thoroughly cleaned at regular intervals to avoid excessive growth.

5.0.0 ◆ WATER TREATMENT IN CLOSED RECIRCULATING WATER SYSTEMS

Hot-water, chilled-water, and dual-temperature systems are examples of closed recirculating systems. Closed recirculating systems are closed loop systems where water is continuously recirculated through sealed components and piping that prevent it from having contact with air (*Figure 3*). As a result, acidic gases are not absorbed

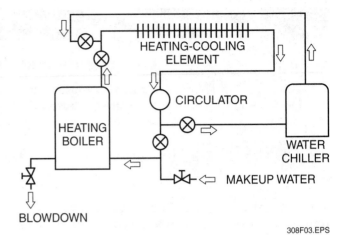

Figure 3 ◆ Simplified dual-temperature closed water system.

308F03.EPS

from the air. In a tight system, the water does not evaporate; therefore, dissolved minerals do not collect. No matter how tight, every closed system will sometimes need the addition of fresh or raw makeup water to compensate for leakage or some other cause of water loss. Dissolved oxygen and other impurities may be introduced into a closed system in this makeup water. Oxygen is one of the main agents of corrosion. Because it is a closed system, the purpose for water treatment is mainly to prevent corrosion and scale formation in the system.

5.1.0 Corrosion

Several chemicals can be used for corrosion control in a closed system. In the past, the use of chromates and hydrazine had been common. However, the EPA has since identified both as suspected carcinogens; thus, they are subjected to more stringent pollution control regulations. Some localities prohibit the use of chromates, while others allow their use only in closed systems. Sodium nitrite is commonly used as an alternative inhibitor for chromates in closed loop systems. Sodium nitrite has been proven nearly as effective as chromate for protecting steel and other ferrous metal components, but it gives little or no protection for components made of nonferrous metals. This requires that other inhibitors be included in the mixture to protect these components. One such nitrite-based inhibitor mixture includes borax as a pH buffer and sodium tolyltriazole as an inhibitor to protect copper, copper alloy, and other nonferrous metals. One drawback of using sodium nitrite is that it will not suppress bacterial growth at temperatures below 120°F.

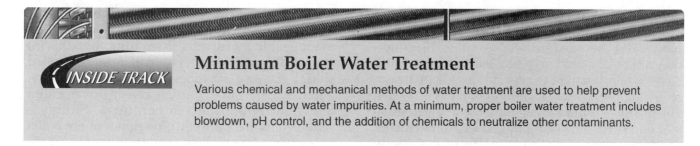

5.2.0 Scale Deposits

In chilled-water systems, scale formation is normally not a problem even if the makeup water is fairly hard. Hot-water systems are more likely to have problems with scale formation due to the higher temperatures. Because a high pH value promotes the buildup of scale, avoid using water with a high pH level whenever possible.

6.0.0 ◆ WATER TREATMENT IN STEAM BOILERS AND SYSTEMS

Steam boilers and steam systems are closed systems. The water used in a boiler must be kept clean for efficient operation. Never add dirty or rusty water to a boiler. Hard water may eventually interfere with the efficient operation of a boiler. For this reason, hard water should be chemically treated with water softeners before being used. Scale, corrosion, fouling, and foaming can all cause boiler problems. Scale deposits on heating surfaces increase boiler temperatures and lower operating efficiency. Corrosion can damage metal surfaces, resulting in metal fatigue and failure. Fouling clogs nozzles and pipes with solid materials, thereby restricting circulation and reducing the heat transfer efficiency. Foaming results in overheating and can cause water impurities to be carried along with the steam into the steam system.

6.1.0 Corrosion

Corrosion erodes away the boiler metal and appears in different forms, including pitting, grooving, and embrittlement. Corrosion can be caused by acidic (low pH) feedwater. Another cause is the dissolved oxygen contained in the boiler feedwater. Oxygen, carbon dioxide, and other gases present in the feedwater are the main causes of corrosion. These gases must be eliminated by the use of a mechanical deaerator (deaerating heater) and/or scavenger chemicals such as sodium sulfite and hydrazine to remove traces of oxygen from the water.

Because makeup water is used, periodic blowdowns are also necessary to control chloride and hardness levels. To minimize corrosion, the boiler water pH should be maintained between 11 and 12 by an alkaline water treatment using soda ash, caustic soda, sodium silicate, or sodium phosphates. Note that the P-alkalinity should be maintained between 300 and 700 ppm. Maintaining high pH levels helps prevent corrosion, but it also promotes the formation of scale; therefore, scale inhibitors must be used in conjunction with high pH levels.

6.2.0 Scale Deposits

When water evaporates into steam in a steam boiler, the steam escapes to the system, while any calcium, magnesium, and other salts that were

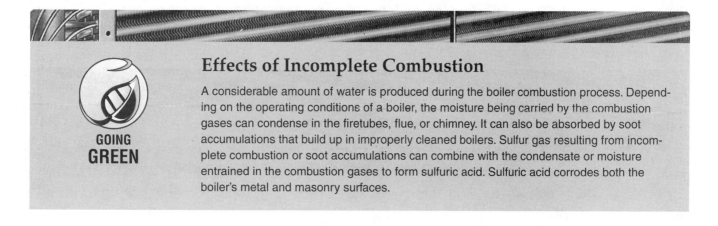

dissolved in the evaporated water remain behind in the water and are deposited on the boiler's tubes and other heat surfaces as scale. This condition can be aggravated, particularly in hard water areas, when steam or condensate losses are heavy, requiring the frequent addition of makeup water. The greater the amount of hardness in the original makeup water, the faster the boiler water reaches the point where scale can form. If a scaling condition becomes bad enough in firetube boilers, the tubes may bulge due to overheating. Tubes can burn out or possibly burst. Scale may also clog the insides of tubes. This can disrupt circulation and raise the boiler pressure, increasing the chance of an explosion.

Scale formation can best be avoided by removing all the calcium, magnesium, and other mineral salts from the feedwater before it enters the boiler. Chemicals commonly used to prevent scale formation are sodium carbonate and phosphate. Another water treatment for scale prevention involves the use of **colloidal substances** such as tannin or lignin. A colloidal substance is a jelly-like material made up of very small, insoluble, and nondiffusible particles that are larger than molecules, but small enough to remain suspended in a fluid without settling to the bottom. When added to the boiler water, these colloidal substances react with and absorb on their surface the calcium, magnesium, and other salts. The entire mass then forms a fluid sludge that is easily removed from the boiler by blowdown. Another common method is to add polymers that keep the sludges in suspension while they flow through the system.

6.3.0 Sludge Deposits

Sludge is a deposit formed by salts and other solids present in boiler water. It appears as lumps or thick masses of material that settle out in the low points of a boiler like a layer of mud. Some sludge in a system occurs naturally. Other sludge is produced on purpose by water treatment. Minerals that exist in water cannot be destroyed, but with the use of proper chemicals, the properties of these minerals can be changed to make them manageable by turning them into nonadhering sludge. Regardless of its source, if not controlled, sludge can insulate tubes, clog boiler circulation, and may bake out or harden, causing overheating and high operating temperatures. Sludge is removed by the periodic blowdown of the boiler. In extreme cases, a thorough cleaning of the boiler with a chemical product may be necessary.

6.4.0 Foaming, Priming, and Carryover

Foaming is a condition caused by high water surface tension due to a scum buildup from oil, grease, and/or sediment on the surface of the boiler water. Foaming prevents steam bubbles from breaking through the water surface and hinders steam production. The trapped bubbles rise and fill the steam space, resulting in impurities being entrained in the steam and then carried over into the steam system. Once impurities enter the steam system, they can form deposits which damage heaters and other system components. Foaming is commonly controlled by filtration, antifoaming agents, and water treatment to remove solid impurities. Common antifoaming chemicals include organic materials such as polymerized esters, alcohols, and amides. Periodic surface blowdown is also done to help minimize foaming. Surface blowdown is covered later in this module.

Both priming and carryover cause boiler water impurities to be entrained in the steam and then carried over into the steam system. Priming occurs when the boiler water level undergoes great changes and/or when violent discharges of bursting bubbles occur. Carryover is a condition in which water solids are entrained in the steam, even if there is no sign of foaming or priming. This is usually indicated by drops of water appearing with the steam. Boiler water solids contained in the steam as a result of either priming or carryover disrupt operation of the equipment coming in contact with the steam and form

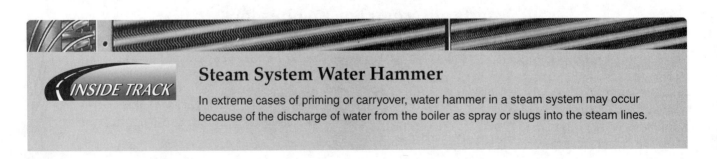

INSIDE TRACK

Steam System Water Hammer

In extreme cases of priming or carryover, water hammer in a steam system may occur because of the discharge of water from the boiler as spray or slugs into the steam lines.

deposits in terminals, valves, piping, and other components. In addition, any moisture carried over with the steam results in heat loss through the steam piping.

Controlling the type and amount of solids in boiler water is the main factor in the chemical control of carryover problems. Small amounts of organic matter and oil will form soap in the boiler and must not be allowed to enter. The amount of solids and alkalinity in the boiler water is also important. *Table 3* lists the maximum allowable total solids and suspended solids in boiler water recommended by the American Boiler Manufacturers Association.

Solids and any slugs of water carried in steam due to carryover may be eliminated by use of a steam separator installed in the supply line. To help prevent carryover problems, do the following:

- Do not allow the boiler water level to get too high.
- If possible, operate the boiler at an even rating.
- Arrange the distribution load so that the rate of steam flow is uniform.

7.0.0 ◆ MECHANICAL WATER TREATMENT EQUIPMENT

In addition to chemical water treatment methods, many mechanical or physical devices are used to filter or otherwise control water quality. These devices may be used alone or in conjunction with a chemical treatment program. Common mechanical water treatment equipment includes the following:

- Filtration equipment including strainers, filters, and centrifugal separators
- Evaporators
- Water softeners
- Deaerators
- Automatic chemical feeder systems
- Blowdown controllers and separators

7.1.0 Filtration Equipment

Mechanical filtration devices are used mainly to trap suspended solids circulating in a system. Strainers, filters, and separators are mechanical filtration devices commonly used to reduce suspended solids to an acceptable level.

Table 3 Maximum Allowable Total Solids in Boiler Water

Boiler Pressure (psi)	Total Solids in Boiler Water (ppm)	Suspended Solids (ppm)
0 to 300	3,500	300
301 to 450	3,000	250
451 to 600	2,500	150
601 to 750	2,000	60
751 to 900	1,500	40

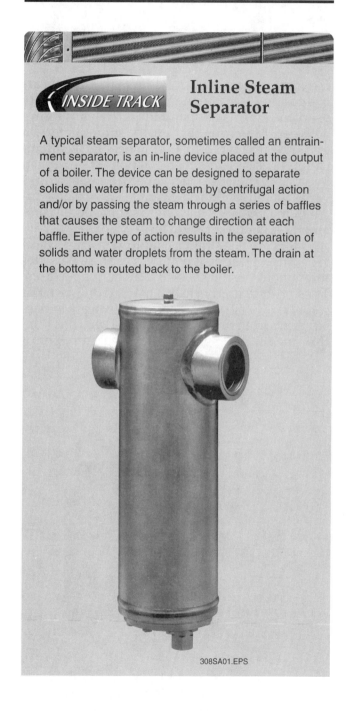

Inline Steam Separator

A typical steam separator, sometimes called an entrainment separator, is an in-line device placed at the output of a boiler. The device can be designed to separate solids and water from the steam by centrifugal action and/or by passing the steam through a series of baffles that causes the steam to change direction at each baffle. Either type of action results in the separation of solids and water droplets from the steam. The drain at the bottom is routed back to the boiler.

308SA01.EPS

7.1.1 Strainers

Strainers (*Figure 4*) are closed containers with a cleanable screen element designed to remove foreign particles as small as 0.01" in diameter. Strainers are available in single or duplex units. They are usually cleaned manually, but some are self-cleaning.

Strainers can be made of cast iron, bronze, stainless steel, alloys, or plastic. Some strainers have magnetic inserts installed in them to catch microscopic iron or steel particles that may be present in the water being strained.

7.1.2 Cartridge Filters

Cartridge filters are closed units that use a cartridge-type filter element. They are commonly used to remove suspended particles ranging between 1 micrometer and 100 micrometers in size. They are typically used where the contamination levels are less than 100 ppm (0.01 percent by mass). The construction of filter cartridges varies widely, and they are made out of many different materials. *Figure 5* shows some of the common types of filter cartridges.

Filters are classified as depth-type or surface-type filters. Depth-type filters capture particles throughout the thickness of the filtering medium. These filters are typically made from yarns or resin-bonded fibers that normally increase in density toward their center. Surface-type filters are thin-media filters made from pleated paper or

MULTI-CARTRIDGE FILTER VESSELS

PLEATED CARTRIDGE FILTERS

ABSORBENT FILTER CARTRIDGES

308F05.EPS

Figure 5 ◆ Multi-cartridge filter vessels and filters.

308F04.EPS

Figure 4 ◆ Duplex strainer.

similar material. These filters are designed to capture particles at or near the surface. Surface-type filters can normally handle higher flow rates because the pleats increase the surface area for filtration and have a better removal efficiency than an equivalent size depth filter.

Once a filter cartridge becomes plugged, as indicated by reduced water flow through the filter, the cartridge normally must be replaced with a new one. Some manufacturers provide guidelines for filter replacement by rating their cartridges according to the number of gallons they can treat. In practice, the frequency of replacement is determined mainly by the size and concentration of the suspended solids in the water that is being filtered.

7.1.3 Multimedia Filters

Multimedia filters consist of a filter tank containing three or four layers of different media. The different grain sizes and types of filtering media provide for depth filtration mainly to increase the capacity of the filter to hold suspended solids. This increases the length of time between **backwashing**. Backwashing is a procedure that reverses the direction of water flow through the filter by forcing the water into the bottom of the filter tank and out the top. Backwashing must be performed on a regular basis to prevent accumulated particles from clogging the filter.

In the multimedia filter shown in *Figure 6*, the coarsest filtering material is at the top with each successively lower layer being of a finer material.

INSIDE TRACK

High-Efficiency and Cleanable Cartridge Filters

Several types of high-efficiency, long-lasting synthetic-membrane cartridges are available that do not release any cartridge fibers into the liquid being filtered. Additionally, some pleated filters used in a duplex filter arrangement can be backwashed (by reversed liquid flow) to remove accumulated particles.

As shown, the filtering media used in this filter are bituminous coal, anthracite coal, sand, and garnet. Untreated water enters the top of the filter tank under pressure and flows down through each of the media layers. At each layer, smaller and smaller particles are trapped as the water works its way from the coarsest to the finest material.

INSIDE TRACK

Duplex Multimedia Filter

This is an example of a duplex multimedia filter. The filtering media used in these filters are anthracite coal, fine garnet, coarse garnet, and crushed rock. The filter assemblies can be configured by their associated valve systems so that they filter the system water simultaneously, or they can be configured so that one is in the backwash mode while the other filters the system water.

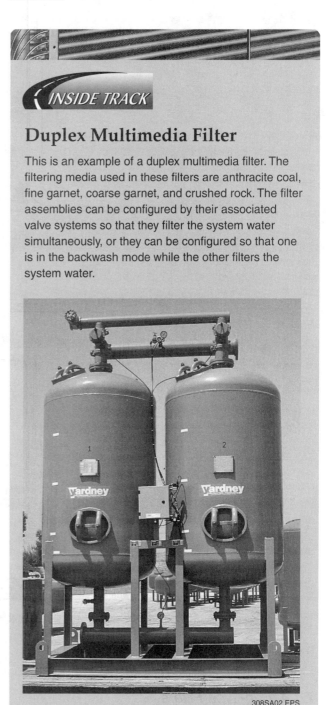

308SA02.EPS

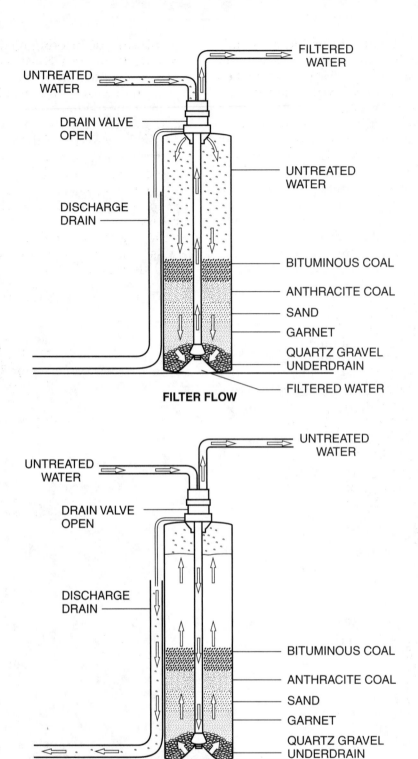

FILTERED WATER

UNTREATED WATER

DRAIN VALVE OPEN

UNTREATED WATER

DISCHARGE DRAIN

BITUMINOUS COAL

ANTHRACITE COAL

SAND

GARNET

QUARTZ GRAVEL UNDERDRAIN

FILTERED WATER

FILTER FLOW

UNTREATED WATER

UNTREATED WATER

DRAIN VALVE OPEN

DISCHARGE DRAIN

BITUMINOUS COAL

ANTHRACITE COAL

SAND

GARNET

QUARTZ GRAVEL UNDERDRAIN

BACKWASH FLOW

308F06.EPS

Figure 6 ◆ Typical multimedia filter operation.

7.1.4 Bag-Type Filters

Bag-type filters consist of a mesh or felt bag supported by a removable metal basket that is enclosed in a tubular housing (*Figure 7*). The bag support basket is usually made from perforated stainless steel. In some units, the basket can be lined with a fine wire mesh and used without a filter bag as a strainer. Some manufacturers make a two-stage bag filter consisting of an inner basket and bag that fits inside a larger outer basket and bag. This provides for a coarse filtering stage (inner bag) followed by a finer one (outer bag).

Untreated water that enters the filter housing is routed to the top of the bag(s) and leaves the unit at the bottom, below the bag(s). The contaminants filtered out of the water as it passes through remain trapped inside the bag.

Filter bags are made from many kinds of materials such as cotton, nylon, polypropylene, and polyester. Bags made from felted material are common because of their finer pores and depth-filtering quality, which provide a high amount of

dirt-loading. When plugged, these bags must be removed and replaced with new ones. Mesh bags are generally coarser, but are lower in cost because they tend to be reusable.

7.1.5 Centrifugal Separators

Centrifugal separators (*Figure 8*) are being used more often than in the past to keep closed and open water systems clean. Generally good for removing suspended particles greater than 45 micrometers in size, they tend to be inadequate at removing smaller ones. To increase their effectiveness, a coagulating compound is sometimes added to the water to make the particles larger, thus making them easier to remove. In the cen-

BAG FILTER VESSELS

FILTER BAGS AND STRAINERS

308F07.EPS

Figure 7 ◆ Typical bag filter vessels and filter inserts.

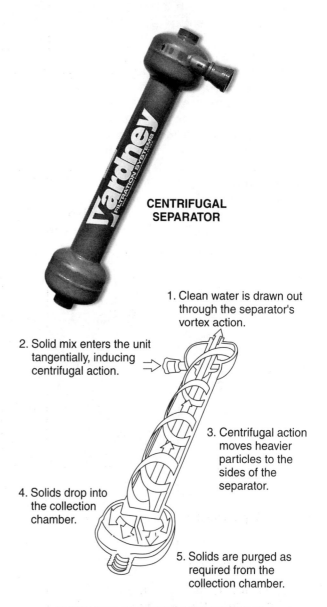

1. Clean water is drawn out through the separator's vortex action.

2. Solid mix enters the unit tangentially, inducing centrifugal action.

3. Centrifugal action moves heavier particles to the sides of the separator.

4. Solids drop into the collection chamber.

5. Solids are purged as required from the collection chamber.

CENTRIFUGAL SEPARATOR DIAGRAM

308F08.EPS

Figure 8 ◆ Typical centrifugal separator.

trifugal separator, which is mounted vertically at a horizontal angle, the untreated water is drawn into the unit, where it is accelerated into a separation chamber. There, centrifugal force causes the heavier solid particles suspended in the water to be tossed to the edges of the separation chamber. From there, they migrate into a collection chamber. The cleaned water is drawn to the low-pressure area of the separation chamber and flows up through the outlet to the system. The solid particles removed from the water are either purged periodically or bled continuously from the separator unit.

7.2.0 Evaporators

Evaporators use a distillation process to remove concentrations of solids. Their use can also remove some organic contaminants and disinfect the water. Evaporators are in-line point-of-entry devices installed in the source water lines that supply makeup or feedwater to various water equipment. There are several types of evaporators. *Figure 9* shows one common type. As shown, it consists of an evaporation (distillation) chamber and an air-cooled condenser section. Untreated supply water enters the evaporation chamber and is heated to boiling. This vaporizes the water, causing all the solids to be retained in the evaporator chamber where they are removed by a drain. The water vapor, free of solid particulate matter,

flows into an air-cooled condenser, where it condenses back into water. This treated water is then collected below the condenser for distribution to the water system. The evaporator unit shown uses an electrical element to heat the water, but other heat sources are often used. Also, a water-cooled condenser is commonly used in some units, instead of an air-cooled condenser.

7.3.0 Water Softeners

Water softeners are used to remove the minerals from hard water that form scale and soap film. They are in-line point-of-entry devices installed in the source water lines that supply makeup or feedwater to the various kinds of water equipment. Many water softeners use a sand-like substance called zeolite that is saturated with sodium (salt) and arranged in a filter bed (*Figure 10*). The zeolite absorbs calcium and magnesium, the main components of hardness, more readily than it does sodium. As the supply of hard water filters through the zeolite, sodium is released and calcium and magnesium are absorbed. A distributor or baffle disperses the untreated water through the zeolite. This ensures that all the untreated water contacts the zeolite instead of channeling or passing through the softener unit without contact. As a result of this process, the water is softened by replacing calcium and magnesium bicarbonate with nonscale-forming sodium bicarbonate.

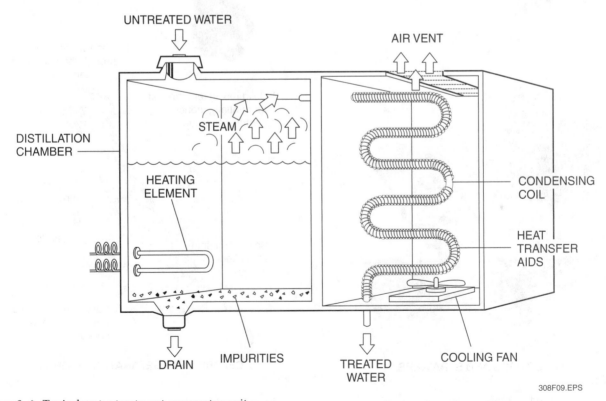

308F09.EPS

Figure 9 ◆ Typical water treatment evaporator unit.

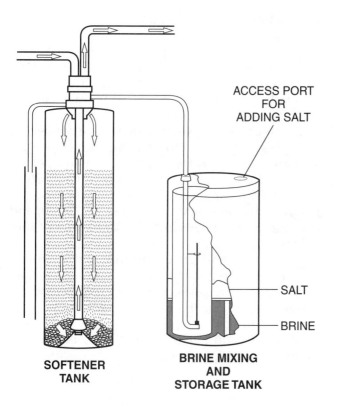

ACCESS PORT
FOR
ADDING SALT

SALT

BRINE

**SOFTENER
TANK**

**BRINE MIXING
AND
STORAGE TANK**

Through extended use, the sodium content of the zeolite bed gradually depletes, thus preventing the unit from softening the water. To prevent this condition, the unit must be regenerated by periodically treating the unit with a strong dose of salt solution. This brine solution is generally mixed and applied from an associated brine tank connected to the softener.

7.4.0 Deaerators

Deaerators (*Figure 11*) are typically used in steam systems to remove air and other noncondensible gases from the system feedwater. Oxygen becomes less soluble as water temperature increases; therefore, removal of oxygen can be done by heating the incoming feedwater to its boiling point. As a result of the boiling, the oxygen and other gases are driven off and vented to the atmosphere. Other impurities that cause hardness precipitate and settle to the bottom of the unit. There, a filtration system traps the solidified impurities.

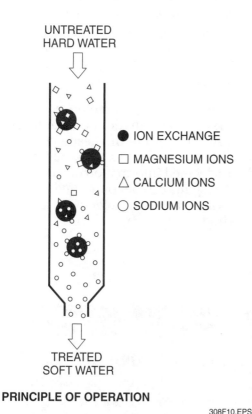

UNTREATED
HARD WATER

● ION EXCHANGE
□ MAGNESIUM IONS
△ CALCIUM IONS
○ SODIUM IONS

TREATED
SOFT WATER

PRINCIPLE OF OPERATION

308F10.EPS

Figure 10 ◆ Typical water softener.

308F11.EPS

Figure 11 ◆ Spray deaerator.

There are several types of deaerators. One type uses system steam as the heat source. In this type, the incoming feedwater is discharged as a fine spray into the upper portion of a closed chamber. The steam enters and mixes with the sprayed water, subjecting the water to an intense scrubbing (cleaning) action. While the cleaning action continues, another traversing stream of steam heats the water to free the oxygen and other noncondensible gases. After coming in contact with the feedwater, most of the steam condenses. Any remaining steam vapor, along with the separated gases, enter a vented condenser where the steam condenses as it comes in contact with tubes that contain cool, incoming feedwater. The liberated oxygen and other gases are vented to the atmosphere.

7.5.0 Automatic Chemical Feed Systems

On many steam and water systems, it is impractical to add chemical treatment manually as it results in inconsistent conditions and spikes in chemical levels which could be detrimental to system components. Automatic chemical feed controllers are often used to supply one or more biocides or other chemicals to the system. Simple controllers can provide pump control to add chemicals based solely on time and programmed dosage, while others have multiple inputs to monitor conductivity, pH, and other related water characteristics. In some cases, they also monitor the volume of make-up water added to the system, which also impacts the amount of chemical treatments to be added and the amount of bleed-off that may be required. Most chemical feed controllers are capable of also controlling a cooling tower bleed-off solenoid valve to maintain proper conductivity and dissolved solid levels, further automating the water treatment and maintenance process.

Pumps used for this application are capable of very precise metering. Coupled with extremely sensitive and accurate controllers, pumps like this help save on chemical costs and maintain consistent water conditions.

Chemical feed systems can also be connected to sophisticated building management systems, allowing the operator to monitor water conditions and chemical usage remotely. Consistent monitoring, especially on larger systems, can result in dramatic savings in chemical costs, and can minimize the environmental impact of biocide use.

Designing and programming the chemical feeder system requires that the water system specialist or other responsible parties understand the types of water problems to expect in the local area and how to correct them. The chemicals used and the dosage levels will vary based on a number of conditions, such as the following:

- The chemistry of the local water source
- Overall system design, including anticipated water temperatures and temperature changes
- The different materials of construction used in system components such as the piping, cooling towers, boilers, heat exchangers, and pumps
- The potential interaction between chosen chemicals

Because these parameters cause such a wide diversity in automatic chemical feeder systems, detailed coverage of these systems and their installation is beyond the scope of this module. More detailed information on automatic chemical feed systems can be found in documents regarding water treatment referenced at the end of this module.

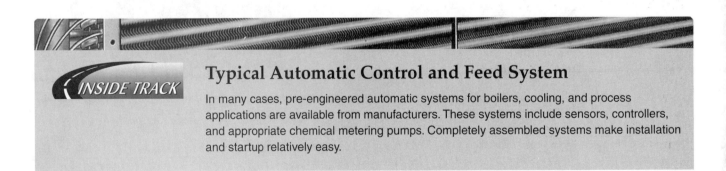

INSIDE TRACK

Typical Automatic Control and Feed System

In many cases, pre-engineered automatic systems for boilers, cooling, and process applications are available from manufacturers. These systems include sensors, controllers, and appropriate chemical metering pumps. Completely assembled systems make installation and startup relatively easy.

7.6.0 Blowdown Controllers and Separators

As previously discussed, steam boilers require blowdown from both the surface and the bottom of the boiler to remove dissolved solids and other contaminants. Although this can be done manually, a substantial cost savings can result from the use of automated blowdown control. Since the liquid and other material discharged during blowdown is at the same temperature as the boiler water, heat energy lost during the blowdown process can be significant. Steam boilers that have little or no condensate return typically need more blowdown due to the increased volume of contaminants entering with the makeup water. In addition, boiler operation and varying loads from one day to the next heavily impact the need for blowdown. As a result, when blowdown is accomplished manually, the operator cannot be certain how much blowdown is required to maintain the necessary TDS levels. An example of a packaged blowdown system is shown in *Figure 12*.

With automated blowdown, probes from the controller monitor the conductivity of the water, which represents the volume of TDS and other contaminants. This information is compared to preset parameters, and a modulating valve is actuated to provide the necessary blowdown. Blowdown can be done from either the surface or the bottom of the boiler.

Steam boilers may also be equipped with a blowdown separator (*Figure 13*). These devices are usually centrifugal-type separators that help reduce corrosion and scaling downstream by separating steam from water and collecting the dissolved solids. Blowdown separators also reduce the temperature of the water being discharged, which is often necessary to comply with environmental regulations and municipal codes.

8.0.0 ◆ GENERAL WATER TREATMENT PROCEDURES AND GUIDELINES

Regardless of water system type, the key to efficient operation is regular and correct performance of water-related maintenance procedures and water treatments. Some general guidelines that pertain to water-related maintenance procedures and water treatments for different types of systems are described in this section.

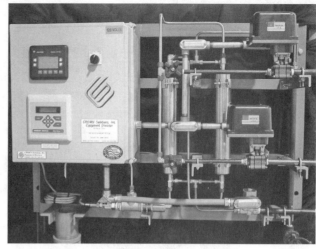

308F12.EPS

Figure 12 ◆ Boiler blowdown control system.

DRAIN CONNECTION

308F13.EPS

Figure 13 ◆ Blowdown separator.

8.1.0 Water Testing

Water systems are most commonly tested using commercially available test kits. Some test kits allow quick on-site analysis of the water (*Figure 14*). Typically, these kits test for hardness, pH, specific treatment chemicals, and other potential contaminants.

INSIDE TRACK

Water Test Report Recommendations

Some reports include recommended EPA ranges for contaminants found and recommendations for any corrective action that may be needed.

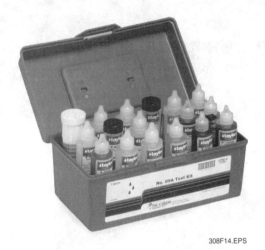

308F14.EPS

Figure 14 ◆ Water analysis test kit.

More comprehensive test kits require that samples be sent back to a test lab for analysis. These kits contain directions for testing the water and all the containers and packaging necessary to ship the samples back to the lab. Most also contain a data sheet (*Figure 15*) that provides the testing facility with important information about the system being tested. For the water samples submitted, the lab sends back a report showing the measured or calculated amount of each contaminant.

HVAC technicians often perform water testing using commercially available test kits. When more extensive or specialized testing is required, a water chemist or other specialist should perform the tests.

When testing water using test kits, remember the following:

- The proper collection, handling, and preservation of water samples is crucial for an accurate water test.
- Use only those containers provided or recommended by a testing laboratory.
- Carefully follow the laboratory and/or the test kit manufacturer's instructions.
- Usually, water must be sampled both before and after it goes through treatment equipment to ensure that a treatment device or program is working properly.

- The type of container needed depends on the test being performed. For microbiological testing, the container must be sterile and contain a chemical to deactivate any residual chlorine.
- Sampling containers used to collect water for chemical analysis often contain a fixing compound that prevents loss or breakdown of the specific chemical. Laboratories specially prepare containers for each category of contaminant. These containers should never be rinsed before use or filled to overflowing.

Water pH can be tested very simply using one of the following four methods:

- *Red and blue litmus paper* – If red litmus paper turns blue, the solution is basic, with a pH above 7.0. If blue litmus paper turns red, the solution is acidic, with a pH below 7.0.
- *pH test papers* – Immerse the test papers in the water, then remove them and compare their color with a color chart to determine the pH.
- *Dye* – Add dye to the water sample and compare the color with a color chart to determine the pH.
- *Electronic pH meter* – Use as directed by the manufacturer's instructions to measure the pH.

8.2.0 Cooling Tower and Open Recirculating System Treatment

Regular maintenance is a must in any water system, but it cannot be stressed enough in open systems because of health and safety issues related to Legionnaires' disease. Poorly maintained cooling towers are known to support the growth of Legionella bacteria. The aerosol produced has the potential of infecting not only the people at the equipment site, but also the surrounding community. Legionnaires' disease can lead to pneumonia, and in some instances is fatal. If cooling towers are not properly maintained, a contractor could be held liable if someone contracts this disease.

DATE OF SAMPLING: _____

WATER ANALYSIS DATA SHEET

STOP! This page must be filled out completely. Failure to do so will prevent an analysis and sample will be discarded.

	NAME	STREET	CITY	STATE

JOB _____

CONTRACTOR/DEALER _____

WHOLESALER _____

REPORT SHOULD BE MAILED TO

COMPANY: _____ FAX NO: _____

ATTENTION: _____ E-MAIL: _____

WATER SAMPLE:

STOP! Make sure that you are sending a make-up (supply) water sample. Do not send recirculating or "system" sample by itself. Failure to send a make-up sample will prevent or delay your analysis report.

WATER:

PROBLEM INVOLVED: Lime Scale ☐   Corrosion ☐   Algae ☐   Slime ☐   Other ☐

SOURCE OF WATER: City ☐   Private Well ☐

SOURCE OF SAMPLE: Make-up ☐   Recirculating* ☐

EQUIPMENT INVOLVED:

1. Evap. Condenser: Make_____ Tonnage_____ Model # _____

   Existing Bleed-off_____gal/hr.

2. Ice Machine: Make_____Type: Cuber ☐   Flaker ☐   Drum ☐

   Model # _____Capacity_____lbs. ice/day

3. Closed System: Chilled ☐   Hot Water Boiler ☐   Both ☐

   Size _____Tons_____Btu/h (output): Capacity_____gals.

4. Steam Heating

   Boiler   Make_____Btu/h _____

5. Evaporative

   (Swamp) Cooler: Make_____ Size _____ cfm

TREATMENT:

1. Chemical: Brand _____ Type or Name_____

   Part/Code No. _____ Results: Good ☐ Bad ☐

2. Feed Equip: Chemical Feed Pump (size_____ GPD) Drip Feed _____

   TDS Monitor/Controller _____ Lockout Timer _____

3. Date System Last Cleaned: _____

TEST RESULTS
(Field or Office Use)

pH:_____

Alk:_____ppm

Hard:_____ppm

Cl:_____ppm

SiO2:_____pprr

Cond:_____mmhos

_____ppm

_____ppm

Nu-Calgon
Wholesaler, Inc.

Calgon is a licensed trade name.

308F15.EPS

Figure 15 ◆ Example of a water sample data sheet.

8.2.1 Periodic Maintenance

The recommended maintenance actions and/or tests related to cooling towers/evaporative condensers and open recirculating water systems are summarized in *Table 4*.

8.2.2 General Cleaning Guidelines

After the installation or major repair of an open system, it is necessary to clean the internal water system of protective oils, films, grease, welding flux, dirt, and other debris. Clean according to the following guidelines:

Step 1   Fill, vent, and leak test the system.

Step 2   Add the chemical cleaning agent prescribed for the system in the amount and manner directed by the manufacturer.

Step 3   Operate all system pumps with their strainers installed for the time interval prescribed by the responsible water treatment specialist.

Step 4   Drain the system; remove, clean, and replace all strainers; and then refill the system.

Step 5   Operate all system pumps for more than 30 minutes. During this time, temporarily open all dead ends, drain valves, and strainer flush valves.

Step 6   Drain and refill the system. Repeat Step 5. On a newly installed system, it may take as many as four flushes to clean the system adequately.

Step 7   Test the system water and certify its cleanliness using a procedure that measures the residual cleaning agent in the water. The final system water pH should be equal to ±0.3 pH units of that of the makeup water.

Step 8   After the system has been tested and certified to be free of cleaning agent, take samples of the system water for future use.

Table 4 Cooling Tower/Open Recirculating Water System Maintenance

Maintenance Action/Test	Frequency
Test and record bacteriological quality of the system water.	Biweekly
Test and record biocide and inhibitor reserves, pH, and conductivity of the system water.	Biweekly
Check that dosing equipment containers are full, pumps are operating properly, and supply lines are not blocked.	Weekly
Check the bleed-off control equipment to make sure it is operating properly and the controller is in calibration. The solenoid valve should be manually operated to confirm that the flow of water to the drain is clear.	Weekly
Check the system for growths and deposits. There should be no algae growth in the towers or slimy feel to the fill pack, tower sump, or side walls.	Weekly
Check the operation of sump immersion heaters. There should be no visible corrosion on the outside of the heater, and the unit should activate when the setpoint on the thermostat is reached.	Monthly
Check the operation of sprays, fans, and drift eliminators. There should be no mechanical damage, and the components should be free of visible deposits. The bypass of aerosol droplets should be minimal when the fans are operating. The distribution system should have no deposits, with an even flow of water to all areas of the tower.	Monthly
Drain, clean, and disinfect cooling towers and associated pipe work in accordance with the method approved for the site. The chlorination period should be a minimum of five hours. Free chlorine residuals should be checked regularly. If possible, the tower pack should be removed for cleaning. Post-cleaning chlorination should be monitored to make sure that free chlorine residuals are maintained.	Semiannually
Review maintenance and water treatment program performance, including the quality of results obtained and the cost of system operation.	Annually

Step 9   Initiate the approved water treatment program specified for use during system operation.

8.2.3 Excessive Scale Formation

If the amount of scale inhibitors prescribed for a water treatment program fails to control the scale buildup as predicted, the following factors should be checked:

Step 1   Check that the prescribed rate of bleed is being maintained.

Step 2   Check to see if the water can be made less scale-forming by a more accurate dosage of the prescribed inhibitor chemicals. Take samples of recirculating and makeup water, and have them analyzed to check for the concentration of scale-forming salts. Look for changes in the concentration levels by comparing them with those previously used as the basis for the water treatment program.

Step 3   Because scale forms more rapidly at higher water temperatures, compare the temperature of the water in the sump with the ambient wet-bulb temperature. If the sump water temperature is more than 10°F above the wet-bulb temperature, possible causes are:

- There is insufficient cooling tower capacity.

- A fan is not working properly due to loose belts, worn bearings, bent blades, or improper shrouding.
- A water-driven fan is operating at too low a speed due to plugged jets, improper water pressure at the fan, worn spindle, plugged louvers, plugged eliminators, or plugged packing in the cooling tower.
- The distribution and breakup of water is inadequate due to insufficient volume of water flow, plugged nozzles, slime-covered water distribution holes, or worn-out tower fill.
- Moist air from the discharge of the cooling tower is returning to the air intake.
- The air intake on an indoor tower is blocked.
- The air current path to the tower is restricted.

Step 4   Because scale may be forming if the temperature rise of the water flowing through the system condenser is too high (more than 10°F), check to see if:

- Slime or debris is covering the pump screen.
- The water lines and/or condenser are undersized or restricted with corrosion or scale.
- There is a worn pump impeller or inadequate pump volume or pressure.
- There is a defective water valve.

8.3.0 Steam Boiler Water Treatment Guidelines

A boiler water treatment program can include pretreatment of the raw makeup water using filters, water softeners, and dealkalizers before it enters the boiler feedwater system. Chemicals can also be used to treat the boiler feedwater and boiler water, and condition any sludge to protect against corrosion and scaling in the boiler. In addition, boiler water treatment includes blow-down—the controlled draining of boiler water out of the boiler in order to maintain the desired concentration of suspended and dissolved solids in the boiler. The water that is removed by blow-down is then replaced with an equal amount of fresh, clean feedwater. Without blowdown, the concentration of suspended and dissolved solids in the boiler water can become excessive and result in the formation of sludge and scale. Priming, foaming, and/or carryover may also occur.

The amount of water removed by blowdown is normally determined by the concentration of total dissolved solids (TDS) contained in the boiler water. Enough water must be removed in order to keep the TDS level below that at which priming, foaming, and/or carryover occurs in the boiler. *Table 3*, presented earlier in this module, lists the limits for total and suspended solids recommended by the American Boiler Manufacturers Association for different boiler operating pressures.

Boilers are normally equipped with two kinds of blowdown valves: a bottom blowdown valve and a surface blowdown valve. The bottom blow-down valve is located at the bottom or lowest part of the boiler. It is used to manually purge the boiler of foreign matter by draining off some amount of the water from the bottom of the boiler in order to control the TDS in the water. This water contains sediment, scale, and other impurities that have settled out of the water as it is heated and have accumulated at the lowest point in the boiler.

The surface blowdown valve is connected to an internal collecting pipe that ends slightly below the boiler's normal operating water level. It is used to skim off impurities on the surface of the water inside the boiler. Surface blowdown usually involves a controlled and continuous draining of boiler water taken off the surface at the water level in order to skim off sediment, oil, and other impurities. The quantity of blowdown water removed from the boiler is controlled by the surface blowdown valve. It is adjusted to provide the flow rate needed to achieve the desired TDS level. Periodic adjustments of the surface blowdown valve must be made in order to increase or decrease the amount of blowdown as determined by water test analysis. Note that some boilers can be equipped with more than one bottom and/or surface blowdown valve. Also, some boilers may not be equipped with a surface blowdown valve.

After a major repair, installation, or when the boiler is extremely dirty, skimming and blow-down may require that the boiler be completely drained and refilled with water one or more times. This process must be continued until the water discharged from the boiler runs clear. Always perform boiler skimming, blowdown, and/or cleaning as directed in the boiler manufacturer's service instructions. General guidelines are given here.

> **NOTE**
> Skimming may not clean the boiler of sediment that has accumulated at the bottom. After skimming, the boiler should be cleaned further by performing the blowdown procedure.

8.3.1 Skimming Off Impurities

Step 1   If not permanently connected to a drain, run a temporary connection from the boiler's skimming valve to a suitable drain.

Step 2   With the boiler empty and cool, slowly begin to add water. After a quantity of water has entered the boiler, never before, turn on the burners and adjust the flame so that the water being added is kept just below the boiling point. Boiling and turbulence must be avoided.

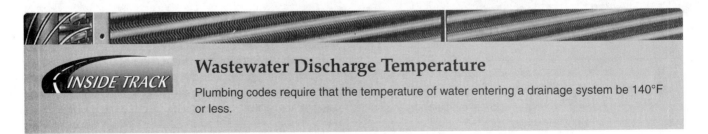

Wastewater Discharge Temperature

INSIDE TRACK

Plumbing codes require that the temperature of water entering a drainage system be 140°F or less.

Step 3 Gradually raise the hot water level in the boiler to the point where the water just flows from the skimming valve, being careful not to raise it above this point.

Step 4 Continue to skim the boiler water until there is no trace of impurities. Water may be checked to make sure it is free from oil by drawing off a sample. If the sample is reasonably free from oil, it will not froth when heated to the boiling point.

NOTE

Only attempt blowdown at a light load. Ideally, temporarily suspend the boiler's heating process to halt water turbulence and allow the solids to settle out.

8.3.2 Bottom Blowdown of the Boiler

Step 1 Check the water level in the boiler.

Step 2 Partially open the bottom blowdown valve. Once the water starts draining, fully open the valve.

Step 3 Remain at the blowdown valve. Monitor the gauge glass during blowdown to make sure that the water level is not lowered to a dangerously low point.

Step 4 When the desired amount of water has been drained from the boiler, close the blowdown valve.

8.3.3 Cleaning

If an exceptional amount of dirt or sludge is present in a boiler, the boiler should be cleaned using an approved boiler cleaning compound according to the manufacturer's instructions. After cleaning, perform the blowdown procedure as needed to thoroughly flush all traces of the cleaning compound out of the boiler.

8.3.4 Monitoring Boiler Parameters

Record key boiler operating parameters and water conditions regularly. Changes in parameter values over time can reveal that the boiler operation is deteriorating and that corrective action should be taken. *Figure 16* shows an example of a form commonly used to record boiler operating and water quality parameters.

9.0.0 ◆ WATER TREATMENT CHEMICAL SAFETY PRECAUTIONS

HVAC technicians using chemicals in water treatment programs must be certain that the chemicals do not harm system components. This includes, but is not limited to, ferrous, nonferrous, plastic, natural rubber, and synthetic rubber items. Of greater concern is how safe a chemical is in terms of health, handling, and spills and the impact of the chemical on the environment. Unless proven otherwise, all chemicals should be treated as hazardous. As such, take the following precautions when working with chemicals:

- Wear appropriate personal safety equipment such as gloves, safety glasses, respirator, protective clothing, and/or aprons to protect your skin, eyes, and respiratory system from contact with chemicals. The exact type of equipment required depends on the potential hazards involved and the local and OSHA rules that apply to the job site.
- Ask for the OSHA material safety data sheet (MSDS), and read and understand it.
- Do not mix or add chemicals other than as directed by the manufacturer or supplier.
- When using a mixture of chemicals, make sure that the chemical reactions in the system are not harmful to the user or the environment.
- Make sure you follow all federal, state, and local rules and regulations that govern the discharge into the environment and/or the proper disposal of all the chemicals used in a system.

WARNING!

Always add chemicals to water in the solution tank. Adding chemicals first and then water can produce a violent reaction and tremendous heat.

RECOMMENDED READINGS

ALKALINiTY. . . 300-500 ppm   CONDUCTANCE:_____ mho

SULFITE. . .   30-60 ppm   CHLORIDE: _____ ppm

CONDENSATE pH. . . 8 to 8.5   BOTTOM
                                   BLOWDOWN SCHEDULE:
                                   _____

BOILER WATER LOG SHEET

DATE	ALKALINITY	SULFITE	CHLORIDE	CONDUCTANCE	CONDENSATE pH	TESTED BY

308F16.EPS

Figure 16 ◆ Example of a boiler survey form.

1. Water that is known to be safe to drink _____.

   a. can be used in HVAC water systems without the need for water treatment
   b. cannot be used in HVAC water systems
   c. should be tested and, if necessary, treated before use in HVAC water systems
   d. normally contains no minerals or gases harmful to HVAC water systems

2. The concentration levels for the various minerals in water is expressed in parts per gallon (ppg).

   a. True
   b. False

3. Water with a pH of 9 is _____.

   a. alkaline
   b. acidic
   c. neutral
   d. very acidic

4. As the hardness level of water decreases, the potential for scaling _____.

   a. increases
   b. decreases
   c. remains the same
   d. fluctuates

5. A cloudy appearance in water is most likely caused by _____ in the water.

   a. suspended solids
   b. dissolved solids
   c. nitrogen
   d. nitrates or silicates

6. Oxygen is one of the main causes of internal corrosion in a water system.

   a. True
   b. False

7. Biological growth is most likely found in _____.

   a. hot-water systems
   b. cooling towers
   c. steam systems
   d. water-to-water heat exchangers

8. Which of the following is unlikely to cause fouling in a water system?

   a. Silt
   b. Bacteria
   c. Acids
   d. Oil

9. If the water recirculating through an open system cooling tower has a pH below 7.0, the most likely cause is absorption of _____.

   a. acidic gases from the air
   b. airborne particles
   c. acidic gases from the makeup water
   d. oxygen from the air

10. The use of chromates to prevent corrosion in closed recirculating water systems _____.

    a. is prohibited in some localities
    b. is permitted in all closed systems
    c. has no restrictions
    d. is prohibited by EPA regulations

11. To minimize corrosion in a steam boiler, the pH of the boiler water should be maintained between _____.

    a. 6.5 and 7.5
    b. 7.5 and 8.5
    c. 9.5 and 10.5
    d. 11 and 12

12. For a given physical size cartridge filter, the surface-type filter _____ than the depth type.

    a. has a higher removal efficiency
    b. has a lower removal efficiency
    c. has about the same removal efficiency
    d. handles lower flow rates

13. Centrifugal separators are used to remove _____.

    a. suspended particles from water
    b. dissolved particles from water
    c. water and impurities from steam
    d. dissolved particles from steam condensate water

14. In the typical water softener _____.
   a. a distillation process is used to remove minerals from hard water
   b. sodium is released and calcium and magnesium are absorbed by a filter bed of zeolite or similar material
   c. the zeolite must be removed and replaced with new zeolite when the sodium content is depleted
   d. the zeolite bed is regenerated by back-washing

15. A device often used in steam systems to remove noncondensible gases from the feedwater is called the _____.
   a. evaporator
   b. deaerator
   c. centrifugal separator
   d. steam separator

Summary

Proper water treatment is an important aspect of HVAC equipment and system operation. It not only protects equipment, it also minimizes maintenance costs, saves energy, and prevents expensive repairs. These are common problems caused by untreated water in heating and cooling systems:

- Corrosion
- Scale formation
- Biological growth
- Suspended solid matter

Water treatment is site specific. The design of a water treatment program depends on the type of water system. The specific program is normally designed by a qualified water specialist. This involves making a detailed analysis of the available water, considering its intended use, then implementing the needed treatment program and/or equipment as required. Water treatment may be done by a physical means such as mechanical filtering, by chemical treatment, or both. Once a water treatment program has been developed, its correct use and program maintenance are often the job of an HVAC technician.

Always remember that the excessive use of chemicals to treat a water system can be just as harmful to the equipment as the use of insufficient chemicals.

Notes

Trade Terms
Introduced in This Module

Alkalinity: A water quality parameter in which the pH is higher than 7. It is also a measure of the water's capacity to neutralize strong acids.

Backwashing: A procedure that reverses the direction of water flow through a multimedia-type filter by forcing the water into the bottom of the filter tank and out the top. Backwashing is performed on a regular basis to prevent accumulated particles from clogging the filter.

Bleed-off: A method used to help control corrosion and scaling in a cooling tower. It involves periodically draining and throwaway (wasting) a small amount of the water circulating in a system. This aids in limiting the buildup of impurities caused by the continuous adding of makeup water to a system.

Colloidal substance: A jelly-like material made up of very small, insoluble, nondiffusible particles larger than molecules, but small enough to remain suspended in a fluid without settling to the bottom.

Concentration: Indicates the strength or relative amount of an element present in a water solution.

Cycles of concentration: A measurement of the ratio of dissolved solids contained in a heating/cooling system's water to the quantity of dissolved solids contained in the related makeup water supply. The term *cycles* indicates when the concentration of an element in the water system has risen above the concentration contained in the makeup water. For instance, if the hardness in the system water is determined to be two times as great as the hardness in the makeup water, then the system water is said to have two cycles of hardness.

Dissolved solids: The dissolved amounts of substances such as calcium, magnesium, chloride, and sulfate contained in water. Dissolved solids can contribute to the corrosion and scale formation in a system.

Electrolysis: The process of changing the chemical composition of a material (called the electrolyte) by passing electrical current through it.

Electrolyte: A substance in which conduction of electricity is accompanied by chemical action.

Fouling: A term used for problems caused by suspended solid matter that accumulates and clogs nozzles and pipes, which restricts circulation or otherwise reduces the transfer of heat in a system.

Grains per gallon (gpg): An alternate unit of measure sometimes used to describe the amounts of dissolved material in a sample of water. Grains per gallon can be converted to ppm by multiplying the value of gpg by 17.

Hardness: A measure of the amount of calcium and magnesium contained in water. It is one of the main factors affecting the formation of scale in a system. As the hardness level increases, the potential for scaling also increases.

Inhibitor: A chemical substance that reduces the rate of corrosion, scale formation, or slime production.

Milligrams per liter (mg/l): Metric unit of measure used to specify exactly how much of a certain material or element is dissolved in a sample of water. One mg/l is equivalent to one ppm.

Parts per million (ppm): Unit of measure used to specify exactly how much of a certain material or element is dissolved in a sample of water. For the purpose of comparison, one ounce of a contaminant mixed with 7,500 gallons of water equals a concentration of about one ppm.

pH: A measure of alkalinity or acidity of water. The pH scale ranges from 0 (extremely acidic) to 14 (extremely alkaline) with the pH of 7 being neutral. Specifically, it defines the relative concentration of hydrogen ions and hydroxide ions. As pH increases, the concentration of hydroxide ions increases.

Scale: A dense coating of mineral matter that precipitates and settles on internal surfaces of equipment as a result of falling or rising temperatures.

Suspended solids: The amount of visible, individual particles in water or those that give water a cloudy look. They can include silt, clay, decayed organisms, iron, manganese, sulfur, and micro-organisms. Suspended solids can clog treatment devices or shield microorganisms from disinfection.

Total solids: Refers to the total amount of both dissolved and suspended solids contained in water.

Additional Resources

This module is intended to be a thorough resource for task training. The following reference works are suggested for further study. These are optional materials for continued education rather than for task training.

ASHRAE Handbook – HVAC Applications. Atlanta, GA: American Society of Heating, Refrigerating, and Air Conditioning Engineers, Inc.

ASHRAE Handbook – HVAC Systems and Equipment. Atlanta, GA: American Society of Heating, Refrigerating, and Air Conditioning Engineers, Inc.

Boilers Simplified. Troy, MI: Business News Publishing Company.

Water Treatment Specification Manual. Troy, MI: Business News Publishing Company.

Figure Credits

Courtesy of SPX Cooling Technologies, Inc., 308F02 (photo)

Clark-Reliance, 308SA01

Hayward Flow Control Systems, 308F04

Parker Hannifin Corporation, 308F05, 308F07

Yardney Water Management Systems, Inc., 308SA02, 308F08

Victory Energy Operations, LLC, 308F11

Crown Solutions, LLC, 308F12

Penn Separator Corp., 308F13

Nu-Calgon, 308F14, 308F15

NCCER CURRICULA — USER UPDATE

NCCER makes every effort to keep its textbooks up-to-date and free of technical errors. We appreciate your help in this process. If you find an error, a typographical mistake, or an inaccuracy in NCCER's curricula, please fill out this form (or a photocopy), or complete the online form at **www.nccer.org/olf**. Be sure to include the exact module ID number, page number, a detailed description, and your recommended correction. Your input will be brought to the attention of the Authoring Team. Thank you for your assistance.

Instructors – If you have an idea for improving this textbook, or have found that additional materials were necessary to teach this module effectively, please let us know so that we may present your suggestions to the Authoring Team.

NCCER Product Development and Revision
13614 Progress Blvd., Alachua, FL 32615

Email: curriculum@nccer.org
Online: www.nccer.org/olf

❏ Trainee Guide ❏ AIG ❏ Exam ❏ PowerPoints Other _____

Craft / Level: _____ Copyright Date: _____

Module ID Number / Title: _____

Section Number(s): _____

Description: _____

_____

_____

_____

Recommended Correction: _____

_____

_____

_____

Your Name: _____

Address: _____

_____

Email: _____ Phone: _____

03309-08

Troubleshooting Electronic Controls

03309-08

Troubleshooting Electronic Controls

Topics to be presented in this module include:

1.0.0   Introduction 9.2
2.0.0   Microprocessor Controls 9.2
3.0.0   Troubleshooting Microprocessor-Controlled Systems 9.3
4.0.0   External Causes of Failure 9.4
5.0.0   Electronic Controls in Heating Systems 9.10
6.0.0   Cooling Systems and Heat Pumps 9.12
7.0.0   Test Instruments 9.18
8.0.0   Standardization 9.18

Overview

Electromechanical controls such as bimetal thermostats and mechanical relays are common in older systems, but today's heating and cooling systems rely on electronic control devices. This is true in applications ranging from the basic residential system to the largest commercial system. Troubleshooting of these modern control systems is very different. Electromechanical control systems use discrete components that can be readily tested and eliminated in case of an operating failure. In an electronic control system, the individual components are not accessible. They are mounted on PC boards, and in some cases are in sealed enclosures. They are treated as "black boxes" for troubleshooting purposes. If there is a control failure, the entire PC board or assembly is replaced. However, the troubleshooter must be able to determine if the control has actually failed before replacing it. This is a skill that requires the ability to analyze wiring diagrams and manufacturer operating sequence literature.

Objectives

When you have completed this module, you will be able to do the following:

1. Describe the similarities and differences between electronic controls and conventional controls.
2. Analyze circuit diagrams and other manufacturers' literature to determine the operating sequence of microprocessor-controlled systems.
3. Use test equipment to diagnose a microprocessor-controlled comfort system.

Trade Terms

Black box
Diagnostics
Enthalpy
Fault message
High-fire
Low-fire
Microprocessor
Radio frequency
  interference (RFI)
Zone damper

Required Trainee Materials

1. Pencil and paper
2. Appropriate personal protective equipment

Prerequisites

Before you begin this module, it is recommended that you successfully complete *Core Curriculum*; *HVAC Level One*; *HVAC Level Two*; and *HVAC Level Three*, Modules 03301-08 through 03308-08.

This course map shows all of the modules in the third level of the *HVAC* curriculum. The suggested training order begins at the bottom and proceeds up. Skill levels increase as you advance on the course map. The local Training Program Sponsor may adjust the training order.

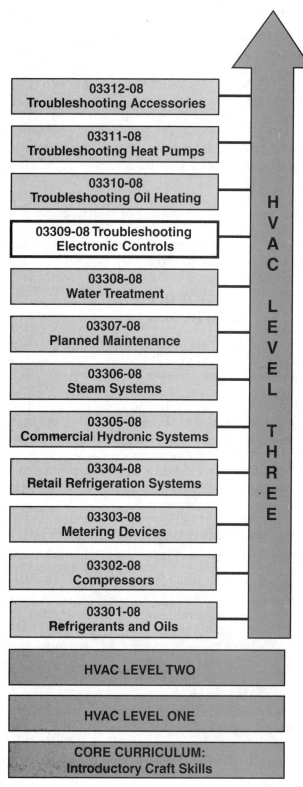

309CMAP.EPS

1.0.0 ◆ INTRODUCTION

In the HVAC Level Two module, *Basic Electronics*, you learned about the operation, usage, and testing of solid-state electronic devices such as thermistors and diodes. With the possible exception of thermistors, you are more likely to find complete, self-contained electronic controls than you are to find discrete electronic components. In this module, you will learn more about troubleshooting **microprocessor**-controlled heating and cooling systems.

The primary objective of this module is to teach you how to troubleshoot electronic controls by performing troubleshooting exercises in the classroom. The information in this module will help prepare you for those exercises.

2.0.0 ◆ MICROPROCESSOR CONTROLS

Electronic controls are packaged in closed modules or on printed circuit boards where the components are visible (*Figure 1*).

An HVAC control system can be compared to a computer (*Figure 2*). If you touch a key on the computer keyboard, you expect a character to appear on the computer screen. The keystroke is the input and the character on the screen is the output. You don't need to understand what happens inside the components in order to understand the cause-and-effect relationship.

If the character fails to appear when you touch the key, you know that something is wrong. If you have a wiring schematic that shows where the input signal goes into the processor and where it comes out on its way to the display monitor, you can use standard test equipment to determine if the problem is in the keyboard, the processor, or the monitor. You may not be able to repair any of these components, any more than you would repair a thermostat, a microprocessor control board, or a hermetic compressor, but you could replace them.

Likewise, if a cooling thermostat calls for cooling, you expect the compressor and fans to turn on. If they don't, you know that something is wrong. It's pretty easy to isolate the problem by making a couple of voltage checks. If it turns out that the control board is receiving the call for cooling, but isn't sending out a call for the compressor and fans to turn on, you simply replace the board. You don't have to know which chip or electrical component failed. All you need to know is what voltage to look for and which pins to test.

309F01.EPS

Figure 1 ◆ Typical HVAC electronic controls.

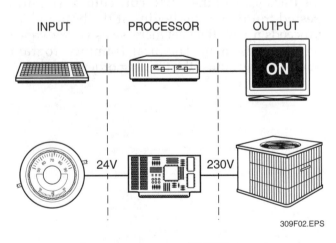

309F02.EPS

Figure 2 ◆ Computer system and HVAC control system.

The microprocessor chip, which is the heart of the personal computer, is also used to create smart control circuits for HVAC systems. It can receive information from sensors, process that information, make decisions based on the information, and issue commands to implement those decisions.

Such controls are more reliable, efficient, accurate, and easier to troubleshoot than conventional controls. The single disadvantage of electronic controls is that you cannot see what goes on inside a microprocessor or integrated circuit. Therefore, it may be more difficult to do circuit tracing in a microprocessor-controlled system. Later in this module, however, you will see how easy it can be to trace a circuit even when you can't follow it with your finger.

3.0.0 ◆ TROUBLESHOOTING MICROPROCESSOR-CONTROLLED SYSTEMS

Figure 3 shows both a microprocessor version and a conventional version of a simplified furnace control circuit. In a conventional control circuit, the schematic diagram shows you all the switches, as well as the coils and contacts of the relays and contactors. Given a little time, you can usually figure out the sequence of operation by studying the circuit and following each load line back to its source.

3.1.0 Troubleshooting Approach

The microprocessor control board is usually represented on the schematic diagram as a **black box**. That is, all you know is what goes into it and what comes out of it; you can't trace the circuit from the loads back to the sources. Even so, it is not that hard to troubleshoot systems with microprocessor controls. In most cases, it is easier because the microprocessor will analyze the information it is receiving, or not receiving, and help you to isolate the fault.

When you are troubleshooting a system that uses microprocessor controls, you will probably have to read the manufacturer's literature to determine the operating sequence of the unit. You

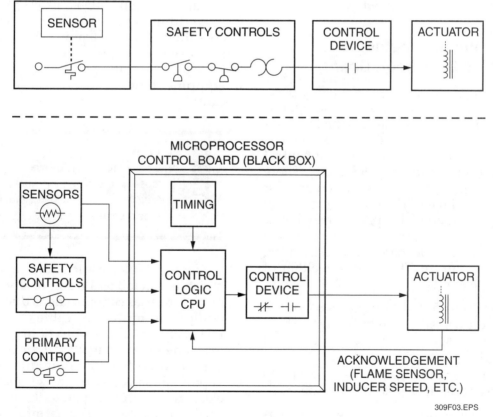

Figure 3 ◆ Conventional versus microprocessor controls.

cannot assume that the electronic control will have the same operating sequence and control features as a similar system with conventional controls. That's because the microprocessor makes operating features available to the designer that were not feasible with conventional circuits.

There are some basic assumptions you can make, however, about the control circuit for any unit you encounter. For example, before activating a compressor or igniting a gas or oil burner, the microprocessor will verify that safety controls such as pressure switches, overloads, and temperature limit switches are in the correct position (usually closed). In a conventional circuit, the same thing is accomplished by putting the safety controls in series between the primary control (such as the thermostat) and the actuator or load. Another reasonable assumption is that when the thermostat is calling for heat or cooling, there will be 24V (or other specified voltage) applied to the control board input. A similar response can be expected at the control board output to energize the compressor or furnace control and activate the fans. You will need to study the literature, however, to determine if there is any built-in delay before these responses appear at the output, such as compressor short-cycle protection, blower-on time delay, or, in an induced-draft furnace, a draft pressure buildup delay.

3.2.0 Determining Whether to Replace the Circuit Board

Electronic controls have almost no moving parts to wear out. Once they are installed and operating, they are unlikely to fail unless the failure is caused by some outside influence. Unless there is an indication of failure, microprocessor control boards and modules should be left alone. There are no preventive maintenance requirements for their modules, and handling or testing them can do more harm than good. In spite of the reliability of electronic controls, service technicians will often replace the control board without isolating the cause of the failure. This can have either of two undesirable effects:

- If the board was not at fault, as is often the case, the fault will still be in the system. There will be a repeat service call and a dissatisfied customer.
- If the board was at fault, whatever caused it to fail may also cause its replacement to fail.

The first step is to determine if the control circuit is really at fault. This can often be done through the built-in **diagnostic** capability, which usually reserves one of the **fault messages** for a failed control board. For example, if a flashing LED is used to announce faults, a complete control failure might be signaled by having the LED remain continuously on (or off). Even then, it is a good idea to check whether the control voltage (24V, for example) is available to the control board. Some controls do not have the ability to distinguish between a complete board failure and the absence of control voltage.

In most cases, you can measure the voltage at the pins on the connector. Some printed circuit boards have test points where the measurements can be made. The wiring schematic will identify the signal that should be available at each test point.

4.0.0 ◆ EXTERNAL CAUSES OF FAILURE

Before replacing a board that appears to have failed, try to determine if an external cause might account for the problem. A variety of such causes are described in the following paragraphs.

4.1.0 Environmental Conditions

Failures in electrically controlled equipment can result from exposure to harmful environmental conditions such as damp or wet locations; exposure to gases, fumes, vapors, or other agents that have a deteriorating effect on the conductors or equipment; or exposure to excessive temperatures.

CAUTION

Equipment must be UL-listed for its intended purpose. The use of equipment not listed for the application may result in equipment failure. It may also be a code violation.

Electrical/electronic equipment subjected to moisture can fail as a result of short circuits and/or high-resistance connections due to corrosion at equipment terminals, cable terminations, and relay contacts. Definitions of dry, damp, and wet locations are given in *NEC Article 100*.

Dirt and other contaminants such as fumes, vapors, abrasives, soot, grease, and oils can cause electronic devices to operate abnormally until they finally break down. The harmful effects of gases, fumes, vapors, and similar agents can cause corro-

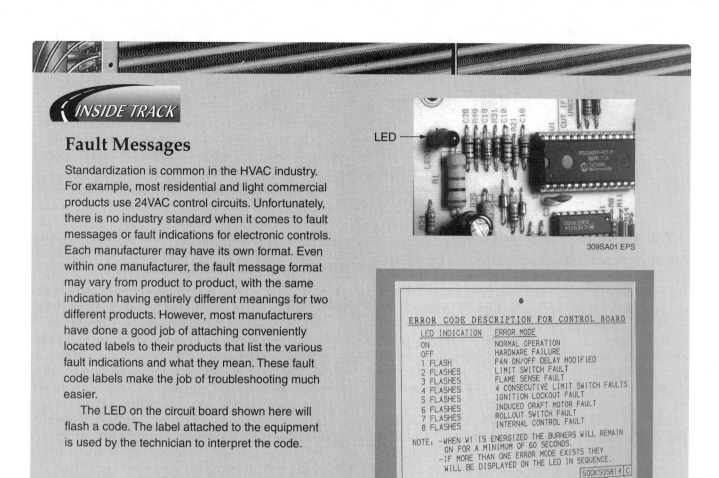

Fault Messages

Standardization is common in the HVAC industry. For example, most residential and light commercial products use 24VAC control circuits. Unfortunately, there is no industry standard when it comes to fault messages or fault indications for electronic controls. Each manufacturer may have its own format. Even within one manufacturer, the fault message format may vary from product to product, with the same indication having entirely different meanings for two different products. However, most manufacturers have done a good job of attaching conveniently located labels to their products that list the various fault indications and what they mean. These fault code labels make the job of troubleshooting much easier.

The LED on the circuit board shown here will flash a code. The label attached to the equipment is used by the technician to interpret the code.

LED

309SA01.EPS

ERROR CODE DESCRIPTION FOR CONTROL BOARD

LED INDICATION	ERROR MODE
ON	NORMAL OPERATION
OFF	HARDWARE FAILURE
1 FLASH	FAN ON/OFF DELAY MODIFIED
2 FLASHES	LIMIT SWITCH FAULT
3 FLASHES	FLAME SENSE FAULT
4 FLASHES	4 CONSECUTIVE LIMIT SWITCH FAULTS
5 FLASHES	IGNITION LOCKOUT FAULT
6 FLASHES	INDUCED DRAFT MOTOR FAULT
7 FLASHES	ROLLOUT SWITCH FAULT
8 FLASHES	INTERNAL CONTROL FAULT

NOTE: –WHEN W1 IS ENERGIZED THE BURNERS WILL REMAIN ON FOR A MINIMUM OF 60 SECONDS.
–IF MORE THAN ONE ERROR MODE EXISTS THEY WILL BE DISPLAYED ON THE LED IN SEQUENCE.

50DK505814 C

309SA02.EPS

sion at equipment terminals, cable terminations, and relay contacts. In some cases, the wiring insulation can be attacked and deteriorate, causing a short circuit or ground fault. Equipment and wiring that is exposed to the harmful effects of gases, fumes, vapors, and similar agents is subject to the requirements for hazardous locations covered in *NEC Articles 500 through 517*.

Equipment that is subjected to unusual amounts of dust and other airborne particles tends to fail due to overheating caused by clogged air filters or the loss of the free air movement required to dissipate the heat generated in the unit. Papers, boxes, and other materials that block free air movement to equipment are another cause of heat-related failures.

Heat increases the resistance of circuits, thereby increasing the current. Heat causes materials to expand, dry out, crack, and/or blister. Equipment that is exposed to temperature extremes often fails as a result of an open circuit in the printed circuit board wiring, an open connection between an integrated circuit chip pin and the printed board wiring, or from a heat-related failure in an integrated circuit. Extreme temperatures can also result in damage to cabling insulation.

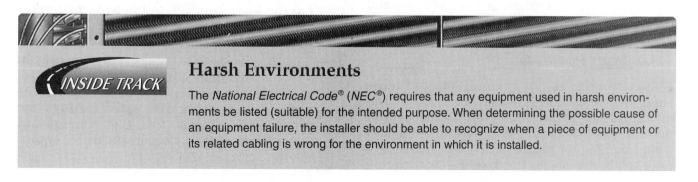

Harsh Environments

The *National Electrical Code®* (*NEC®*) requires that any equipment used in harsh environments be listed (suitable) for the intended purpose. When determining the possible cause of an equipment failure, the installer should be able to recognize when a piece of equipment or its related cabling is wrong for the environment in which it is installed.

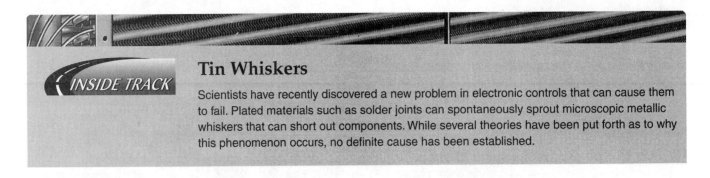

Tin Whiskers

Scientists have recently discovered a new problem in electronic controls that can cause them to fail. Plated materials such as solder joints can spontaneously sprout microscopic metallic whiskers that can short out components. While several theories have been put forth as to why this phenomenon occurs, no definite cause has been established.

4.2.0 Improper Installation

Improper installation of equipment by an unqualified or careless installer can result in failures. All systems and equipment should be installed in accordance with the manufacturer's instructions and the current national and local codes. As previously mentioned, installation of the wrong type of equipment for a particular environment can result in failures. Incorrect mounting or support of equipment enclosures can cause the equipment to be susceptible to damage or loose connections resulting from vibration or shock. Vibrations and physical abuse often cause breakdowns. Failure to properly tighten terminal lugs or to properly terminate a connection can result in an electrical/electronic device failing prematurely.

Cable shorts or opens can occur as a result of staples installed too tightly in cable runs, or cables being dragged over sharp edges during installation. Cables installed in high-traffic areas can be damaged by being stepped on. The impedance or resistance of some cables installed near electrical equipment can be affected if the cable is nicked, cut, or crimped. Cables installed alongside or over fluorescent lights or too close to electrical junction boxes, hot water pipes, or heating ducts can fail due to overheating. Similarly, cables improperly run in the same raceway with power circuits can cause electrical noise to be induced into data-carrying cables.

Manufacturing defects are also common. For example, you may find a loose circuit board after delivery or installation. Shipping and transporting can also loosen or damage circuit boards and components.

4.3.0 Poor Power Quality

Abnormal primary power conditions can cause equipment to operate erratically or become damaged and fail if not protected by the proper devices. The waveform characteristics of the utility power can become distorted by severe weather, utility power fluctuations, or changing building load conditions, all of which can result in poor power quality. Some of the common types of power abnormalities include:

- *Sags* – Sags, also called brownouts, are short-term decreases in voltage levels. Sags are typically caused by the startup power demands of many electrical devices such as motors, compressors, elevators, and shop tools. Sags can also be deliberately caused by the power company as a way of coping with extraordinary power demands. In a procedure called rolling brownouts, the power company systematically lowers voltage levels in certain areas for hours or days at a time. Typically, this occurs in urban areas in the summer when air conditioning requirements are at their peak. Sags can starve electronic equipment of the power needed to function properly. In computer systems, sags can cause frozen keyboards, unexpected system crashes, and the loss or corruption of data. In other electronic equipment, sags may stress components, causing them to fail prematurely.

- *Blackouts* – A blackout is a total loss of utility power. Blackouts are caused by excessive demands on the power grid, lightning storms, and ice on power lines. In computer systems, this can cause the loss of work stored in RAM or cache, and possible loss of the data stored on the hard drive.

- *Spikes or transients* – Spikes or transients are instantaneous, dramatic increases in voltage typically caused by a nearby lightning strike or when utility power comes back online after having been knocked out in a storm. Spikes can cause damage to hardware and loss of data.

- *Surges* – Surges are short-term increases in voltage. They typically occur when large electrical motors are switched off because of the extra voltage being dissipated through the power lines. Surges stress components and cause premature failures in electronic equipment.

- *Electrical noise* – Electrical noise, also referred to as electromagnetic interference (EMI) and **radio frequency interference (RFI)**, distorts the smooth utility power sine wave. Electrical noise is caused by many factors, including lightning strikes, load switching, generators, radio transmitters, and industrial equipment. Noise introduces glitches and errors into computer software programs and data files.

- *Harmonics* – A harmonic is a sinusoidal waveform with a frequency that is an integral multiple of the fundamental system frequency. For example, the harmonics of a 60Hz sine wave would be 120Hz (second harmonic), 180Hz (third harmonic), 300Hz (fifth harmonic), and so on. When one or more harmonic components are added to the fundamental frequency, a distorted or nonlinear waveform is produced. Harmonics are especially prevalent whenever there are several types of equipment, such as switching power supplies, personal computers, adjustable speed drives, and medical test equipment, that draw current in short pulses. Electronic ballasts used in fluorescent and high-intensity discharge (HID) lighting fixtures are also common sources of harmonic generation. The presence of harmonics can result in overheated transformers and neutrals, and tripped circuit breakers.

4.4.0 Electrostatic Discharge

Electrostatic discharge (ESD) is the charge produced by the transfer of electrons from one object to another. Static electricity is generated as the result of electron movement between two different materials that come in contact with each other and are then separated. When the two materials are good conductors, the excess electrons in one will return to the other before the separation is complete. However, if one of them is an insulator, both will become charged by the loss or gain of electrons (unless grounded). An example of this is when you get an electric shock from touching a doorknob after walking across a carpet.

Static electricity charges can have potentials that exceed several thousand volts. Some examples are:

- Walking across a carpet – 35,000V
- Walking over a vinyl floor – 2,000V
- Handling vinyl envelopes, such as for work instructions – 7,000V
- Picking up a common plastic bag – 20,000V
- Touching office furniture that is padded with polyurethane foam – 18,000V

Electrostatic discharges that occur at or near electronic equipment can damage sensitive semiconductor circuits and devices. Most static electricity shocks that humans can feel have a voltage level in excess of 2,000V. Static electricity charges lower than this level are normally below the threshold of human sensation. However, these lower-voltage static electricity charges can damage sensitive semiconductor devices. Some semiconductor devices can be damaged by electrostatic charges as low as 10V.

When subjected to ESD, a sensitive electronic component can experience a catastrophic failure, meaning it is immediately damaged to the point at which it is totally inoperative. ESD can also cause a component to fail in a way that allows it to continue to work properly for a while, but results in poor system performance and eventual system failure. This type of failure is called a latent or hidden failure. Another type of failure caused by ESD, called an upstart failure, can cause a current flow in an integrated circuit that is not significant enough to cause a catastrophic failure, but may result in gate leakage that causes the intermittent loss of software or the incorrect storage of information.

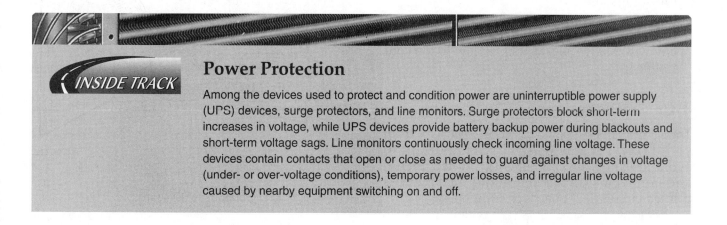

Power Protection

Among the devices used to protect and condition power are uninterruptible power supply (UPS) devices, surge protectors, and line monitors. Surge protectors block short-term increases in voltage, while UPS devices provide battery backup power during blackouts and short-term voltage sags. Line monitors continuously check incoming line voltage. These devices contain contacts that open or close as needed to guard against changes in voltage (under- or over-voltage conditions), temporary power losses, and irregular line voltage caused by nearby equipment switching on and off.

As an HVAC technician, you must be continually alert to the threat of ESD and understand how to control it. Control of ESD can be accomplished through grounding and isolation.

Some guidelines that will help prevent failures associated with ESD include the following:

- When working with ESD-sensitive devices, make sure you ground, isolate, and neutralize each device.

- Keep the work area clean and clear of unnecessary materials, particularly common plastics.

- Test grounding devices, such as wrist straps, before use to make sure they have not become loose or intermittent.

- Always ship and handle ESD-sensitive components in metallized shielding bags, conductive bags, or conductive tote boxes.

- When handling ESD-sensitive printed circuit (PC) boards or components, avoid touching the components, printed circuit, and connector pins.

- When handing an ESD-sensitive PC board to another person, make sure both individuals are grounded to the same ground point or potential.

- Maintain the room humidity at a minimum of 40 percent, if possible.

- Ensure that all carpets and clothing of operating personnel are made of natural materials.

- If carpets are made of synthetic materials, make sure they have a conductive backing.

- Make sure that equipment operators wear conductive footwear.

- Use anti-static materials whenever possible to increase conductivity.

- Voltage surges and excessive voltage can also cause control circuit failures. You should always make sure the control voltage is within the allowed limits, usually 18V to 30V. Also, make sure the unit is properly grounded at the time of installation.

- Excessive heat can also damage electronic controls. The equipment should not be altered in any way that would restrict airflow to the control board.

4.4.1 Grounding

Transients can be created by static electricity from your body. For this reason, it is important to avoid touching the components, printed circuit, and connector pins when handling sensitive printed circuit (PC) boards or components.

Always ground yourself before touching the PC board or components. This is normally done by using a grounding device that directly connects you to ground, thereby eliminating any static charge from being generated by your body.

Using ESD grounding wrist straps (*Figure 4*) is the most common grounding method, especially for field work. To provide protection from ESD, the wrist strap must be properly grounded to either the utility ground or to a common ground point on a properly grounded work surface. It must also make good contact with your skin. When working on HVAC equipment containing electronic controls, grounding can be accomplished by touching grounded bare metal in the cabinet before touching or handling an electronic control.

Other devices used for ESD protection in a shop or lab environment can include conductive footwear and workstations equipped with properly grounded conductive or dissipative work surfaces used in conjunction with a grounded conductive floor or floor mat.

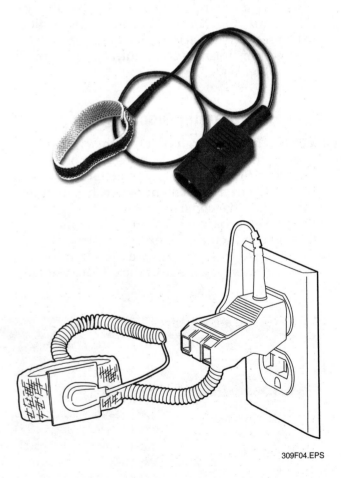

309F04.EPS

Figure 4 ◆ ESD grounding wrist straps.

4.4.2 Isolation

Isolation of sensitive components and assemblies during storage and transportation is another way to prevent ESD-related component failures. Static charges cannot penetrate containers that are made of conductive materials. For this reason, sensitive devices are normally contained in metallized shielding bags (*Figure 5*), conductive bags, and/or conductive tote boxes during storage and/or transportation. It is important to point out that the outside surfaces of these conductive containers can carry static charges that must be removed before the container is opened. This discharge can only be safely accomplished by a properly grounded person opening the container at a static-safe location.

4.5.0 Thermistor Failure

Thermistors, which are used as sensors in microprocessor-controlled systems, will sometimes fail or provide incorrect information. The built-in diagnostic feature of many systems will evaluate information coming from the thermistors and report a fault when it senses an invalid input. A thermistor is not likely to fail on its own. If a thermistor fault is reported, it could mean that the thermistor has been damaged, that the thermistor lead is damaged or loose, or that the thermistor has come loose from its mounting and is therefore not providing an accurate indication.

The resistance of a thermistor itself can be tested with an ohmmeter. The manufacturer's literature will usually contain a graph or table that shows expected resistance readings for a range of temperatures, similar to the information shown in *Table 1*.

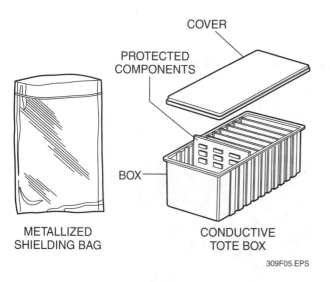

COVER

PROTECTED COMPONENTS

BOX

METALLIZED SHIELDING BAG

CONDUCTIVE TOTE BOX

309F05.EPS

Figure 5 ◆ ESD isolation containers.

Table 1 Thermistor Values

TEMPERATURE (°F)	RESISTANCE (Ohms)	TEMPERATURE (°F)	RESISTANCE (Ohms)	TEMPERATURE (°F)	RESISTANCE (Ohms)
-60	362,640	45	11,396	150	1,020
-55	297,140	50	9,950	155	929
-50	245,245	55	8,709	160	844
-45	202,841	60	7,642	165	768
-40	168,250	65	6,749	170	699
-35	139,960	70	5,944	175	640
-30	116,820	75	5,249	180	585
-25	98,420	80	4,644	185	535
-20	82,665	85	4,134	190	490
-15	69,685	90	3,671	195	449
-10	58,915	95	3,265	200	414
-5	50,284	100	2,913	205	380
0	42,765	105	2,600	210	350
5	36,475	110	2,336	215	323
10	31,216	115	2,092	220	299
15	26,786	120	1,879	225	276
20	23,164	125	1,689	230	255
25	19,978	130	1,527	235	236
30	17,276	135	1,377	240	219
35	14,980	140	1,244		
40	13,085	145	1,126		

309T01.EPS

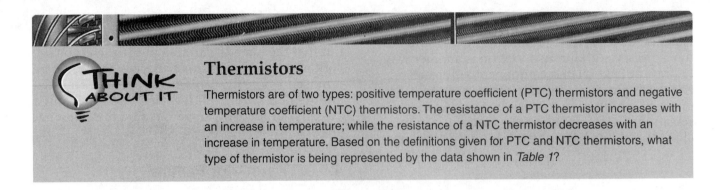

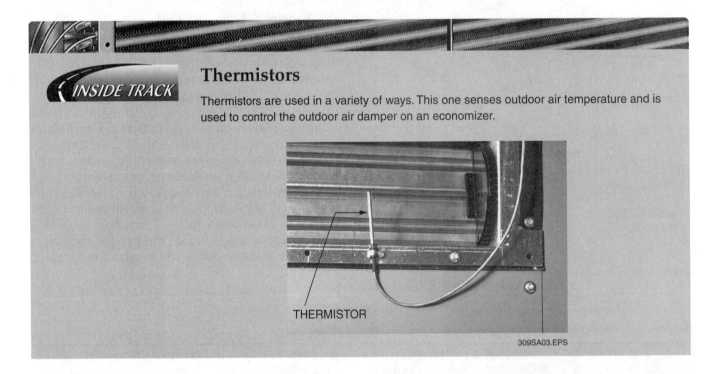

5.0.0 ◆ ELECTRONIC CONTROLS IN HEATING SYSTEMS

Figure 6 shows a printed circuit board containing the microprocessor control circuits for a packaged gas heating and cooling unit. The schematic diagram for the board is shown in *Figure 7*. Following are the highlights of the operating sequence for this unit.

- As in all furnaces, the sequence is initiated by a call for heat from the room thermostat, in the form of a 24V signal. This starts the induced-draft motor.

309F06.EPS

Figure 6 ◆ Gas unit control board.

Small Units (Blower <10 Amps)

TYPICAL SCHEMATIC

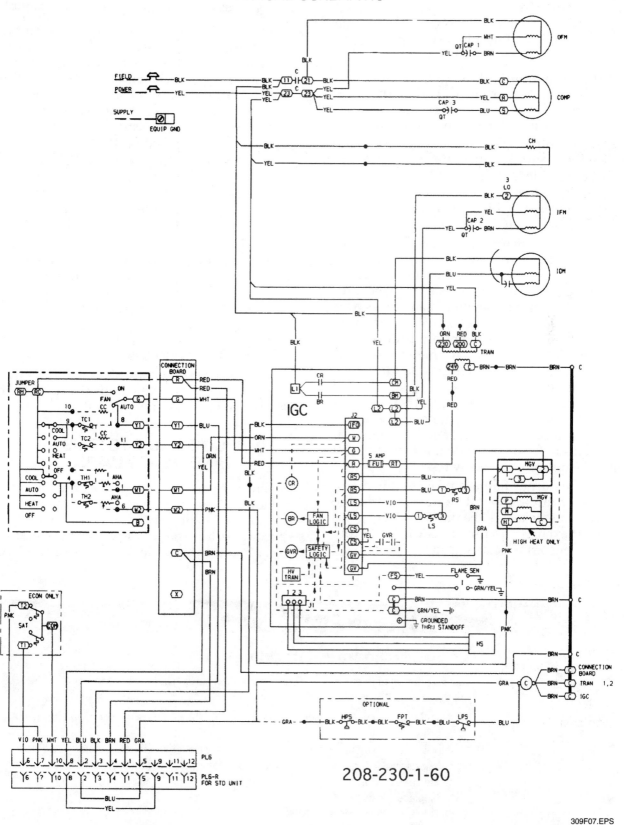

208-230-1-60

309F07.EPS

Figure 7 ◆ Gas unit control board wiring schematic.

- Before the microprocessor sends a signal to initiate combustion or start the compressor, it will first verify that each safety control is in its correct position. In a gas furnace control, for example, the microprocessor might ask itself the following questions before turning on the gas valve and sending a high-voltage pulse to the ignitor:

  – Is there continuity through the rollout and limit switches, indicating that they are closed?
  – Is a switch closed message coming from the pressure switch or other sensor to indicate that the induced-draft motor is running?

  If those conditions are met, the microprocessor will energize the gas valve relay and activate the high-voltage ignition transformer.

- Once the gas valve and ignitor have been turned on, the microprocessor uses the current signal from the flame sensor to verify that there is a steady, strong flame. If so, the ignitor is shut off. If not, the microprocessor may be programmed to retry for ignition many times over a period of several minutes. After each attempt, the microprocessor will impose a delay that gives the induced-draft fan time to purge the heat exchangers to prevent gas buildup. One electronic furnace control on the market will retry for ignition 33 times over a 15-minute period before it gives up. At that point, the microprocessor locks out ignition and a fault indicator begins to flash. The system is then disabled until the power is manually reset.

- When the burner flame is proven, the microprocessor fan logic will be notified and the blower time delay will begin. This delay gives the heat exchangers time to warm up before the blower starts circulating air.

The operating sequence just described is not very different from that of any furnace with an electronic ignition. The difference lies in the special features that can be programmed into the microprocessor-controlled system. For example:

- Some systems have a feature to prevent nuisance trips of the limit switch. The microprocessor will automatically adjust the blower-off delay if the unit is tripping off on the limit switch after the thermostat opens. It will also flash a fault message to notify someone that this has happened because the problem could be caused by a lack of airflow.

- Variable-speed blowers and induced-draft fan motors are common features in high-efficiency furnaces. Some are programmed to automatically adapt to changing conditions such as the opening and closing of **zone dampers** in a zoned system.

- Most microprocessor-controlled systems can diagnose and isolate faults. Most automatic diagnostics include a failure message that indicates if the control board/module itself has failed. This must be checked before the board is replaced because other problems, such as a power loss, might also cause this message.

- Some gas heating controls can automatically adjust the length of the **high-fire** and **low-fire** cycles to maintain indoor comfort.

6.0.0 ◆ COOLING SYSTEMS AND HEAT PUMPS

Cooling systems do not require a great deal of control logic unless they have variable-capacity compressors, variable-speed blowers, or some other special feature. The primary advantages of electronic controls in cooling systems for residential

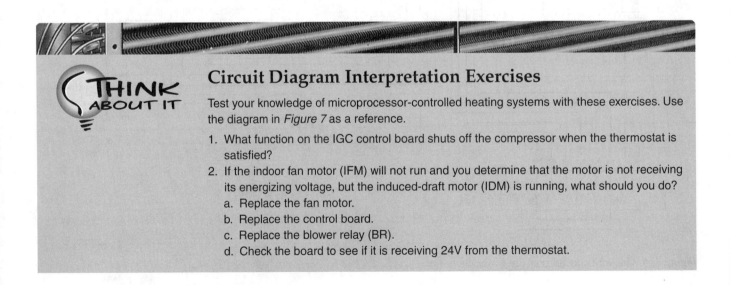

THINK ABOUT IT

Circuit Diagram Interpretation Exercises

Test your knowledge of microprocessor-controlled heating systems with these exercises. Use the diagram in *Figure 7* as a reference.

1. What function on the IGC control board shuts off the compressor when the thermostat is satisfied?
2. If the indoor fan motor (IFM) will not run and you determine that the motor is not receiving its energizing voltage, but the induced-draft motor (IDM) is running, what should you do?
   a. Replace the fan motor.
   b. Replace the control board.
   c. Replace the blower relay (BR).
   d. Check the board to see if it is receiving 24V from the thermostat.

and light commercial use are their reliability and accuracy. However, in large cooling systems, such as those used in office buildings, microprocessor controls and digital control systems are very important.

Packaged electronic controls have become common in heat pumps because the need for defrost makes the logic of heat pump controls more complicated than those of straight cooling systems. Even when conventional controls are used, a packaged electronic control just for the

defrost mode may be provided. An example of this application is shown in the HVAC Level Two module, *Heat Pumps*.

6.1.0 Cooling Systems

Figure 8 shows the control setup for the air side of a commercial heating/cooling system with an economizer. The circuit diagram in *Figure 9* is an example of the electronic control circuit for such a system.

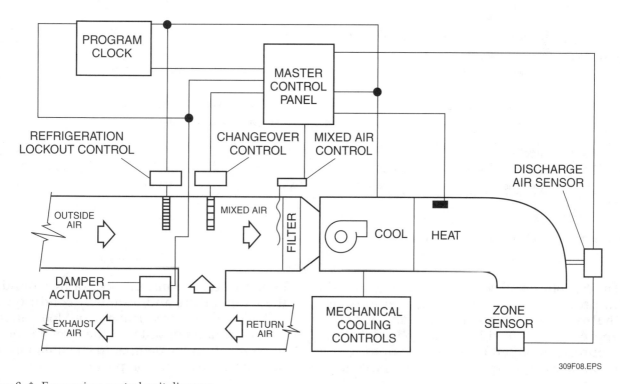

Figure 8 ◆ Economizer control unit diagram.

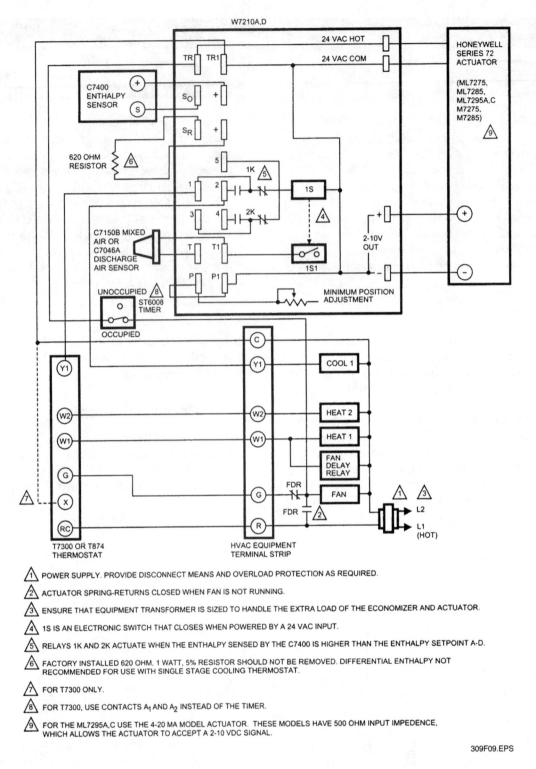

Figure 9 ◆ HVAC electronic control system diagram.

The economizer allows outdoor air to be mixed with processed air when the temperature or **enthalpy** of the outdoor air is suitable for this purpose. Sensors feed information back to the microprocessor about the condition of the outdoor air, discharge air, and the air in the conditioned space. The microprocessor then decides if the temperature and humidity conditions are suitable for the introduction of outdoor air. If the conditions are not suitable, the economizer control will return the damper to its minimum position and cycle on the compressor.

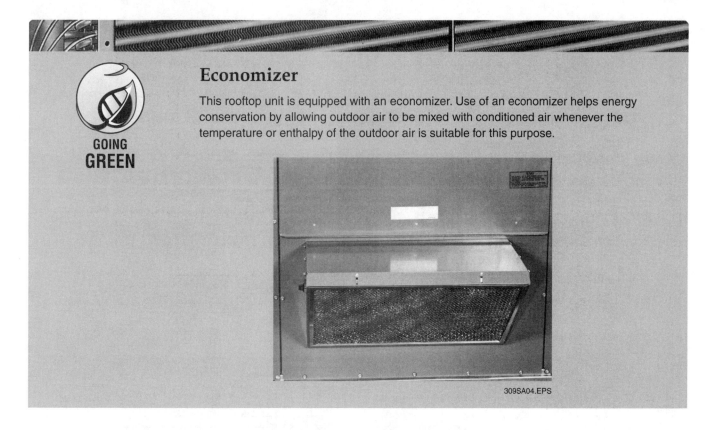

Economizer

This rooftop unit is equipped with an economizer. Use of an economizer helps energy conservation by allowing outdoor air to be mixed with conditioned air whenever the temperature or enthalpy of the outdoor air is suitable for this purpose.

GOING GREEN

309SA04.EPS

In large systems, the controls can be very complex. Outdoor air (free cooling) might be the first stage of cooling. As the enthalpy of the outdoor air changes, the control system will begin to cycle on mechanical cooling and reduce the amount of outdoor air.

6.2.0 Heat Pumps

Figure 10 shows the integrated electronic control board for a heat pump. *Figure 11* is the wiring schematic for a packaged heat pump. Even though you can't see everything that is going on inside the control boards, you can determine a lot about the sequence and timing of operations. The unit contains two control boards, a defrost board (DB) and an ICM board that controls the operation of the variable-speed indoor blower motor. Looking at the defrost board helps reveal how it functions. A timing (logic) circuit is in series with the compressor contactor coil and the low- and high-pressure switches in the Y circuit. This indicates that the defrost board contains a compressor start-delay timer. Another timing (logic) circuit is in series with the defrost thermostat. This indicates that defrost is based on elapsed time and outdoor coil temperature. Elapsed time is field-selectable and is done based on the position of dip switches, as shown on the schematic.

A direct connection between the room thermostat cooling contacts and O on the defrost board indicates the reversing valve is energized in cooling. Open relay contacts within the defrost board imply that during defrost, they close to energize the reversing valve and apply power to the electric heat relays (HR) by providing a current path from R to W2. Closed relay contacts across OF1 and OF2 open during defrost to stop outdoor fan operation. The same contacts can be seen in series with the outdoor fan motor (OFM) in the high-voltage side (upper part) of the schematic.

309F10.EPS

Figure 10 ◆ Heat pump control unit.

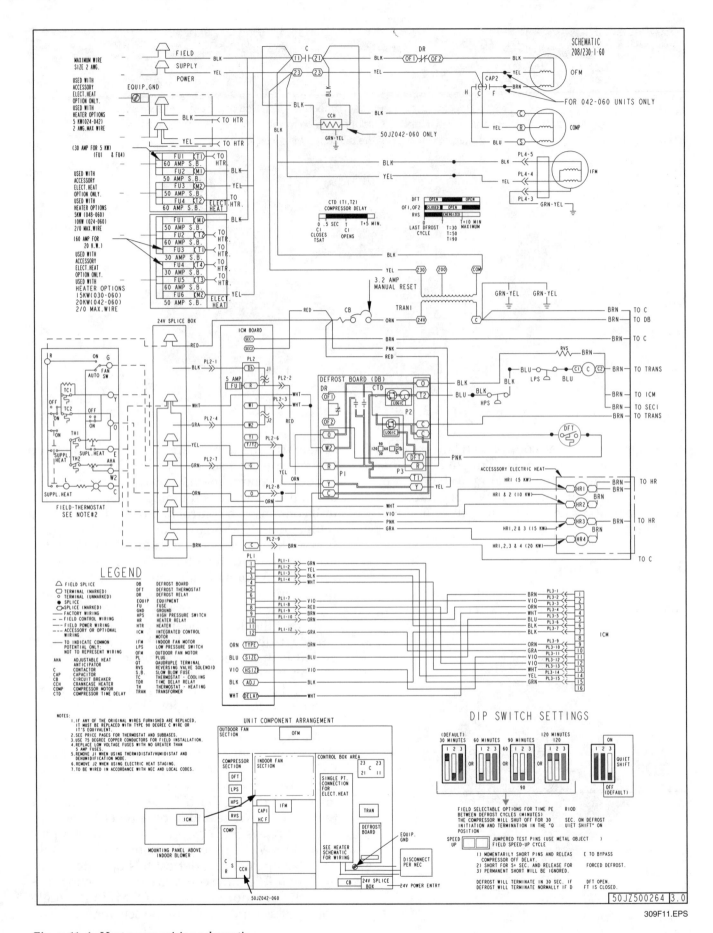

Figure 11 ◆ Heat pump wiring schematic.

The ICM board is used to control the variable-speed indoor fan motor (IFM). Looking at the schematic, one might conclude that there are four electric motors in the unit; three in the high-voltage side (outdoor fan motor, compressor motor, and indoor fan motor), and one in the low-voltage side (ICM). In fact, the indoor fan motor and the ICM motor are one and the same. The indoor fan motor has high-voltage power applied at all times, even when the contactor is de-energized. This is a characteristic of electronically commutated motors. In order to run, it must also receive various low-voltage input signals from the ICM board through the 16-pin connector on the motor. This results in the indoor fan motor having two connections, one for high-voltage (upper part of schematic), and another that carries the low-voltage input signals (lower part of schematic).

The ICM board generates the low-voltage signals based on the inputs it receives from the room thermostat. For example, during cooling operation, signals from the thermostat generate output signals from the ICM board that tell the fan to run at its highest speed. If continuous fan operation is required, the lack of a Y signal (cooling) and the presence of a G signal (fan operation only) cause the ICM board to signal the fan to run at a much lower speed.

The ICM board generates the appropriate speed control signals based on the position of jumper wires on pins on the board. For example, motor speed is partially based on system capacity. Cooling fan speed of a 3-ton system would be higher than cooling fan speed of a 2-ton system. Heater size in kilowatts also determines fan speed in the heating mode. Larger heaters require more airflow. When properly configured, the ICM board will allow the fan to operate at a number of different speeds based on the equipment's operating mode. Configuration procedures are not readily apparent on the schematic diagram and must be obtained by reading the manufacturer's installation instructions.

For the special features offered by an electronic control, you must read the manufacturer's literature. The training programs available from manufacturers also are very helpful in learning about the specifics of a particular product or family of products. One example of information that may be included in the product literature is shown in *Figure 12*. Having this diagram when checking out the control circuit would eliminate a lot of the guesswork because it shows what signal should appear at each pin of the board connector.

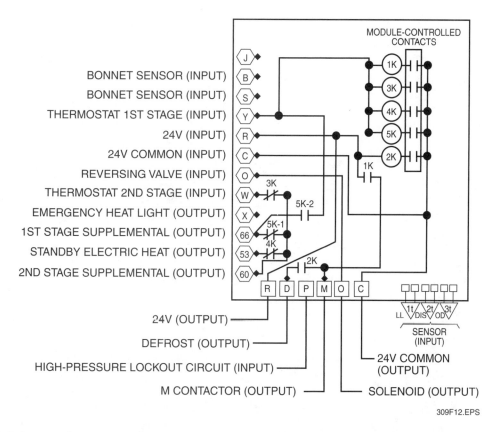

Figure 12 ◆ Input and output signals.

7.0.0 ◆ TEST INSTRUMENTS

The same test equipment and tools used to test and troubleshoot systems with conventional controls are also used on systems with electronic control circuits. No special test instruments are needed to troubleshoot electronic control circuits. Using a VOM to check the voltage at the inputs and outputs of the control board should be sufficient.

Whenever you check resistance or continuity in an electronic circuit, make sure the component you are testing is disconnected or isolated from the board.

CAUTION

Be cautious when using an ohmmeter to test components that are connected to the control board. Because the ohmmeter generates its own voltage, it might damage the board-mounted circuits.

Many manufacturers market special diagnostic devices to test microprocessor-controlled systems. These devices help to isolate hard-to-solve problems by allowing the technician to exercise the various system operating functions and then display the results. The manufacturer of a multi-stage, variable-speed gas furnace provides interface hardware (*Figure 13*) and software that enable a personal computer or hand-held device to be connected to the furnace control board so that vital information can be obtained and displayed to aid in troubleshooting. *Figure 14* shows a typical screen and the type of information available.

8.0.0 ◆ STANDARDIZATION

It is common to see standardization of control circuits within a particular manufacturer's products. It is more cost effective for the manufacturer to use the same control board or module in every furnace or air conditioning system than it is to design a new control circuit for each product, as was common with conventional circuits.

However, the use of a standard control does not imply that the operating sequences will be identical from one product to another. The microprocessor control allows designers to change operating features by changing the programming of the microprocessor. Therefore, even though the same module is used on two products, they may have a few different operating features.

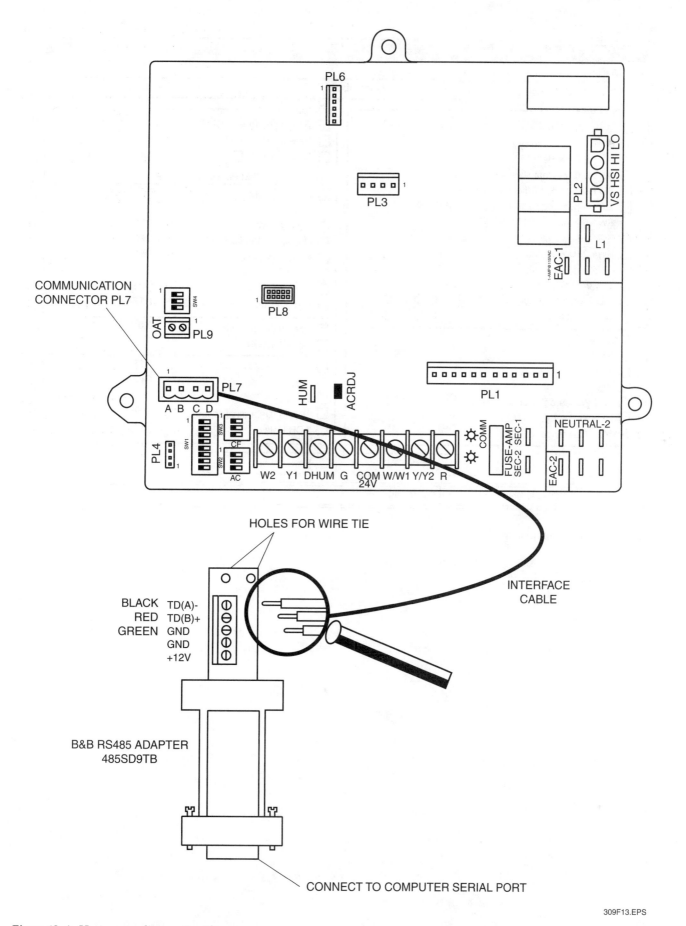

Figure 13 ◆ Heat pump diagnostic aid.

309F13.EPS

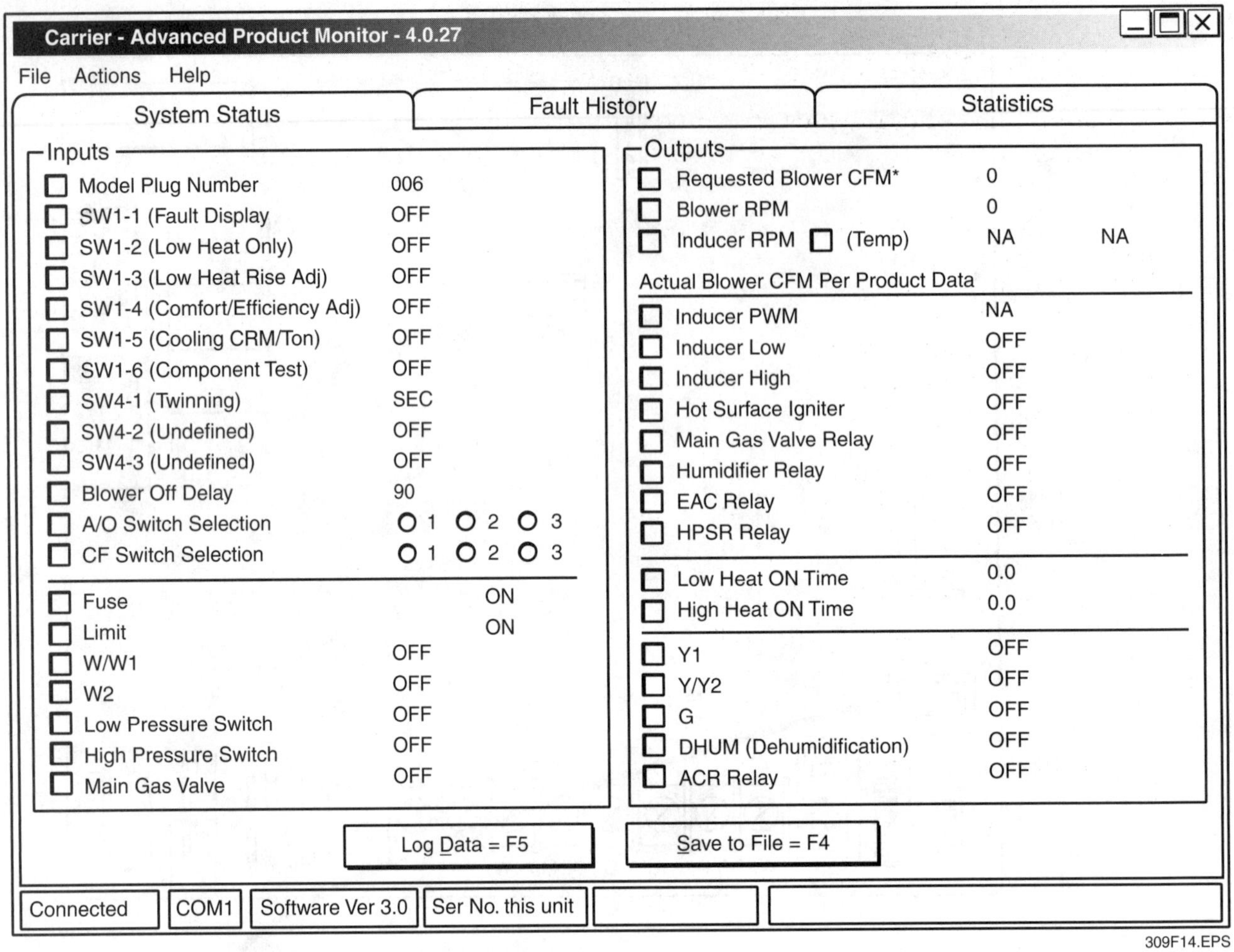

Carrier - Advanced Product Monitor - 4.0.27

File   Actions   Help

| System Status | Fault History | Statistics |

Inputs

Model Plug Number	006	
SW1-1 (Fault Display)	OFF	
SW1-2 (Low Heat Only)	OFF	
SW1-3 (Low Heat Rise Adj)	OFF	
SW1-4 (Comfort/Efficiency Adj)	OFF	
SW1-5 (Cooling CRM/Ton)	OFF	
SW1-6 (Component Test)	OFF	
SW4-1 (Twinning)	SEC	
SW4-2 (Undefined)	OFF	
SW4-3 (Undefined)	OFF	
Blower Off Delay	90	
A/O Switch Selection	○ 1 ○ 2 ○ 3	
CF Switch Selection	○ 1 ○ 2 ○ 3	

Fuse	ON
Limit	ON
W/W1	OFF
W2	OFF
Low Pressure Switch	OFF
High Pressure Switch	OFF
Main Gas Valve	OFF

Outputs

Requested Blower CFM*	0	
Blower RPM	0	
Inducer RPM ☐ (Temp)	NA	NA

Actual Blower CFM Per Product Data

Inducer PWM	NA
Inducer Low	OFF
Inducer High	OFF
Hot Surface Igniter	OFF
Main Gas Valve Relay	OFF
Humidifier Relay	OFF
EAC Relay	OFF
HPSR Relay	OFF

Low Heat ON Time	0.0
High Heat ON Time	0.0

Y1	OFF
Y/Y2	OFF
G	OFF
DHUM (Dehumidification)	OFF
ACR Relay	OFF

| Log Data = F5 | Save to File = F4 |

| Connected | COM1 | Software Ver 3.0 | Ser No. this unit | | |

309F14.EPS

Figure 14 ◆ Information display.

1. Which of the following is generally safe to assume about a microprocessor-controlled comfort system?

   a. The compressor will turn on as soon as there is a call for cooling.
   b. When the thermostat calls for heat or cooling, the specified voltage will be applied to the control board input.
   c. If the diagnostic program signals a board failure, replacing the board will fix the problem.
   d. The diagnostic program will indicate a complete board failure in the absence of control voltage.

2. If a microprocessor-controlled heat pump works in the heating mode but will not provide cooling, the logical first step is to ____.

   a. replace the control board
   b. use the built-in diagnostic capability to help in isolating the fault
   c. replace the reversing valve
   d. check the compressor and its protective devices

3. All of the following are likely external sources of problems in a microprocessor control circuit except ____.

   a. moisture
   b. ozone depletion
   c. current surges
   d. excessive heat

4. The approved way to handle a control device containing a microprocessor is to ____.

   a. wear work gloves
   b. hold the board with your fingertips
   c. pick it up with pliers or tweezers
   d. ground yourself before touching it

5. A thermistor should be tested with a(n) ____.

   a. voltmeter
   b. ohmmeter
   c. ammeter
   d. megger

6. You would expect to find all of the following features on a gas heating unit with conventional (electromechanical) controls except a(n) ____.

   a. flame rollout switch
   b. intermittent pilot
   c. automatic adjustment of the blower delay period
   d. induced-draft fan

Refer to the input and output signals diagram in *Figure 12* to answer Questions 7 through 9.

7. To verify that the control board is sending out a signal to activate the compressor, you should check ____.

   a. pins M and B on the board
   b. pins M and P on the board
   c. across the transformer secondary
   d. pins O and B on the board

8. Check that a defrost output is present by placing a voltmeter across ____.

   a. R and D
   b. O and B
   c. D and B
   d. R and O

9. In the defrost mode, the energizing voltage for the reversing valve solenoid comes from pin ____.

   a. O
   b. P
   c. R
   d. B

10. Which of the following statements about electronically commutated motors is *incorrect*?

   a. Continuous blower speed and cooling fan speed are always the same.
   b. Power is always applied to the motor.
   c. Low-voltage inputs determine motor speed.
   d. Unit capacity and electric heater size can be factors affecting motor speed.

Summary

Microprocessor controls improve the efficiency, reliability, and accuracy of HVAC systems. Although they may seem more complex than conventional control systems, they are often easier to troubleshoot because of their built-in diagnostic and testing features.

The same test equipment and tools used to test systems with conventional controls are used to test systems with microprocessor controls. Many manufacturers also offer diagnostic devices specific to microprocessor-controlled systems.

An important thing to remember when troubleshooting systems that use a microprocessor control is that the control board should not be replaced until you have definitely determined that it has failed. If the board has failed, do everything you can to eliminate external problems that could have caused the failure. External causes of failure include environmental conditions, improper equipment installation, poor power quality, and damage caused by electrostatic discharge.

Notes

Black box: A term given to a solid-state, hermetically sealed electronic assembly that is replaced rather than field-repaired.

Diagnostics: A term used to identify the built-in testing and fault isolation capability of a microprocessor-controlled system.

Enthalpy: The total heat content of a substance. In HVAC, it is the total heat content of the air and water vapor mixture as measured from an established point.

Fault message: A coded message delivered by a flashing light, display readout, or printout to indicate the existence of a fault and (usually) its probable source.

High-fire: The maximum output stage of a two-stage gas furnace control.

Low-fire: The reduced gas pressure stage of a two-stage gas furnace control.

Microprocessor: A tiny computer made from a single integrated circuit chip.

Radio frequency interference (RFI): Electronic noise caused by equipment operating at a very high frequency.

Zone damper: A damper used to control airflow to a zone in a zoned comfort system.

Additional Resources and References

Additional Resources

This module is intended to be a thorough resource for task training. The following reference works are suggested for further study. These are optional materials for continued education rather than for task training.

Electronic Devices, 2002. Thomas L. Floyd. Upper Saddle River, NJ: Prentice Hall, Inc.

Electronics Fundamentals, 2001. Thomas L. Floyd. Upper Saddle River, NJ: Prentice Hall, Inc.

Figure Credits

Emerson Climate Technologies, 309F01

Carrier Corporation, 309SA01, 309SA03, 309F06, 309F07, 309F11, 309F13, 309F14

Topaz Publications, Inc., 309SA02, 309F04 (photo), 309SA04

Courtesy of Honeywell International Inc., 309F09

ICM Controls, 309F10

NCCER CURRICULA — USER UPDATE

NCCER makes every effort to keep its textbooks up-to-date and free of technical errors. We appreciate your help in this process. If you find an error, a typographical mistake, or an inaccuracy in NCCER's curricula, please fill out this form (or a photocopy), or complete the online form at **www.nccer.org/olf**. Be sure to include the exact module ID number, page number, a detailed description, and your recommended correction. Your input will be brought to the attention of the Authoring Team. Thank you for your assistance.

Instructors – If you have an idea for improving this textbook, or have found that additional materials were necessary to teach this module effectively, please let us know so that we may present your suggestions to the Authoring Team.

NCCER Product Development and Revision

13614 Progress Blvd., Alachua, FL 32615

Email: curriculum@nccer.org
Online: www.nccer.org/olf

❏ Trainee Guide   ❏ AIG   ❏ Exam   ❏ PowerPoints   Other _____

Craft / Level: _____ Copyright Date: _____

Module ID Number / Title: _____

Section Number(s): _____

Description: _____

_____

_____

_____

_____

Recommended Correction: _____

_____

_____

_____

_____

Your Name: _____

Address: _____

_____

Email: _____ Phone: _____

03310-08

Troubleshooting
Oil Heating

03310-08
Troubleshooting Oil Heating

Topics to be presented in this module include:

1.0.0	Introduction	10.2
2.0.0	Typical Operation	10.2
3.0.0	Oil Burner Troubleshooting	10.8
4.0.0	Troubleshooting Controls	10.10
5.0.0	System Troubleshooting	10.16
6.0.0	Condensing Oil Furnaces	10.28

Overview

In northern sections of the U.S., many homes are heated with oil furnaces. There was a time when oil was a very inexpensive way to heat, so oil furnaces became very popular. In other instances, natural gas just wasn't available to the home when it was built, so oil was the best alternative. Although the end result is the same as that of a gas furnace, the method of producing heat is significantly different. One of the main differences is that the oil must be pumped to the point of combustion, whereas natural gas is supplied under pressure. Because the components of oil furnaces are different from those of gas-fired furnaces, troubleshooters must learn to interpret wiring diagrams that are significantly different from those of gas furnaces. In addition, troubleshooting and maintenance of oil furnaces requires different test equipment, such as gas analyzers and draft gauges.

Objectives

When you have completed this module, you will be able to do the following:

1. Describe the basic operating sequence for oil-fired heating equipment.
2. Interpret control circuit diagrams for an oil heating system.
3. Develop a troubleshooting chart for an oil heating system.
4. Identify the tools and instruments used in troubleshooting oil heating systems.
5. Correctly use the tools and instruments required for troubleshooting oil heating systems.
6. Isolate and correct malfunctions in oil heating systems.
7. Describe the safety precautions that must be taken when servicing oil heating systems.

Trade Terms

Cad cell
Flame-retention burner

Primary control
Stack control

Required Trainee Materials

1. Pencil and paper
2. Appropriate personal protective equipment

Prerequisites

Before you begin this module, it is recommended that you successfully complete *Core Curriculum*; *HVAC Level One*; *HVAC Level Two*; and *HVAC Level Three*, Modules 03301-08 through 03309-08.

This course map shows all of the modules in the third level of the *HVAC* curriculum. The suggested training order begins at the bottom and proceeds up. Skill levels increase as you advance on the course map. The local Training Program Sponsor may adjust the training order.

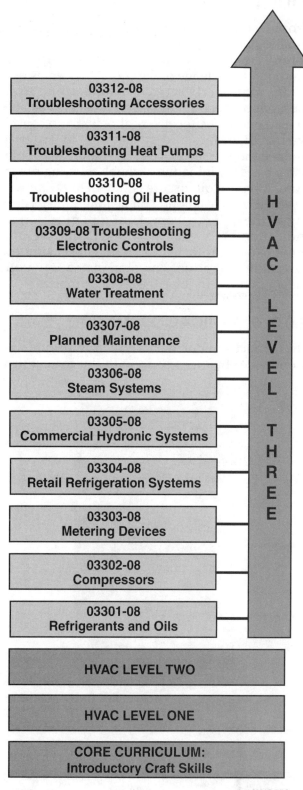

310CMAP.EPS

1.0.0 ◆ INTRODUCTION

Troubleshooting oil heating units is similar to troubleshooting gas heating units, except that the control circuitry may not be quite as complex. A thorough knowledge of the control system is necessary to properly service, maintain, and troubleshoot oil heating units. You must also be able to read and interpret electrical power and control wiring diagrams and schematics in order to understand the sequence of operation and establish a basis for initiating a troubleshooting procedure. A typical oil furnace is shown in *Figure 1*.

When troubleshooting oil heating units, listening to the owner's complaint will give you an indication as to what the problem may be. You must use this information while observing the system in operation. You then may apply your theory in order to solve the service problem.

2.0.0 ◆ TYPICAL OPERATION

Oil furnaces in current production provide an atomized oil vapor mixed with the correct portion of air to the combustion chamber (see *Figure 2*). An ignition transformer provides the spark at a set of electrodes to ignite the fuel-air mixture in the combustion chamber. A **primary control** locks the system out on a flame failure. The lockout must be manually reset. **Cad cells** are commonly used to sense the burner light to verify the presence of a flame for continued operation.

> **NOTE**
>
> Before the introduction of the cad cell, a **stack control** with a bimetal element was used as a safety control. The bimetal element responded to a rise in flue gas temperature. That technology is now obsolete.

The oil pump and burner assembly pressurize the fuel oil and force it through a calibrated nozzle (*Figure 3*), which projects a cone-shaped, atomized spray into the combustion chamber of the heat exchanger. A blower provides the combustion air for the fuel mixture. The electrodes ignite the mixture. In some installations, the spark is on whenever the burner is running and stays on until the burner is extinguished.

310F01.EPS

Figure 1 ◆ Oil furnace.

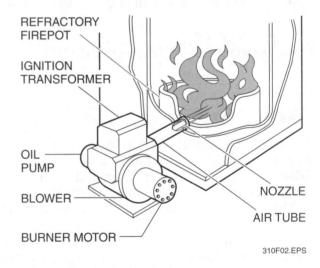

310F02.EPS

Figure 2 ◆ Pressure oil burner.

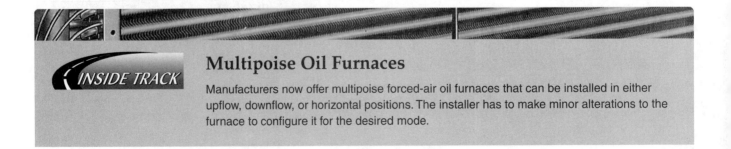

INSIDE TRACK

Multipoise Oil Furnaces

Manufacturers now offer multipoise forced-air oil furnaces that can be installed in either upflow, downflow, or horizontal positions. The installer has to make minor alterations to the furnace to configure it for the desired mode.

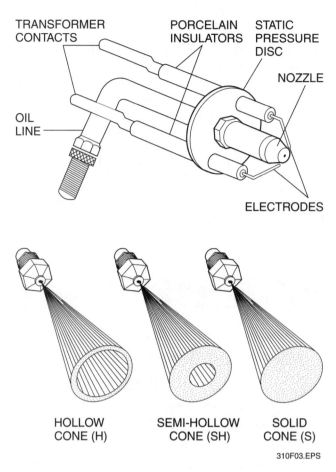

TRANSFORMER CONTACTS

PORCELAIN INSULATORS

STATIC PRESSURE DISC

NOZZLE

OIL LINE

ELECTRODES

HOLLOW CONE (H)

SEMI-HOLLOW CONE (SH)

SOLID CONE (S)

310F03.EPS

Figure 3 ◆ Oil burner nozzle and spray patterns.

The heat exchanger of an oil-fired furnace has primary and secondary heating surfaces. The primary heating surface surrounds the flame, whereas the secondary heating surface is often a series of steel sections over which the hot combustion gases pass before leaving the heat exchanger on their way to the chimney flue.

2.1.0 Typical Safety Devices

All oil-fired furnaces are equipped with a central electrical control connected in series with the room thermostat. The central or primary control, which is a safety-control device, senses the presence or absence of flame and shuts down the burner if it fails to function properly.

Another safety device is the limit control that is actuated by a bimetal element placed in the discharge air plenum. It is usually calibrated to shut off the burner if the discharge air temperature reaches or exceeds 200°F. This limit control will usually reset when the temperature drops 25°F below the cutout point. A secondary limit control is an additional safety control used on some downflow furnaces to protect the blower motor against excessively high temperatures. If a limit switch or other safety control trips more than once, the cause should be determined.

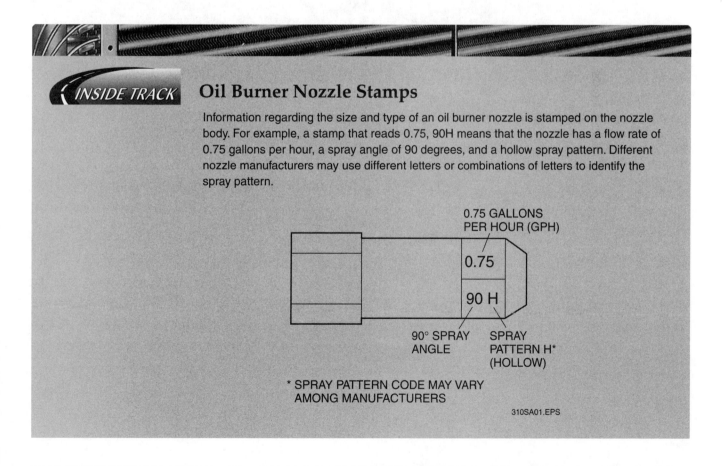

INSIDE TRACK

Oil Burner Nozzle Stamps

Information regarding the size and type of an oil burner nozzle is stamped on the nozzle body. For example, a stamp that reads 0.75, 90H means that the nozzle has a flow rate of 0.75 gallons per hour, a spray angle of 90 degrees, and a hollow spray pattern. Different nozzle manufacturers may use different letters or combinations of letters to identify the spray pattern.

0.75 GALLONS PER HOUR (GPH)

0.75

90 H

90° SPRAY ANGLE

SPRAY PATTERN H* (HOLLOW)

* SPRAY PATTERN CODE MAY VARY AMONG MANUFACTURERS

310SA01.EPS

2.2.0 Operational Effectiveness

The operational effectiveness of an oil-fired furnace is measured by combustion efficiency.

The electrical operating sequence of an oil furnace must be determined by analyzing the schematic wiring diagram and the product literature. As you develop familiarity with oil furnaces in general, you will be increasingly able to determine quickly the operating sequence of an unfamiliar furnace when you encounter it in your work or in your training.

Figure 4 is the schematic diagram for an oil furnace and add-on air conditioner with a cad cell flame detector. The operating sequence is included on the diagram. *Figure 5* is the wiring diagram for the same furnace. In a troubleshooting situation, the wiring diagram will help you determine where to connect your test meter. *Figures 6* and *7* are the schematic and wiring diagrams for a similar unit. Although these diagrams represent only two of the many oil-fired systems you may encounter, they will help you gain more insight into the operation of these systems. Study the operating sequences of these units, and then test your knowledge by completing the exercises.

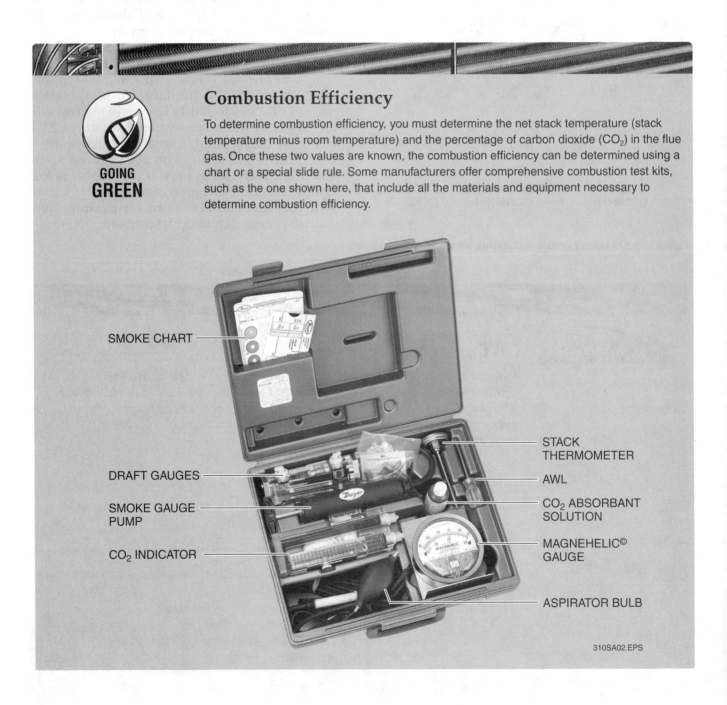

GOING GREEN

Combustion Efficiency

To determine combustion efficiency, you must determine the net stack temperature (stack temperature minus room temperature) and the percentage of carbon dioxide (CO_2) in the flue gas. Once these two values are known, the combustion efficiency can be determined using a chart or a special slide rule. Some manufacturers offer comprehensive combustion test kits, such as the one shown here, that include all the materials and equipment necessary to determine combustion efficiency.

SMOKE CHART

DRAFT GAUGES

SMOKE GAUGE PUMP

CO_2 INDICATOR

STACK THERMOMETER

AWL

CO_2 ABSORBANT SOLUTION

MAGNEHELIC© GAUGE

ASPIRATOR BULB

310SA02.EPS

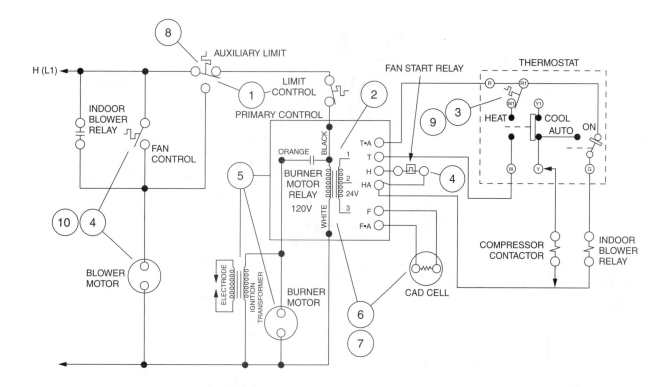

1. Line potential feeds through the secondary limit and limit controls to power primary control.

2. The primary control provides 24-volt control circuit.

3. On a heating demand, the thermostat heating bulb makes.

4. The fan control heater is energized through primary control. After a short period, the heater provides sufficient heat to close the fan contacts. This energizes the blower motor.

5. The primary control simultaneously energizes the burner motor and ignition transformer at the oil burner. The burner motor operates the oil pump and combustion blower to feed air and oil vapor into the combustion chamber. The fuel mixture should ignite with the spark furnished by ignition transformer.

6. If combustion does not take place within the time specified on the primary control, as detected by cad cell, the primary control locks itself out.

7. Should a flame failure occur during the "on" cycle, the primary control locks itself out in response to the cad cell.

8. The secondary limit opens at temperatures above setpoint. This de-energizes the primary control but still allows blower motor operation through the NO contacts.

9. As the heating demand is satisfied, the thermostat heating bulb breaks. This de-energizes the oil burner circuits.

10. The blower motor continues running until furnace temperature drops below fan control setpoint.

310F04.EPS

Figure 4 ◆ Operating sequence number one.

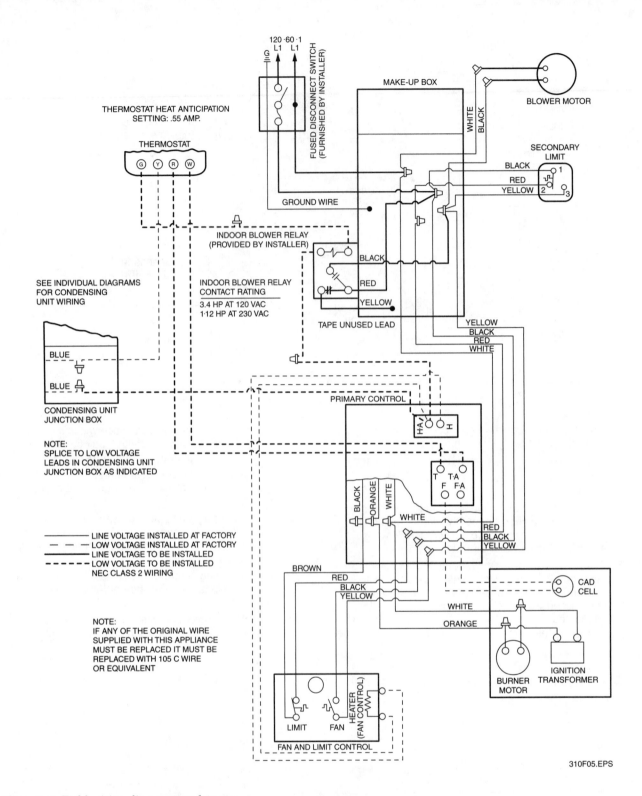

Figure 5 ◆ Field wiring diagram number one.

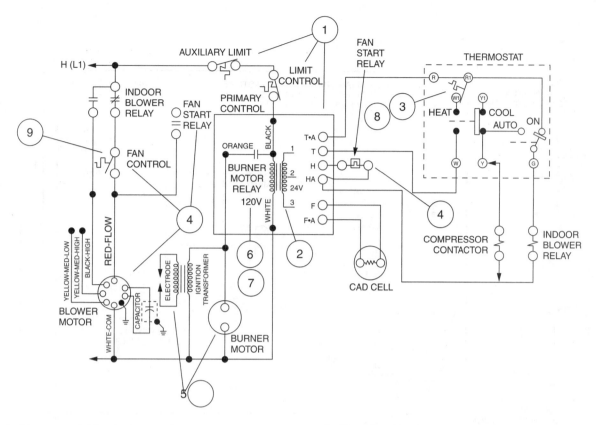

1. Line potential feeds through the secondary limit and limit controls to power primary control.

2. The primary control provides 24-volt control circuit.

3. On a heating demand, the thermostat heating bulb makes.

4. On some units the fan start relay is energized. In approximately 30 seconds the relay contacts make to energize the indoor blower motor on low speed. On other units, the fan control initiates blower motor operation.

5. The primary control simultaneously energizes the burner motor and ignition transformer at the oil burner. The burner motor operates the oil pump and combustion blower to feed air and oil vapor into the combustion chamber. The fuel mixture should ignite with the spark furnished by ignition transformer.

6. If combustion does not take place within the time specified on the primary control, as detected by cad cell, the primary control locks itself out.

7. Should a flame failure occur during the heating cycle, the primary control locks itself out in response to the cad cell.

8. As the heating demand is satisfied, the thermostat heating bulb breaks. This de-energizes the oil burner circuits.

9. The blower motor continues running until furnace temperature drops below fan control setpoint.

310F06.EPS

Figure 6 ◆ Operating sequence number two.

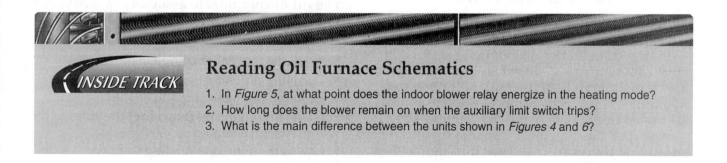

Reading Oil Furnace Schematics

1. In *Figure 5*, at what point does the indoor blower relay energize in the heating mode?
2. How long does the blower remain on when the auxiliary limit switch trips?
3. What is the main difference between the units shown in *Figures 4* and *6*?

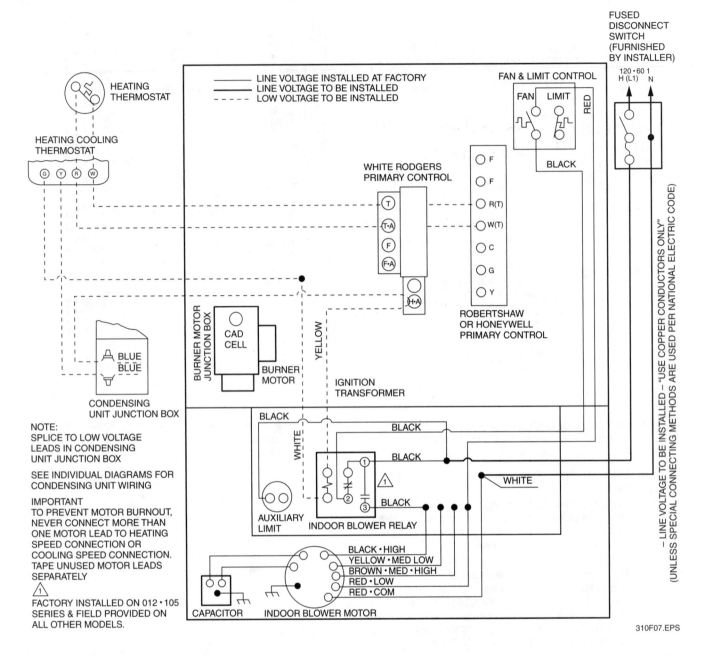

Figure 7 ◆ Field wiring diagram number two.

3.0.0 ◆ OIL BURNER TROUBLESHOOTING

The components of the burner system (*Figure 8*) include the oil supply and the burner. The gun-type atomizing burner, shown in *Figure 9*, is the most commonly used unit. This type of burner uses a fuel pump to deliver oil under pressure (typically 100 psig, but as much as 150 psig) to a calibrated nozzle where it is broken into a fine, cone-shaped mist. An electric spark provided by an ignition transformer and electrode ignite the mixture.

3.1.0 Oil Burner

The oil burner nozzle assembly is a common source of problems. The following conditions can be traced to a faulty nozzle:

- No flame (partial or complete blockage)
- Pulsating pressure
- Flame changing in size and shape
- Flame impinging on (touching) the sides of the combustion chamber
- Low carbon dioxide reading in flue gases (less than 8 percent)

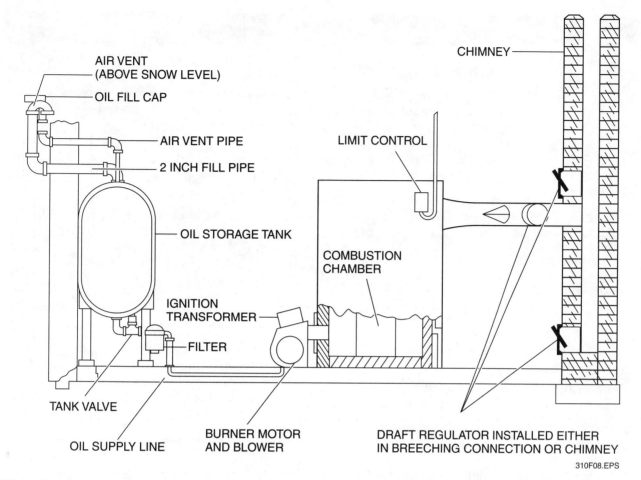

Figure 8 ◆ System components.

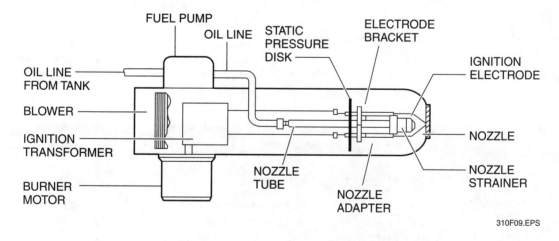

Figure 9 ◆ Gun-type burner.

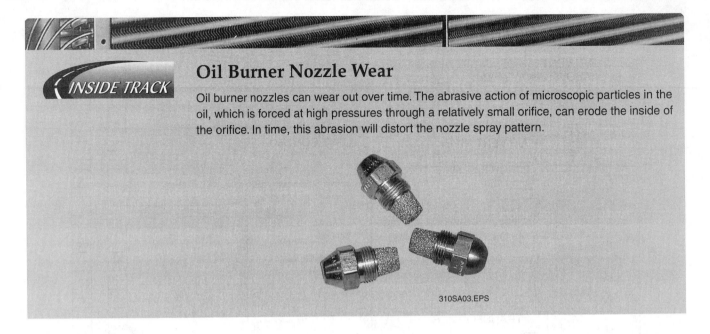
- Delayed ignition
- Oil odors
- Carbon formation or burning oil at the nozzle (may be caused by a bent or warped nozzle)

If the nozzle appears to be faulty, it should be replaced. In general, it is a good idea to replace it annually.

If the nozzle strainer is clogged, determine the cause. The fuel oil line or the oil pump strainer may be plugged, or there may be water or other contaminants in the fuel tank.

If a varnish-like substance has formed on the nozzle, it indicates overheating. There are four major causes of overheating:

- An overfired burner (nozzle too large)
- The fire burning too close to the nozzle
- Too small a firebox
- The nozzle being positioned too far forward

If the fire is too close to the nozzle, a small increase in oil pump pressure may correct the problem. When the nozzle is too far forward, it may be moved back by using a short adapter or by shortening the oil line.

NOTE

A special nozzle wrench makes removing and installing oil burner nozzles much easier.

4.0.0 ◆ TROUBLESHOOTING CONTROLS

The control system (*Figure 10*) turns the heating system on and off in response to temperature changes in the conditioned space. It also provides a means of shutting down the system if an unsafe condition arises.

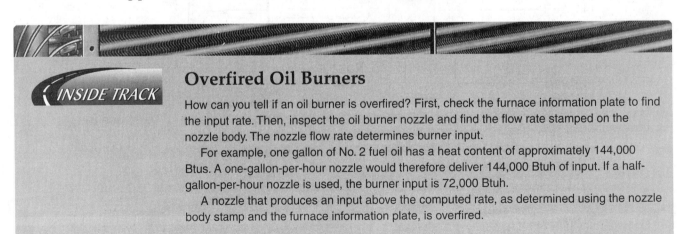

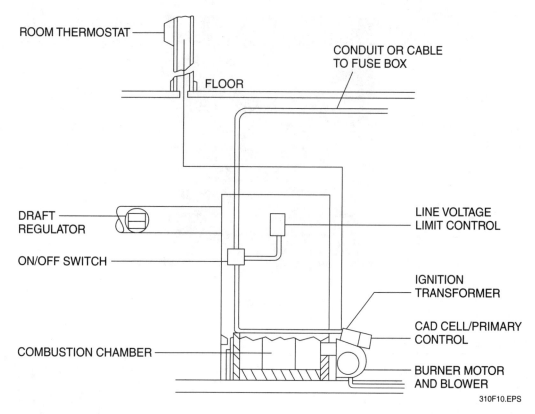

ROOM THERMOSTAT

CONDUIT OR CABLE
TO FUSE BOX

FLOOR

DRAFT
REGULATOR

ON/OFF SWITCH

LINE VOLTAGE
LIMIT CONTROL

IGNITION
TRANSFORMER

CAD CELL/PRIMARY
CONTROL

COMBUSTION CHAMBER

BURNER MOTOR
AND BLOWER

310F10.EPS

Figure 10 ◆ Oil burner control system.

An actual diagram of a control circuit is shown in *Figure 11*. The following are operating features of this control circuit:

- The room thermostat senses changes in the space temperature and then signals the primary control to energize or de-energize the burner.
- The primary control acts as the nerve center. It receives signals from the thermostat and flame detector and sends out signals to control the oil burner.
- The flame detector has a similar function to that of a thermocouple or flame rod in a gas furnace. If a flame is not detected within a prescribed time period, it acts to shut off the burner. In modern oil furnaces, the flame detector is usually a cad cell detector, which responds to the light created by the flame in the combustion chamber. In older furnaces, a bimetal detector in the stack was used for this purpose. You may encounter either type when working with residential and light commercial systems. Many large gas and oil heating systems use an ultraviolet flame detector.
- The line voltage limit control de-energizes the burner if overheating occurs.

- Auxiliary controls may include devices to control the blower or fan, low-limit controls, zone valve controls, air cleaner controls, humidifier controls, time delay controls, and motor control relays.

4.1.0 Primary Control

The primary control (*Figure 11*) is a combination control and safety device. It contains a relay or electronic switching circuit that turns on the system by transferring line voltage to the burner motor and the ignition transformer when a call for heat is received from the thermostat. The safety feature is controlled by the combustion detector, which is a cad cell flame detector. The flame detector must be matched to the primary control. If the primary control does not receive a continuous signal that combustion is taking place, its relay will be de-energized, turning off the burner motor. The primary control allows a prescribed time to elapse before it shuts off and locks out the burner. If a flame is established within the prescribed time, burner operation will continue. Different primary controls have different lockout times. The equipment manufacturer selects the lockout time based on individual product design.

(VIEW A) PRIMARY CONTROL

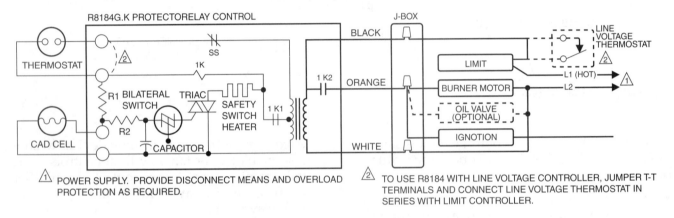

(VIEW B) PRIMARY CONTROL WIRING DIAGRAM

310F11.EPS

Figure 11 ◆ Primary control wiring diagram.

WARNING!

The primary control contains a reset button. When the button is pushed, and the thermostat is calling for heat, the ignition sequence will start. Each time this occurs, oil is sprayed into the combustion chamber. If ignition does not occur, the oil will accumulate in the chamber, creating an explosion hazard. Therefore, the button should not be pressed more than once. All accumulated oil must be cleaned up before the furnace is started.

The primary control shown in *Figure 11* is a solid-state control. Earlier types used relays or contactors for switching.

4.2.0 Cad Cell Flame Detector

The cad cell (photocell) flame detector consists of a plug-in, light-sensitive cell and a socket with a mounting bracket and lead wire (one-piece units are also available). It is mounted in the burner unit (*Figure 12*) and wired into the control circuitry as shown in *Figure 11*. The location of the cad cell is critical. It must be positioned so that it can detect the light from the flame in order to function properly.

Before checking the cad cell and primary control, check the following:

• Main power supply (unit disconnect) on
• Electrodes properly positioned and with proper gap

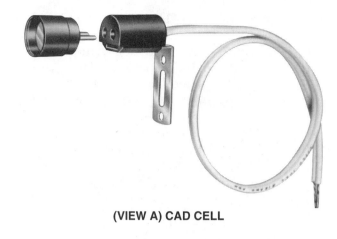

(VIEW A) CAD CELL

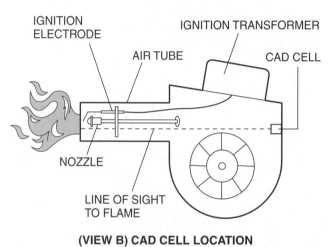

(VIEW B) CAD CELL LOCATION

310F12.EPS

Figure 12 ◆ Cad cell location.

- Contact between ignition transformer and electrodes is good
- Oil pump pressure adequate
- Oil burner nozzle clean and correctly sized
- Oil supply adequate and piping correctly installed

The following is an example of control circuit troubleshooting based on the circuit shown in *Figure 11*.

When checking the cad cell resistance with an ohmmeter, the resistance should range from 300 to 1,000 ohms (Ω) when the burner is operating (light showing). With the cad cell face covered (no light available), the resistance should be roughly 100,000Ω. This is high enough to block any current flow in the circuit. If the cad cell resistance is above 1,600Ω during the burner run cycle, the cad cell may need cleaning or aligning, or the burner flame may need to be adjusted. The equipment required for checking the cad cell and primary control includes:

- 0 to 150VAC voltmeter
- 1,500Ω resistor
- Insulated jumper wires
- Ohmmeter

NOTE

Clean the soot or oil film that can form on the face of a cad cell flame detector with a cotton-tipped swab.

4.2.1 Troubleshooting a Primary Control if Burner Starts and Locks Out

WARNING!

Checking the primary control and cad cell must be done with live circuits; therefore, the troubleshooter must observe all the precautions necessary to avoid the danger of electrical shock and equipment damage.

To troubleshoot a primary control if the burner starts and locks out, proceed as follows:

Step 1 Make sure the limit switches are closed and line voltage at the black and white leads of the primary control is 120 volts. Switch off the power at the furnace disconnect.

Step 2 Disconnect at least one room thermostat lead from terminals T-T. Place a jumper wire across terminals T-T.

Step 3 Disconnect the two cad cell leads from terminals F-F. Connect one end of a 1,500Ω resistor to one of the F-F terminals. This resistor will simulate a cad cell that senses light. Make sure that there is enough lead length to connect quickly the other end of the resistor to the other F-F terminal.

Step 4 Close the furnace disconnect, press the reset button on the primary control, and quickly (within 5 to 10 seconds) connect the other end of the 1,500Ω resistor to the remaining F-F terminal.

Step 5 If the primary locks out with the resistor across F-F, the primary is defective. When in doubt, wait five minutes and repeat the test, making sure to connect the resistor quickly. If the primary does not lock out with the resistor in place, the control is functioning properly.

4.2.2 Troubleshooting a Cad Cell

To troubleshoot a cad cell, proceed as follows:

Step 1 Switch off the furnace power. Disconnect at least one lead of the cad cell from terminals F-F and check using an ohmmeter. Resistance should be around 100,000Ω.

Step 2 Reconnect the cad cell to the F-F terminals, disconnect at least one lead of the room thermostat from the T-T terminals, and place a jumper wire across the T-T terminals.

Step 3 Press the reset button on the primary control and switch on the furnace power to initiate burner operation.

Step 4 After the burner ignites the fuel, disconnect both leads of the cad cell from the F-F terminals. The burner should stop after a short interval (15 to 45 seconds). If the burner does not shut off, the primary control is probably defective.

Step 5 Switch off the furnace power and connect a jumper wire to one of the F-F terminals.

Step 6 Press the reset button on the primary control and switch on the primary power. The burner should start and run. As soon as the burner starts, connect the other end of the jumper wire to the other F-F terminal.

Step 7 With the burner operating, measure the resistance across the two cad cell leads. The resistance should not exceed 1,600Ω. Ideally, it should be between 600Ω and 700Ω.

Step 8 If the resistance is 1,500Ω or less, the cad cell is functioning properly. If the resistance is higher than 1,600Ω, the cad cell face may be dirty, cracked, broken, or misaligned such that it does not completely sense the light from the flame.

4.3.0 Ignition Components

You may run across any one of several types of ignition controls:

- *Constant ignition* – The constant ignition control comes on when the burner is energized and stays energized as long as the burner is firing.

- *Intermittent ignition* – The intermittent ignition comes on when the burner is energized, but is cut off after the main burner flame is established. It may also de-energize after a preset ignition timing period has elapsed.

- *Nonrecycling control* – The nonrecycling control attempts to restart the burner immediately upon a loss of flame. It will attempt to restart the burner until it locks out on safety.

- *Recycling control* – The recycling control shuts down the burner immediately upon a loss of flame. It will attempt to restart the burner once before going into a safety lockout mode.

The ignition system consists of the ignition transformer (*Figure 13*) and the spark electrodes. The correct positioning and adjustment of the electrodes is covered in the *Planned Maintenance* module. The ignition transformer can be checked as follows:

Step 1 Turn off power to the unit.

Step 2 Shut off the fuel supply or disconnect the burner motor lead.

 WARNING!
Follow the manufacturer's instructions and be extremely careful when performing this procedure. The ignition transformer produces a 10,000V pulse.

310F13.EPS

Figure 13 ◆ Ignition transformer.

Step 3 Connect a voltmeter capable of measuring high voltage across the ignition transformer output terminals, and then turn the power back on. The reading should be at least 10,000V. If this result is not obtained, replace the transformer.

4.4.0 Auxiliary Components

Auxiliary components include solenoid oil flow valves, electronic time delays, and aquastats. These components are used to perform various control functions.

4.4.1 Solenoid Oil Flow Valves

Solenoid on-off flow valves can be mounted directly in the pipeline or on support brackets. These magnetic valves provide fuel oil to the burner only when the burner operates. They close immediately upon a loss of power. Some models have an integral thermistor that delays opening until the burner motor reaches full speed. They are wired into the line voltage supply to the burner motor (*Figure 14*).

Under normal operation, the valve opens when the thermostat calls for heat and closes immediately when the call for heat is ended.

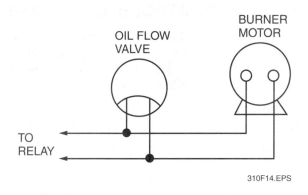

Figure 14 ◆ Oil valve circuit.

These valves usually offer reliable service and should not be replaced until all other sources of trouble have been eliminated. To check magnetic oil flow valve operation, proceed as follows:

Step 1 Check the valve operation by listening for an audible click when it opens and closes.

Step 2 If the valve does not open when the thermostat calls for heat:
- Check for normal fuel pressure.
- Make sure that the bleed line is not obstructed.
- Check the power supply at the valve. If no voltage is available, check the power source and circuit controls (relays and primary control). If the proper voltage is available, replace the coil or the entire valve.

4.4.2 Electronic Time Delay

The electronic time delay allows the burner motor to establish draft. It also can be used to establish first-stage flame before the opening of the fuel valve on a two-stage oil burner. It is connected in the control circuit as illustrated in the schematic diagram in *Figure 15*. On a call for heat, the electronic circuit in the time delay device (ST 70) delays the opening of the oil valve for approximately five seconds. Troubleshooting the electronic time delay consists of checking for supply voltage. If voltage is present and the delay does not function, it must be replaced.

4.4.3 Aquastat

The aquastat (*Figure 16*) is an immersion control used in conjunction with a solid-state primary control on an oil-fired boiler. This device can provide combinations of limit and circulation control in a hydronic heating system.

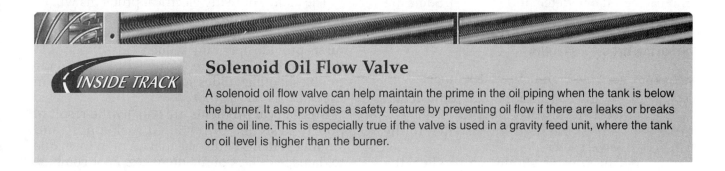

Solenoid Oil Flow Valve

A solenoid oil flow valve can help maintain the prime in the oil piping when the tank is below the burner. It also provides a safety feature by preventing oil flow if there are leaks or breaks in the oil line. This is especially true if the valve is used in a gravity feed unit, where the tank or oil level is higher than the burner.

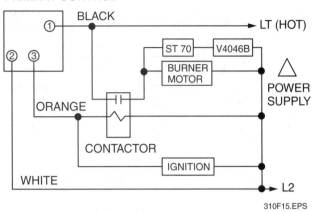

PRIMARY CONTROL

310F15.EPS

Figure 15 ◆ Time delay schematic.

310F16.EPS

Figure 16 ◆ Aquastat controller.

All aquastats require a 24V zone thermostat and cad cell flame detector (*Figure 17*). Some are designed for junction box mounting and incorporate a remote sensor. To check the operation of an aquastat, proceed as follows:

Step 1   Check the high-limit switch mode:
- Set the thermostat to call for heat and close the disconnect switch.
- Adjust the high-limit switch to the lowest setting.
- The burner should stop when the water temperature reaches the setting on the high-limit switch.

NOTE

If the water temperature is above the low-limit setting on the low-limit switch, the circulator will continue operation as long as the thermostat is calling for heat.

Step 2   Check the low-limit switch and circulator switch:
- Adjust the low-limit switch to the highest setting.
- The burner should start. The circulator should not start if the water temperature is at least 10°F below the setpoint of the device.
- Adjust the low-limit switch to the lowest setting; the burner should stop.
- Adjust the thermostat to call for heat; the burner and circulator should come on.
- While the burner and circulator are operating, raise the low-limit setting. The circulator should stop. Lower the low-limit setting; the circulator should resume operation.
- Return all switches to their normal settings before leaving the site.

5.0.0 ◆ SYSTEM TROUBLESHOOTING

Problems in oil-fired heating systems can be electrical, or they can be caused by a number of mechanical problems. They often manifest themselves in one of the following ways:

- No heat
- Not enough heat
- Too much heat
- Noise
- Odor
- Soot buildup
- Excessive cost of operation

The source of many electrical problems will be found in the control circuits. In addition to schematic and wiring diagrams, most manufacturers provide operating sequence descriptions in their product literature. Many will also provide troubleshooting guides such as the ones shown in *Appendix A* and *Appendix B*.

Mechanical problems are usually the result of poorly designed or installed fuel supply and venting systems or defects that occur in them due to damage or lack of maintenance. Poor combustion efficiency, often a result of neglected maintenance, can also lead to customer complaints.

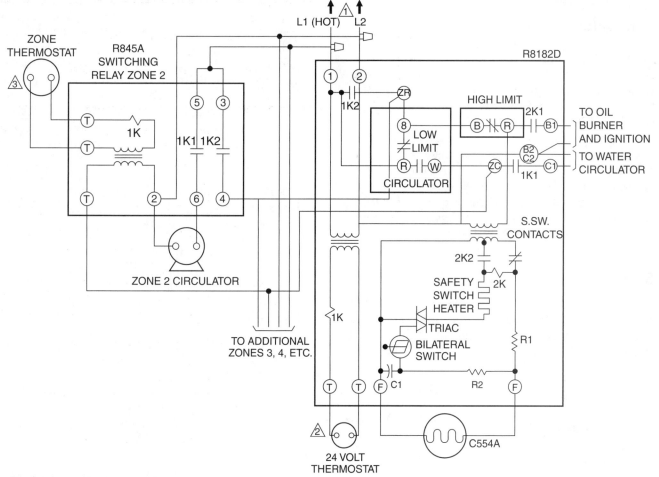

Figure 17 ◆ Schematic for a solid-state primary control with an aquastat.

⚠ 120V AC POWER SUPPLY PROVIDES DISCONNECT MEANS AND OVERLOAD
PROTECTION AS REQUIRED.

⚠ THERMOSTAT HEAT ANTICIPATOR SETTING. 0.2 AMP FOR R8182D.

⚠ THERMOSTAT HEAT ANTICIPATOR SETTING. 0.4 AMP FOR R845A.

310F17.EPS

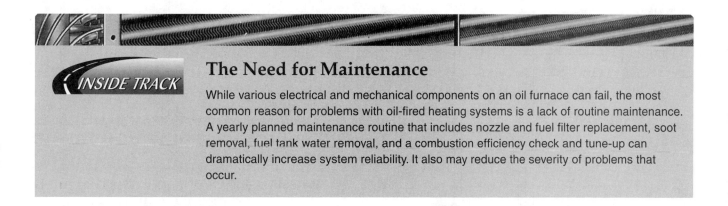

INSIDE TRACK

The Need for Maintenance

While various electrical and mechanical components on an oil furnace can fail, the most
common reason for problems with oil-fired heating systems is a lack of routine maintenance.
A yearly planned maintenance routine that includes nozzle and fuel filter replacement, soot
removal, fuel tank water removal, and a combustion efficiency check and tune-up can
dramatically increase system reliability. It also may reduce the severity of problems that
occur.

5.1.0 Fuel Supply

If the burner loses flame but continues to run with a jumper wire in place across the flame detector terminals, then the fault is probably in the fuel system. Listen for a whine at the pump, which indicates high suction and a supply restriction. Prior to checking the pump, valves, and controls, assess the fuel supply for proper installation and piping.

Oil supply lines in residential installations usually use ⅜" (9.53mm) outside diameter (OD) soft copper tubing. To prevent oil and air leaks in fuel lines, always use flare fittings on copper tubing connections. Compression fittings should never be used on copper fuel supply lines.

5.2.0 Piping Systems

A single-line (one-pipe) system is simply one line from the tank to the fuel-burning unit. A two-line (two-pipe) system has two lines from the tank to the burner. This type should always be used whenever the oil supply level is below the burner. It eliminates the necessity of bleeding air from the lines because the air is returned to the tank.

Figures 18 through *21* are examples of oil furnace installations that have been used in the past. National, state, and local codes covering oil furnace installations, especially underground tanks, are under constant review and are subject to change. Before beginning any installation, consult the manufacturer's installation instructions, as well as applicable codes. If an underground tank is used, make sure it meets current construction standards. The EPA has published special requirements for commercial installations with underground tanks that are different from those for residential installations.

Figure 18 illustrates a single-line system with the supply tank in the same space as the burner. The single-line system should only be used when the burner is located on the same level or above the bottom of the oil supply tank. Single-pipe systems often depend on gravity to feed oil to the

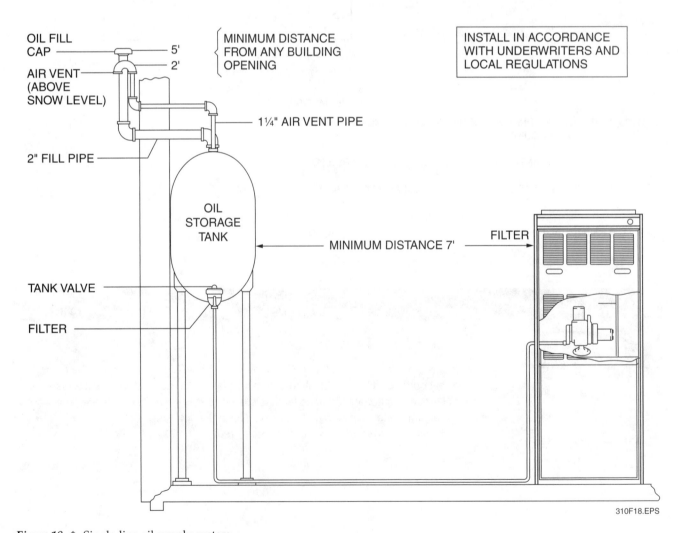

Figure 18 ◆ Single-line oil supply system.

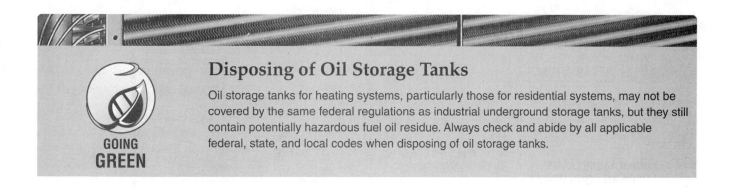

Disposing of Oil Storage Tanks

Oil storage tanks for heating systems, particularly those for residential systems, may not be covered by the same federal regulations as industrial underground storage tanks, but they still contain potentially hazardous fuel oil residue. Always check and abide by all applicable federal, state, and local codes when disposing of oil storage tanks.

GOING GREEN

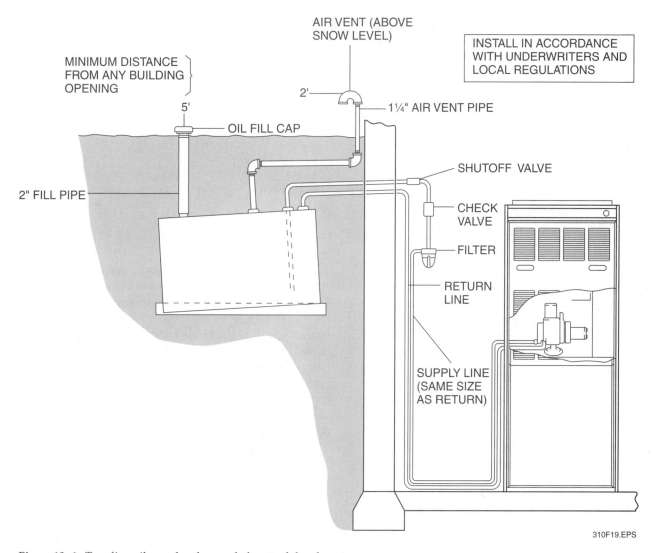

Figure 19 ◆ Two-line oil supply—burner below tank level.

burner. In gravity-feed installations, a solenoid oil flow valve may be appropriate to help prevent oil leakage from a damaged oil line.

The piping on this application must be leak-free. No air should be allowed to enter the oil line and pump after the supply system has been purged.

One-pipe and two-pipe systems require different placement of the bypass plug on the pump. Be sure to follow the manufacturer's instructions carefully.

Figure 19 depicts a two-line system with the burner located below or on the same level as the supply tank. Note that the storage tank is buried

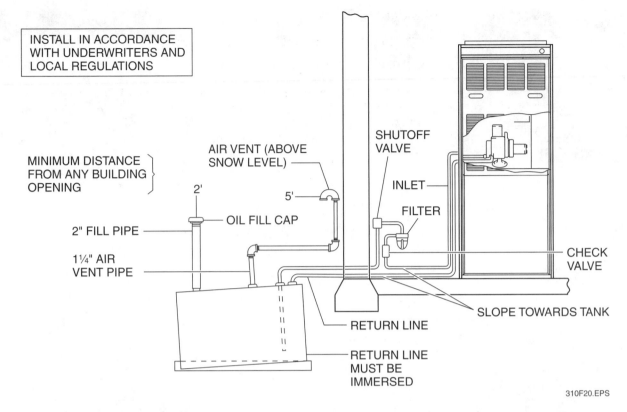

Figure 20 ◆ Two-line oil supply—burner above tank level.

outside the structure. An all-brass check valve should be installed in the suction line as close to the tank as possible. It is also acceptable to install a globe valve in the supply line so the oil can be shut off for burner service or removal.

Figure 20 shows the two-line system with the burner located above the tank when a lift of less than 15' is required. A check valve and globe shutoff valve should also be included in this piping network.

Figure 21 shows a piping system required where a lift of more than 15' is necessary and multiple units are tied together with a single oil system. A boost pump must be used in this application. The schematic in the illustration portrays a series-loop system where all the oil is pumped through individual smaller tanks on each unit in series and then returned directly to the main supply through a separate return line.

Some manufacturers recommend that the piping system use two pipe connections at the boost pump and at each oil rooftop unit. The supply line from the main tank to the boost pump should include a fusible valve as well as oil filter(s) of adequate size.

To guarantee a trouble-free supply system for multiple units, the following factors should be considered:

• A rooftop system should be limited to a maximum of eight oil-fired units.

• The fittings and lengths of supply and return lines must be accurately calculated to guarantee proper delivery of fuel to all units.

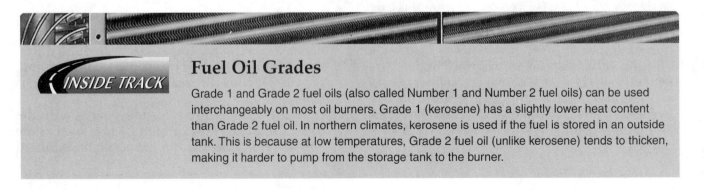

Fuel Oil Grades

Grade 1 and Grade 2 fuel oils (also called Number 1 and Number 2 fuel oils) can be used interchangeably on most oil burners. Grade 1 (kerosene) has a slightly lower heat content than Grade 2 fuel oil. In northern climates, kerosene is used if the fuel is stored in an outside tank. This is because at low temperatures, Grade 2 fuel oil (unlike kerosene) tends to thicken, making it harder to pump from the storage tank to the burner.

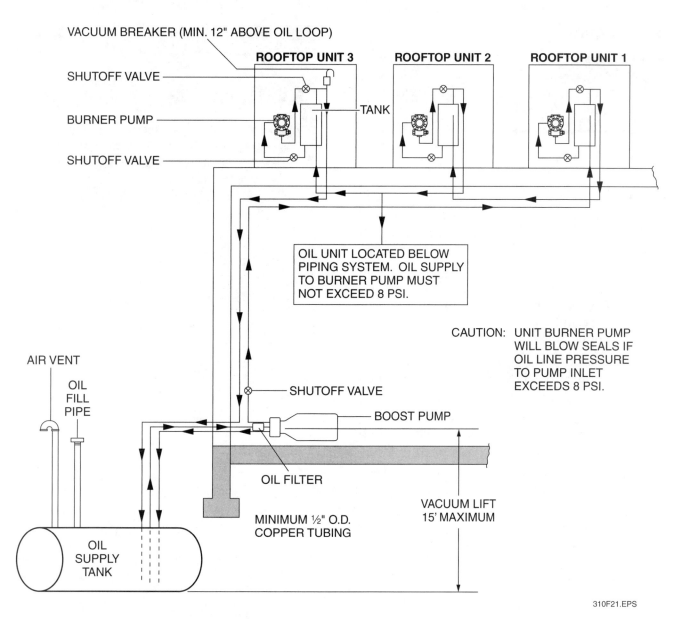

VACUUM BREAKER (MIN. 12" ABOVE OIL LOOP)

SHUTOFF VALVE

BURNER PUMP

SHUTOFF VALVE

ROOFTOP UNIT 3   ROOFTOP UNIT 2   ROOFTOP UNIT 1

TANK

OIL UNIT LOCATED BELOW
PIPING SYSTEM. OIL SUPPLY
TO BURNER PUMP MUST
NOT EXCEED 8 PSI.

CAUTION: UNIT BURNER PUMP
WILL BLOW SEALS IF
OIL LINE PRESSURE
TO PUMP INLET
EXCEEDS 8 PSI.

AIR VENT

OIL
FILL
PIPE

SHUTOFF VALVE

BOOST PUMP

OIL FILTER

MINIMUM ½" O.D.
COPPER TUBING

VACUUM LIFT
15' MAXIMUM

OIL
SUPPLY
TANK

310F21.EPS

Figure 21 ◆ Oil piping system for rooftop units.

- Supply lines should be run to the farthest burner first and then piped toward the boost pump so that the last unit is the closest to the boost pump. The manufacturer's recommendations as to supply pump pressure must be followed to eliminate seal failure.

- The return lines should never include smaller pipe sizes than the supply lines. One size larger is sometimes recommended on the vertical return line to permit free flow of fuel back to the supply tank.

- A vacuum breaker should be installed at the highest point in the system. It should be located on the last unit in the supply run, closest to the return line.

- All lines that are exposed to low ambient outside temperature must be insulated.

- Local and other applicable codes must be followed and will take precedence over any other design and construction considerations for the fuel oil supply system.

NOTE

Figures 19, *20*, and *21* show underground fuel storage tanks. These figures are for reference only and should not be used as a guide for the actual installation of an underground fuel storage tank. Underground installation of fuel storage tanks is strictly governed by a host of national and local codes. Consult and follow all applicable codes when installing an underground tank.

5.3.0 Odors

To check for odor caused by fuel oil, proceed as follows:

Step 1 Check for leaks in the oil supply and return piping systems. Tighten or replace connections or ruptured lines as necessary.

Step 2 Check the fuel tank for proper venting to the outdoors if it is a basement tank. Correct venting problems where necessary and securely tighten or seal the vent connections.

Step 3 Check for seepage around the fill or vent connections if there is an oil film on the tank.

Step 4 Check for seepage through the oil pump cover gasket or the pump shaft bushings. If necessary, replace the pump.

Step 5 Check for loose high-pressure oil line connections to the gun assembly, and tighten the connections or replace the oil line as necessary.

Step 6 Check for fuel filter leakage, correct the cause, and replace the filter.

5.4.0 Flue and Chimney Exhaust

The flue pipe and chimney exhaust can also become a source of odor or improper burning operation. As with any other HVAC system, the proper design and installation of the flue pipe and chimney exhaust system eliminates problems that could be encountered by the HVAC troubleshooter.

Refer to *Figure 22* and the numbered list below for an explanation of some of the areas that might lead to chimney problems:

1. Downdraft might be caused by the top of the chimney being lower than surrounding objects. The correction requires extending the chimney above all objects within 30' or as recommended by local building codes.

2. A restricted opening may be caused by problems with the chimney cap, ventilator, or coping.

Chimney Liners

Correcting masonry-related chimney problems is much easier today due to advances in technology. If a chimney is oversized or is improperly lined, a flexible chimney liner kit rated for use with fuel oils can be installed to correct the problem.

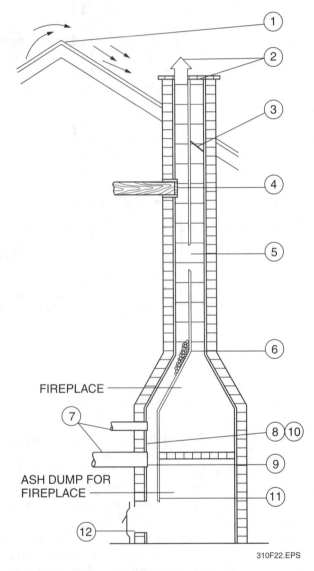

310F22.EPS

Figure 22 ◆ Common chimney problems.

3. Obstructions in the chimney can be located by using a light and a mirror.

4. A joist projecting into the chimney can cause problems.

5. A break in the chimney lining or tile may be present.

6. A collection of soot at a narrow opening or turn may be present.

7. Two or more openings may exist in the same chimney. Refer to local codes for recommendations.

8. Loose or open connections at joints in the vent pipe may be present.

9. A burner smoke pipe that extends into the chimney can cause problems.

10. A loosely seated vent pipe in the chimney can cause problems.

11. Failure to extend the length of the flue pipe (drip leg) properly (a loosely fitting clean-out door) can cause problems.

12. An oversized chimney might not be able to establish an adequate draft.

NOTE
Birds often build nests in chimneys during summer months, and small animals can fall into chimneys and die. Both situations can block or restrict the chimney.

To check for combustion problems caused by improper flue piping and/or the chimney exhaust system, proceed as follows:

Step 1 Check the combustion for carbon dioxide. Check stack temperature and smoke with the blower off and with the blower running. A different reading with the blower running indicates a possible heat exchanger leak.

Step 2 Inspect all joints in the vent pipe for loose or open connections.

Step 3 Check for proper draft with a draft gauge:
• Drill two ¼" holes in the flue between the heating unit and the barometric damper (*Figure 23*).

NOTE
Two holes are drilled to speed up testing and reduce instrument handling (draft gauge stack thermometer).

• Drill one ¼" hole in the fire inspection door cover.
• Check the draft reading over the fire (above the flame level, if possible). Adjust the barometric damper to assure a negative pressure within the combustion chamber of about 0.02

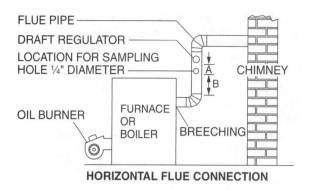

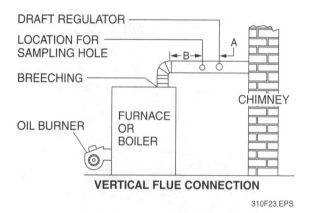

Figure 23 ◆ Flue sampling hole.

inches of water column (in. w.c.). If it is not possible to take draft readings over the fire, adjust the draft regulator to give a breech draft reading between 0.04 and 0.06 in. w.c., or as recommended by the burner manufacturer.

• Seal the fire door draft sampling hole with a metal plug or high-temperature sealant. It may not be necessary to seal the vent pipe sampling holes.

Step 4 Check the smoke readings:
• After the burner has been operating for five or ten minutes, take a smoke measurement (*Figure 24*) in the vent pipe following the test instrument manufacturer's procedure. Oily or yellow smoke spots on the filter paper are a sign of unburned fuel. This condition can sometimes be caused by too much air.
• Record the smoke and carbon dioxide measurements (*Figure 25*) from the vent pipe and compare to the specified readings. Adjust the air setting to within specifications and repeat the draft, carbon dioxide and/or oxygen, and smoke measurements to ensure they remain within the limits.

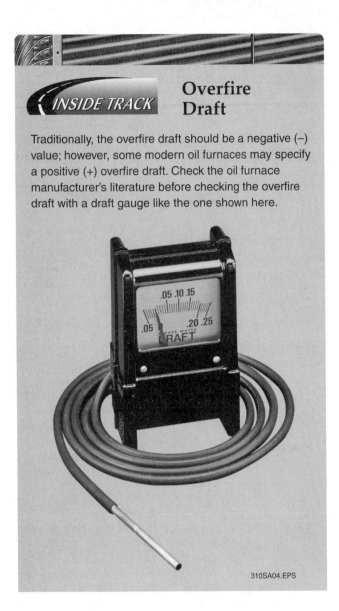

310F25.EPS

Figure 25 ◆ Gas analyzer kit for measuring carbon dioxide.

310F24.EPS

Figure 24 ◆ Smoke sampling device.

- If the combustion performance specifications cannot be met, check for air leaks into the combustion chamber and repair the leaks with furnace cement or high-temperature sealant.
- To check for dilution by leakage into the combustion chamber, measure the carbon dioxide content at as high a point as possible over the fire. Use a stainless steel tube inserted through the fire door sample hole and compare this with the carbon dioxide measured in the vent pipe. A differential of more than one percent carbon dioxide between the vent pipe and the overfire readings will usually indicate air entry into the combustion chamber.
- If the carbon dioxide level indicated in *Table 1* cannot be reached without exceeding No. 1 smoke, poor mixing of air and fuel is likely. This could be caused by an improper match between air pattern and nozzle spray pattern. This problem can frequently be corrected by replacing the nozzle with one having a different spray angle and pattern.

To measure stack temperature, proceed as follows (*Figure 26*):

Table 1 Air Adjustments

High-Pressure Gun-Type Burner	Typical Carbon Dioxide in Tuned Flue Gas*
Older gun burners – No internal air handling parts other than end cone and stabilizer	8 percent
Newer gun burners – Special internal air handling parts	9 percent
Flame retention gun burners – Flame retention heads	10 percent

* Based on acceptable smoke test; generally a trace, but not usually exceeding No. 1.

Figure 26 ◆ Stack thermometers.

310F26.EPS

Step 1 Insert the flue temperature thermometer into the second hole in the vent pipe.

Step 2 Subtract the room air temperature from the stack thermometer reading and record the net temperature. If the net stack

temperature exceeds 400°F to 600°F during steady operation for matched-packaged oil-burner units (600°F to 700°F for conversion burners), excessive stack loss is indicated.

Step 3 Check the combustion efficiency:

- Determine the net stack temperature and the measured carbon dioxide level after the system has been adjusted as close as possible.
- Use the instrument manufacturer's calculator or consult *Table 2*. Generally, combustion efficiency should not be less than 75 percent for standard, non-flame-retention burners and higher for flame-retention burners.

Step 4 If the proper combustion efficiency cannot be achieved, check for an unbalanced flame. If the flame appears unbalanced (*Figure 27*) and the fuel input may be too low, check the nozzle size and oil pressure.

- If the nozzle is partially plugged or defective, replace it with another as recommended by the manufacturer or choose one from an interchange chart.

Table 2 No. 2 Fuel Oil Efficiency Table

% CO_2	NET STACK TEMPERATURE (°F)												
	300°	350°	400°	450°	500°	550°	600°	650°	700°	750°	800°	850°	900°
15	87½	86½	85¼	84¼	84¼	82	81	79¾	78¾	77½	76½	75½	74¼
	87½	86¼	85	84	83	81¾	80¾	79¼	78½	77¼	76	75	73¾
14	87¼	86	84¾	83¾	82¾	81¾	80¼	79	78	76¾	75½	74½	73
	87	85¾	84½	83½	82½	81¼	80	78¾	77½	76¼	75¼	74	72¼
13	87¾	85½	84¼	83¼	82	80¾	79½	78¼	77	75¾	74½	73½	71¾
	86½	85¼	84	83	81½	80¼	79	77¾	76½	75¼	73¾	72¾	71
12	86¼	85	83¾	82¼	81¼	79¾	78½	77¼	75¾	74½	73	71½	70¼
	86	84¾	83½	82	80¾	79¼	78	76	75¼	73¾	72¼	70¾	69½
11	85¾	84½	83	81½	80¼	78¾	77¼	75¼	74½	73	71½	70	68½
	85½	84	82½	81	79½	78	76½	75½	73¾	72	70½	69	67½
10	85	83½	82	80½	78¾	77¼	75¾	74¾	72¾	71	69½	68	66¼
	84½	83	81½	79¾	78	76½	75	73	71¾	70	68¼	66¾	65
9	84	82¼	80¾	79	77¼	75¾	74	72	70¾	68¾	67¼	65¼	63½
	83½	81¾	80	78¼	76½	74¾	73	71	69½	67½	65½	63¾	62
8	83	81	79¼	77½	75½	73¾	71¾	70¾	68	66	64	62	60
	82¼	80¼	78½	76½	74½	72½	70½	68½	66½	64¼	62¼	60	58
7	81½	79½	77¼	75¼	73¼	71	69	67	64¾	62½	60¼	57¾	55½
	80¾	78½	76¼	74	71¾	69½	67¼	65¼	62¾	60¼	57¾	55½	53
6	79¾	77¼	75	72½	70	67¾	65¼	62¼	60¼	57½	55	52½	50
	78½	76	73½	71	68	65½	63	60	57½	54½	51¾	49	46½
5	77¼	74½	71¾	69	65¾	63	60	57	54	51	48	45½	42½
	75½	72½	69½	66¼	63	60	56¾	53¾	50¼	47	43½	40¼	36¾
4	73¼	69¾	66¼	62¾	59¼	55¾	52	48	45	41¼	37½	33¾	30

PERCENT CO_2 (vertical axis) — **CO_2 MEASUREMENTS**

310T02.EPS

Figure 27 ◆ Unbalanced flame.

- If the oil pressure is too low (less than 100 psig, or as recommended by the manufacturer), readjust it or replace the burner or pump.

NOTE

Water can form in the fuel tank due to condensation. This water can make its way to the pump where it can corrode and eventually ruin the pump. Fuel tanks should be periodically checked for the presence of water. If water is found, it must be removed.

- Using a smoke spot comparison chart (*Figure 28*), make final burner adjustments and recheck the steady-state efficiency (15 minutes of burner operation or when the stack temperature rises less than 5°F in one minute). The efficiency should be greater than the efficiency prior to modification, with the smoke reading preferably between 0 and 1, but never greater than 2. The steady-state stack temperature should be 350°F or more.

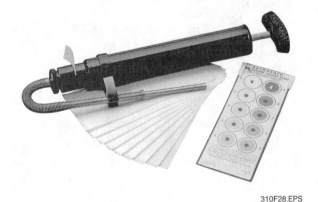

Figure 28 ◆ Smoke spot comparison chart.

If the owner still complains of not enough heat after all the adjustments are made and the burner seems to be firing properly, it may be appropriate to check the heat rise across the furnace plenum.

NOTE

If an oil pump is properly adjusted, the needle on its pressure gauge should be steady. If the needle pulsates, check for a partially clogged fuel filter or pump screen and/or air leaks in the fuel supply lines.

The vacuum gauge on a single-stage pump should read less than 10 inches of vacuum. The vacuum gauge on a two-stage pump should read less than 15 inches of vacuum. If a high vacuum is detected, check for kinks or restrictions in the fuel supply lines and/or a plugged fuel filter.

To check the temperature (heat) rise and limit control, proceed as follows:

Step 1   Drill access holes in the supply and return air ductwork (*Figure 29*) and insert a plenum thermometer in each opening. Make sure that the thermometer is properly calibrated.

Step 2   The thermometer in the supply duct must be located out of the line of sight of the furnace heat exchanger to prevent radiant heat from affecting the thermometer.

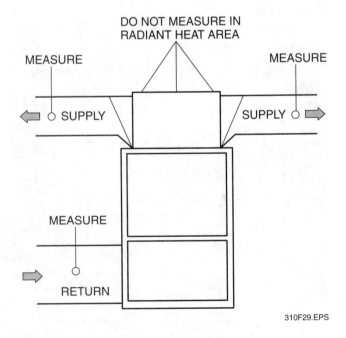

Figure 29 ◆ Measuring temperature rise.

Step 3 Jumper the thermostat terminals and allow the furnace to run for about 15 minutes.

Step 4 Check the temperature rise differential between the supply air and return air.

Step 5 The temperature rise should be slightly above the midpoint of the rise range listed on the furnace information plate. Adjust the blower speed to achieve the reading.

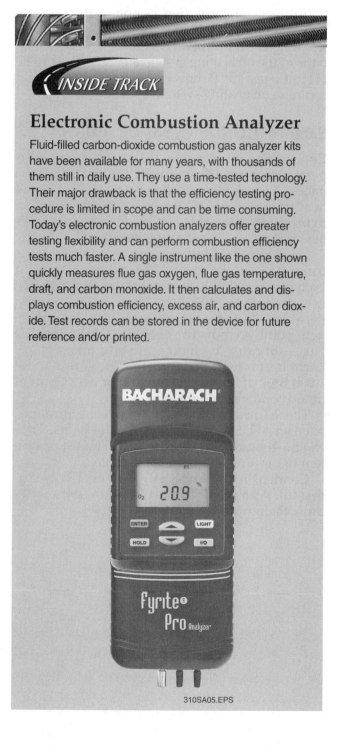

Electronic Combustion Analyzer

Fluid-filled carbon-dioxide combustion gas analyzer kits have been available for many years, with thousands of them still in daily use. They use a time-tested technology. Their major drawback is that the efficiency testing procedure is limited in scope and can be time consuming. Today's electronic combustion analyzers offer greater testing flexibility and can perform combustion efficiency tests much faster. A single instrument like the one shown quickly measures flue gas oxygen, flue gas temperature, draft, and carbon monoxide. It then calculates and displays combustion efficiency, excess air, and carbon dioxide. Test records can be stored in the device for future reference and/or printed.

310SA05.EPS

Step 6 Allow the burner to operate with the blower disconnected until the limit control opens and turns the burner off. Most forced-air oil furnace limit switches are set to open at 200°F. Check the plenum temperature, but do not allow it to exceed 250°F. Turn off the disconnect switch if it reaches that temperature. If the supply temperature reaches 250°F and does not trip the limit control, replace the limit control and return the blower circuit to the system.

Proper ductwork design is as important to oil furnaces as it is to gas furnaces. An improperly designed duct system can cause blower motor overloading or inadequate airflow that could open the limit control.

5.5.0 Flame-Retention Burners

Flame-retention burners (*Figure 30*) are used in modern oil-burning appliances and are designed to produce significant improvements in overall heating plant efficiency while simultaneously reducing the number of burner-related service calls. Flame-retention burners use motors that operate at 3,450 rpm instead of the conventional 1,725 rpm. This higher speed is needed to overcome the greater resistance to air flow that is created by the retention ring.

310F30.EPS

Figure 30 ◆ Flame-retention oil burner.

The advantages of a flame-retention burner include the following:

- Better air-oil mixing and flame retention within the air pattern
- A higher flame temperature with less excess air
- More usable energy produced for the amount of oil consumed
- Less effect on flame from stack draft variations
- Fewer products of incomplete combustion
- During the off-cycle, the retention head reduces the flow of air through the burner, over the heat exchanger, and out the stack, thereby removing less heated air from the structure

5.6.0 Troubleshooting Common Oil Heating Problems

This section provides a troubleshooting guide to common causes of the primary complaints associated with oil heating service calls. The following checks should be made on a no-heat complaint:

Step 1   Check for power at the main disconnect switch.

Step 2   Check the burner motor fuse.

Step 3   Check the burner on-off switch.

 WARNING!
If the switch is off, make sure the combustion chamber is free of oil, or oil vapor, before turning the switch on. Failure to do so could result in an explosion.

Step 4   Check the oil supply.

Step 5   Make sure all manual oil valves are open.

Step 6   Check the limit switches. They must be closed.

Step 7   Reset the safety switch, set the thermostat to call for heat, and then perform the following checks:

- Simulate flame failure by shutting off the oil supply hand valve while the burner is operating. The burner should lock out.
- Simulate ignition or fuel failure by shutting off the power supply while the burner is operating normally. The system should shut down immediately. Wait a few minutes for the stack to cool, then restore power. The system should restart immediately if the thermostat is calling for heat.

If the system still does not operate properly, note the point at which it fails. Use the troubleshooting chart in *Appendix A* to determine which parts of the oil heating system are most likely to be at fault. *Appendix B* shows the most likely causes of common oil heating system failures.

6.0.0 ◆ CONDENSING OIL FURNACES

Condensing oil furnaces, like their gas counterparts, are more efficient than conventional furnaces. A condensing furnace extracts additional heat from flue gases by passing the combustion byproducts through additional heat exchangers before they are vented to the outdoors.

Condensing oil furnaces may have as many as three heat exchangers. The condensing heat exchanger, which is the final stage, is a coil-type heat exchanger. In this condensing coil, the temperature of the flue gases is reduced below the dew point, so moisture condenses from the flue gases. Because a change of state takes place, latent heat is extracted from the flue gases. About 1,000 Btus of heat are extracted for every pint of liquid condensed in this process.

These furnaces, like gas condensing furnaces, have an AFUE rating of 90 percent or better. They can be vented with PVC pipe and require a system to capture and drain condensate.

Condensing oil furnaces were never very successful. As a result, some manufacturers withdrew them from the market.

1. The device most commonly used in oil furnaces to verify a continuous flame is the _____.
   a. flame rectifier
   b. thermopile
   c. cad cell
   d. hot surface ignitor

2. The air needed to support combustion in an oil furnace is obtained from _____.
   a. a blower in the oil burner assembly
   b. holes in the heat exchanger
   c. natural convection occurring in the vicinity of the furnace
   d. supply and return air

3. Which of the following receives a signal from the thermostat and switches primary power to the burner and other components?
   a. Cad cell
   b. Limit switch
   c. Primary control
   d. Master cylinder

4. A two-line fuel pipe system is required when the _____.
   a. oil supply level is above the burner
   b. oil supply level is below the burner
   c. distance from the tank to the furnace is more than 15'
   d. distance from the tank to the furnace is more than 30'

5. Return lines in a two-pipe system should be _____.
   a. smaller than the supply lines
   b. the same size as the supply lines
   c. the same size as the supply lines, except for vertical runs
   d. always larger than the supply lines

6. An unbalanced (off-center) flame is usually caused by _____.
   a. too much air
   b. a leaking heat exchanger
   c. a defective nozzle
   d. a defective cad cell

7. A downdraft in a chimney could be caused by _____.
   a. the top of the chimney being too high
   b. the top of the chimney being lower than surrounding surfaces
   c. an obstruction in the chimney
   d. a crack in the chimney lining

8. The combustion efficiency of a standard oil furnace should be at least _____ percent.
   a. 50
   b. 75
   c. 90
   d. 100

9. If the stack temperature is different with the blower running than it is with the blower off, it indicates _____.
   a. heat exchanger leakage
   b. a leak in the stack
   c. the thermostat heat anticipator is not properly adjusted
   d. an unbalanced flame

10. A major difference between the flame-retention burner and a conventional oil burner is _____.
    a. the conventional burner is more efficient
    b. with a flame-retention burner, combustion occurs inside the burner, rather than in a combustion chamber
    c. the flame-retention burner operates at a much lower speed
    d. the flame-retention burner produces a higher temperature flame

Summary

Oil furnace troubleshooting usually starts with an owner's complaint, which must be analyzed in a logical manner to arrive at the solution to the problem. One of the critical tasks of the HVAC technician is to check a system thoroughly before condemning any control or component.

An oil burner system consists of the oil supply and the burner. Gun-type atomizing units are the most common burners used. Oil burner nozzle assemblies are a common source of problems.

Problems may also occur in oil furnace control systems. Components to check include thermostats, primary controls, flame detectors, limit controls, auxiliary controls, and ignition components.

Auxiliary components such as solenoid oil flow valves, electronic time delays, and aquastats may also cause problems.

At the time of the troubleshooting call, it is a good idea to replace filters and similar devices that will contribute to the efficient operation of the system.

The responsible troubleshooter will not leave the site until the reason for the system failure has been corrected. To make sure of this, the system should be cycled several times and then monitored for a reasonable amount of time after startup.

Notes

Trade Terms Introduced in This Module

Cad cell: A light-sensitive device (photocell) in which the resistance reacts to changes in the amount of light. In an oil furnace, it acts as a flame detector.

Flame-retention burner: A high-efficiency oil burner.

Primary control: A combination switching and safety control that turns the oil burner on and off in response to signals from the thermostat and safety controls.

Stack control: A type of obsolete flame detector consisting of a bimetal switch inserted into the flue stack.

Appendix A

Troubleshooting Summary

No-Heat Complaints

When troubleshooting a no-heat complaint, always check the following basic points first:

- Make sure the power is on at the main switch.
- Make sure the burner motor fuse is not blown.
- Check the reset on the primary control. If the switch reset is tripped, make sure the combustion chamber is free of oil or oil vapor before pressing the reset button.
- Check the oil supply. Make sure fuel is available and there are no blockages or restrictions in the fuel line.
- Make sure the manual oil valves are open.
- Make sure the limit switches are closed.
- Make sure the room thermostat is calling for heat.

After completing these checks, run through the starting procedure. If the system still does not operate properly, note the point at which it fails.

Check the appropriate table to determine which parts of the system are most likely to be the cause of the trouble.

Miscellaneous Problems

Other complaints, such as smoke, soot, odors, or combustion equipment noise are generally caused by defective burner operation. Check the burner manufacturer's instructions for possible causes and corrective actions.

PROBLEM	POSSIBLE CAUSE	CORRECTIVE ACTION
Burner motor does not start		
Trouble in primary control	Check cad cell	Replace if resistance is incorrect
Trouble in thermostat	Broken wires or loose connections Defective thermostat	Replace wires or tighten connections Replace thermostat
Faulty burner components	Broken wires or loose connections Motor start switch or thermal overload switch open Defective motor Defective pump	See manufacturer's instructions
Burner motor starts – no flame established		

Trouble in primary control	Check cad cell or stack relay
	Loose connections or broken wires
	  between ignition transformer and
	  primary control
		Replace wires or tighten connections

Trouble in ignition system	Defective transformer Ignition electrodes: • Improperly positioned • Spaced too far apart • Loose • Dirty Dirty or damaged ceramic insulators	Replace transformer Check manufacturer's instructions Replace or clean electrode assembly
Faulty burner components	Dirty nozzle Loose, misaligned, or worn nozzle Clogged oil pump strainer Clogged oil line Air leak in suction line Defective pressure regulator valve Defective pump Improper draft Water in oil Oil too heavy	Correct per manufacturer's instructions
Burner motor starts – flame goes on and off after start-up		
Trouble in primary control	Cad cell problems	Check and/or clean cad cell
Trouble in ignition system	Limit settings too low Dirty oil filter	Readjust setting or differential Replace filter
Trouble in thermostat	Broken wires or loose connections Defective anticipator Defective thermostat	Replace broken wires or tighten connections Adjust anticipator Replace thermostat
Faulty burner components	Low oil pressure Defective pump, clogged or dirty oil lines Fluctuating water level (hydronic systems) Weak transformer Cracked electrodes	See manufacturer's instructions Replace transformer Replace porcelain electrodes

310A01.EPS

PROBLEM	POSSIBLE CAUSE	CORRECTIVE ACTION
System overheats house		
Trouble in primary control	Defective primary	With burner running, disconnect one low-voltage thermostat lead from primary control terminal; if relay does not drop out, replace primary
Trouble at thermostat	Wiring shorted	Repair or replace wiring
	Thermostat out of calibration	Recalibrate thermostat with an accurate thermometer
	Defective anticipator or anticipator improperly set	Replace or readjust anticipator; replace thermostat, if necessary
	Thermostat stuck in call-for-heat	Repair or replace thermostat
	Thermostat improperly located in area under control, or draft on stat through wall	Relocate thermostat: must be out of drafts, away from radiating surfaces, ducts, steam pipes, and sunny locations
	Thermostat not level	Level thermostat
Trouble in distribution system	Ductwork close to thermostat	Relocate thermostat
	Flow control valve (hot water) stuck in ON position	Repair or replace valve
	Circulator does not stop running	Check circulator circuits, replace switching device
Faulty burner components	Extremely oversized system	Replace system
	System control valves stuck in ON position	Repair or replace control valve
System underheats house		
Trouble at thermostat	Open or loose wiring	Repair or replace wiring
	Thermostat out of calibration	Recalibrate thermostat with an accurate thermometer
	Defective anticipator or anticipator improperly set	Replace or readjust anticipator; replace thermostat, if necessary
	Thermostat stuck in not calling for heat	Repair or replace thermostat
	Thermostat improperly located	Relocate to sense more accurately
	Thermostat not level	Level thermostat
Trouble at high- or low-limit controller	Controller set too low	Raise setpoint
	Slow to return to ON position	Adjust differential
	Controller defective	Replace controller
Trouble at blower (for forced-air system)	Cutting out or cycling on overload	Check air filters
	Burned out blower motor	Replace motor
	Running too slow or inadequate capacity	Check/replace motor
	Fan belt broken or loose	Replace or tighten belt
System defects	Undersized furnace or boiler system (possible if addition has been made to house)	Replace furnace or boiler system
	Distribution system inadequate	Correct problem
	Sooted heat exchanger	Clean heat exchanger
	Dirty warm-air filters	Replace filters
	Dirty boiler water (steam system)	Drain/refill boiler

310A02.EPS

PROBLEM	POSSIBLE CAUSE	CORRECTIVE ACTION
System underheats house (continued)		
Faulty burner components	Fuel line clogged or strainer plugged Air in fuel line or excess air in system Dirty or improperly sized nozzle Low oil pressure or defective pump Motor does not come up to proper speed Inoperative or inadequate circulator (hydronic system)	Clear clog Check fuel lines for leaks Replace nozzle Adjust or replace pump Replace motor Replace circulator
System cycles too frequently (this problem is outside the primary control; the most common problems are listed below)		
Trouble at thermostat	Poor location Poorly adjusted anticipator Loose connections or defective wiring	Relocate thermostat Readjust anticipator; replace thermostat, if necessary Tighten connections or replace broken wires
Trouble at high limit	Limit set too low Differential too narrow	Reset limit Reset differential; replace control, if necessary
Faulty burner system	Dirty or clogged air filter Loose connections Fluctuating voltage Intermittent shorts	Replace filter Repair connections Contact utility Find short circuit
Relay chatters after pulling in		
Trouble at thermostat	Open contacts	Replace thermostat
Trouble at primary control	Supply voltage too low Defective load relay	Check for loose connections Replace primary

310A03.EPS

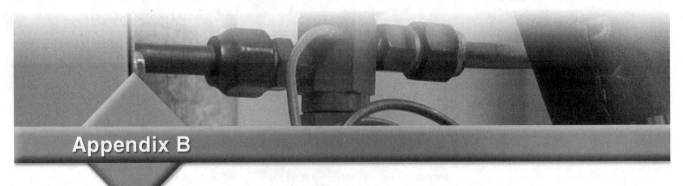

Appendix B

Troubleshooting Chart

GENERALLY THE CAUSE	OCCASIONALLY THE CAUSE	RARELY THE CAUSE
Make these checks first.	Make these checks only if the first checks failed to locate the trouble.	Make these checks only if other checks failed to locate the trouble.

PROBLEM: No heat – burner fails to start

Power failure Blown fuses or tripped circuit breaker Open disconnect switch or blown fuse Thermostat set too low Thermostat switch in improper position Primary or safety controls Flame detector	Control relay or contactor Limit controls Thermostat not level Fuel pump Burner motor	Faulty wiring Loose terminals Low voltage Thermostat faulty

PROBLEM: No heat – burner starts but fails to ignite

Electrodes Ignition transformer Improper burner adjustment Oil supply and oil pressure Manual fuel valve	Fuel pump Nozzle Burner motor Oil line filter Fuel lines	Low voltage Water in oil

PROBLEM: No heat – burner starts and fires, then loses flame before locking out on safety

Fuel pump Nozzle Oil supply and oil pressure Oil line filters Water in oil Primary or safety controls Flame detector Improper burner adjustment	Fuel lines Nozzle Displaced or damaged baffles	Faulty wiring Loose terminals Vent or flue problem

PROBLEM: Not enough heat – burner cycle too short

Control relay or contactor Thermostat heat anticipator setting Thermostat not level Thermostat location Dirty filters	Limit controls Fan controls Blower bearings Blower motor Defective blower wheel Dirty blower wheel	Faulty wiring Loose terminals Low voltage Low air volume Ductwork small or restricted

PROBLEM: Not enough heat – burner cycle too long

Blower belt broken or slipping Dirty filters Dirty or plugged heat exchanger	Fuel pump Low air volume Dirty blower wheel Oil line filter	Ductwork small or restricted Oil supply and oil pressure

310A04.EPS

GENERALLY THE CAUSE Make these checks first.	OCCASIONALLY THE CAUSE Make these checks only if the first checks failed to locate the trouble.	RARELY THE CAUSE Make these checks only if other checks failed to locate the trouble.
PROBLEM: Too much heat – burner cycles too long		
Thermostat heat anticipator setting too high Thermostat out of calibration	Thermostat not level Thermostat location	
PROBLEM: Too much heat – burner runs continuously		
Faulty wiring Control relay or contactor	Thermostat faulty	Thermostat not level Thermostat location
PROBLEM: Combustion noise		
Nozzle Electrode Dirty or plugged heat exchanger Fuel pump Burner air adjustment	Burner motor Ignition transformer	Faulty wiring Loose terminals Vent or flue problem Oil supply
PROBLEM: Air noise		
Blower Ductwork small or restricted	Cabinet	Blower wheel Dirty filters
PROBLEM: Cost of operation		
Burner motor Barometric control Blower motor Dirty filters Improper burner adjustment Inefficient burner	Low air volume Insufficient insulation Excessive infiltration	Low voltage Fan control Nozzle Ignition transformer Blower belt broken or slipping Dirty blower wheel Ductwork small or restricted Vent or flue problem Input too low
PROBLEM: Mechanical noise		
Blower bearings Blower motor Blower belt broken or slipping Blower wheel out of balance	Cabinet	Low voltage Control relay or contactor Fuel pump Burner motor Oil valve Ruptured heat exchanger Displaced or damaged baffles Fuel lines
PROBLEM: Odor		
Barometric control Vent or flue problems Flue gas spillage Fuel lines Oil supply Oil pump shutoff valve	Low air volume Improper burner adjustment Ruptured heat exchanger Displaced or damaged baffles Blocked heat exchanger Stagnant water in humidifier Water or moisture in air system	Faulty wiring or loose terminals Control transformer Fuel pump Dirty filters Dirty heat exchanger Input too high Input too low Outdoor odors

310A05.EPS

Additional Resources and References

Additional Resources

This module is intended to be a thorough resource for task training. The following reference work is suggested for further study. This is optional material for continued education rather than for task training.

Heating, Ventilating, and Air Conditioning Fundamentals, 1995. Raymond A. Havrella. Englewood Cliffs, NJ: Prentice Hall, Inc.

Figure Credits

Photo courtesy of Rheem Manufacturing Company, 310F01

Dwyer Instruments, Inc., 310SA02

Topaz Publications, Inc., 310SA03

Emerson Climate Technologies, 310F11 (photo)

Courtesy of Honeywell International Inc., 310F12(A), 310F16

France/A Scott Fetzer Company, 310F13

Bacharach, 310SA04, 310F24, 310F25, 310F26, 310F28, 310SA05

R.W. Beckett Corporation, 310F30

NCCER CURRICULA — USER UPDATE

NCCER makes every effort to keep its textbooks up-to-date and free of technical errors. We appreciate your help in this process. If you find an error, a typographical mistake, or an inaccuracy in NCCER's curricula, please fill out this form (or a photocopy), or complete the online form at **www.nccer.org/olf**. Be sure to include the exact module ID number, page number, a detailed description, and your recommended correction. Your input will be brought to the attention of the Authoring Team. Thank you for your assistance.

Instructors – If you have an idea for improving this textbook, or have found that additional materials were necessary to teach this module effectively, please let us know so that we may present your suggestions to the Authoring Team.

NCCER Product Development and Revision
13614 Progress Blvd., Alachua, FL 32615

Email: curriculum@nccer.org
Online: www.nccer.org/olf

❏ Trainee Guide ❏ AIG ❏ Exam ❏ PowerPoints Other _____

Craft / Level: _____ Copyright Date: _____

Module ID Number / Title: _____

Section Number(s): _____

Description: _____

_____

_____

_____

Recommended Correction: _____

_____

_____

_____

Your Name: _____

Address: _____

_____

Email: _____ Phone: _____

03311-08

Troubleshooting Heat Pumps

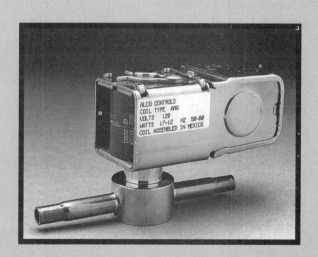

03311-08

Troubleshooting Heat Pumps

Topics to be presented in this module include:

1.0.0	Introduction	11.2
2.0.0	Heat Pump Operation	11.4
3.0.0	Electrical Operating Sequence	11.5
4.0.0	Troubleshooting	11.16

Overview

This module covers heat pump operation and the specialized troubleshooting methods associated with heat pumps. A heat pump is more complex than a cooling-only system. It contains more control devices and has a more complex wiring diagram than a cooling-only system. For example, a heat pump contains a reversing valve and a defrost control circuit, along with electric heating elements that provide supplementary and emergency heat and their associated controls. Despite the increased complexity, however, it is sometimes easier to isolate a heat pump malfunction. This is because the cooling and heating functions can be examined separately. If both modes are affected, the problem is in a component that is common to both functions, the compressor contactor, for example. If only one function is affected, only the components unique to that function must be examined.

Objectives

When you have completed this module, you will be able to do the following:

1. Describe the basic operating sequence for an air-to-air heat pump.
2. Interpret control circuit diagrams for heat pumps.
3. Develop a checklist for troubleshooting a heat pump.
4. Identify the tools and instruments used in troubleshooting heat pumps.
5. Correctly use the tools and instruments required for troubleshooting heat pumps.
6. Isolate and correct malfunctions in heat pumps.
7. Describe the safety precautions associated with servicing heat pumps.

Trade Terms

Direct-acting valve
Field wiring
Pilot-operated valve
Transformer phasing

Required Trainee Materials

1. Pencil and paper
2. Appropriate personal protective equipment

Prerequisites

Before you begin this module, it is recommended that you successfully complete *Core Curriculum*; *HVAC Level One*; *HVAC Level Two*; and *HVAC Level Three*, Modules 03301-08 through 03310-08.

This course map shows all of the modules in the third level of the *HVAC* curriculum. The suggested training order begins at the bottom and proceeds up. Skill levels increase as you advance on the course map. The local Training Program Sponsor may adjust the training order.

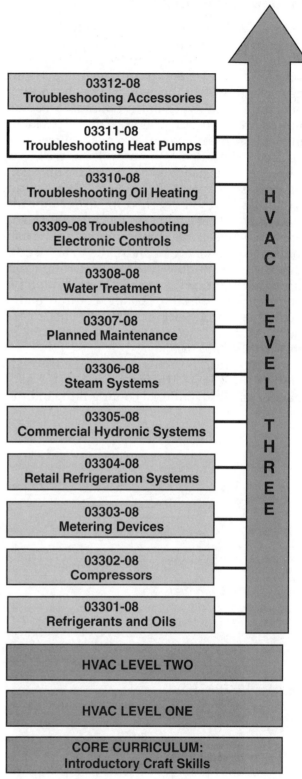

03312-08
Troubleshooting Accessories

03311-08
Troubleshooting Heat Pumps

03310-08
Troubleshooting Oil Heating

03309-08 Troubleshooting
Electronic Controls

03308-08
Water Treatment

03307-08
Planned Maintenance

03306-08
Steam Systems

03305-08
Commercial Hydronic Systems

03304-08
Retail Refrigeration Systems

03303-08
Metering Devices

03302-08
Compressors

03301-08
Refrigerants and Oils

HVAC LEVEL THREE

HVAC LEVEL TWO

HVAC LEVEL ONE

CORE CURRICULUM:
Introductory Craft Skills

311CMAP.EPS

1.0.0 ◆ INTRODUCTION

A service call for a heat pump is likely to be more complicated than one for a cooling-only system or a furnace. Heat pump control circuits are more complex because they control heating and cooling cycles as well as the defrost cycle. In addition, there are several types of heat pumps. Most heat pumps are air-to-air split systems or packaged units, but some use water as a heat source. Some are used to heat domestic hot water as well.

On the other hand, a heat pump is still a cooling unit, so many of the same troubleshooting principles that you learned in earlier modules also apply to heat pumps. In this module, the differences between heat pumps and other types of comfort systems are explained. If you need to review the techniques for troubleshooting cooling equipment, refer to the HVAC Level Two module, *Troubleshooting Cooling*. For information on troubleshooting electrical components, refer to the HVAC Level Two module, *Introduction to Control Circuit Troubleshooting*.

A heat pump is a combination heating and cooling unit. It produces cooling in the same manner as a conventional cooling unit, and then reverses the cycle to produce heat (*Figure 1*). Both systems have these same components:

- Compressor
- Condenser and evaporator coils
- Blower fan
- Condenser fan
- Service valves or service ports
- Pressure and temperature controls
- Refrigerant and refrigerant lines

In addition to the typical air conditioning components listed, the heat pump requires a reversing valve as well as additional metering devices and control circuits.

Because heat pumps perform both heating and cooling functions, the coils are identified by location: the condensing coil and evaporator coil on an air conditioner are identified as the outdoor coil and indoor coil on a heat pump. Like cooling equipment, a heat pump can be a single packaged unit or it can be a matched split system.

An advantage of the heat pump is that it is less expensive to operate than electric resistance heat. Plus, it provides cooling. For heating purposes, a

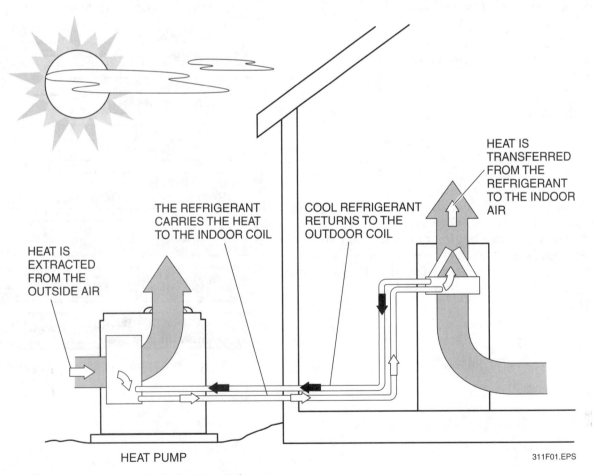

Figure 1 ◆ Heat pump operation in the heating mode.

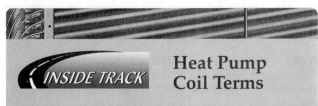
heat pump may or may not be less costly to operate than gas or oil heat, depending on the climate and fuel costs. It is, however, likely to be more cost effective for combined heating and cooling than other methods.

1.1.0 Heat Pump Classifications

Heat pumps are classified according to their heat source and the medium to which the heat is transferred (heat sink). A water-to-air heat pump, for example, picks up heat from water flowing through a heat exchanger and transfers it to the air flowing over another coil.

The more common types of heat pumps are: air-to-air, water-to-water, air-to-water, and water-to-air. In the air-to-air system, air is the external heat source and the medium (heat sink) that comes in contact with the indoor refrigerant coil. Some larger heat pump systems may have more

than one heat source and may also use both water and air as heat sinks.

An air-to-water system also uses air as the heat source, but the heating and cooling are provided by a water distribution system within the structure. The air-to-water system uses convection devices to transfer the heat between the water and the air. These systems are used primarily to heat water.

Ground water, from the standpoint of temperature, is an excellent heat source but is not available in all areas, and in some areas is in limited supply during certain periods. Therefore, air is the predominant heat source in residential and small commercial installations.

On occasion, heat pump systems may use a coil buried in the ground as a heat source (ground source or geothermal heat pump). Fluid is circulated through the buried coil (*Figure 2*) and the earth is used as a heat sink to absorb or dissipate heat. The fluid (typically a nontoxic anti-freeze solution) is circulated to a fluid-to-refrigerant heat exchanger located in the air handler inside the structure. The refrigerant then flows to the indoor coil in the air handler where the heat is rejected (heating mode) or absorbed (cooling mode). A typical residential geothermal heat pump is a packaged product that contains the compressor, refrigerant-to-fluid heat exchanger, indoor coil, and blower motor. The pump that circulates the fluid through the ground coil is often a separate component.

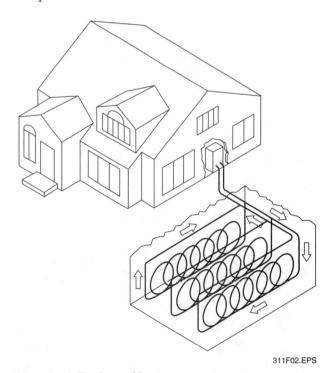

311F02.EPS

Figure 2 ◆ Geothermal heat pump system.

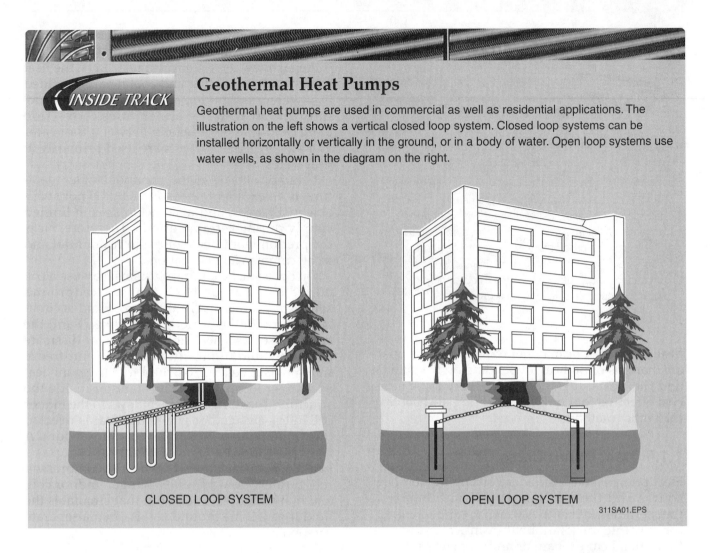

Geothermal Heat Pumps

Geothermal heat pumps are used in commercial as well as residential applications. The illustration on the left shows a vertical closed loop system. Closed loop systems can be installed horizontally or vertically in the ground, or in a body of water. Open loop systems use water wells, as shown in the diagram on the right.

CLOSED LOOP SYSTEM

OPEN LOOP SYSTEM

311SA01.EPS

Other heat sources besides air and water are sometimes used. Some examples are waste heat from selected industrial processes, exhaust air from ventilation, solar energy, and heat extracted from refrigerated spaces. These sources are commonly used in addition to, rather than as replacements for, the basic heat source.

2.0.0 ◆ HEAT PUMP OPERATION

Even at very cold outdoor temperatures, there is some heat in the air. The air-to-air heat pump operates on the principle of extracting this heat from the outdoor air and transferring it to the indoor air. The reversing valve is the key to this phenomenon. In the heating mode, it causes the refrigerant to flow in the opposite direction from that of the cooling mode. The outdoor coil then acts as the evaporator, while the indoor coil becomes the condenser. The control circuits are usually arranged so that the reversing valve is de-energized in the heating mode. That way, if there is a control failure, the heating function will continue to operate. Defrost will not work, however, so the unit will eventually freeze up.

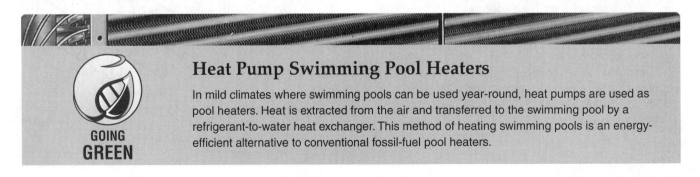

Heat Pump Swimming Pool Heaters

In mild climates where swimming pools can be used year-round, heat pumps are used as pool heaters. Heat is extracted from the air and transferred to the swimming pool by a refrigerant-to-water heat exchanger. This method of heating swimming pools is an energy-efficient alternative to conventional fossil-fuel pool heaters.

GOING GREEN

Another key difference between cooling-only systems and heat pumps is the metering device arrangement. As you can see in *Figure 3*, the heat pump has two metering devices. In the arrangement shown, the devices are fixed-orifice metering devices. Built-in or separate check valves are used to block or allow refrigerant flow, depending on the mode of operation. In other applications, fixed-orifice, piston-type metering devices are used. When this type is used, check valves are not required because the piston acts as a check valve.

The following sections are intended to serve as a review of basic air-to-air heat pump operation. For a more detailed presentation covering all types of heat pumps, refer to the HVAC Level Two module, *Heat Pumps*.

2.1.0 Cooling Cycle

In most heat pumps, the reversing valve is energized when the thermostat calls for cooling and de-energized when it calls for heating. This guarantees that heat is available even if there is a failure in the reversing valve control circuits.

When the unit is in the cooling mode (*Figure 3, View A*), the refrigerant flow (see arrows) is the same as that of any cooling unit. Notice the path through the reversing valve. The check valve used with the indoor coil metering device is closed, forcing the refrigerant to be metered. The cold, low-pressure refrigerant flowing through the indoor coil (evaporator) absorbs heat from the conditioned space and is boiled into a superheated vapor. Hot, high-pressure refrigerant gas leaving the compressor is pumped through the outdoor coil (condenser), where the heat is rejected. The check valve used with the outdoor coil metering device is open, allowing refrigerant to bypass the metering device. The refrigerant pressures and temperatures shown are typical of those covered in the discussion of cooling systems.

2.2.0 Heating Cycle

In the heating mode (*Figure 3, View B*), the reversing valve changes position. The refrigerant leaving the compressor is routed in the opposite direction from that of the cooling cycle; that is, instead of flowing through to the outdoor coil, the hot, high-pressure refrigerant vapor leaving the compressor flows to the indoor coil, which is now acting as a condenser. The check valve used with the indoor coil metering device is open, allowing refrigerant to bypass the metering

device. Heat is extracted from the refrigerant at that point because the air in the conditioned space is cool in relation to the refrigerant in the coil. The cooled refrigerant then flows to the outdoor coil. The check valve used with the outdoor coil metering device is closed, forcing the refrigerant to be metered. Acting as an evaporator, the outdoor coil absorbs heat from the relatively warmer outdoor air. Heat provided in this way is known as reverse cycle heat or compression heat. Note the significant differences in the pressures and temperatures at key points in the system as compared with the cooling mode.

An accumulator and crankcase heater, which are optional accessories on cooling units, are essential on most heat pumps. The crankcase heater keeps the liquid refrigerant from migrating back to the compressor, where it could cause slugging at startup. The accumulator traps liquid refrigerant that gets into the suction line during low-temperature operation in the heating mode. The liquid refrigerant boils into a vapor in the accumulator.

2.3.0 Defrost Cycle

At outdoor temperatures below about 50°F, frost will form on the outdoor coil of the heat pump while it is running in the reverse cycle heating mode. As the frost builds up, it will restrict airflow and significantly reduce heat transfer. The frost is usually melted by reversing the flow of refrigerant so that the hot, high-pressure refrigerant from the compressor discharge flows through the outdoor coil. The defrost cycle is automatic. The start and termination of the cycle may be based on elapsed time, coil temperature, or other factors.

While the unit is in defrost, a supplementary electric heater or the gas burner of a furnace is usually turned on to prevent the occupied areas from becoming cold. Because heat pumps are not especially efficient at very low outdoor temperatures, electric heaters are used to supplement reverse cycle heat in all-electric heat pumps.

3.0.0 ◆ ELECTRICAL OPERATING SEQUENCE

The key to troubleshooting any heat pump is understanding the operating sequence. The control circuits for a heat pump are generally more complicated than those of a cooling-only or heating-only system because there are more controls and more operating modes.

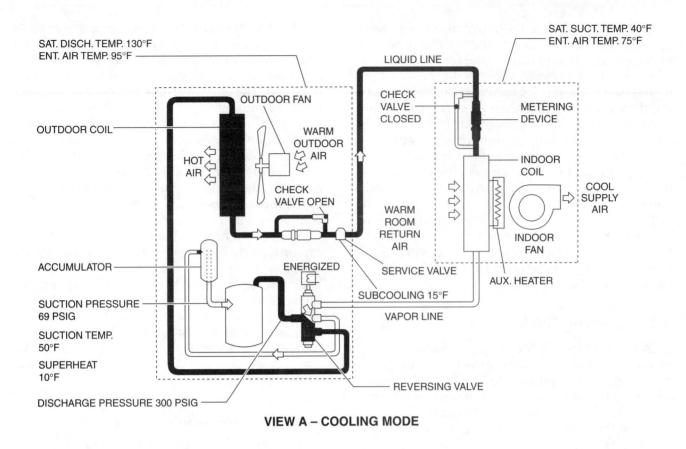

VIEW A – COOLING MODE

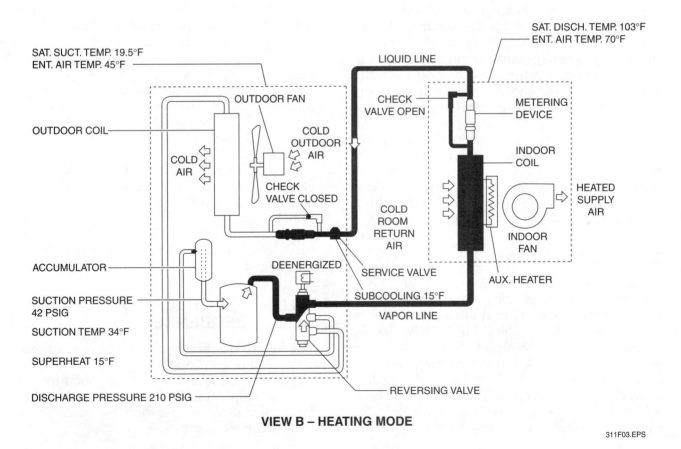

VIEW B – HEATING MODE

311F03.EPS

Figure 3 ◆ Refrigerant flow in an air-to-air heat pump.

For most heat pumps, there are five different operating modes:

- Cooling mode
- Heating cycle with compression heat only
- Heating cycle with compression heat and supplemental heat
- Defrost cycle
- Emergency heating (compression heating disabled)

Figures 4 through *7* show the schematic wiring diagrams for a 3½-ton split system heat pump working in tandem with an electric furnace. Each of the diagrams shows the operating sequence of a different operating mode.

- *Figure 4* illustrates the cooling sequence of operation.

- *Figure 5* depicts the same unit in the compression heating cycle with supplementary heat turned on.
- *Figure 6* shows the defrost cycle.
- *Figure 7* shows the sequence of operation for emergency heat.
- *Figure 8* will help you interpret the abbreviations you find on these and other diagrams. Often, the name assigned to a control can tell you what it does in the circuit.
- *Figure 9* defines the symbols used on these and other diagrams.

Study each of these diagrams and follow the operating sequence listed at the bottom of the diagram.

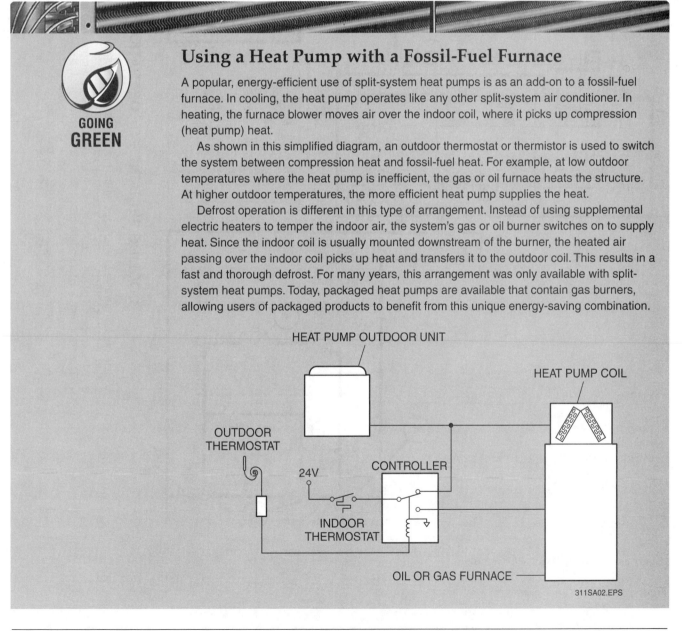

GOING GREEN

Using a Heat Pump with a Fossil-Fuel Furnace

A popular, energy-efficient use of split-system heat pumps is as an add-on to a fossil-fuel furnace. In cooling, the heat pump operates like any other split-system air conditioner. In heating, the furnace blower moves air over the indoor coil, where it picks up compression (heat pump) heat.

As shown in this simplified diagram, an outdoor thermostat or thermistor is used to switch the system between compression heat and fossil-fuel heat. For example, at low outdoor temperatures where the heat pump is inefficient, the gas or oil furnace heats the structure. At higher outdoor temperatures, the more efficient heat pump supplies the heat.

Defrost operation is different in this type of arrangement. Instead of using supplemental electric heaters to temper the indoor air, the system's gas or oil burner switches on to supply heat. Since the indoor coil is usually mounted downstream of the burner, the heated air passing over the indoor coil picks up heat and transfers it to the outdoor coil. This results in a fast and thorough defrost. For many years, this arrangement was only available with split-system heat pumps. Today, packaged heat pumps are available that contain gas burners, allowing users of packaged products to benefit from this unique energy-saving combination.

HEAT PUMP OUTDOOR UNIT

HEAT PUMP COIL

OUTDOOR THERMOSTAT

24V

CONTROLLER

INDOOR THERMOSTAT

OIL OR GAS FURNACE

311SA02.EPS

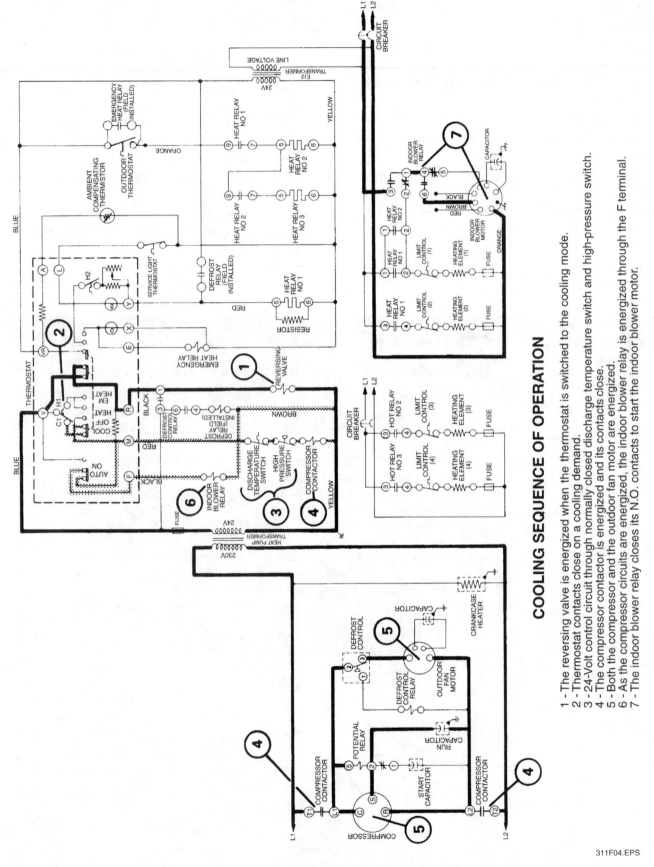

COOLING SEQUENCE OF OPERATION

1 - The reversing valve is energized when the thermostat is switched to the cooling mode.
2 - Thermostat contacts close on a cooling demand.
3 - 24-Volt control circuit through normally closed discharge temperature switch and high-pressure switch.
4 - The compressor contactor is energized and its contacts close.
5 - Both the compressor and the outdoor fan motor are energized.
6 - As the compressor circuits are energized, the indoor blower relay is energized through the F terminal.
7 - The indoor blower relay closes its N.O. contacts to start the indoor blower motor.

Figure 4 ◆ Cooling sequence diagram.

311F04.EPS

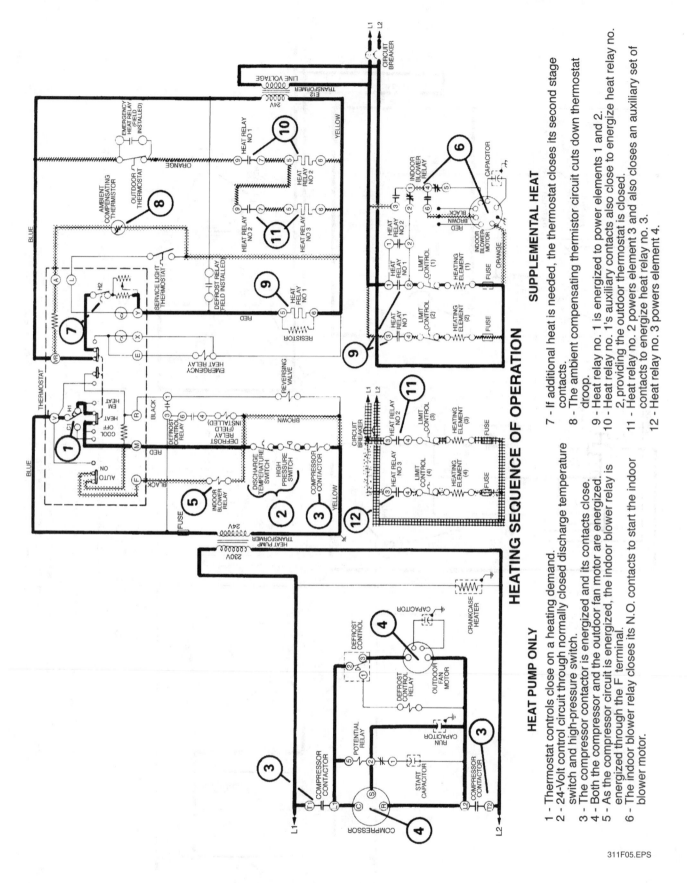

HEATING SEQUENCE OF OPERATION

HEAT PUMP ONLY

1 - Thermostat controls close on a heating demand.
2 - 24-Volt control circuit through normally closed discharge temperature switch and high-pressure switch.
3 - The compressor contactor is energized and its contacts close.
4 - Both the compressor and the outdoor fan motor are energized.
5 - As the compressor circuit is energized, the indoor blower relay is energized through the F terminal.
6 - The indoor blower relay closes its N.O. contacts to start the indoor blower motor.

SUPPLEMENTAL HEAT

7 - If additional heat is needed, the thermostat closes its second stage contacts.
8 - The ambient compensating thermistor circuit cuts down thermostat droop.
9 - Heat relay no. 1 is energized to power elements 1 and 2.
10 - Heat relay no. 1's auxiliary contacts also close to energize heat relay no. 2, providing the outdoor thermostat is closed.
11 - Heat relay no. 2 powers element 3 and also closes an auxiliary set of contacts to energize heat relay no. 3.
12 - Heat relay no. 3 powers element 4.

Figure 5 ♦ Heating sequence diagram.

311F05.EPS

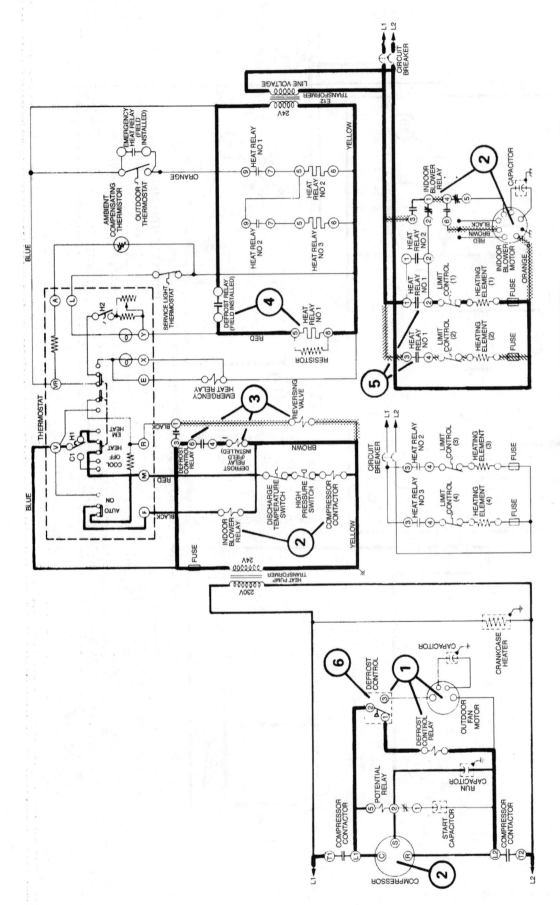

DEFROST CYCLE SEQUENCE OF OPERATION

1 - The defrost control switches position under defrost conditions. This turns off the outdoor fan and energizes the defrost control relay.

2 - The compressor and indoor blower motor continue to operate.

3 - The defrost control relay closes its N.O. contacts to energize the reversing valve and the defrost relay.

4 - The defrost relay closes its contacts to energize heat relay no. 1.

5 - Heat relay no. 1 closes its contacts to power elements 1 & 2.

6 - The defrost control will return to its normal position after the defrost cycle is complete.

311F06.EPS

Figure 6 ◆ Defrost cycle sequence diagram.

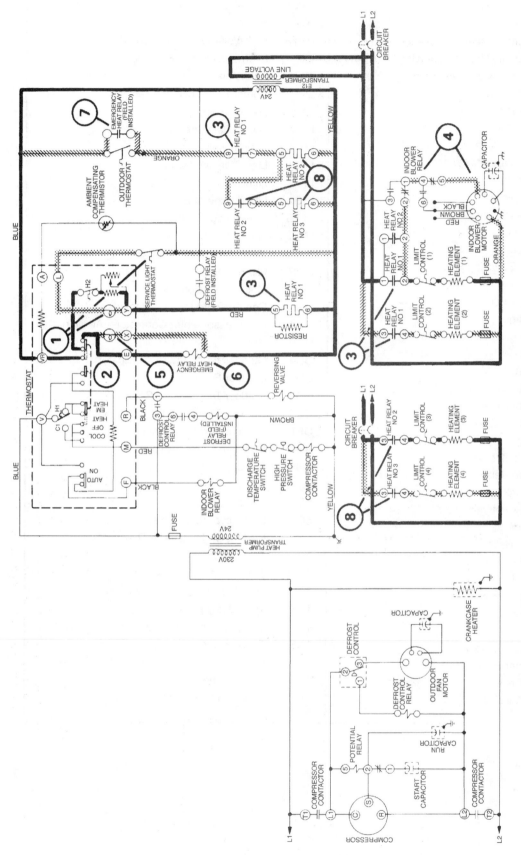

EMERGENCY HEAT SEQUENCE OF OPERATION

1 - Should a heat pump malfunction occur, the service light comes on.
2 - Place the thermostat in the emergency heat position.
3 - Heat relay no. 1 remains energized as before and elements no. 1 and no. 2 are energized.
4 - Switching the thermostat deenergizes the indoor blower relay, which consequently changes the blower motor speed.
5 - Switching the thermostat also brings on another service light to remind the homeowner that the system is on emergency heat.
6 - The emergency heat relay is energized through the E terminal.
7 - The relay closes its contacts to bypass the outdoor thermostat.
8 - Heat relays no. 2 and no. 3 are energized to bring on elements (3) and (4).

311F07.EPS

Figure 7 ◆ Emergency heat sequence diagram.

AH	Supplementary Heater	HPCO	High-Pressure Cutout
AR	Auxiliary Relay	HTR	Heater
BH	Supplementary Heater Contactor	HVTB	High-Voltage Terminal Board
BHA	Supplementary Heater Contactor – Auxiliary	IOL	Internal Overload Protector
BLR	Balancing Relay	1K	Economizer Relay
CA	Cooling Anticipator	2K	Compressor Lock-out Relay
CB	Circuit Breaker	LA	Lock-in Relay
CF	Fan Capacitor	LAS	Low Airflow Sensing
CFA	Outdoor Fan Capacitor	LB	Lock-in Relay
CH	Supplementary Htr. Contactor	LPCO	Low-Pressure Cutout
CL	Contactor Relay	LT	Light
CN	Wire Connector	LTS	Low-Temperature Switch
COC	Changeover Control	LVTB	Low-Voltage Terminal Board
CR	Run Capacitor	MAC	Mixed Air Control
CPR	Compressor	MPM	Motor Protection Module
CSR	Capacitor Switching Relay	MPP	Min. Position Potentiometer
D	Defrost Relay	MPS	Motor Protection Sensor
DFC	Defrost Control	MS	Compressor Motor Contactor
DFT	Defrost Timer	MSA	Compressor Motor Contactor – Auxiliary
DH	Supplementary Htr. Contactor	MTR	Motor
DR	Defrost Relay	ODA	Outdoor Temperature Anticipator
DS	Diaphragm Switch	ODF	Outdoor Fan Relay
DT	Defrost Termination Switch	ODS	Outdoor Temperature Sensor
EDC	Evaporator Defrost Control	ODT	Outdoor Thermostat
EDR	Evaporator Defrost Relay	OFT	Outdoor Fan Thermostat
ETB	Economizer Terminal Board	RH	Emergency Heat Relay
ER	Electric Heater Relay	RHS	Resistance Heat Switch
F	Indoor Fan Relay	S-FU	$\frac{1}{16}$ Amp Fuse
FL	Flexible Link	SEN	Sensor
FP	Feed-Back Potentiometer	SC	Switchover Valve Solenoid
FR	Contactor Relay	SM	System Switch (Room Thermostat)
FA	Contactor Relay	T	Thermistor
FM	Manual Fan Switch	TCO	Temperature Limit Switch
FS	Fan Signal Relay	TDL	Discharge Line Thermostat
FST	Fan Switch Terminal	TDR	Time Delay Relay
FTB	Fan Terminal Board	TM	Compressor Motor Thermostat
FU	Fuse	TNS	Transformer
FUA	Compressor Fuse	TS	Heating & Cooling Thermostat
FUB	Transformer Fuse	TSC	Cooling Thermostat
H	Heating Relay	TSH	Heating Thermostat
HA	Heating Anticipator		

311F08.EPS

Figure 8 ◆ HVAC abbreviations.

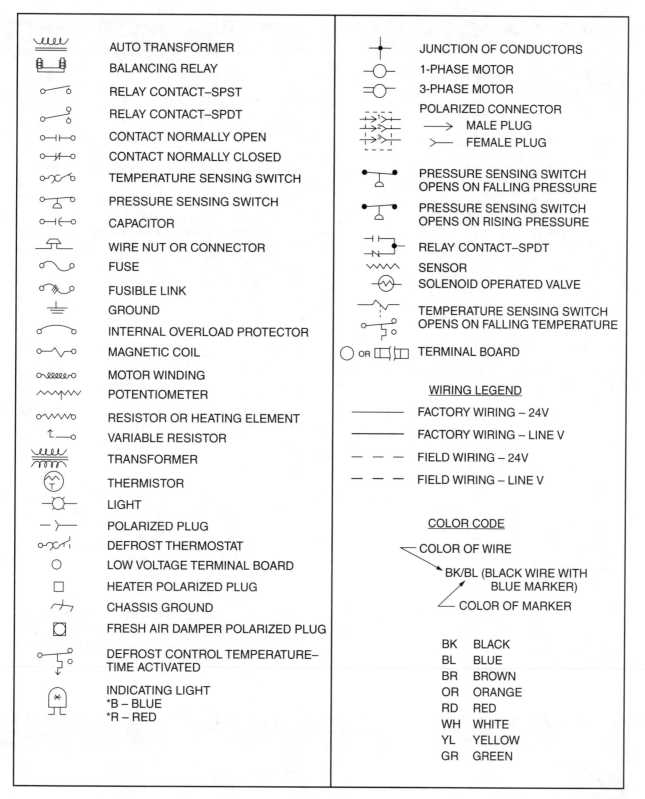

AUTO TRANSFORMER	JUNCTION OF CONDUCTORS
BALANCING RELAY	1-PHASE MOTOR
RELAY CONTACT–SPST	3-PHASE MOTOR
RELAY CONTACT–SPDT	POLARIZED CONNECTOR
CONTACT NORMALLY OPEN	MALE PLUG
CONTACT NORMALLY CLOSED	FEMALE PLUG
TEMPERATURE SENSING SWITCH	PRESSURE SENSING SWITCH OPENS ON FALLING PRESSURE
PRESSURE SENSING SWITCH	PRESSURE SENSING SWITCH OPENS ON RISING PRESSURE
CAPACITOR	RELAY CONTACT–SPDT
WIRE NUT OR CONNECTOR	SENSOR
FUSE	SOLENOID OPERATED VALVE
FUSIBLE LINK	TEMPERATURE SENSING SWITCH OPENS ON FALLING TEMPERATURE
GROUND	TERMINAL BOARD
INTERNAL OVERLOAD PROTECTOR	
MAGNETIC COIL	
MOTOR WINDING	WIRING LEGEND
POTENTIOMETER	FACTORY WIRING – 24V
RESISTOR OR HEATING ELEMENT	FACTORY WIRING – LINE V
VARIABLE RESISTOR	FIELD WIRING – 24V
TRANSFORMER	FIELD WIRING – LINE V
THERMISTOR	
LIGHT	
POLARIZED PLUG	COLOR CODE
DEFROST THERMOSTAT	COLOR OF WIRE
LOW VOLTAGE TERMINAL BOARD	BK/BL (BLACK WIRE WITH BLUE MARKER)
HEATER POLARIZED PLUG	COLOR OF MARKER
CHASSIS GROUND	
FRESH AIR DAMPER POLARIZED PLUG	

DEFROST CONTROL TEMPERATURE–
TIME ACTIVATED

INDICATING LIGHT
*B – BLUE
*R – RED

BK	BLACK
BL	BLUE
BR	BROWN
OR	ORANGE
RD	RED
WH	WHITE
YL	YELLOW
GR	GREEN

311F09.EPS

Figure 9 ◆ HVAC symbols.

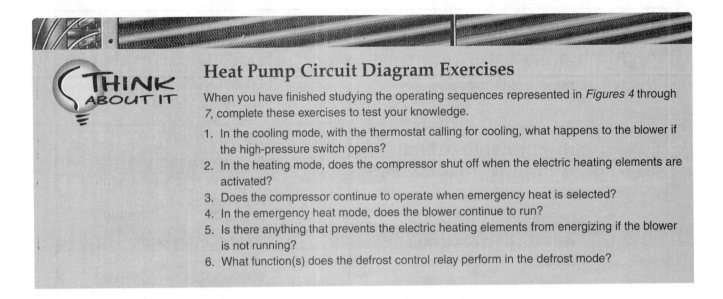

Heat Pump Circuit Diagram Exercises

When you have finished studying the operating sequences represented in *Figures 4* through *7*, complete these exercises to test your knowledge.

1. In the cooling mode, with the thermostat calling for cooling, what happens to the blower if the high-pressure switch opens?
2. In the heating mode, does the compressor shut off when the electric heating elements are activated?
3. Does the compressor continue to operate when emergency heat is selected?
4. In the emergency heat mode, does the blower continue to run?
5. Is there anything that prevents the electric heating elements from energizing if the blower is not running?
6. What function(s) does the defrost control relay perform in the defrost mode?

3.1.0 New Technology

The operating sequences just described are typical for a heat pump using electromechanical controls. They are ideal for understanding basic electrical operating sequences. While millions of electromechanical heat pumps are still in widespread use, they are rapidly being replaced with heat pumps that use electronic controls and variable-speed motor technology. *Figure 10* is the schematic diagram of a modern packaged heat pump that contains electronic controls and a variable-speed blower motor. Features of interest in this unit include the following:

- An electronic defrost board (DB) that contains a compressor start delay timer and timing circuit that controls the defrost interval. Dip switches on the control allow the installer to select a defrost interval suited to local climate conditions. Another dip switch enables a feature that allows for quieter defrost operation.

Test pins are available to allow a speed-up of timing functions.

- An ICM board (right of DB) acts as an interface between the room thermostat and the electronically commutated indoor blower motor (ICM). Based on inputs from the room thermostat, the ICM board tells the blower how fast to run based on the operating mode of the system.
- Like the schematic diagrams in the previous figures, this diagram contains features to help technicians understand the operating sequence and troubleshoot problems more easily. Features include a legend, component locator, and graphs depicting the various timing sequences of the electronic control.

If you study *Figure 10*, you will see that although the unit contains electronic controls, those controls perform the same functions as the electromechanical controls in the heat pump shown in *Figures 4* through *7*.

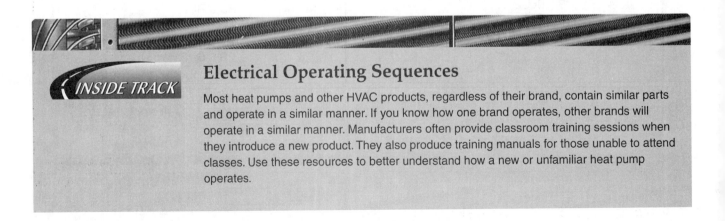

Electrical Operating Sequences

Most heat pumps and other HVAC products, regardless of their brand, contain similar parts and operate in a similar manner. If you know how one brand operates, other brands will operate in a similar manner. Manufacturers often provide classroom training sessions when they introduce a new product. They also produce training manuals for those unable to attend classes. Use these resources to better understand how a new or unfamiliar heat pump operates.

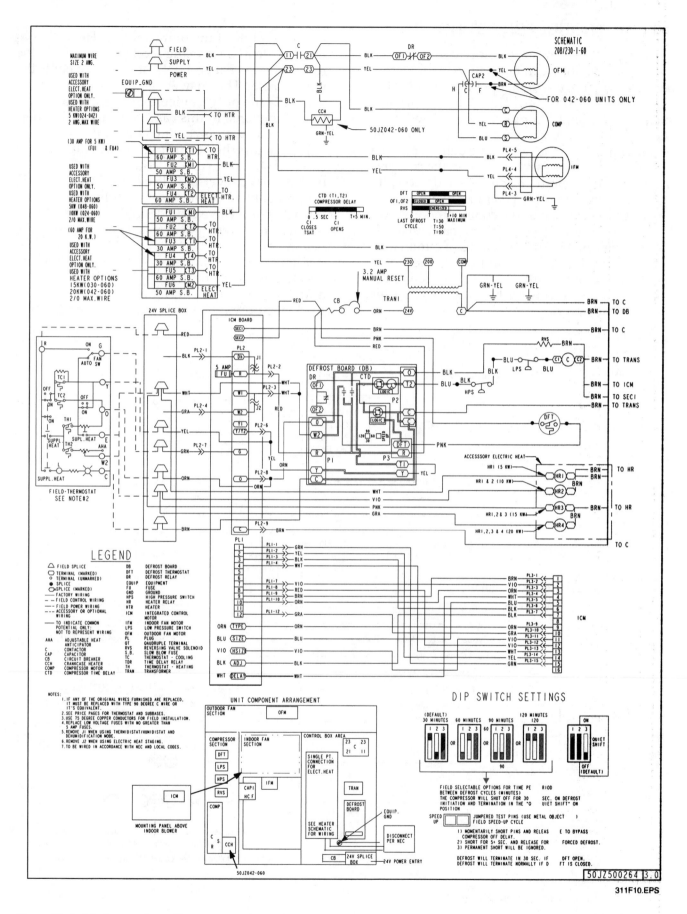

Figure 10 ◆ Modern packaged heat pump schematic.

4.0.0 ◆ TROUBLESHOOTING

Although heat pumps are more complex than cooling units, they can be easy to troubleshoot because they have several operating modes that share common functions. Therefore, if a problem such as a disabled compressor is observed in the cooling mode, you can check to see if the same problem occurs in the heating or defrost mode. If it does, the problem is most likely in the compressor or one of its control or protective devices. If it doesn't, then the problem is most likely in some device that is unique to the cooling mode. If you know the operating sequence of a particular heat pump, you can, by logical process of elimination, often isolate to a failed component merely by observing how the unit operates in each mode. You can often do this without even using test equipment.

For example, if the unit shown in *Figure 4* does not operate in the cooling mode, but works in the reverse cycle heat and defrost modes, what is the likely cause? The answer is: a bad room thermostat. It is the only device that is common to all three functions. In this case, the part of the thermostat that controls cooling is at fault.

Here's another example: what is the likely cause if the unit runs in cooling, but does not operate in the reverse cycle heating mode? Again, it's probably a bad thermostat. Although a defective reversing valve is a possibility, it is less likely because the reversing valve is de-energized in the heating mode.

Many manufacturers provide troubleshooting instructions in their product literature. *Figures 11* and *12* show an example of a troubleshooting road map provided by one manufacturer. Another

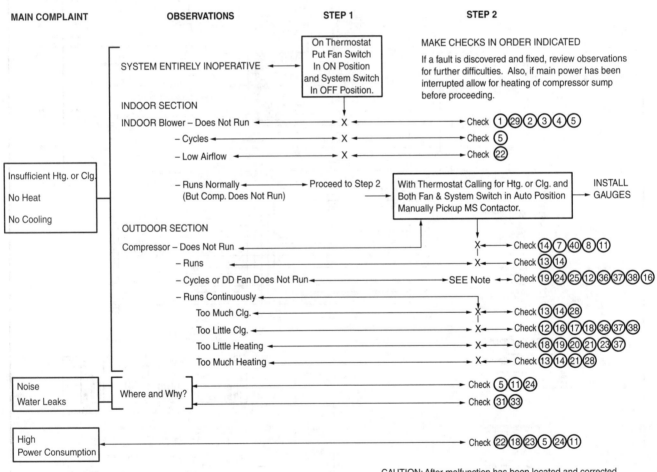

Figure 11 ◆ Troubleshooting procedures.

example is provided in the *Appendix*. Trouble-shooting aids like these can be helpful, but they are often either too general or too complicated. There is no substitute for knowing the operating sequence of the product you are troubleshooting. A heat pump troubleshooting chart may identify many possible causes of a particular symptom; or,

you may have to spend a lot of time reading through the chart to find the set of symptoms that matches the problem you are trying to solve. If you know the circuit, and you think about the conditions created by the problem, you can usually get pretty close to the failed component just by flipping a switch or two.

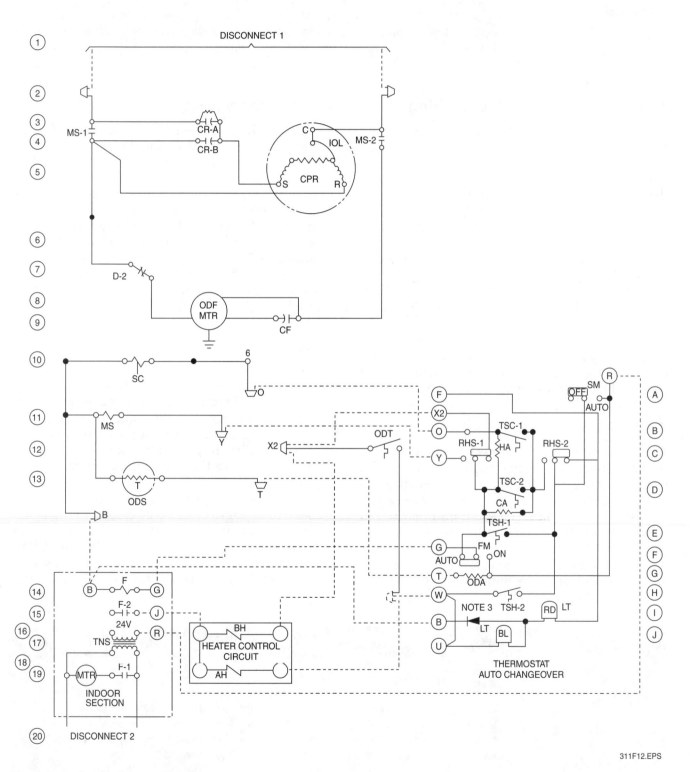

Figure 12 ◆ Troubleshooting diagram.

311F12.EPS

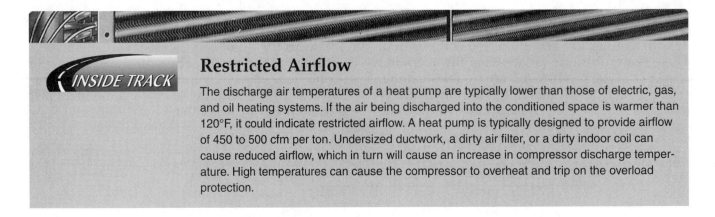

4.1.0 Control Circuit Field Wiring

Control circuit **field wiring** must be connected by at least 18 AWG copper wire, or the relays and contactors may chatter or fail to operate. If the wiring circuits are longer than 100', wire heavier than 18 AWG copper will have to be used in order to prevent excessive voltage drop.

Low primary voltage at the transformer may cause relays and contactors to chatter or fail to operate; therefore, the primary voltage must be checked. If the primary voltage is below 10 percent of the rating stamped on the nameplate, erratic control operation may result. Check for loose connections at the terminals and wire nut connections, because the additional resistance can also cause erratic control device operation. Control circuit fuses or breakers (if any) should also be checked before checking the field wiring.

4.1.1 Field Wiring

To check field wiring, first check the terminals for proper connection. Then disconnect each end of the wire in a specific circuit and measure the resistance with an ohmmeter. Excessive resistance will indicate loose connections or corroded terminals. Open circuits will be indicated by a reading of infinity on the ohmmeter. To check thermostat wiring in the wall, join two wires with a wire nut and check continuity at the other end.

4.1.2 Factory Wiring

Factory wiring in the unit can be checked in the same manner as field wiring. Field wiring should be checked before factory wiring. When factory miswiring occurs, the malfunction should show up when the unit is checked out during installation. Factory wiring problems seldom occur after the system has been in operation. Check the equipment wiring using the wiring diagram attached to the unit, and trace the circuits by observing the actual location of the electrical devices, terminal designations, and wire color codes as designated on the wiring diagram.

4.2.0 Thermostats

Heat pump thermostats (*Figure 13*) should be checked to ensure that the thermostat is the correct model number to match the system. Thermostats should not be mounted where extraneous cooling or heating sources will affect the sensing bulb. They must be mounted level and the wire hole in the wall behind them must be plugged.

Today's heat pump room thermostats are likely to be digital electronic controls rather than electromechanical mercury-bulb thermostats. There are several reasons for this. First and foremost is the fact that mercury is a toxic substance. If the glass capsule containing the mercury

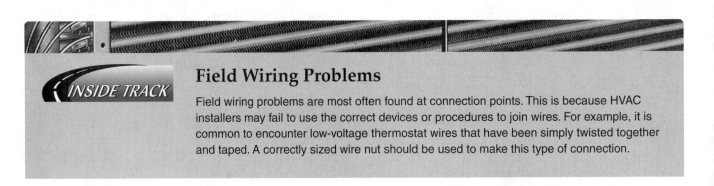

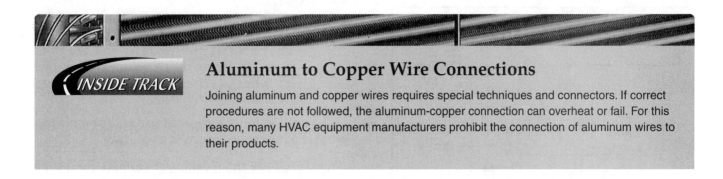

311F13.EPS

Figure 13 ◆ Heat pump room thermostat.

breaks, people could be exposed to the mercury. In replacement situations, old mercury-bulb thermostats must be treated as a hazardous waste and disposed of properly. Digital electronic room thermostats offer more precise temperature control and allow the full range of features designed into modern heat pumps using electronic controls to be fully utilized. The digital room thermostat shown in *Figure 13* is typical of the newest generation of controls. It offers the versatility of being used for heating, cooling, heat/cool, and heat pump operation. The installer configures the thermostat for the desired mode of operation. Within the heat pump mode, some thermostats offer the versatility of programming for heat pump operation with auxiliary electric heaters or with a gas or oil furnace. Often, the traditional mode selection subbase switches are removed and replaced with a mode selection button. When the button is pressed, the mode is displayed on the digital display. Within the thermostat itself, the traditional hard-wired electromechanical circuits are replaced with electronic circuits. Miniature relays provide the switching that was previously done by the subbase switches. Because the thermostats are electronic devices, they must be powered to operate. This can be done

with batteries and/or the 24 volts available in the control circuit. If control circuit power is used, a 24-volt common lead must be provided and connected to the common (C) terminal on the thermostat.

To check this type of thermostat, it is best to begin by determining the function controlled by each terminal. In the case of the unit shown in *Figure 12*:

- *Terminal B* – Wire color is blue: it is the 24V common lead from the control transformer secondary circuit.
- *Terminal R* – Wire color is red: it is the 24V power lead from the control transformer secondary circuit.
- *Terminal Y* – Wire color is yellow: it connects to the compressor contactor identified on a schematic wiring diagram as MS or MSA.
- *Terminal G* – Wire color is green: it controls the indoor fan relay (F or FA).
- *Terminal O* – Wire color is orange: it controls the reversing valve on a heat pump system.
- *Terminal W* – Wire color is white: it controls the heat source relay for supplementary heat.
- *Terminal T* – Wire color is brown: it connects to the anticipator circuit of the outdoor sensor (heat pump application).
- *Terminal X2* – Any color wire: it connects to the emergency heat circuit (heat pump).
- *Terminal F* – Any color wire: it connects to the low airflow switch.

This wiring description applies to the specific heating/cooling unit shown in *Figure 12*. It does not apply to all heat pumps.

Figure 14 is a detailed schematic of an electromechanical mercury-bulb room thermostat similar to that depicted in *Figure 12*. Although these thermostats are no longer manufactured, there are still millions of them in use.

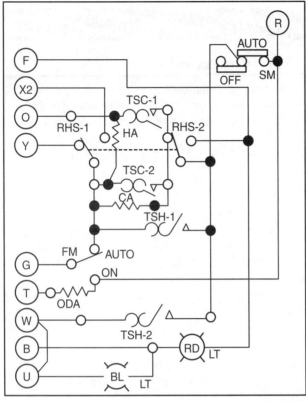

311F14.EPS

Figure 14 ◆ Thermostat wiring diagram.

4.2.1 Thermostat Troubleshooting Sequence

To check a thermostat for a short circuit when the power is on and the control devices do not de-energize, proceed as follows:

Step 1 Remove the thermostat from the subbase and disconnect the field wiring from the subbase. If the control devices remain energized, check for a short in the thermostat wires between the thermostat and the equipment connections.

Step 2 If the control device is de-energized with the thermostat removed, inspect the thermostat, the subbase, and the field wire leads for shorts (contact with one another). If that sequence checks out, replace the thermostat.

To check the thermostat for an open circuit with the power on, proceed as follows:

Step 1 Check the thermostat subbase mounting screws (if any); they must be tight.

Step 2 Remove the thermostat from the base and measure the voltage between R and B (switching lead to common) to check the 24V supply from the transformer.

Step 3 Check the subbase pressure contacts.

Step 4 Check all the field wiring connections and terminals at the subbase.

Step 5 Connect a jumper at the subbase terminals with the power on at the following terminals:

- R to G will energize the indoor fan.
- R to Y will energize the compressor.
- R to O will energize the reversing valve.
- R to W and G will energize the resistance auxiliary heat if the indoor fan relay contacts are closed.
- R to X2 and G will energize the emergency heat controls (heat pump) if the indoor fan relay contacts are closed.

Step 6 If the components operate as they are energized, replace the thermostat. If the components do not operate with the jumper wire checks, the problem is not in the thermostat; the field wiring and control circuit must be checked.

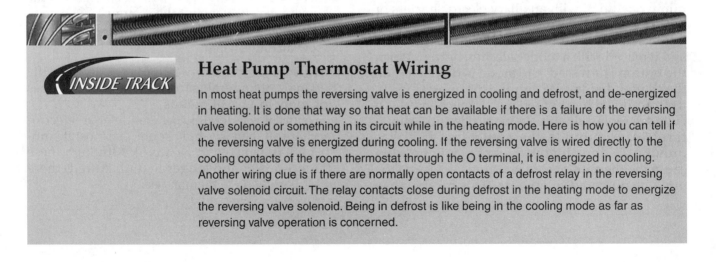

Heat Pump Thermostat Wiring

In most heat pumps the reversing valve is energized in cooling and defrost, and de-energized in heating. It is done that way so that heat can be available if there is a failure of the reversing valve solenoid or something in its circuit while in the heating mode. Here is how you can tell if the reversing valve is energized during cooling. If the reversing valve is wired directly to the cooling contacts of the room thermostat through the O terminal, it is energized in cooling. Another wiring clue is if there are normally open contacts of a defrost relay in the reversing valve solenoid circuit. The relay contacts close during defrost in the heating mode to energize the reversing valve solenoid. Being in defrost is like being in the cooling mode as far as reversing valve operation is concerned.

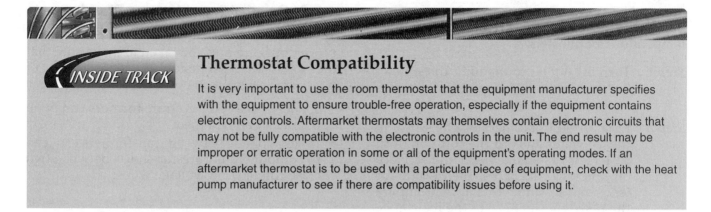

An outdoor thermostat for a heat pump with auxiliary heat can be checked as follows:

Step 1   Connect the ohmmeter as shown in *Figure 15*. Set the sensing bulb in a container with ice, salt, and water. Adjust the dial on the outdoor thermostat to its highest or warmest setting. The ohmmeter should indicate an open circuit.

Step 2   To test the calibration of the outdoor thermostat, place an electronic thermometer in the solution and stir the solution until 32°F is registered on the thermometer. Slowly rotate the outdoor thermostat dial until the ohmmeter indicates that the contacts are closed. The dial should indicate about 32°F. If more than a ±5° variation is found to exist between the thermometer and the dial, replace the outdoor thermostat.

Thermistors are used in modern heat pumps to sense outdoor temperature and coil temperature.

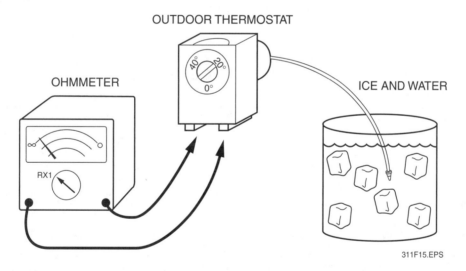

Figure 15 ◆ Outdoor thermostat check.

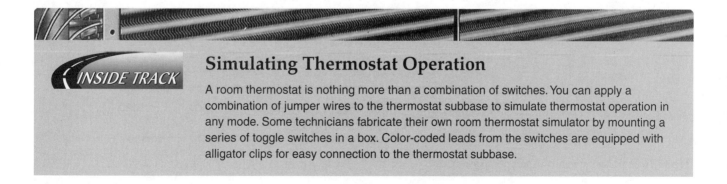

Their calibration can be checked in a manner similar to that used to check conventional outdoor temperature thermostats.

Step 1 Turn off power to the unit, disconnect the thermistor, and immerse it in an ice water solution.

Step 2 Place an electronic thermometer sensor in the ice water and stir the solution until 32°F registers on the thermometer. Place ohmmeter leads across the thermistor leads and read the resistance. Compare the resistance at 32°F to the resistance that the thermistor should read at that temperature by consulting a resistance versus temperature chart (*Figure 16*). In the example shown, the resistance measured at that temperature should be about 33,000 ohms. If the temperature that the thermistor is attempting to measure is known (air temperature or the surface temperature of a refrigerant line), enter that value into the chart to find the resistance at that temperature. Then compare the measured resistance to the value that the chart says it should be. Check the manufacturer's literature to determine if the resistance measured is within acceptable limits.

4.3.0 Control Transformer Phasing

Transformers connected in parallel can increase power, but care must be taken to make sure that they are in phase. Transformers connected out of phase can lead to such problems as:

- A short between two transformers can burn out one or both of them.
- Excess voltage between transformers can lead to component failures or shorts between two close-running low-voltage wires.

Transformers and Heat Pumps

Most heat/cool units (that aren't heat pumps) can operate effectively with a 24V transformer that has a power rating of 40VA. Heat pumps have more electrical components than most heat/cool units, however, so a 40VA transformer may be unable to handle the load. This is especially true of fossil-fuel furnaces equipped with add-on heat pumps. These systems may require transformers rated for 60VA or higher.

THERMISTOR CURVE

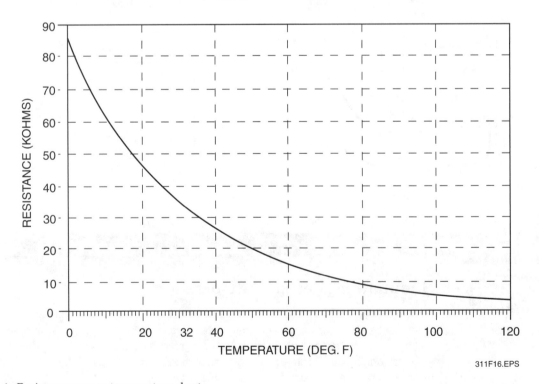

311F16.EPS

Figure 16 ◆ Resistance versus temperature chart.

Manufacturers will usually label built-in transformers with marked terminals and provide wiring diagrams. When separate transformers are connected, the terminals are not usually marked for **transformer phasing**, and they should be checked. All three-phase transformers should be tested for correct phasing. Phasing of transformers is covered in the HVAC Level Two module, *Alternating Current*.

4.4.0 Speed Controller

In the past, direct-drive blower motors used speed controllers that varied voltage to the motor as a way to change motor speed. That technology is now obsolete. If variable-speed blower operation is required today, electronically-commutated motors (ECM) are used. When used with electronic controls, ECM motors deliver precise air quantities based on the input signals they receive. The motor speeds are determined by the application and other factors. The equipment installer can either set dip-switches or moves jumpers on an interface board (*Figure 17*) to allow the motor to operate at different speeds under different circumstances. On the interface board shown, equipment capacity and the size of auxiliary electric heaters are factors that will determine how fast the blower motor will run. The amplitude of the voltage signal and the terminals on the motor to which they are applied, are based on the signals received from

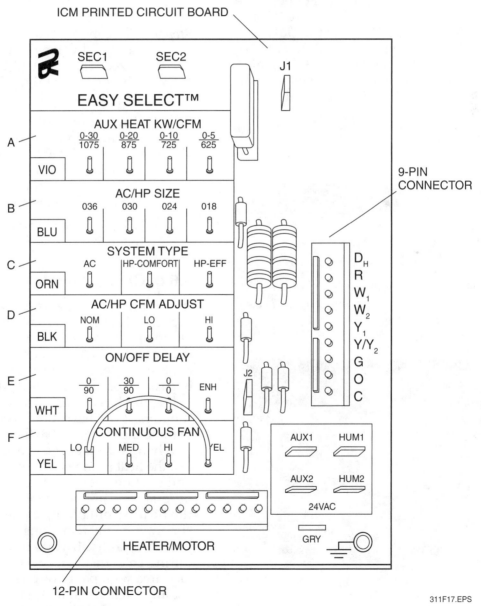

311F17.EPS

Figure 17 ◆ Speed controller schematic.

the room thermostat. Different combinations of voltage applied to different terminals of the motor cause it to operate at different speeds during different equipment operating modes.

4.5.0 Magnetic Relays and Solenoid

When the coil is energized in a typical solenoid, a magnetic field is produced, which pulls the valve stem, controlling the flow of a liquid or gas. There are two types of solenoid valves:

- **Direct-acting valves** – The solenoid coil pulls the valve port open directly by lifting the pin from the valve seat. Power limitations of the solenoid restrict the port size of a given operation.

- **Pilot-operated valves** – The solenoid coil operates a plunger covering a small pilot port hole. When the port hole is open, pressure trapped between a larger piston and the port hole escapes. The resulting pressure differential on the two sides of the piston causes it to move toward the pilot port hole; this opens the main port hole on the other side of the piston. To close the main port hole, the process is reversed. Heat pump reversing valves are usually pilot-operated valves.

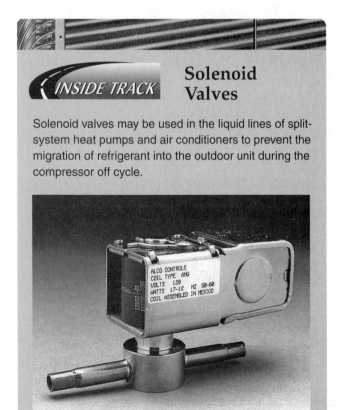

INSIDE TRACK

Solenoid Valves

Solenoid valves may be used in the liquid lines of split-system heat pumps and air conditioners to prevent the migration of refrigerant into the outdoor unit during the compressor off cycle.

311SA03.EPS

When troubleshooting solenoid valves, look for these major categories of problems:

- Valve does not open.
- Valve does not close.
- Valve produces excess noise.
- Coil is burned out.

4.5.1 Testing a Solenoid Valve

When testing solenoid valves, perform the following:

Step 1   Inspect for loose or improperly installed wires.

Step 2   Test supply voltage to the coil with a voltmeter. The voltage should fall within +10 percent and –15 percent of the rated voltage.

Step 3   Place the tip of a steel screwdriver on the solenoid valve plunger when it is energized. The magnetic force of the solenoid should attract the screwdriver, indicating the magnetic attraction of the coil.

Step 4   Inspect the coil for burning, which can result from too high an ambient temperature, too much moisture, or too much humidity. If this has caused the coil to fail, provide proper ventilation or drip guards when installing the next coil.

Step 5   Check the coil resistance while it is disconnected from the circuit.

4.6.0 Check Valves

If a unit has two expansion valves or two sets of capillary tubes, there will be two check valves. During the heating cycle, one check valve is used to bypass the indoor metering device; during the cooling cycle, the other check valve is used to bypass the outdoor metering device.

If a unit has one expansion valve and two distributors, it will have four check valves. These help to route the refrigerant to the proper distributor.

When you suspect a check valve is defective, test the heat pump equipment in both the heating and cooling cycles. If the equipment maintains normal head and suction pressure during either the heating or cooling cycle, but pressures are abnormal in the remaining cycle, a check valve is probably faulty. If you are testing check valves during cold weather, cover the outdoor coil surface area with paper or plastic while conducting cooling cycle tests. If you block the outdoor coil until the head pressure is around 250 psig, cooling cycle conditions can be simulated.

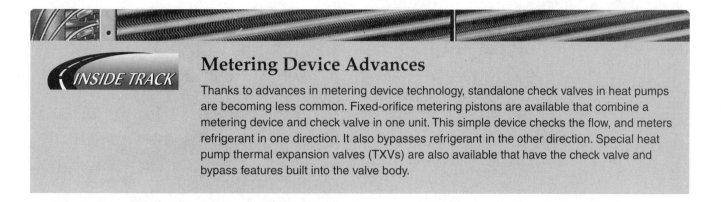

Metering Device Advances

Thanks to advances in metering device technology, standalone check valves in heat pumps are becoming less common. Fixed-orifice metering pistons are available that combine a metering device and check valve in one unit. This simple device checks the flow, and meters refrigerant in one direction. It also bypasses refrigerant in the other direction. Special heat pump thermal expansion valves (TXVs) are also available that have the check valve and bypass features built into the valve body.

4.6.1 Check Valve Problems

Figure 18 illustrates typical check valve construction. When a check valve becomes faulty, the following problems may be encountered:

- If the outdoor check valve sticks closed during a cooling cycle, there will be low suction pressure, high discharge pressure, and high superheat at the indoor coil. The compressor will be hot and may trip the internal overload protector. However, if the indoor unit has a thermostatic expansion valve, a defective thermal element on this valve can lead to similar symptoms. If the outdoor check valve is stuck closed during the cooling mode, the equipment will still operate at normal pressures in the heating cycle.

- If the indoor check valve sticks closed during a heating cycle, there will be low suction pressure, high discharge pressure, and high superheat at the outdoor coil. Again, the compressor will be hot and may trip the internal overload protector. If the outdoor unit has a thermostatic expansion valve, a defective thermal element will lead to similar symptoms. If the indoor check valve is stuck closed during the heating cycle, the equipment will still operate at normal pressures in the cooling cycle.

- If the indoor check valve sticks open during the cooling cycle, there will be high suction pressure and liquid flooding back to the compressor. During the heating cycle, however, the equipment should be operating at normal pressures.

- If the outdoor check valve sticks open during a heating cycle, there will be high suction pressure, low superheat, and liquid flooding back to the compressor. The compressor will be cool or cold. During the cooling cycle, equipment should operate at normal pressures.

- If the indoor check valve leaks during the cooling cycle, there will be a slightly high suction

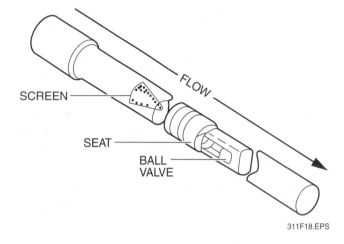

Figure 18 ◆ Check valve construction.

pressure and a slightly low head pressure. Depending upon the rate of the leak, there could be cool or normal temperatures in the compressor. In the heating cycle, however, the equipment would operate at normal pressures.

- If the outdoor check valve is leaking during the heating cycle, there will be slightly high suction pressure and slightly low head pressure. The compressor will be near normal temperatures or slightly cool and the equipment should operate at normal pressures during the cooling cycle.

One additional test should be run before replacing a suspected stuck open/closed check valve (*Figure 19*):

NOTE

This test assumes that the valve is a steel ball valve.

Step 1   Shut down the equipment, allowing head and suction pressures to equalize.

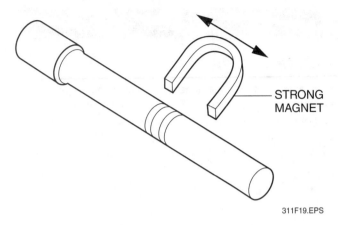

Figure 19 ◆ Magnet test.

Once the pressures are equal, slide a strong magnet back and forth along the check valve body.

Step 2 If the ball valve is stuck, you will not hear anything. If the ball valve is free, you will detect a clear clicking sound as the magnet moves.

Step 3 If you hear the clicking, the check valve is not stuck and does not need to be replaced.

4.6.2 Replacing Check Valves

If you determine the check valve must be replaced, follow these steps:

Step 1 Properly recover or isolate the refrigerant.

Step 2 Remove the faulty check valve.

Step 3 Avoid bending or deforming the valve body when installing the replacement. Use care when handling the valve.

Step 4 Avoid overheating the check valve or applying excessive brazing material when brazing.

NOTE

When installing a replacement check or reversing valve, wrap the valve body with a wet rag to keep the valve from overheating during brazing.

4.7.0 Reversing Valves

A reversing valve (*Figure 20*) has a four-way slide mechanism that directs the refrigerant in one of four ways, depending on the system needs. Here is one way to test a reversing valve:

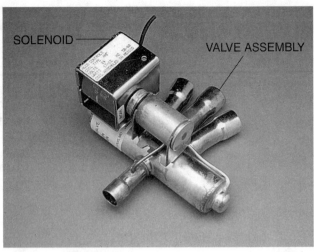

Figure 20 ◆ Reversing valve.

Step 1 Inspect the valve and solenoid coil for dents, deep scratches, or cracks. These may indicate failure.

Step 2 Check the electrical system to be sure the solenoid has power and can be energized. You should hear the plunger click.

Step 3 If the coil cannot be energized, check other components in the electrical system for the source of the problem. Check the coil resistance with the wiring disconnected.

Step 4 Check the system for the proper amount of charge. If the charge is too low, the reversing valve will be affected.

Step 5 Check the system pressure differential. For a valve to operate there must be a minimum operating pressure differential. While some valves will shift when the differential is as low as 10 psig, others may require up to 100 psig. To be sure the valve is defective, test at the higher pressure before condemning it. A low charge will prevent the valve from shifting position.

The reversing valve may also be tested by measuring temperatures. *Figure 21* shows the temperature measurement points for testing a reversing valve while it is in the heating position.

NOTE

To avoid error, temperature readings should be made at least five inches from the valve body. The temperature probe should be insulated.

Newer Refrigerants and Higher Pressures

Newer refrigerants operate at much higher pressures than older refrigerants (such as HCFC-22). If a heat pump containing a new refrigerant (such as HCFC-410A) requires a replacement component (such as a reversing valve), make sure you install an exact replacement as specified by the pump manufacturer.

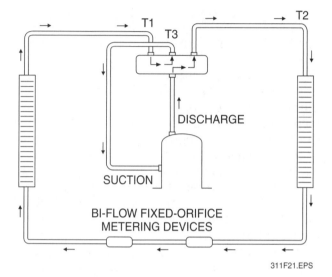

311F21.EPS

Figure 21 ◆ Reversing valve test.

Step 1   Set the thermostat to turn on cooling.

Step 2   Attach the thermometers at T1, T2, and T3.

Step 3   If the temperature difference between thermometers T2 and T3 is greater than 3°F, the valve is leaking. Replace the reversing valve.

Step 4   Set the thermostat to turn on heating.

Step 5   If the difference between T1 and T3 is greater than 3°F, the valve is leaking and requires replacement.

CAUTION

While reversing valves are rugged, they can still suffer from mishandling. A dropped valve may easily be ruined. Because it is sometimes awkward to install, you may be tempted to tap it a few times with a hammer to ensure a tight fit. Avoid this temptation; you may tap the copper tubing, but never the valve body.

Also, you may need to twist the valve gently to straighten it, but avoid using the valve as a lever to straighten itself. It is easy to bend the solenoid stem or warp the valve body slightly, which will render it useless. When installing a valve, the soldering temperature must not exceed 250°F. Protect the valve body from overheating. A nitrogen purge will keep the valve cool and prevent the formation of copper oxide. A wet rag or heat paste must be used on the valve in conjunction with the purge.

THINK ABOUT IT

More Troubleshooting Exercises

Refer to the heat pump wiring diagram in *Figure 12* to solve the following problems. A list of HVAC abbreviations is provided in *Figure 8*.

1. Is the reversing valve (switchover valve) energized or de-energized in the heating mode? What is the energizing path?
2. If the outdoor fan (ODF) continues to run during defrost, what component should be checked? What symptoms would you be likely to see if the fan continues running during defrost?

4.8.0 Defrost Control

Most of the heat pump defrost controls you encounter will be solid-state packaged controls that will be replaced if they are determined to be defective. These controls usually rely on a combination of compressor run time and outdoor coil temperature. However, there are some that use only outdoor coil temperature to initiate and terminate defrost.

Most solid-state defrost controls have a self-test feature that allows the technician to initiate defrost without waiting for time or temperature demand. Initiation procedures vary from manufacturer to manufacturer.

If defrost activates, but does not readily melt the frost buildup, the problem could be that the outdoor fan is not being turned off during defrost. If that is the case, check the switch or relay contacts that control the outdoor fan. An incorrect refrigerant charge can also affect defrost performance.

If the unit is blowing cold air into the conditioned space during defrost, it means the electric heater or burner flame is not turning on when it should. If the heater works when staged on by the thermostat, it means the problem is in the defrost relay or the defrost control board, depending on the design. If the heater does not work in either situation, check the heating element and its associated thermal protection or the burner control circuit.

If an electromechanical type of defrost control is used, you can wait for the cycle to initiate on its own (assuming the conditions are right) or manipulate the unit into starting defrost. If an elapsed-time control is used, you can simply reset the timer. If the control requires temperature as well as run time to initiate defrost, you may have to get around the control by placing the sensor in ice water.

Because switching the unit into defrost is like initiating a call for cooling, one way to eliminate the reversing valve as a possible source of trouble in defrost is to see if the system operates normally in cooling. Heat pump defrost control operating cycles are covered in detail in the HVAC Level Two module, *Heat Pumps*.

4.9.0 Refrigerant Charge

The tolerance on the refrigerant charge for a heat pump with fixed-orifice metering is very tight. On some systems, it must be within ½ ounce of the manufacturer's specification for the heat pump to operate properly. Systems that use fixed metering devices are critically charged because the charge is so specific. For that reason, you should use a short connection, rather than connecting a standard manifold gauge hose to the pressure gauge, when checking the charge. Using hoses equipped with check valves will reduce the loss of refrigerant experienced when disconnecting from the service valve.

If the charge is low, it indicates a refrigerant leak. The leak must be located and repaired before refrigerant is added. Check the product literature. Some manufacturers recommend that the unit be evacuated to a deep vacuum before adding the new charge using the charging charts provided. Other manufacturers will provide procedures for adding a partial charge. At least one major manufacturer requires that the charge be weighed during heating operation.

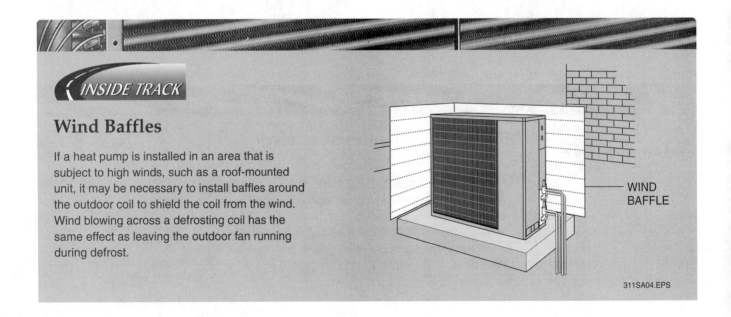

INSIDE TRACK

Wind Baffles

If a heat pump is installed in an area that is subject to high winds, such as a roof-mounted unit, it may be necessary to install baffles around the outdoor coil to shield the coil from the wind. Wind blowing across a defrosting coil has the same effect as leaving the outdoor fan running during defrost.

WIND
BAFFLE

311SA04.EPS

Defrost Options

Many time-temperature defrost controls allow the installer to select the time between defrosts based on the local climate. In areas of freezing fog or drizzle, for example, a shorter interval would be selected because frost would build up quickly under these conditions. In drier climates, a longer interval would be more suitable, because frost would accumulate more slowly.

The time-temperature defrost method is used in many heat pumps, but other methods may be used. For example, some heat pumps use a pressure switch that checks the pressure drop across a coil. As frost builds on the coil and less air flows through it, the pressure drop increases until it reaches a preset point. At that point, the defrost cycle is triggered.

Most electronic defrost controls provide a feature that allows the technician to initiate the defrost cycle in order to avoid delays while troubleshooting. In this case, a coin is used to jumper across the defrost speed-up terminals on the defrost control board.

311SA05.EPS

Checking Superheat

Check the superheat on the suction line between the accumulator and the compressor. An acceptable alternative is to check the superheat entering the accumulator in the cooling mode. Checking superheat entering the accumulator in the heating mode may result in low superheat values, because liquid is returned in the heating and defrost modes.

If the heat pump has a suction line accumulator, which most do, it may contain refrigerant. You can usually tell because the accumulator will have frost or sweat on it if it contains refrigerant. Give the system time to boil out the refrigerant before checking the charge.

CAUTION

Newer HFC refrigerants, such as HFC-410A, have to be charged into a system as a liquid to prevent fractionation. If the system is operating, liquid can be added into the suction side by using a metering orifice in the refrigerant charging hose. This ensures that the liquid being added into the suction line goes into the system as a gas and not as a liquid. Adding liquid directly into the suction side of an operating system could damage or ruin the compressor.

1. Heat pumps produce heat by _____.
   a. circulating indoor air through an electric heating element
   b. circulating low-pressure, low-temperature refrigerant through the indoor coil
   c. extracting heat from the outdoor air and transferring it to the indoor air
   d. circulating hot, high-pressure refrigerant through the outdoor coil

2. In addition to the reversing valve, the other major difference between a heat pump and a cooling-only system is that _____.
   a. the heat pump requires a head pressure control
   b. the cooling unit requires a filter-drier
   c. heat pumps use only fixed-orifice metering devices
   d. heat pumps require two metering devices

3. If a heat pump works normally in the heating mode but the defrost mode will not cycle on, the _____ is most likely defective.
   a. compressor
   b. thermostat
   c. reversing valve
   d. compressor contactor

4. In the cooling mode, the reversing valve is normally _____.
   a. energized
   b. de-energized
   c. bypassed
   d. back-seated

5. When a heat pump is operating in the reverse cycle heating mode, the refrigerant bypasses the _____.
   a. indoor coil metering device
   b. outdoor coil metering device
   c. indoor coil
   d. outdoor coil

6. In the reverse cycle heating mode, _____.
   a. the indoor coil acts as an evaporator
   b. the indoor coil acts as a condenser
   d. the outdoor coil acts as a condenser
   c. defrost is only needed in very cold climates

7. Each of these is a condition commonly found in the defrost mode, *except* _____.
   a. the outdoor fan shuts off
   b. the indoor fan runs
   c. an electric heater is energized
   d. the compressor shuts off

8. In the diagram in *Figure 7*, what voltage should you read across the coil of the indoor blower relay when the unit is in the emergency heat mode?
   a. 0 volts
   b. 24 volts
   c. 120 volts
   d. A small voltage representing the voltage drop across the coil

9. Terminal R of a heat pump thermostat is normally connected to the _____.
   a. secondary of the control transformer
   b. primary of the control transformer
   c. reversing valve
   d. compressor control circuit

10. Connecting a jumper between R and Y on a heat pump thermostat will _____.
    a. turn on the indoor fan
    b. turn on the compressor
    c. energize the reversing valve
    d. turn on the supplementary heaters

11. Which of the following is true of a pilot-operated valve?
    a. It is non-magnetic.
    b. The solenoid opens the main valve port.
    c. The solenoid operates a plunger, which in turn opens the valve port.
    d. It must be manually operated.

12. If the outdoor check valve sticks closed during the cooling cycle, there will be _____.
    a. high suction pressure and high superheat at the outdoor coil
    b. low suction pressure and low superheat at the outdoor coil
    c. low suction pressure and high superheat at the indoor coil
    d. abnormal temperatures and pressures in the heating cycle

13. The function of a reversing valve will be affected by all of the following, *except* _____.

    a. a defective solenoid
    b. low system charge
    c. a scratch or crack in the valve
    d. a plunger that clicks when the solenoid is energized

14. If a heat pump will not switch into defrost, and you suspect the reversing valve, the quickest way to verify your suspicion is to _____.

    a. see if the reversing valve switches when there is a call for cooling
    b. see if the reversing valve switches when there is a call for heating
    c. bypass the reversing valve and see if the unit switches into defrost
    d. wait for the next defrost cycle

15. If a heat pump periodically blows cold air into the conditioned space during the heating season, a likely cause is _____.

    a. low refrigerant charge
    b. the auxiliary electric heater or furnace burner is not turning on during defrost
    c. the outdoor fan is not turning off during defrost
    d. the outdoor thermostat is defective

Summary

The control circuits for a heat pump are more complex than those of a cooling-only system because the heat pump performs more functions: cooling, heating, and defrost. If you consider that many heat pumps also have supplementary electric heaters and emergency heat functions, it makes the control circuits seem even more complex. The added complexity is often an advantage to a troubleshooter, however, because the ability to switch between functions makes it easy to eliminate large portions of the control circuits when isolating a problem.

Technicians troubleshooting heat pumps should have a good understanding of the techniques and tools used in troubleshooting electrical devices and cooling equipment. For effective troubleshooting of a heat pump, it is important to learn the electrical operating sequence. Mechanical troubleshooting is similar to that of a cooling system. Temperature and pressure readings will be different in the heating mode than they are in cooling, however. A heat pump is also much less able to tolerate a variation in refrigerant charge, so a refrigerant leak or a slight charging error can have a significant performance impact on the heat pump.

Notes

Trade Terms Introduced in This Module

Direct-acting valve: A valve in which the solenoid operates the valve port.

Field wiring: Wiring that is installed during installation of the system; for example, such as a thermostat hookup to unit.

Pilot-operated valve: A valve in which the solenoid operates a plunger, which in turn creates or releases pressure that operates the valve.

Transformer phasing: Making sure that parallel transformers are operating in phase.

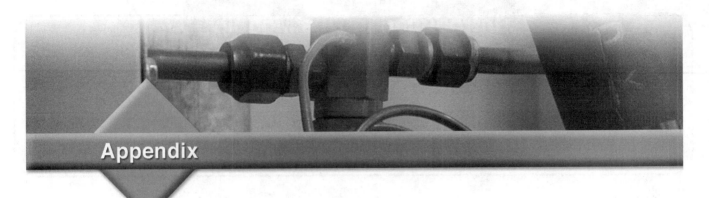

Troubleshooting Chart

GENERALLY THE CAUSE	OCCASIONALLY THE CAUSE	RARELY THE CAUSE
Make these checks first.	Make these checks only if the first checks failed to locate the trouble.	Make these checks only if other checks failed to locate the trouble.

PROBLEM: Compressor and outdoor fan motor do not start

Power failure Blown fuse Faulty wiring – Motor to line side of contactor – Load side, contactor to motor terminal – Control circuit Loose terminals – Motor to line side of contactor – Load side, contactor to motor terminal – Control circuit Compressor overload Contactor coil Pressure control – Internal overload in compressor Refrigerant charge low High head pressure Low suction pressure Run capacitor Start capacitor – Compressor motor – Compressor stuck	Low voltage – Motor to line side of contactor – Load side, contactor to motor terminal – Control circuit – Defective contacts in contactor Control circuit thermostat	Unbalanced power supply Outdoor fan motor Indoor fan motor Refrigerant overcharge Noncondensibles Defective power element – Outdoor expansion valve – Indoor expansion valve Thermostat setting

PROBLEM: Outdoor fan motor will not start

Faulty wiring – Load side, contactor to motor terminal Loose terminals – Load side, contactor to motor terminal Defrost relay Outdoor fan motor Outdoor fan motor capacitor	Defective control, timer, or relay	

PROBLEM: Compressor hums but will not start

Faulty wiring – Motor to line side of contactor – Load side, contactor to motor terminal Loose terminals – Motor to line side of contactor – Load side, contactor to motor terminal Compressor overload Potential relay Start capacitor	Unbalanced power supply Low voltage – Load side, contactor to motor terminal Defective contacts in contactor Compressor motor Compressor bearing defective Compressor stuck	Refrigerant overcharge High head pressure Noncondensibles Restrictions

311A01.EPS

GENERALLY THE CAUSE	OCCASIONALLY THE CAUSE	RARELY THE CAUSE
Make these checks first.	Make these checks only if the first checks failed to locate the trouble.	Make these checks only if other checks failed to locate the trouble.

PROBLEM: Compressor cycles on overload

Faulty wiring – Motor to line side of contactor – Load side, contactor to motor terminal Loose terminals – Motor to line side of contactor – Load side, contactor to motor terminal Low voltage – Motor to line side of contactor – Load side, contactor to motor terminal Unbalanced power supply Potential relay Run capacitor Start capacitor	Defective contacts in contactor Compressor overload Compressor motor Refrigerant charge low Refrigerant overcharge High head pressure High suction pressure Noncondensibles Excessive load cooling Defective reversing valve	Compressor valve defective Compressor oil level Fins dirty or plugged – Indoor section of system – Outdoor section of system Fan belt slipping Coil air short circuiting Air volume low (for cooling) Blower belt slipping Air volume low Air filters dirty

PROBLEM: Compressor off or cycling on low-pressure control

Refrigerant charge low Low suction pressure Defective power element Push rod packing leak – Outdoor expansion valve Indoor fan motor Outdoor fan motor Air volume low Air filters dirty Small or restricted ductwork Restrictions	Indoor fan relay Indoor fan motor Sensing element loose or poorly located Cycle too long (clock timer) Defective power element Fins dirty or plugged Fan belt slipping Low-temperature coil air (for cooling)	Pressure control Out of adjustment (de-ice control) Dirty internally Push rod packing leak – Indoor expansion valve Coil air short circuiting Air stratification in conditioned space

PROBLEM: Compressor off on high-pressure control

Defrost relay Outdoor fan motor Indoor fan relay Indoor fan motor Refrigerant overcharge Cycle too long (clock timer) Coil Defective control timer or relay Fins dirty or plugged Fan belt slipping Coil air short circuiting Blower belt slipping Air volume low Air filters dirty	Out of adjustment (de-ice control) Defective reversing valve Air volume low (for cooling) Fins dirty or plugged Small or restricted ductwork	Pressure control Temperatures Noncondensibles Air stratification in conditioned space Auxiliary heat upstream from indoor fan

311A02.EPS

GENERALLY THE CAUSE	OCCASIONALLY THE CAUSE	RARELY THE CAUSE
Make these checks first.	Make these checks only if the first checks failed to locate the trouble.	Make these checks only if other checks failed to locate the trouble.
PROBLEM: Compressor noisy		
Hold down bolts Compressor valves defective Compressor oil level Refrigerant overcharge	Run capacitor Leaking or defective check valve – Indoor section of system – Outdoor section of system Superheat setting TXV stuck open Loose thermal bulb – Indoor section of system	Compressor bearings defective
PROBLEM: Compressor loses oil		
Refrigerant leak Low suction pressure Refrigeration piping incorrect	TXV stuck open Restrictions	Superheat setting Refrigeration lines too long
PROBLEM: Unit runs in heat cycle—pumps down in cooling or defrost cycle		
Defective reversing valve Defective power element Restrictions	Check valve sticking closed Dirty internally	Push rod packing leak – Indoor section of system
PROBLEM: Unit runs normal in one cycle—high suction pressure in other cycle		
Leaking reversing valve Leaking or defective check valve – Outdoor section of system Loose thermal bulb – Outdoor section of system	Superheat setting Leaking or defective check valve – Indoor section of system Loose thermal bulb – Indoor section of system	TXV stuck open – Outdoor section of system – Indoor section of system
PROBLEM: Head pressure too high		
Refrigerant overcharge Temperatures Noncondensibles Fins dirty or plugged Fan belt slipping Air volume too low (for cooling) – Outdoor coil Blower belt slipping Air volume too low – Indoor coil Air filters dirty	Excessive load cooling Coil air short circuiting Fins dirty or plugged Refrigeration piping incorrect	Compressor oil level Small or restricted ductwork Thermostat setting
PROBLEM: Head pressure too low		
Refrigerant charge low Low suction pressure Leaking reversing valve Low-temperature coil air (for cooling)	Compressor valves defective	

311A03.EPS

GENERALLY THE CAUSE	OCCASIONALLY THE CAUSE	RARELY THE CAUSE
Make these checks first.	Make these checks only if the first checks failed to locate the trouble.	Make these checks only if other checks failed to locate the trouble.
PROBLEM: Suction pressure too high		
Compressor valves defective High head pressure Excessive load cooling Reversing valve leaking	Refrigerant overcharge Temperatures Leaking or defective check valve – Indoor section of system – Outdoor section of system Loose thermal bulb – Outdoor section of system TXV stuck open – Indoor section of system	TXV stuck open – Outdoor section of system Superheat setting Loose thermal bulb – Indoor section of system
PROBLEM: Suction pressure too low		
Refrigerant charge low Fan belt slipping Blower belt slipping Air volume low Air filters dirty Indoor fan motor Indoor fan relay	Temperatures Leaking or defective check valve – Outdoor section of system – Indoor section of system Superheat setting Dirty internally Defective power element Fins dirty or plugged Coil air short circuiting Small or restricted ductwork Check valve sticking closed Restrictions	Low head pressure Push rod packing leak – Outdoor section of system – Indoor section of system Air stratification in conditioned space Refrigeration piping incorrect
Power failure Blown fuse Run capacitor Faulty wiring – Motor to line side of contactor – Load side of contactor to motor – Control circuit Indoor fan motor	Indoor fan relay	Low voltage – Motor to line side of contactor – Load side, contactor to motor terminal – Control circuit Thermostat
PROBLEM: Indoor coil frosting or icing		
Low suction pressure Blower belt slipping Air volume low Air filters dirty – Indoor section of system Indoor fan motor Indoor relay Small or restricted ductwork	Refrigerant charge low Temperatures Reversing valve sticking closed Dirty internally	

311A04.EPS

GENERALLY THE CAUSE	OCCASIONALLY THE CAUSE	RARELY THE CAUSE
Make these checks first.	Make these checks only if the first checks failed to locate the trouble.	Make these checks only if other checks failed to locate the trouble.

PROBLEM: Compressor runs continuously—no cooling

Compressor valve defective Refrigerant charge low Reversing valve leaking Reversing valve defective Fins dirty or plugged (outdoor) Blower belt slipping Air volume low Air filters dirty Small or restricted ductwork – Indoor section of system	High suction pressure Noncondensibles Superheat setting Dirty internally Push rod packing leak – Indoor section of system Restrictions	Fins dirty or plugged (indoor) Coil air short circuiting Air volume low (for cooling) Low-temperature coil air Defective power element Check valve sticking closed – Indoor section of system Check valve leaking or defective Air stratification in conditioned space

PROBLEM: Compressor runs continuously—cooling

Faulty wiring (control circuit) Thermostat (control circuit)	Thermostat location	Thermostat setting Air duct not insulated

PROBLEM: Liquid refrigerant flooding back to compressor in cooling cycle—TXV system

Superheat setting – Indoor section of system Loose thermal bulb – Indoor section of system	Refrigerant overcharge Temperatures TXV stuck open – Indoor section of system Check valve leaking or defective – Indoor section of system	Fins dirty or plugged Blower belt slipping Air volume low Air filters dirty Small or restricted ductwork Air stratification in conditioned space

PROBLEM: Liquid refrigerant flooding back to compressor in cooling cycle—cap tube system

Refrigerant overcharge Indoor fan motor Indoor fan relay Blower belt slipping Air volume low Air filters dirty	High head pressure Fins dirty or plugged Small or restricted ductwork Check valve leaking or defective	Temperatures Air stratification in conditioned space

PROBLEM: Compressor runs continuously—no heating

Refrigerant charge low Reversing valve leaking Contactor stuck	Defective control, timer, or relay	Superheat setting Dirty internally Push rod packing leak – Outdoor section of system Air stratification in conditioned space

PROBLEM: Compressor runs continuously—heating

Faulty wiring (control circuit) Thermostat Thermostat location Contactor stuck	Thermostat setting	Air duct not insulated

311A05.EPS

GENERALLY THE CAUSE Make these checks first.	OCCASIONALLY THE CAUSE Make these checks only if the first checks failed to locate the trouble.	RARELY THE CAUSE Make these checks only if other checks failed to locate the trouble.
PROBLEM: Unit runs in cooling cycle—pumps down in heating cycle		
Defective reversing valve Defective power element	Dirty internally Fins dirty or plugged Fan belt slipping Coil air short circuiting Check valve sticking closed Restrictions	Push rod packing leak – Outdoor section of system
PROBLEM: Unit cycles on low-pressure control when switching cycle or at end of defrost cycle		
Defective low pressure control Refrigerant charge low Defective reversing valve Defective power element Push rod packing leak – Outdoor section of system – Indoor section of system		
PROBLEM: Defrost cycle initiates—no ice on coil		
Refrigerant charge low De-ice control out of adjustment Sensing element loose or poorly located Defective control, timer, or relay	Faulty wiring (control circuit) Fins dirty or plugged Fan belt slipping Coil air short circuiting Air volume low (for cooling)	Cycle too long (clock timer) Defective reversing valve Superheat setting
PROBLEM: Defrost cycle initiates but won't terminate defrost relay		
Defrost relay Refrigerant charge low De-ice control out of adjustment Defective control, timer, or relay	Faulty wiring (control circuit) Compressor valves defective Reversing valve defective	
PROBLEM: Reversing valve will not shift		
Faulty wiring (control circuit) Loose terminal (control circuit) Defective reversing valve Refrigerant charge	Low voltage (control circuit) Defrost relay Compressor stuck	
PROBLEM: Ice buildup on lower part of outdoor coil		
Defrost relay Refrigerant charge low De-ice control out of adjustment Sensing element loose or poorly located	Compressor valves defective Cycle too long (clock timer) Reversing valve leaking Reversing valve defective Superheat setting – Outdoor section of system	Loose thermal bulb – Outdoor section of system
PROBLEM: Liquid refrigerant flooding back to compressor in heating cycle—TXV system		
Superheat setting – Outdoor section of system Loose thermal bulb – Outdoor section of system	Leaking or defective check valve TXV stuck open – Outdoor section of system	Refrigerant overcharge Temperatures

311A06.EPS

GENERALLY THE CAUSE Make these checks first.	OCCASIONALLY THE CAUSE Make these checks only if the first checks failed to locate the trouble.	RARELY THE CAUSE Make these checks only if other checks failed to locate the trouble.
PROBLEM: Liquid refrigerant flooding back to compressor in heating cycle—cap tube system		
Refrigerant overcharge High head pressure	Leaking or defective check valve	Temperatures
PROBLEM: Auxiliary heat on—indoor blower off		
Faulty wiring – Load side, contactor to motor terminal – Control circuit Loose terminals – Load side, contactor to motor terminal – Control circuit Fan relay Indoor fan motor	Thermostat	
PROBLEM: Excessive operating costs		
Compressor stuck Refrigerant charge low De-ice control out of adjustment Reversing valve leaking Reversing valve defective Blower belt slipping Air volume low Air filters dirty Air duct not insulated No outdoor thermostat Outdoor thermostat adjustment	Refrigerant overcharge Fins dirty or plugged Fan belt slipping Coil air short circuiting Thermostat location Air stratification in conditioned space Refrigeration piping incorrect Unit incorrect size for application	Faulty wiring – Motor to line side of contactor Loose terminals – Motor to line side of contactor Noncondensibles Superheat setting TXV stuck open Dirty internally Defective power element Thermostat setting
PROBLEM: Unit short cycles on defrost control		
Refrigerant charge low De-ice control out of adjustment Defective control, timer, or relay Fan belt slipping	Defective power element Fins dirty or plugged	
PROBLEM: Outdoor blower does not stop in defrost cycle		
Fan relay IOD or ID section Defrost relay		

Additional Resources and References

Additional Resources

This module is intended to be a thorough resource for task training. The following reference work is suggested for further study. This is optional material for continued education rather than for task training.

Troubleshooting Heat Pumps (Residential-Light Commercial), 1999. Syracuse, NY: Carrier Corporation.

Figure Credits

Carrier Corporation, 311F10, 311F12, 311F16, 311F17, 311SA05

Courtesy of Honeywell International Inc., 311F13

Emerson Climate Technologies, 311SA03, 311F20, 311SA04

NCCER CURRICULA — USER UPDATE

NCCER makes every effort to keep its textbooks up-to-date and free of technical errors. We appreciate your help in this process. If you find an error, a typographical mistake, or an inaccuracy in NCCER's curricula, please fill out this form (or a photocopy), or complete the online form at **www.nccer.org/olf**. Be sure to include the exact module ID number, page number, a detailed description, and your recommended correction. Your input will be brought to the attention of the Authoring Team. Thank you for your assistance.

Instructors – If you have an idea for improving this textbook, or have found that additional materials were necessary to teach this module effectively, please let us know so that we may present your suggestions to the Authoring Team.

NCCER Product Development and Revision

13614 Progress Blvd., Alachua, FL 32615

Email: curriculum@nccer.org
Online: www.nccer.org/olf

❏ Trainee Guide   ❏ AIG   ❏ Exam   ❏ PowerPoints   Other _____

Craft / Level: _____ Copyright Date: _____

Module ID Number / Title: _____

Section Number(s): _____

Description: _____

_____

_____

_____

_____

Recommended Correction: _____

_____

_____

_____

_____

Your Name: _____

Address: _____

_____

Email: _____ Phone: _____

03312-08

Troubleshooting Accessories

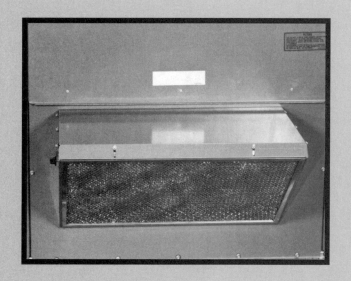

03312-08
Troubleshooting Accessories

Topics to be presented in this module include:

1.0.0	Introduction	.12.2
2.0.0	Troubleshooting Approach	.12.2
3.0.0	Humidifiers	.12.3
4.0.0	Electronic Air Cleaners	.12.8
5.0.0	Ultraviolet Lamps	.12.12
6.0.0	Economizers, Zone Control, and Heat Recovery Ventilators	.12.13

Overview

Residential furnaces, especially in cold, dry northern climates, are often equipped with humidifiers to maintain a healthy moisture level in the indoor air. Electronic air cleaners may also be installed on residential heating and cooling equipment. The majority of specialized accessories are found on commercial systems, however. These accessories include economizers, energy/ heat recovery ventilators, and zone dampers. Servicing of equipment containing these devices requires specialized knowledge. This is especially true of economizers, because there are several types, each with its own special operating characteristics.

Objectives

When you have completed this module, you will be able to do the following:

1. Describe a systematic approach for troubleshooting HVAC system accessories.
2. Isolate problems with electrical and/or mechanical functions of HVAC system accessories.
3. Use equipment manufacturer's troubleshooting aids to troubleshoot HVAC system accessories.
4. Identify and properly use the service instruments needed to troubleshoot HVAC system accessories.
5. Troubleshoot problems in selected HVAC system accessories.
6. State the safety precautions associated with the troubleshooting of HVAC accessories.

Trade Terms

Dump zone
Free cooling

Mechanical cooling
Water hammer

Required Trainee Materials

1. Pencil and paper
2. Appropriate personal protective equipment

Prerequisites

Before you begin this module, it is recommended that you successfully complete *Core Curriculum*; *HVAC Level One*; *HVAC Level Two*; and *HVAC Level Three*, Modules 03301-08 through 03311-08.

This course map shows all of the modules in the third level of the *HVAC* curriculum. The suggested training order begins at the bottom and proceeds up. Skill levels increase as you advance on the course map. The local Training Program Sponsor may adjust the training order.

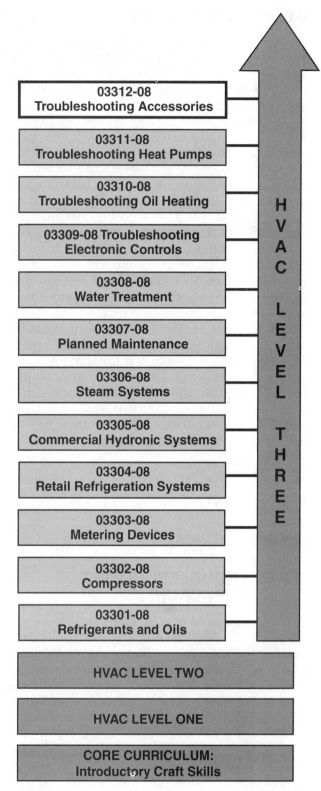

03312-08
Troubleshooting Accessories

03311-08
Troubleshooting Heat Pumps

03310-08
Troubleshooting Oil Heating

03309-08 Troubleshooting
Electronic Controls

03308-08
Water Treatment

03307-08
Planned Maintenance

03306-08
Steam Systems

03305-08
Commercial Hydronic Systems

03304-08
Retail Refrigeration Systems

03303-08
Metering Devices

03302-08
Compressors

03301-08
Refrigerants and Oils

HVAC LEVEL TWO

HVAC LEVEL ONE

CORE CURRICULUM:
Introductory Craft Skills

HVAC LEVEL THREE

312CMAP.EPS

1.0.0 ◆ INTRODUCTION

Troubleshooting the accessories and optional equipment used with HVAC systems is done in the same way as troubleshooting basic heating, cooling, or air distribution equipment. It involves close examination of the equipment to detect telltale signs of what may be wrong. There are standard procedures to follow that make troubleshooting accessories easier than the trial-and-error method. By using the manufacturer's wiring diagrams, specifications, and troubleshooting aids, the cause of almost any malfunction can be quickly isolated. The HVAC technician must be able to listen to the owner's complaint, observe the system in operation, and then identify the most probable cause of the malfunction. Accurate diagnosis and repair of HVAC accessories also requires proper selection and safe use of test equipment and instruments.

There are many kinds of accessories that can be used with HVAC equipment. In addition, accessories used with residential equipment are often different than those used with commercial equipment. This module will cover troubleshooting of the common accessories and optional equipment introduced in the HVAC Level Two module, *Air Quality Equipment*. They include the following:

- Humidifiers
- Electronic air cleaners
- Heat recovery ventilators
- Economizers
- Zoned control systems

2.0.0 ◆ TROUBLESHOOTING APPROACH

For the purpose of this module, problems are divided into two categories: electrical and mechanical. This does not mean that all malfunctions fit easily into one of these two categories. Obviously, there are problems where one affects the other. Whether troubleshooting an electrical or mechanical problem, the use of a logical, structured approach to troubleshooting is essential. You must know how to diagnose specific control devices and components used with accessories. A good set of service tools and the ability to read electrical wiring diagrams is a must; otherwise, you are only guessing at the cause of a problem. It is also recommended that you review the material previously studied in the HVAC Level Two module, *Introduction to Control Circuit Troubleshooting*. It describes the elements necessary to perform a systematic approach to troubleshooting. It also discusses the different kinds of troubleshooting aids provided by equipment manufacturers, both in their equipment and in their service literature.

2.1.0 Electrical Troubleshooting

The electrical troubleshooting information covered in this module is limited to those electrical devices that are specific to certain accessories. Troubleshooting most of the electrical problems and components in HVAC equipment is performed in the same way regardless of the type of equipment being serviced. Troubleshooting also requires the use of various pieces of test equipment (*Figure 1*). Because of this, fault isolating and troubleshooting data common to all HVAC equipment, including accessories, is described in *Introduction to Control Circuit Troubleshooting*. For quick reference, the troubleshooting topics covered that are relevant to accessories are listed here:

- Isolating to a faulty circuit via the process of elimination
- Isolating to a faulty component
- Single-phase and three-phase input voltage measurements
- Fuse and circuit breaker checks
- Resistive and inductive load checks
- Switch and relay/contactor checks
- Control transformer checks

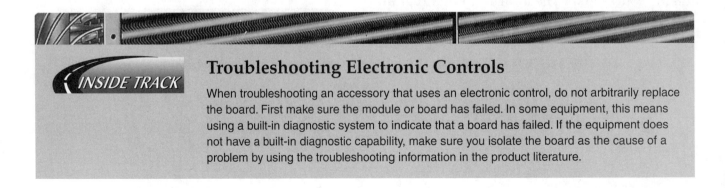

INSIDE TRACK

Troubleshooting Electronic Controls

When troubleshooting an accessory that uses an electronic control, do not arbitrarily replace the board. First make sure the module or board has failed. In some equipment, this means using a built-in diagnostic system to indicate that a board has failed. If the equipment does not have a built-in diagnostic capability, make sure you isolate the board as the cause of a problem by using the troubleshooting information in the product literature.

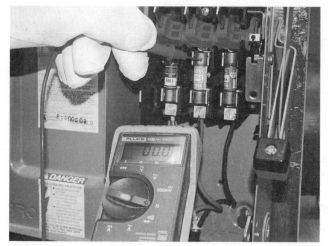

Figure 1 ◆ Technician troubleshooting with a multimeter.

General procedures used to troubleshoot equipment containing electronic control modules and boards are covered in the HVAC Level Three module, *Troubleshooting Electronic Controls*.

2.2.0 Troubleshooting Aids

The first thing to do when troubleshooting an accessory is consult the product literature. This literature usually describes the mechanical and electrical operation of the equipment, including the sequence of operation. Voltage and/or signal inputs and outputs related to the operation of the accessory are also defined.

Service literature almost always includes troubleshooting aids such as wiring diagrams and troubleshooting charts. These aids will help you isolate an equipment problem to the specific failed component.

3.0.0 ◆ HUMIDIFIERS

A humidifier is a device used to add humidity. All humidifiers introduce water vapor into a building's conditioned air at a predetermined rate; therefore, they must be connected to a water supply. Some humidifiers also require a drain connection.

Humidifiers must be carefully installed and mounted as directed in the manufacturer's installation instructions. Some are motor-driven and controlled by a humidistat.

3.1.0 General Humidifier Troubleshooting

Most humidifier problems can be prevented with periodic maintenance. Evaporative humidifiers remove all minerals from the water that is evaporated. These minerals are left on the media and water reservoir. The float, float-valve orifice, and valve seat for most humidifiers need periodic cleaning to remove mineral deposits. If so equipped, water solenoid valve orifices and strainers should also be cleaned. In addition, algae, bacteria, and virus growth can cause problems. Algaecide can be used to help neutralize algae growth.

Most manufacturers recommend that the water reservoir be drained and the humidifier components cleaned about every two months. In hard water areas, more frequent cleaning may be necessary. *Figures 2, 3,* and *4* show typical control circuits for three different humidifiers.

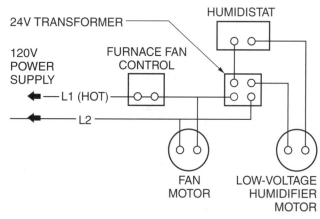

Figure 2 ◆ Control circuit schematic.

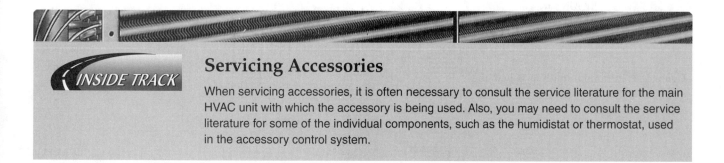

Servicing Accessories

When servicing accessories, it is often necessary to consult the service literature for the main HVAC unit with which the accessory is being used. Also, you may need to consult the service literature for some of the individual components, such as the humidistat or thermostat, used in the accessory control system.

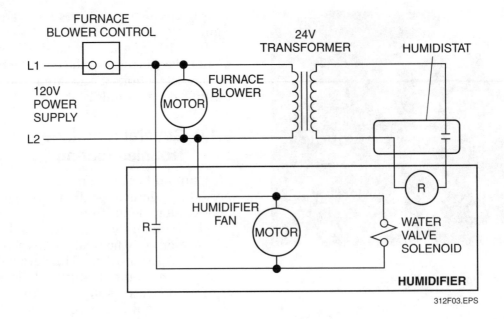

Figure 3 ◆ Schematic of a typical fan-powered humidifier circuit.

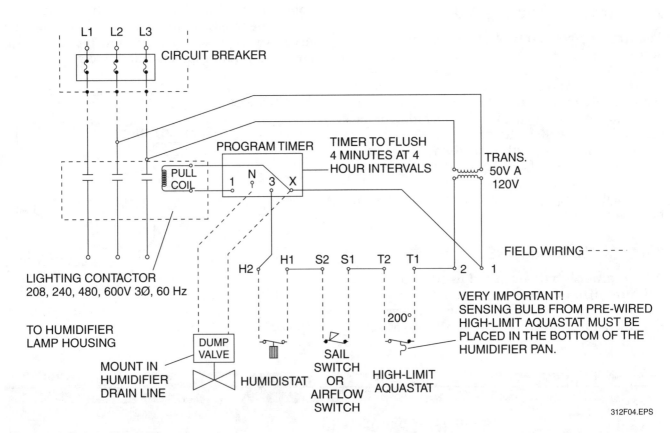

Figure 4 ◆ Infrared humidifier schematic.

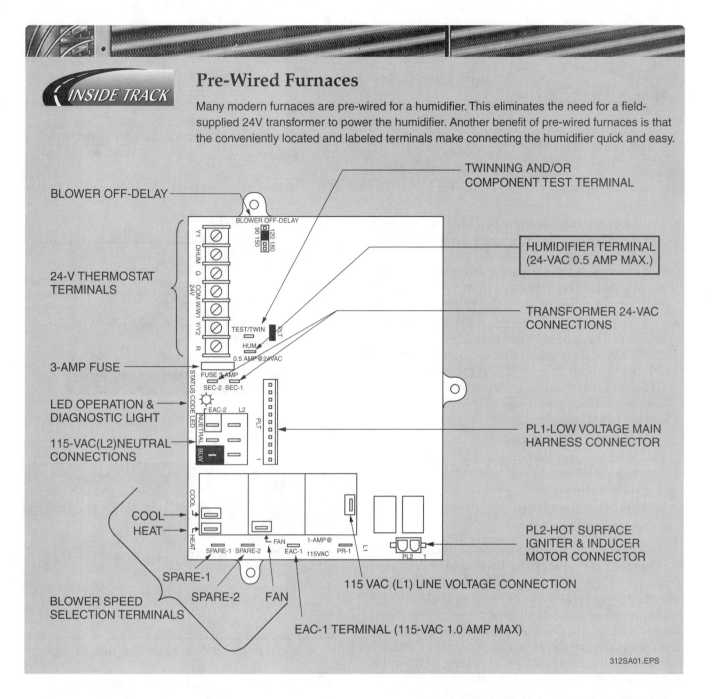

Pre-Wired Furnaces

Many modern furnaces are pre-wired for a humidifier. This eliminates the need for a field-supplied 24V transformer to power the humidifier. Another benefit of pre-wired furnaces is that the conveniently located and labeled terminals make connecting the humidifier quick and easy.

INSIDE TRACK

TWINNING AND/OR COMPONENT TEST TERMINAL

BLOWER OFF-DELAY

HUMIDIFIER TERMINAL (24-VAC 0.5 AMP MAX.)

24-V THERMOSTAT TERMINALS

TRANSFORMER 24-VAC CONNECTIONS

3-AMP FUSE

LED OPERATION & DIAGNOSTIC LIGHT

PL1-LOW VOLTAGE MAIN HARNESS CONNECTOR

115-VAC(L2)NEUTRAL CONNECTIONS

COOL
HEAT

PL2-HOT SURFACE IGNITER & INDUCER MOTOR CONNECTOR

SPARE-1

115 VAC (L1) LINE VOLTAGE CONNECTION

BLOWER SPEED SELECTION TERMINALS

SPARE-2   FAN

EAC-1 TERMINAL (115-VAC 1.0 AMP MAX)

312SA01.EPS

Common problems encountered with humidifiers are as follows:

- *Humidifier not running* – This is usually caused by an electrical control circuit problem. As applicable, check the overload protection, humidistat, and control circuit components. Also check to see if the humidifier motor has failed. Electrical failures such as a tripped overload or failed motor can result from the rotating media being bound because of a mineral buildup or mechanical failure.

- *Water overflow* – Check for a dirty, misadjusted, or faulty float valve. If equipped with a solenoid valve, check for dirt or scale lodged be-

tween the valve plunger and seat. The valve may need cleaning, adjusting, or replacement. Also check for water backing up the drain line as a result of the line being plugged or restricted.

NOTE

Water overflowing from a humidifier can corrode metal parts, including furnace heat exchangers, leading to premature failure of the part. Water can also damage electrical and electronic components.

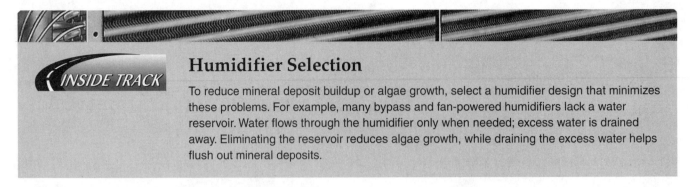

Humidifier Selection

To reduce mineral deposit buildup or algae growth, select a humidifier design that minimizes these problems. For example, many bypass and fan-powered humidifiers lack a water reservoir. Water flows through the humidifier only when needed; excess water is drained away. Eliminating the reservoir reduces algae growth, while draining the excess water helps flush out mineral deposits.

- *Noise* – If the unit is equipped with a solenoid valve and noise occurs each time the valve opens and closes, check to make sure the unit is snug against the plenum. If **water hammer** occurs, it may be necessary to place a length of high-pressure hose before the valve.
- *Low or high levels of humidity* – The recommended level of relative humidity (RH) that should be maintained in a building during the winter is about 30 percent. Symptoms of high RH are condensation on windows and inside exterior walls. Symptoms of low RH are:
  - Dry, itchy skin
  - Static electricity shocks
  - Clothing static cling
  - Sinus problems
  - Chilly feeling
  - Sickly pets and plants
  - Loose furniture joints

The first thing to check when the humidity is too high or too low is the humidistat setpoint (*Figure 5*). *Figure 6* lists typical recommended levels of indoor relative humidity that should be maintained at various outdoor winter temperatures. If the humidistat setpoint is correct, check the calibration of the humidistat. This is done by comparing the humidistat dial setting to the relative humidity level in the conditioned space as measured with a sling psychrometer. First, turn the humidistat dial to its lowest setting, then slowly increase the setting until a light click is heard. The relative humidity setting indicated by the dial should be close to the relative humidity measured with the sling psychrometer. If the humidistat is out of calibration, adjust it, if possible. Excessive humidity may be caused by continuous operation of the humidifier as a result of a failed humidistat or a defective or sticking contact on the humidifier relay.

Modern room thermostats can contain a built-in humidistat that measures and controls humidity electronically. The desired humidity level is entered into the control by pressing a button and

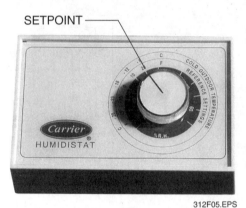

SETPOINT

Figure 5 ◆ Humidistat.

312F05.EPS

AT OUTDOOR TEMPERATURE (°F)	RECOMMENDED (SAFE) INDOOR RH (%)
–20	15
–10	20
0	25
10	30
20	35
30	40

BASED ON AN INDOOR TEMPERATURE OF 72°F

312F06.EPS

Figure 6 ◆ Recommended indoor relative humidity in the winter.

observing the desired humidity level on a digital display. The control will then operate the humidifier to maintain the desired humidity level during heating operation in much the same way as done with a conventional humidistat. If an outdoor temperature sensor is used with the combination thermostat/humidistat, the control automatically adjusts the indoor humidity level based on outdoor temperature. This type of humidistat also is used to control humidity (dehumidify) in the cooling mode. Normal air conditioning operation removes humidity from the air. However, there

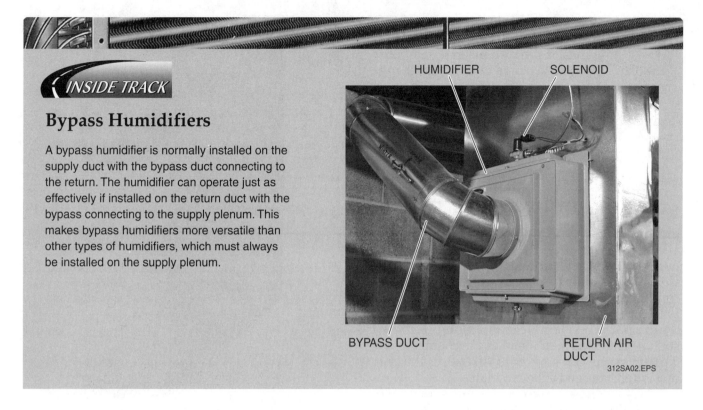

INSIDE TRACK

Bypass Humidifiers

A bypass humidifier is normally installed on the supply duct with the bypass duct connecting to the return. The humidifier can operate just as effectively if installed on the return duct with the bypass connecting to the supply plenum. This makes bypass humidifiers more versatile than other types of humidifiers, which must always be installed on the supply plenum.

HUMIDIFIER   SOLENOID

BYPASS DUCT   RETURN AIR DUCT

312SA02.EPS

may be times when humidity is high, but the temperature in the space is not high enough to initiate a call for cooling. In these cases, the humidistat calls for compressor operation. Typically this arrangement is used with a variable-speed blower and an electronic control in the air handler. When a call for dehumidification is present without a call for cooling, the blower operates at a much lower speed to maximize moisture removal. The software in the humidistat and the air handler electronic control allow dehumidification without overcooling the space. To enable this feature, the room thermostat and/or the air handler have to be properly configured according to the manufacturer's instructions.

Some reasons for low humidity not directly related to the humidifier are as follows:

- The water supply feed valve/saddle supply valve is closed or not fully open.
- Excessive air infiltration from permanently opened windows, exhaust fans, or open fireplace dampers can cause low humidity.
- An oversized furnace can cause low humidity because burner cycles are too short for the humidifier to do its job. With this condition, it may be necessary to set the blower for continuous operation so that the humidifier can run over a longer period of time.

Low or high humidity levels can also be caused by using a humidifier with the wrong

capacity. Humidifiers are typically rated in gallons of water per day. Humidifier selection depends on the volume of the building or area in square feet ($ft.^2$). It also depends on the building's air tightness. *Figure 7* shows a typical graph used for the selection of residential humidifiers. Similar graphs and/or charts are available for commercial and industrial humidifiers.

The terms loose, average, and tight are defined as follows:

- *Loose* – The building has little insulation, no vapor barriers, and no storm doors or windows. In homes, it can also mean an undampered fireplace. The air exchange rate is about 1.5 changes per hour.

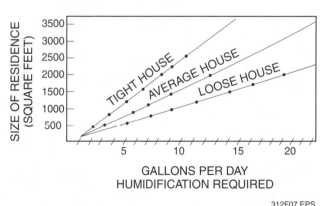

312F07.EPS

Figure 7 ◆ Humidifier capacity chart.

- *Average* – The building is insulated, has vapor barriers, and has loose storm doors or windows. In homes, it can also mean a dampered fireplace. The air exchange is about 1.0 change per hour.
- *Tight* – The building is well-insulated, has vapor barriers, and tight storm doors or windows. In homes, it can also mean a dampered fireplace. The air exchange is about 0.5 change per hour.

4.0.0 ◆ ELECTRONIC AIR CLEANERS

Electronic air cleaners (EACs) (*Figure 8*) remove airborne particles and odors from the air circulated to the conditioned space. They can be stand-alone units or may be mounted in the air conditioning system. EACs create an electrostatic field that produces a positive or negative charge on the particles being carried along in the air stream. The EAC then removes the charged particles from the air stream by drawing them to oppositely charged collector plates.

EACs have a high-voltage, solid-state power supply. The high voltage produced by the power supply, which can range from 6,000 to 10,000 volts DC, is used first to electrically charge (ionize) all

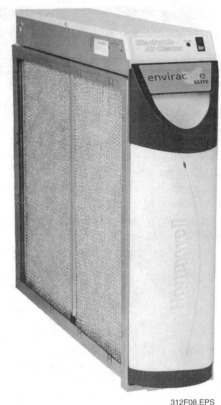

312F08.EPS

Figure 8 ◆ Electronic air cleaner (EAC).

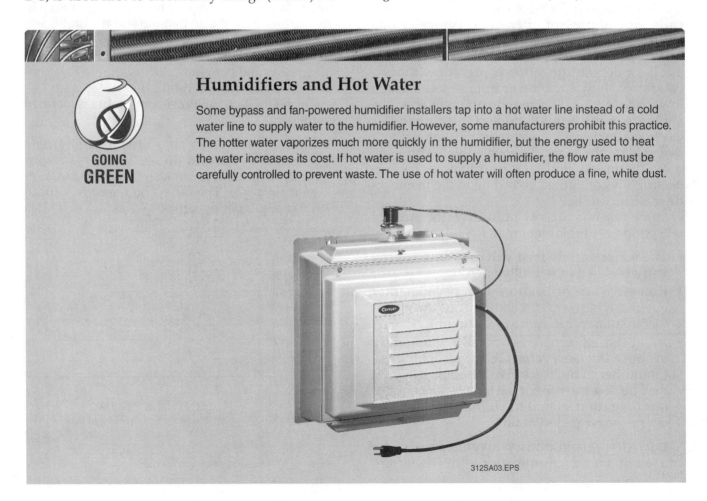

GOING GREEN

Humidifiers and Hot Water

Some bypass and fan-powered humidifier installers tap into a hot water line instead of a cold water line to supply water to the humidifier. However, some manufacturers prohibit this practice. The hotter water vaporizes much more quickly in the humidifier, but the energy used to heat the water increases its cost. If hot water is used to supply a humidifier, the flow rate must be carefully controlled to prevent waste. The use of hot water will often produce a fine, white dust.

312SA03.EPS

particles in the air that pass through the filter. After the particles are ionized, the high voltage is then used to attract, collect, and hold the particles in the filter. *Figure 9* shows a typical EAC.

As shown, the filter portion of an EAC consists of a prefilter, ionizer section, collector section, and charcoal filter section. As the air enters the EAC, larger particles are trapped by the prefilter section. Smaller particles pass through the prefilter to the ionizing section.

This section consists of a fine tungsten wire grid (ionizer grid) which is connected to the high-voltage DC power supply. An electrostatic ionizing field created by the high voltage on the wire grid gives particles a positive or negative charge as they pass through the grid. These charged particles are then drawn into the collector section.

The collector section consists of a series of equally spaced, parallel collector plates connected to the high-voltage DC power supply so that the even- and odd-numbered plates are at a positive and negative DC voltage, respectively. As the ionized particles flow between the plates, they are attracted and held on oppositely charged collector plates. The air, cleaned of pollutants,

then passes through the charcoal filter section where odors are absorbed. From there, it passes on to the conditioned space. The pollutants remain held in the collector section until they are removed when the filter is cleaned.

Figure 10 shows the schematic diagram for a typical EAC. In the unit shown, the high-voltage power supply is made from discrete components such as diodes, capacitors, and resistors. In newer modules, the power supply may be a solid-state circuit that is fully enclosed in a non-repairable, sealed assembly.

The EAC is normally electrically connected to the furnace fan circuit and operates only when the furnace blower motor is energized. On some units, this control is through contacts of a sail switch installed in the air stream. Others may use contacts of the fan control relay.

Some EACs are electrically interlocked so that the unit shuts off automatically when the service doors are opened to gain access to the unit.

EACs do not clean air effectively when dirty. It is recommended that the electronic cells and protective screens be cleaned every two to three months.

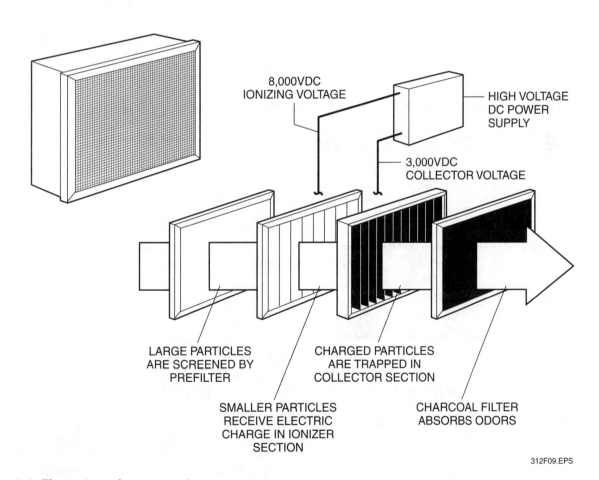

Figure 9 ◆ Electronic air cleaner operation.

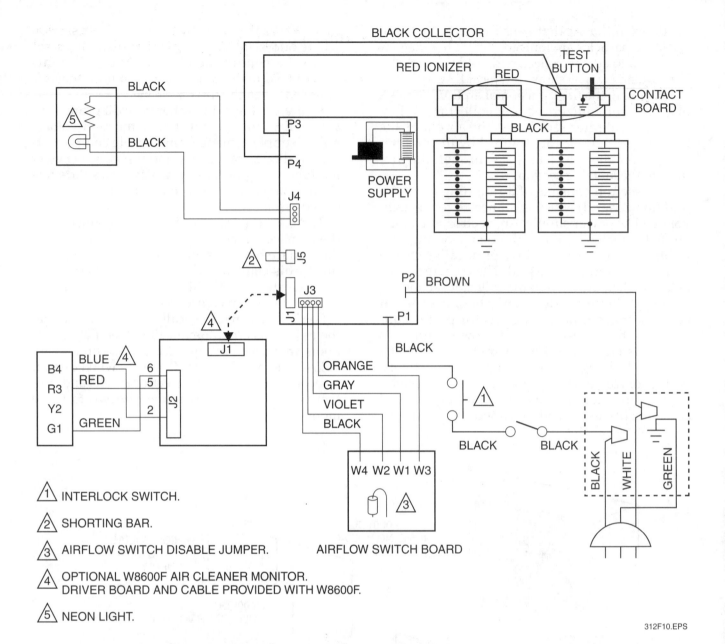

Figure 10 ◆ Electronic air cleaner schematic.

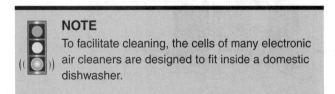

NOTE

To facilitate cleaning, the cells of many electronic air cleaners are designed to fit inside a domestic dishwasher.

Some manufacturers have designed their equipment so that the power indicator flashes when the EAC needs cleaning. If cleaned regularly, many EAC problems can be avoided. However, some conditions can exist that cause an EAC to become dirty and overloaded in an abnormally short time. These conditions are:

- New carpeting has been installed.
- Untreated concrete floors exist.
- Dusty construction work is being performed in the area.

EAC operation is sensitive to airflow. When operated at airflow rates above those recommended by the manufacturer, they can become quite inefficient. When the airflow is reduced below the manufacturer's recommended minimum, enough ozone can be generated to cause an annoying odor. If the EAC is operating and is clean, but is not cleaning the air, or if the smell of ozone is present, check to make sure that the system airflow matches the EAC specifications.

INSIDE TRACK

Connecting Electronic Air Cleaners

An electronic air cleaner must operate any time the furnace or air handler blower operates, regardless of the operating mode or fan speed. Many HVAC equipment manufacturers pre-wire their products with a set of terminals to which an air cleaner may be connected. This eliminates the need for field-supplied air cleaner wiring.

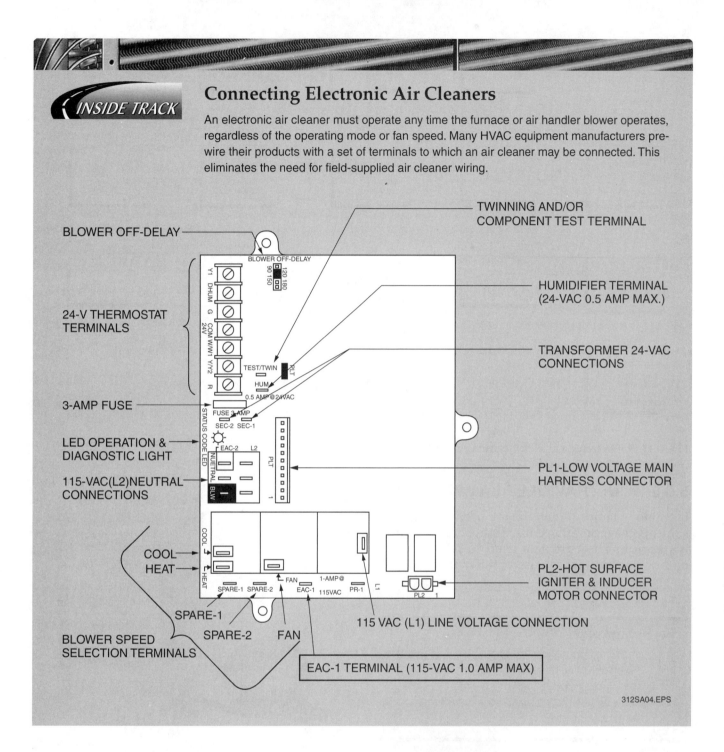

BLOWER OFF-DELAY

TWINNING AND/OR COMPONENT TEST TERMINAL

24-V THERMOSTAT TERMINALS

HUMIDIFIER TERMINAL (24-VAC 0.5 AMP MAX.)

TRANSFORMER 24-VAC CONNECTIONS

3-AMP FUSE

LED OPERATION & DIAGNOSTIC LIGHT

115-VAC(L2)NEUTRAL CONNECTIONS

PL1-LOW VOLTAGE MAIN HARNESS CONNECTOR

COOL
HEAT

PL2-HOT SURFACE IGNITER & INDUCER MOTOR CONNECTOR

SPARE-1

SPARE-2   FAN

BLOWER SPEED SELECTION TERMINALS

115 VAC (L1) LINE VOLTAGE CONNECTION

EAC-1 TERMINAL (115-VAC 1.0 AMP MAX)

312SA04.EPS

The recommended procedure for troubleshooting problems in an EAC is to first perform a visual inspection to identify any mechanical damage that might be the cause of the problem. For electrical problems, use the troubleshooting aids in the manufacturer's service literature to fault-isolate the unit. *Figure 11* shows a typical manufacturer's troubleshooting flow chart.

WARNING!

If you must measure high voltage when troubleshooting an electronic air cleaner, use a voltmeter that is designed and rated for measuring the high voltages encountered in the unit.

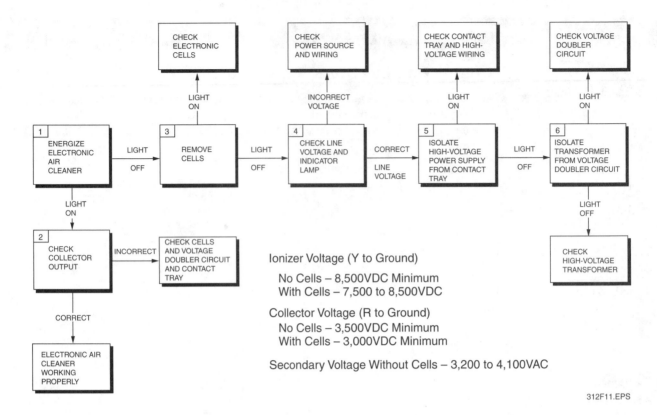

Figure 11 ◆ Electronic air cleaner manufacturer's troubleshooting flow chart.

5.0.0 ◆ ULTRAVIOLET LAMPS

Ultraviolet lamps (*Figure 12*) are used for microbe control on evaporative coils, drain pans, and duct insulation. They are also used to kill airborne pathogens and may be used in conjunction with an electronic air cleaner.

> **WARNING!**
> The light from ultraviolet lamps can damage eyes. When servicing these lamps, follow all the manufacturer's safety warnings and make sure the lamp is switched off and disconnected from the power supply.

There is not much troubleshooting the HVAC technician can do for ultraviolet lamps. If a lamp is not working, check its power supply. Check the lamp itself for physical damage, and make sure there is no debris in the lamp socket.

Ultraviolet lamps must be replaced after a certain number of hours of operation. Consult the manufacturer's instructions for information about the life span of the lamps and at what intervals they should be changed.

Figure 12 ◆ Ultraviolet lamp.

> **CAUTION**
> Ultraviolet lamps contain a small quantity of mercury, a hazardous substance. If a lamp breaks during handling, use caution when cleaning up the debris and dispose of it according to local regulations.

6.0.0 ◆ ECONOMIZERS, ZONE CONTROL, AND HEAT RECOVERY VENTILATORS

Economizers, zone control, and heat recovery ventilators help to provide conditioned air in a way that results in the greatest human comfort and economical system operation. Even though these systems perform somewhat different functions, they use most of the same components. These components typically include thermostats, humidistats, electronic controls, and motor-operated dampers. Because of the similarity in components and in troubleshooting the components, this section will first briefly review the basic operation of each type of system. This is followed by the troubleshooting information related to the common components.

6.1.0 Economizers

An economizer (*Figure 13*) is an accessory used on commercial units that reduces operating cost during cooling by using outdoor air for cooling (**free cooling**) and, whenever possible, reduces compressor (**mechanical cooling**) operation. It does this by controlling the amount of outside air that is brought into a conditioned space. *Figure 14* shows a basic economizer system. It consists of a damper actuator assembly and related economizer control module.

Control signals applied to the economizer control module come from the thermostat located in the conditioned space, an enthalpy (total heat) sensor located in the outdoor air duct, and from the discharge air sensor located on the discharge side of the system evaporator coil. The enthalpy sensor responds to changes in the air dry-bulb temperature and humidity, allowing the use of outdoor air at higher temperatures for free cooling as long as its humidity is low. The setpoint controls the changeover from cooling by compressor operation to cooling with outside air. The discharge air sensor monitors the average air temperature on the face of the system indoor coil and compares it to the predetermined setpoint. It also senses the temperature of the building air, which is a mixture of return air coming from the conditioned space and outdoor air.

312F13.EPS

Figure 13 ◆ Rooftop unit with an economizer installed.

Economizers

In addition to energy savings and reduced long-term maintenance costs, economizers provide the natural advantage of good building ventilation. While in the free-cooling mode, the system is also using anywhere from 50 to 100 percent outside air. This level typically exceeds the code-required ventilation rates for most building applications. Economizers also provide the option of ventilating a building at the needed rate based on occupancy. This is normally achieved through the use of CO_2 sensors mounted in rooms that might have high occupancy rates; conference rooms, assembly halls, restaurants, training rooms, and meeting centers are examples. These rooms are often used on a sporadic scheduling basis and therefore spend much of the time less than fully occupied. By interlocking a CO_2 sensor with the economizer, the ventilation rate can be controlled when the system is either in a design cooling or design heating mode when 100 percent outside air is not desirable. The CO_2 sensor only increases the rate of ventilation to match the human occupancy load, thus maximizing both energy savings and human comfort with superior ventilation. This control sequence is called demand control ventilation (DCV) and is becoming much more of an expected feature in high-occupancy applications.

GOING GREEN

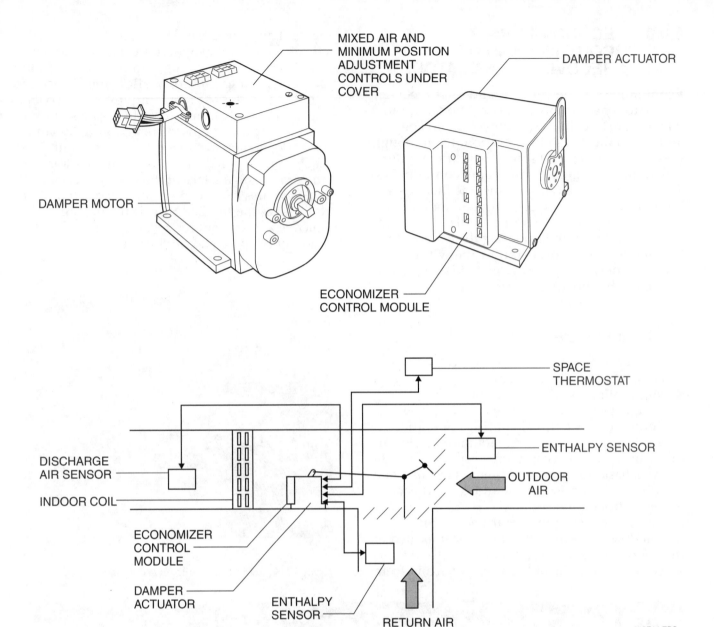

Figure 14 ◆ Basic economizer system.

There are many types of economizers. A simplified schematic for an economizer is shown in *Figure 15*. The operation of this unit, as briefly described here, is similar to that of many economizers found in the field.

On a call for cooling by the space thermostat, the system operates as follows:

When the enthalpy of the outdoor air is below the setpoint, the outdoor air damper is proportioned open and the return air damper is proportioned closed to maintain a temperature selected by the setpoint of the mixed/discharge air sensor. Typically, this is between 50°F and 56°F.

During economizer operation, mechanical cooling is operated by stage 2 cooling on the space thermostat. When the enthalpy of the outdoor air is above the setpoint, the outdoor air damper closes to its minimum position. A call for cooling from the space thermostat turns on mechanical cooling. During the unoccupied period, the damper actuator spring returns the outdoor damper to the fully closed position.

Some economizer systems use a second enthalpy sensor located in the return duct. Use of two sensors is called differential enthalpy. When used, the economizer control selects the air (outside or return) with the lower enthalpy for cooling. If the outdoor air has lower enthalpy than the return air, then the outdoor air damper will be opened to bring in outdoor air for free cooling.

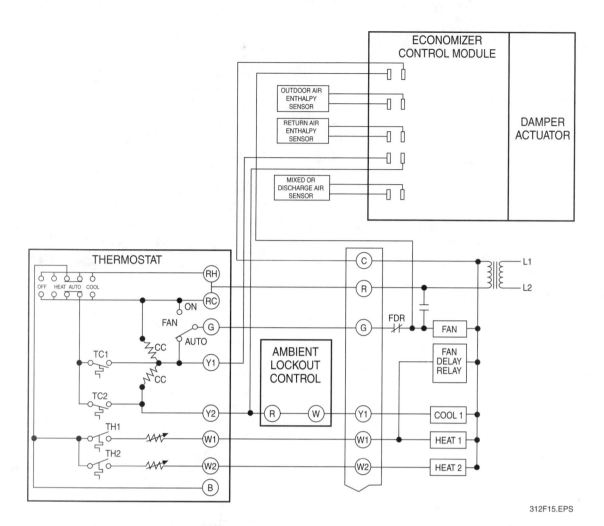

Figure 15 ◆ Simplified schematic of an economizer.

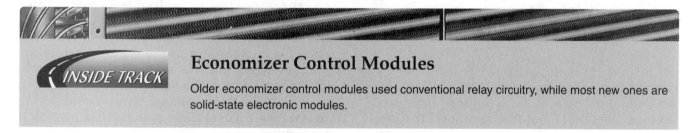

Economizer Control Modules

Older economizer control modules used conventional relay circuitry, while most new ones are solid-state electronic modules.

6.1.1 Economizer Types

When troubleshooting economizers, you will encounter three basic types: dry-bulb economizers, enthalpy economizers, and integrated economizers. The dry-bulb economizer (*Figure 16*) uses an outdoor air temperature sensor (OAT) to activate a control relay whenever the outdoor temperature falls below a user-specified value (typically 55°F). The control relay (CR) locks the compressor out and space cooling is accomplished by modulating the amount of outdoor air entering the return side of the supply air fan. As the outdoor air temperature falls below 55°F,

more return air is mixed with outdoor air to maintain a 55°F supply air temperature.

The enthalpy economizer (*Figure 17*) uses an outdoor air temperature sensor (OAT) wired in series with an enthalpy sensor (hOA) and a control relay (CR). When the outdoor air temperature is below 60°F and the outdoor enthalpy is below 24.7 Btu/lb, the control relay is activated. The control relay prevents compressor operation, and zone cooling is accomplished by modulating the amount of outdoor air entering the return side of the supply air fan.

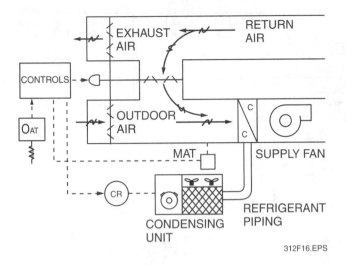

Figure 16 ◆ Dry-bulb economizer.

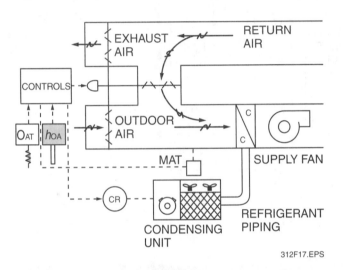

Figure 17 ◆ Enthalpy economizer.

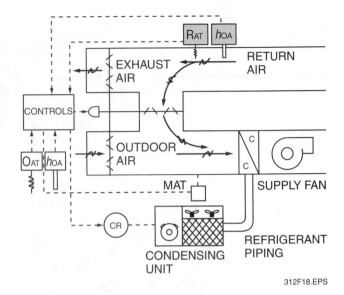

Figure 18 ◆ Integrated economizer.

6.2.0 Zone Control

Zone control provides a number of separately controlled spaces or zones within a building where different heating or cooling temperatures can be controlled at the same time. This provides a way to overcome variations in cooling or heating loads that occur in different areas of the building at different times.

6.2.1 Typical Zoned System

A typical zoned system (*Figure 19*) consists of an electronic control panel that is connected to a thermostat and a motorized damper that controls the air supplied to each zone. The thermostat controls the temperature of the zone while the damper, which is installed in the supply duct to each zone, controls the flow of conditioned air to the zone. The control panel is also connected to the control circuits of the building heating/cooling equipment.

There are many variations in zoned systems, but the principles of operation are basically the same. The operation of the system shown in *Figure 19* is like that of many zoned systems.

Each zone is controlled by its own thermostat which, via the control panel, operates the zone automatic damper, register, or diffuser. When any zone thermostat calls for heating (or cooling), the control panel turns on the heating (or cooling) equipment, and closes the dampers to all the zones that are satisfied. When all the zone thermostats are satisfied, the control panel zone relays are de-energized. This causes all the zone dampers to go to the position (open or closed) as

The integrated economizer (*Figure 18*) uses two sensors to measure enthalpy: a temperature sensor and humidity sensor. An outdoor air temperature sensor (OAT), outdoor air humidity sensor (hOA), return air temperature sensor (RAT), and return air humidity sensor (hOA) are wired to the controls of the central unit. The controller calculates outdoor air and return air enthalpy values. Below a pre-determined outdoor air temperature (55°F in this example), the economizer provides cooling with the compressor off, just like a dry-bulb economizer. For outdoor air temperatures above 55°F, when the outdoor enthalpy is less than the return air enthalpy, 100 percent outdoor air is used to assist compression cooling. When the outdoor air enthalpy is above the return air enthalpy, the outdoor air dampers are set to the minimum ventilation position, and the compressor provides the needed cooling capacity.

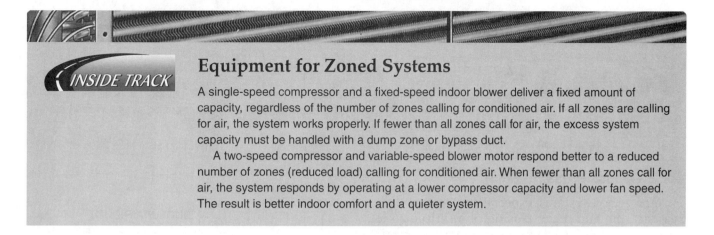

Equipment for Zoned Systems

A single-speed compressor and a fixed-speed indoor blower deliver a fixed amount of capacity, regardless of the number of zones calling for conditioned air. If all zones are calling for air, the system works properly. If fewer than all zones call for air, the excess system capacity must be handled with a dump zone or bypass duct.

A two-speed compressor and variable-speed blower motor respond better to a reduced number of zones (reduced load) calling for conditioned air. When fewer than all zones call for air, the system responds by operating at a lower compressor capacity and lower fan speed. The result is better indoor comfort and a quieter system.

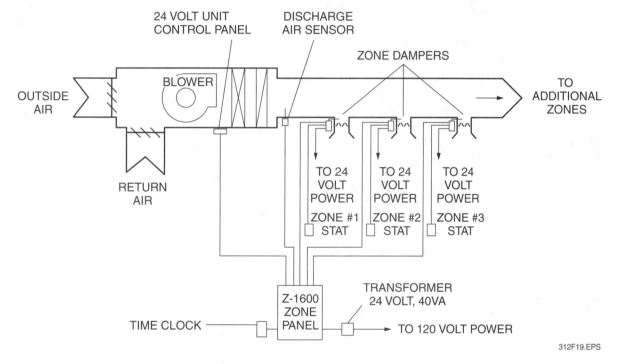

312F19.EPS

Figure 19 ◆ Zoned system schematic.

selected by a switch provided on the control panel for each zone. When the zone switch is set to the open position, fan operation is allowed to continue in order to provide for continuous air circulation in any or all of the zones.

Some zoned system control panels provide for automatic changeover of the system from the heating mode to the cooling mode. Other zoned systems must be manually switched between operating modes. Typically, this is accomplished at the zone 1 (master) thermostat. There, the thermostat is switched to the desired HEAT, OFF, or COOL mode. The operation of the fan, either ON or AUTO, is also controlled from the zone 1 thermostat. The other zone thermostats may not have these control features.

In the system shown in *Figure 19*, a fresh air intake damper is also used to control the intake of fresh outdoor air into the forced ventilation system. The fresh air intake damper shown is controlled by a fresh air control. Typically, this control can be set to drive the fresh air damper to any one of five positions: closed, 25 percent, 50 percent, 75 percent, and open. In small systems, only one damper is normally used to provide fresh air ventilation. On larger homes or commercial installations, one fresh air damper may be installed in the fresh air duct and another to close off the return. The return air damper would be adjusted to operate in reverse of the fresh air damper. As the fresh air damper opens, the return air damper will close, and vice versa.

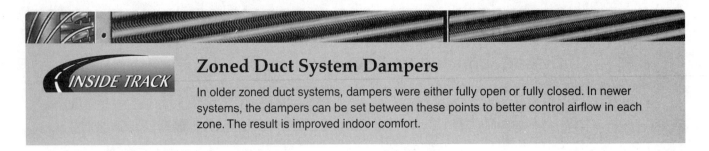

Zoned Duct System Dampers

In older zoned duct systems, dampers were either fully open or fully closed. In newer systems, the dampers can be set between these points to better control airflow in each zone. The result is improved indoor comfort.

6.2.2 Bypass Pressure Relief

In some zoned systems, especially in retrofit duct systems, the automatic closing of multiple zone dampers can create an excessive increase in air pressure and velocity throughout the duct system. Sometimes, this increase can overcome the static pressure rating of the fan motor, causing a reduction in the volume of air through the unit. To avoid such problems, bypass pressure relief may need to be added to the system. Bypass pressure relief can be added by using a control damper installed in a **dump zone** bypass or direct return bypass duct configuration in the system. *Figure 20* shows typical dump zone and direct bypass damper setups.

A dump zone is an area located within a zoned building that does not require precise temperature control, such as a basement, large foyer, or hall. As the zone thermostats in the controlled areas are satisfied and the related zone dampers close, any excess system airflow is routed via the bypass pressure relief damper into the dump zone. The bypass pressure relief damper acts as a pressure regulator to maintain the design pressure in the supply trunk for the operating zones. It also bypasses enough air to maintain full airflow through the heating or cooling equipment. Two types of bypass dampers are in common use: barometric dampers and motorized dampers. The barometric damper is generally used because it does not require electrical connections and control wiring. The barometric damper's balancing mechanism is adjusted to open the damper when the system pressure gets too high. A return grille normally is located in a dump zone area to avoid high pressure buildup in the space, and to prevent over-conditioning of the space. One benefit of using a dump zone bypass is that the heating or cooling capacity supplied by the bypassed air will pick up part of the building load. The disadvantage is that the area where the air is discharged is not controlled, and the temperature may become too hot or too cold.

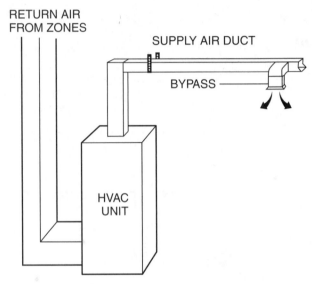

DUMP ZONE BYPASS IN BASEMENT

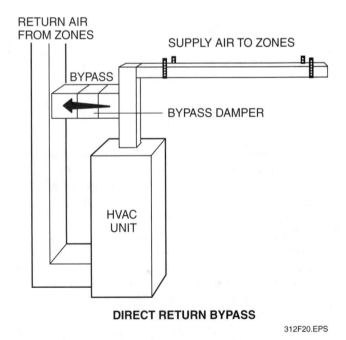

DIRECT RETURN BYPASS

312F20.EPS

Figure 20 ◆ Dump zone and direct return bypass zones.

The direct-return bypass uses a bypass pressure relief damper installed in a bypass duct around the furnace or fan coil. Since the air discharged by the damper is bypassed directly to the return inlet of the furnace or air handler, it has no chance to pick up any part of the building heating or cooling load. This is a disadvantage because when the bypass damper is wide open, the air temperature returning to the equipment will increase during heating and decrease during cooling. Bypassed air that is too hot can cause the furnace to shut off on high limit. Air that is too cold can cause the air conditioning coil to freeze up. To avoid excessive temperature swings in the heating or cooling equipment, a direct-return bypass is usually sized to handle about 25 percent of the total system cfm when the bypass damper is fully open.

Often, additional protection is provided by using an anti-freezeup control to prevent the cooling coil from freezing, while a single-pole bulb thermostat control may be used to protect the heating equipment.

NOTE

In commercial applications, the ceiling plenum is often used as the dump zone.

6.3.0 Recovery Ventilators

Energy recovery ventilators (ERVs) and heat recovery ventilators (HRVs) draw fresh outdoor air into the home. At the same time, they expel stale indoor air to the outside (*Figure 21*). The expelled air takes with it odors and other unwanted contaminants.

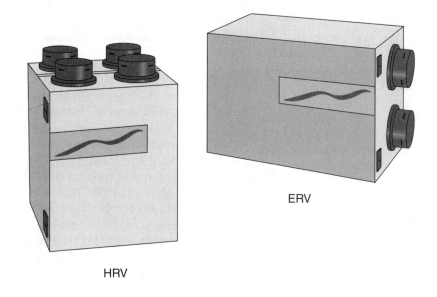

ERV

HRV

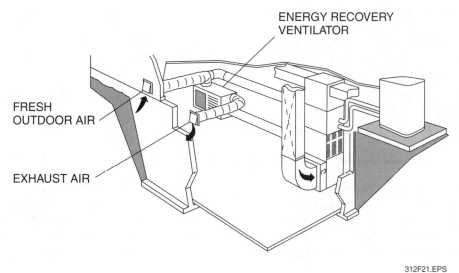

ENERGY RECOVERY VENTILATOR

FRESH OUTDOOR AIR

EXHAUST AIR

312F21.EPS

Figure 21 ◆ Recovery ventilators.

While removing the stale air from the home, recovery ventilators also recover much of the heat or cooling energy from the outgoing indoor air and transfer it to the fresh incoming air.

ERV and HRV units are similar in construction and in operation. The heart of both units is a special energy recovery core. In the ERV, this core transfers both sensible and latent heat between the fresh incoming air and the stale air, allowing it to remove moisture from the air during the summer cooling season. The HRV energy recovery core operates in a similar manner, except that it is designed to transfer only sensible heat between the air streams. ERVs provide more efficient year-round operation than HRVs. For this reason, they are used in most areas of the United States and some parts of Canada. HRVs are used to supply fresh air, quickly remove excessive moisture, and recover heat energy during the heating season. They normally are used in areas with colder climates and extended heating seasons.

An ERV/HRV is normally installed as part of a forced-air heating/cooling system. In order for the ventilator to operate properly, the related furnace or fan coil blower and duct system must be adequate to handle the airflow requirements for the ventilator being used. The ERV/HRV is controlled independently by a simple wall-mounted control. A blower interlock relay incorporated in the furnace/fan coil control circuit ties the operation of the ERV/HRV unit to the furnace or fan coil blower. When the ERV/HRV is in operation, the interlock relay is energized, causing the furnace or fan coil blower to operate. This circulates both fresh air and return air throughout the duct system.

NOTE

To maintain a good transfer of energy and proper airflow, the air system must be properly balanced and the ERV/HRV components clean of dust and dirt. The system must be balanced so intake and exhaust airflows are approximately equal. Typically, the ventilator is set up to bring in slightly more air than it exhausts. Balancing is done with a balancing kit at installation, and any time an airflow problem is suspected.

The ERV/HRV air filters, screens, and recovery core must be cleaned to remove dust and dirt. These tasks should be performed at the intervals recommended by the ventilator manufacturer. Typically, the recovery core of an ERV should be

cleaned about every three months. The core of an HRV should be cleaned about once per year, following the end of the heating season.

NOTE

Some cores are made of cardboard and cannot be washed with water. Follow the manufacturer's instructions.

Troubleshooting an inoperative or malfunctioning ERV/HRV involves isolating the problem to one of its three main parts: the wall control, electronic control board, and fan motors. Normally, the wall control is either good or bad, which can be determined with a VOM. To troubleshoot the electronic control board, the wall control must be attached before the unit will operate properly. Any configuration jumpers located on the board must match the configuration specified in the service literature. The easiest way to check their operation is to use the wall control to initiate low-speed and high-speed operation.

Troubleshooting electrical problems in a ventilator must always be done in accordance with the detailed troubleshooting procedures and diagrams provided in the service literature supplied with the unit.

6.4.0 Troubleshooting Economizers and Zoned Control Systems

Because there is such diversity in the types of economizers and zoned control systems, it is essential that the product service literature be used to troubleshoot problems in these devices. The remainder of this section gives some general guidelines that might prove helpful when troubleshooting thermostats, electronic controls, and dampers associated with these systems.

6.4.1 Troubleshooting Thermostats

Thermostats (*Figure 22*) used with economizers and zoned systems are the same as those used to control basic heating/cooling equipment. They can be of the cooling-only, heating-only, or heating/cooling type. The operation of each of these thermostats is covered in the HVAC Level Two module, *Introduction to Control Circuit Troubleshooting*. Additional information pertaining to troubleshooting non-electronic thermostats is given in the HVAC Level Two module, *Troubleshooting Cooling*.

Electronic programmable thermostats are commonly used in zoned systems. These thermostats use microprocessors and integrated circuits to provide a wide variety of control and energy-saving features. Different thermostats offer different features; the more sophisticated the thermostat, the more features it offers. In some zoned systems, one electronic programmable room thermostat placed in a central location acts as a master. Sensors in the other zones are often simple passive devices (thermistors) that sense temperature only and send that information back to the master thermostat. In some systems, that information is sent wireless. The master thermostat acts as a central point to program all functions of the main equipment and zone system components. It displays information from the room sensors and other information fed to the main control panel from other sensors in the system. The main control panel uses inputs from all sensing devices in all locations to operate the zoned system.

Regardless of the type of thermostat used, the approach to troubleshooting a thermostat for a heating/cooling system problem is basically the same. If a unit is not running at all, one of the first things to check is that the thermostat function switch is set to the desired position for heating/cooling or fan operation, and the thermostat setpoint is above or below room temperature, depending on the operating mode. If set correctly, the next thing to do is check that control power is applied to the thermostat. If not, check for missing control voltage or an open circuit. A visual check should also be made to look for the following conditions that can cause improper thermostat or system operation:

- To prevent chattering, the thermostat should be mounted on an inside wall in a location that is free of vibration. Electronic programmable room thermostats are not as susceptible to vibration as electro-mechanical room thermostats.
- The thermostat should not be subjected to drafts or dead air spots, hot or cold air from ducts or diffusers, or radiant heat from direct sunlight or hidden heat from appliances.
- Make sure the thermostat is level (mercury bulb thermometers).
- If applicable, make sure the mercury bulb or bimetal element is not sticking.

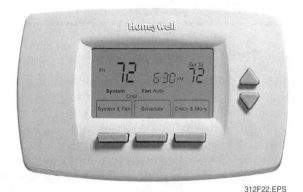

312F22.EPS

Figure 22 ◆ Electronic room thermostat.

CAUTION

The mercury used in older room thermostats is considered a hazardous material. When replacing a mercury bulb room thermostat, dispose of the old thermostat in accordance with local regulations.

INSIDE TRACK

Electronic Programmable Thermostats

Manufacturers of electronic programmable thermostats generally provide troubleshooting aids in their service literature that should be used to troubleshoot a particular thermostat.

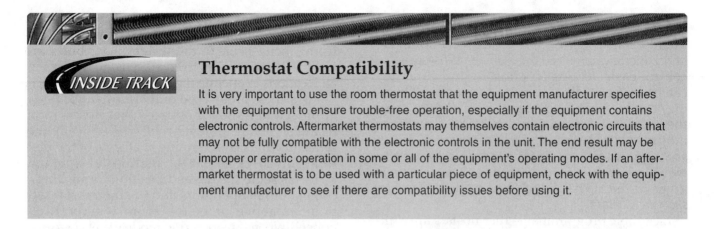

Thermostat Compatibility

It is very important to use the room thermostat that the equipment manufacturer specifies with the equipment to ensure trouble-free operation, especially if the equipment contains electronic controls. Aftermarket thermostats may themselves contain electronic circuits that may not be fully compatible with the electronic controls in the unit. The end result may be improper or erratic operation in some or all of the equipment's operating modes. If an aftermarket thermostat is to be used with a particular piece of equipment, check with the equipment manufacturer to see if there are compatibility issues before using it.

6.4.2 Troubleshooting Electronic Controls

When troubleshooting an accessory that uses an electronic control board/panel, do not arbitrarily replace the board because there is a problem with the equipment. Use an intelligent approach to first make sure the board or panel has failed. Know the sequence of operation for the equipment, then check for the presence of proper input and output voltages and/or signals applied to and from the board. Random testing of the input and output signals to any electronic control, without regard to the sequence of operation, will most likely result in the needless replacement of a good control. Use any available diagnostic features built into the equipment or the troubleshooting information in the manufacturer's service literature.

When troubleshooting electrical problems in a zoned system, the problem can often be isolated to the control board or zoned control circuit by switching zone connections at the control board. For example, assume that all zones are working correctly for cooling when the thermostat calls for cooling, except for zone 1. In zone 1, the temperature has risen above the setpoint of the zone thermostat, but the compressor does not run. The cause of this problem can be in the zone 1 control circuit, or it can be in the zone 1 circuits on the control panel. If the connections for the zone 1 and zone 2 thermostat/damper control circuits are switched at the control board, and the problem still exists in zone 1, then you know that the problem exists in the zone 1 control circuit, and not the control panel. If the problem moves from zone 1 to zone 2 when the zone 2 thermostat calls for cooling, then the zone 1 circuits on the control panel are defective.

CAUTION

Microprocessors and integrated circuits used in electronic controls are sensitive to, and can be damaged by, static electricity. Before handling the board or touching its component parts, always discharge your body's static electricity by grounding yourself to a metal object, such as the equipment chassis, a water or gas line, or a metallic electrical conduit. It is also good practice to never touch the connector pins of the microprocessor and/or integrated circuit. Before troubleshooting an electronic control, refer to the manufacturer's service literature to make yourself aware of any other special precautions about handling the control.

6.4.3 Troubleshooting Dampers

Leaky or inoperative dampers waste energy. When the outside air is too hot or too cold, it is necessary to keep unwanted, untreated air from entering the building. Also, when used with automated systems, it's important to make sure the dampers operate properly when the system controller sends a signal. Leaking or sticking zone dampers can also unbalance the system air distribution. Customer complaints of stuffiness, constant unit operation, or no cooling can all be the result of a problem with the unit's damper(s). To troubleshoot dampers (*Figure 23*), you should make the following checks:

- Check that all voltages and electrical signals used to actuate the damper are within specifications.

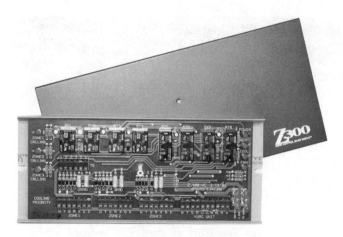

ZONE CONTROL PANEL

RECTANGULAR DAMPER

312F23.EPS

Figure 23 ◆ Zone control panel and damper.

- Observe damper motors and actuators through an operating cycle to check for defects or binding. Make sure that all mounting bolts are securely fastened.
- Check that the linkage from the actuators is adjusted to be sure that the blades of the damper fully open or close within the stroke or travel of the actuator arm.
- Check that all blades close tightly in the closed position. If necessary, adjustments should be made to damper linkage(s) to close any partially open blades.

> **NOTE**
>
> Some dampers have a minimum position setting and will not close completely when the system is calling.

- Make sure that all damaged blades are replaced. Dirt, soot, and lint should be removed, especially around the moving parts.
- Check the blade edge and side seals. Replace where necessary.
- Check the pins, straps, and bushings/bearings for wear, rust, or corrosion and replace any as required.
- Inspect the condition of caulking where used to make damper frames tight to the structure.

1. Humidifiers are controlled by a(n) _____.
   a. check valve
   b. economizer
   c. ventilator
   d. humidistat

2. If a humidifier is not running, the problem is usually caused by a(n) _____.
   a. faulty electrical control circuit
   b. dirty float valve
   c. high RH condition
   d. oversized furnace

3. When a humidifier experiences water overflow, the problem is often caused by a(n) _____.
   a. faulty electrical control circuit
   b. dirty float valve
   c. high RH condition
   d. oversized furnace

4. Symptoms of high RH include _____.
   a. dry, itchy skin
   b. loose furniture joints
   c. static electricity shocks
   d. condensation on windows

5. The first thing to check when the humidity in a conditioned space is too high or too low is _____.
   a. humidistat calibration
   b. humidistat setpoint
   c. furnace size
   d. water supply feed valve

6. A tight building is characterized by the _____.
   a. lack of vapor barriers
   b. presence of an undampered fireplace
   c. lack of insulation
   d. presence of tight storm doors and windows

7. In an electronic air cleaner, the output of the high-voltage power supply is applied to the _____.
   a. ionizer section
   b. collector section
   c. collector and ionizer
   d. prefilter

8. When an electronic air cleaner is operated at an airflow above that recommended by the manufacturer, the filter's cleaning ability is _____.
   a. slightly better
   b. worse
   c. the same
   d. doubled

9. The enthalpy sensor(s) used with an economizer control _____.
   a. room temperature
   b. room humidity
   c. cooling system changeover from free cooling to mechanical cooling
   d. both room temperature and humidity

10. An economizer uses a mixed/discharge air sensor to control the open or closed position of _____.
    a. both the outdoor and return dampers during free cooling
    b. the outdoor damper during free cooling
    c. the outdoor damper during mechanical cooling
    d. both the outdoor and return dampers during both free and mechanical cooling

11. In an economizer that uses differential enthalpy sensing, the economizer control module selects _____ air for cooling.
    a. only outside
    b. only return
    c. either outside or return
    d. mixed/discharge

12. A device that uses a heat exchanger to transfer heat from warmed indoor air to ventilation air entering a building is a(n) _____.
    a. economizer
    b. zone damper control
    c. heat recovery ventilator
    d. mixed air thermostat

13. During the summer months, cooling supplied to the lobby zone of a zoned system operates normally during the morning and evening hours, but seems to run continuously during the afternoon hours. The problem can be caused by _____.
    a. a thermostat subjected to radiant heat from direct sunlight
    b. the zone damper being stuck open
    c. the thermostat setpoint being too high
    d. a thermostat that is subjected to cold air from ducts or diffusers

14. When troubleshooting a zoned system for a problem of no heat in all zones, the problem can be caused by a defective _____.
    a. zone control panel
    b. fresh air control
    c. zone thermostat
    d. zone damper

15. When troubleshooting a problem suspected to be caused by a malfunctioning damper, one of the first things to do is _____.
    a. check the control voltage applied to the compressor contactor
    b. observe the operation of the damper for an operating cycle to check for defects and binding
    c. jumper the contacts on the zone thermostat
    d. replace the damper with a known good one

Summary

Effective troubleshooting of HVAC accessories is a process by which the HVAC technician listens to a customer's complaint, performs an independent analysis of a problem, and then initiates and performs a systematic, step-by-step approach to troubleshooting that results in the correction of the problem. The technician must understand the purpose and principles of operation of each component in the accessory being serviced. Also, the relationship between the operation of the accessory and the main HVAC equipment that it is used with must also be understood. The technician must be able to tell whether a given device is functioning properly and be able to recognize the symptoms arising from the improper operation of any part of the equipment. Based on the symptoms revealed by the analysis of system operation, the technician should use the troubleshooting aids provided by the system manufacturer to identify and repair the cause of the problem.

Notes

Trade Terms Introduced in This Module

Dump zone: An uncontrolled area in a zoned system that is used to avoid low airflow problems by routing the excess airflow into the dump zone to maintain system balance.

Free cooling: A mode of economizer operation. During cooling operation, the system compressor is turned off and the indoor fan is used to bring outside air into the building to provide cooling.

Mechanical cooling: The cooling provided in the conventional manner by the compressor and mechanical refrigeration system.

Water hammer: A banging noise in water pipes that occurs when the water supply solenoid valve on a humidifier abruptly shuts off water to the humidifier.

Additional Resources and References

Additional Resources

This module is intended to be a thorough resource for task training. The following reference works are suggested for further study. These are optional materials for continued education rather than for task training.

Refrigeration and Air Conditioning, An Introduction to HVAC/R, Fourth Edition. Larry Jeffus. Air Conditioning and Refrigeration Institute. Prentice Hall.

Figure Credits

Tim Dean, 312F01

Carrier Corporation, 312SA01, 312SA03, 312SA04

Topaz Publications, Inc., 312F05, 312SA02, 312F13

Courtesy of Honeywell International Inc., 312F08, 312F10, 312F22

American Air & Water, Inc., 312F12

Jackson Systems, 312F19, 312F23

NCCER CURRICULA — USER UPDATE

NCCER makes every effort to keep its textbooks up-to-date and free of technical errors. We appreciate your help in this process. If you find an error, a typographical mistake, or an inaccuracy in NCCER's curricula, please fill out this form (or a photocopy), or complete the online form at **www.nccer.org/olf**. Be sure to include the exact module ID number, page number, a detailed description, and your recommended correction. Your input will be brought to the attention of the Authoring Team. Thank you for your assistance.

Instructors – If you have an idea for improving this textbook, or have found that additional materials were necessary to teach this module effectively, please let us know so that we may present your suggestions to the Authoring Team.

NCCER Product Development and Revision

13614 Progress Blvd., Alachua, FL 32615

Email: curriculum@nccer.org
Online: www.nccer.org/olf

❏ Trainee Guide   ❏ AIG   ❏ Exam   ❏ PowerPoints   Other _____

Craft / Level: _____   Copyright Date: _____

Module ID Number / Title: _____

Section Number(s): _____

Description: _____

_____

_____

_____

_____

Recommended Correction: _____

_____

_____

_____

_____

Your Name: _____

Address: _____

_____

Email: _____   Phone: _____

Glossary of Trade Terms

Alkalinity: A water quality parameter in which the pH is higher than 7. It is also a measure of the water's capacity to neutralize strong acids.

Alkylbenzene: A type of synthetic hydrocarbon oil that is compatible with mineral oil but has better oil return properties. It is widely used in low-temperature applications. It is not compatible with synthetic oils.

Azeotrope: A blended refrigerant that behaves like a pure refrigerant. The refrigerant evaporates and condenses at one given pressure and temperature. From the technician's standpoint, working with an azeotropic blend is like working with a pure refrigerant.

Backwashing: A procedure that reverses the direction of water flow through a multimedia-type filter by forcing the water into the bottom of the filter tank and out the top. Backwashing is performed on a regular basis to prevent accumulated particles from clogging the filter.

Binary blend: A blended refrigerant consisting of two refrigerants.

Black box: A term given to a solid-state, hermetically sealed electronic assembly that is replaced rather than field-repaired.

Bleed-off: A method used to help control corrosion and scaling in a cooling tower. It involves periodically draining and throwaway (wasting) a small amount of the water circulating in a system. This aids in limiting the buildup of impurities caused by the continuous adding of makeup water to a system.

Blowdown: The process of purging a boiler of foreign matter.

Boiler horsepower (Bohp): One Bohp is equal to 9.803 kilowatts. Also, it is the heat necessary to vaporize 34.5 pounds of water per hour at 212°F. This is equal to a heat output of 33,475 Btus per hour or about 140 square feet of steam radiation.

Bubble point: The point at which refrigerant starts evaporating when heat is added to a zeotropic refrigerant. It is also the point that refrigerant finishes condensing when heat is removed from the refrigerant. See *dew point.*

Burnout: A condition in which the breakdown of the motor winding insulation causes the motor to short out or ground electrically.

Cad cell: A light-sensitive device (photocell) in which the resistance reacts to changes in the amount of light. In an oil furnace, it acts as a flame detector.

Capacity control: Methods used in cooling systems to adjust system operation to match changes in the system cooling load.

Capillary tube: A copper tube with a fixed length and fixed diameter, usually with an inside diameter of $\frac{1}{16}$" to $\frac{1}{8}$". Used as a metering device.

Cavitation: The result of air formed due to a drop in pressure in a pumping system.

Chlorofluorocarbons (CFCs): The most damaging refrigerants to the environment since they contain more chlorine atoms in their structure. Production of CFCs in the United States ended in 1995.

Clearance volume: The amount of clearance between a piston at the top dead-center position of travel and the cylinder head.

Colloidal substance: A jelly-like material made up of very small, insoluble, nondiffusible particles larger than molecules, but small enough to remain suspended in a fluid without settling to the bottom.

Combustion efficiency: Producing the most heat with the least amount of energy.

Compliant scroll compressor: A version of the scroll compressor that allows the orbiting scroll to temporarily shift from its normal operating position if liquid refrigerant enters the compressor.

Compound: A substance made up of different elements. A refrigerant that is not a blend is a compound.

Compression: The reduction in volume of a vapor or gas by mechanical means.

Concentration: Indicates the strength or relative amount of an element present in a water solution.

Cycles of concentration: A measurement of the ratio of dissolved solids contained in a heating/cooling system's water to the quantity of dissolved solids contained in the related makeup water supply. The term *cycles* indicates when the concentration of an element in the water system has risen above the concentration contained in the makeup water. For instance, if the hardness in the system water is determined to be two times as great as the hardness in the makeup water, then the system water is said to have two cycles of hardness.

Design load: The maximum load at which a system is designed to operate.

Dew point: The point at which refrigerant stops evaporating when heat is added to a zeotropic refrigerant. It is also the point that refrigerant starts condensing when heat is removed from the refrigerant. See *bubble point.*

Diagnostics: A term used to identify the built-in testing and fault isolation capability of a microprocessor-controlled system.

Dielectric strength: The ability of refrigerant oil or any material to resist breaking down in the presence of voltage.

Direct-acting valve: A valve in which the solenoid operates the valve port.

Direct-expansion (DX) evaporator: An evaporator in which liquid is completely converted to vapor; also known as dry-expansion.

Dissolved solids: The dissolved amounts of substances such as calcium, magnesium, chloride, and sulfate contained in water. Dissolved solids can contribute to the corrosion and scale formation in a system.

Distributor line: Tubing between the distributor and evaporator or the line between the metering device and evaporator.

Distributor: A special fitting, generally machined, containing multiple passageways that distribute liquid refrigerant evenly to the evaporator circuits.

Drip leg: A drain for condensate in a steam line placed at a low point or change of direction in the line and used with a steam trap.

Dry steam: The steam that exists after all the water has been vaporized into steam at its saturation temperature.

Electrolysis: The process of changing the chemical composition of a material (called the electrolyte) by passing electrical current through it.

Electrolyte: A substance in which conduction of electricity is accompanied by chemical action.

Electronic expansion valve (EEV): A microprocessor-controlled expansion valve in which the orifice is controlled by a precision DC motor. The EEV maintains a constant superheat.

Enthalpy: The total heat content of a substance. In HVAC, it is the total heat content of the air and water vapor mixture as measured from an established point.

Fault message: A coded message delivered by a flashing light, display readout, or printout to indicate the existence of a fault and (usually) its probable source.

Field wiring: Wiring that is installed during installation of the system; for example, such as a thermostat hookup to unit.

Firetube boiler: A boiler in which the flue gases are contained inside the tubes, with the heated water on the outside.

Fixed-orifice metering device: A device in which the metering orifice is located in a replaceable piston. The term *fixed-orifice metering device* may also be used to refer to a capillary tube.

Flame-retention burner: A high-efficiency oil burner.

Flash economizer: An energy-conservation process used in chillers that accomplishes intercooling of the refrigerant between the condenser and evaporator using a flash chamber or similar device. Flash gas of coolant produced in the flash chamber flows directly to the second stage of the chiller compressor, bypassing the evaporator and the first compressor stage.

Flash gas: Vapor refrigerant that is formed as the liquid refrigerant is squeezed through the metering orifice. Flash gas is produced when some of the liquid refrigerant boils off to cool the remaining liquid as the liquid passes through the metering device.

Flash point: The temperature at which heated oil vapors burst into flame. It is a measure of the flammability of the lubricant. Ideally, compressor lubricant oils should have a high flash point for safety reasons.

Glossary of Trade Terms

Flash steam: Formed when hot condensate is released to a lower pressure and re-evaporated.

Floc point: The temperature at which a 90/10 mixture of oil/refrigerant forms a cloudy or flocculent precipitate of wax in the mixture.

Flood-back: Liquid refrigerant that makes its way past the evaporator and returns to the compressor. It often causes dilution of the oil in the crankcase and a more severe condition known as slugging.

Flooded evaporator: An evaporator in which the refrigerant remains in liquid form as it leaves the evaporator.

Flooded starts: A condition in which slugging, foaming, and inadequate lubrication occur at compressor startup as a result of the oil in the compressor crankcase having absorbed refrigerant during shutdown.

Flooding: A condition in which there is a continuous return of liquid refrigerant in the suction vapor being returned to the compressor during operation.

Flow meter: A venturi tube or orifice plate measuring device installed in a water or other fluid system for the purpose of making pressure drop and velocity measurements.

Fouling: A term used for problems caused by suspended solid matter that accumulates and clogs nozzles and pipes, which restricts circulation or otherwise reduces the transfer of heat in a system.

Fractionation: A process in which the component refrigerants of a blended refrigerant boil off into a vapor state at different temperatures.

Glide: The temperature range in which a zeotropic refrigerant blend evaporates and condenses.

Global warming potential (GWP): A measure of a substance's ability to contribute to global warming that is expressed as a number. The higher the number, the greater the warming potential.

Grains per gallon (gpg): An alternate unit of measure sometimes used to describe the amounts of dissolved material in a sample of water. Grains per gallon can be converted to ppm by multiplying the value of gpg by 17.

Greenhouse effect: An effect in which atmospheric gases such as carbon monoxide trap solar heat in the atmosphere in the same way that a greenhouse captures and holds solar heat.

Halogen: A class of elements that include chlorine and fluorine that are used in the manufacture of refrigerants. A fully halogenated refrigerant is one in which all hydrogen atoms of a hydrocarbon molecule are replaced by halogen atoms.

Hantavirus: Any of several viruses that cause hantavirus pulmonary syndrome (HPS), a potentially fatal respiratory disease.

Hardness: A measure of the amount of calcium and magnesium contained in water. It is one of the main factors affecting the formation of scale in a system. As the hardness level increases, the potential for scaling also increases.

Head pressure: A measure of pressure drop, expressed in feet of water or psig. It is normally used to describe the capacity of circulating pumps. It indicates the height of a column of water that can be lifted by the pump, reflecting friction losses in piping.

High-fire: The maximum output stage of a two-stage gas furnace control.

Hunting: A condition in which an expansion valve alternately underfeeds and overfeeds the evaporator.

Hydrocarbon: A compound composed only of carbon and hydrogen atoms. When some or all of the hydrogen atoms in a hydrocarbon are combined with a halogen such as chlorine or fluorine, the end result is a compound known as a halogenated hydrocarbon or halocarbon. Most common refrigerants are halocarbons.

Hydrogenated chlorofluorocarbon (HCFC): A refrigerant that contains chlorines atoms in its structure, but fewer than found in CFCs. They are kinder to the environment than CFCs, allowing them to be phased out over a longer period of time. By 2030, all HCFC refrigerants will no longer be manufactured.

Hydrogenated fluorocarbon (HFC): A refrigerant that contains no chlorine atoms in its structure, which makes this group of refrigerants the most environmentally friendly of the three classifications of halocarbon refrigerants.

Glossary of Trade Terms

Hydronic system: A system that uses water or water-based solutions as the medium to transport heat or cold from the point of generation to the point of use.

Hygroscopic: The ability of a substance to absorb moisture from the air. Many synthetic refrigerant oils are very hygroscopic.

Inhibitor: A chemical substance that reduces the rate of corrosion, scale formation, or slime production.

Ionizing wires: The fine wires in the ionizing section of an electronic air cleaner that apply a positive charge to the airborne particles.

Latent heat of condensation: The heat given up or removed from a gas in changing back to a liquid state (steam to water).

Latent heat of fusion: The heat that is gained or lost in changing to or from a solid (ice to water or water to ice).

Latent heat of vaporization: The heat that is gained in changing from a liquid to a gas (water to steam).

Latent heat: The heat energy absorbed or rejected when a substance is changing state (solid to liquid, liquid to gas, or vice versa) and there is no change in the measured temperature.

Line duty device: A motor protection device that senses current flow and temperature in the motor winding. If an overload occurs, it opens the motor winding circuit to remove the line voltage.

Lithium bromide: A salt solution commonly used in air conditioning absorption chillers.

Low-fire: The reduced gas pressure stage of a two-stage gas furnace control.

Microprocessor: A tiny computer made from a single integrated circuit chip.

Milligrams per liter (mg/l): Metric unit of measure used to specify exactly how much of a certain material or element is dissolved in a sample of water. One mg/l is equivalent to one ppm.

Miscible: The desirable property that allows oil to dissolve in refrigerant and refrigerant to dissolve in oil.

Mixture: The combination of two or more compounds. Blended refrigerants are mixtures.

Near-azeotrope: A zeotrope with azeotropic properties. The refrigerant HFC-410A is an example of a near-azeotrope. Like zeotropes, near-azeotropes exhibit glide characteristics but they are usually minimal to the point that they are not a factor for technicians to consider when servicing a unit containing such a refrigerant.

Orifice: A tiny opening designed to pass liquid refrigerant.

Ozone depletion potential (ODP): A measure of a substance's ability to deplete atmospheric ozone that is expressed as a number. CFC refrigerants have greater ozone depletion potentials than HFC refrigerants.

Parts per million (ppm): Unit of measure used to specify exactly how much of a certain material or element is dissolved in a sample of water. For the purpose of comparison, one ounce of a contaminant mixed with 7,500 gallons of water equals a concentration of about one ppm.

pH: A measure of alkalinity or acidity of water. The pH scale ranges from 0 (extremely acidic) to 14 (extremely alkaline) with the pH of 7 being neutral. Specifically, it defines the relative concentration of hydrogen ions and hydroxide ions. As pH increases, the concentration of hydroxide ions increases.

Phase change: The conversion of refrigerant from one state to another, such as from liquid to vapor.

Pilot duty device: A motor protection device that senses current overload or temperature within the motor. If an overload occurs, it opens the motor contactor control circuit to remove power from the motor.

Pilot-operated valve: A valve in which the solenoid operates a plunger, which in turn creates or releases pressure that operates the valve.

Polyalkylene glycol (PAG): Synthetic refrigerant oil used with HFC refrigerants. It is very hygroscopic.

Polyolester (POE): Synthetic refrigerant oil used with HFC refrigerants. It is very hygroscopic.

Polyvinyl ether (PVE): A synthetic refrigerant oil with properties similar to mineral oils which can be used with HFC refrigerants. It has an advantage over POE and PAG oils in that it is not hygroscopic.

Glossary of Trade Terms

Positive-displacement compressor: Any compressor where the pumping action is created by pistons or moving chambers.

Pour point: An oil quality related to viscosity. It can be defined as the temperature at which oil first starts to flow. See *viscosity.*

Pressure (force-feed) lubrication system: A method of compressor lubrication that uses an oil pump mounted on the end of the crankshaft to force oil to the compressor main bearings, lower connecting rod bearings, and piston pins.

Pressure drop: The difference in pressure between two points. In a steam system, it is the result of power being consumed as the steam moves through pipes, heating units, and fittings. It is caused by the friction created between the inner walls of the pipe or device and the moving steam.

Pressure drop: The difference in pressure between two points. In a water system, it is the result of power being consumed as the water moves through pipes, heating units, and fittings. It is caused by the friction created between the inner walls of the pipe or device and the moving water.

Primary control: A combination switching and safety control that turns the oil burner on and off in response to signals from the thermostat and safety controls.

Radio frequency interference (RFI): Electronic noise caused by equipment operating at a very high frequency.

Safety drop time: The time it takes for the gas valve to shut off the gas supply after the pilot is disabled.

Saturated steam: The pure or dry steam produced at the temperature that corresponds to the boiling temperature of water at the existing pressure.

Scale: A dense coating of mineral matter that precipitates and settles on internal surfaces of equipment as a result of falling or rising temperatures.

Secondary coolant: Any liquid, such as water, that is cooled by a system refrigerant and then used to absorb heat from the conditioned space without a change in state. Also known as indirect coolant.

Sensible heat: Heat that can be measured by a thermometer or sensed by touch. The energy of molecular motion.

Short cycling: A condition in which the compressor is restarted immediately after it has been turned off.

Single phasing: A condition in which a three-phase motor continues to run after losing one of the three input phases while operating.

Skimming: The process of removing impurities and foam from the water surface of a steam boiler; also commonly referred to as surface blowdown.

Slinger ring: A ring attached to the outer edge of the fan that picks up water as the fan rotates and slings it onto the condenser coil.

Specific heat: The amount of heat required to raise the temperature of one pound of a substance one degree Fahrenheit. Expressed as Btu/lb/°F. At sea level, water has a specific heat of 1 Btu/lb/°F. At sea level, air has a specific heat of 0.24 Btu/lb/°F.

Splash lubrication system: Method of compressor lubrication in which the crankcase oil is splashed onto the cylinder walls and bearing surfaces during each revolution of the crankshaft while the compressor is running.

Stack control: A type of obsolete flame detector consisting of a bimetal switch inserted into the flue stack.

Static pressure drop: The pressure differential between the air entering the evaporator and the air leaving it.

Static pressure: In a water system, static pressure is created by the weight of the water in the system. It is referenced to a point such as a boiler gauge. Static pressure is equal to 0.43 pounds per square inch, per foot of water height.

Subcooling: The reverse of superheat. It is the temperature of a liquid when it has cooled below its condensing temperature.

Superheat: The measurable heat added to a vapor or gas above its boiling point as a liquid at that pressure; or heat added to the refrigerant above the refrigerant's boiling point.

Superheated steam: Steam which has been heated above its saturation temperature.

Surge chamber: A device that separates liquid and vapor refrigerant. Used with flooded evaporators. Liquid is recirculated back to the chiller or evaporator and vapor returns to the compressor.

Suspended solids: The amount of visible, individual particles in water or those that give water a cloudy look. They can include silt, clay, decayed organisms, iron, manganese, sulfur, and microorganisms. Suspended solids can clog treatment devices or shield microorganisms from disinfection.

Ternary blend: A blended refrigerant consisting of three refrigerants.

Thermal economizer: An energy-conservation process used in chillers in which warm condensed refrigerant is brought into contact with the coldest (inlet) water tubes in a condenser or heat exchanger. This causes the condensed refrigerant to subcool to a temperature below the condensing temperature.

Thermal expansion valve (TEV or TXV): A metering device with an external sensing bulb that senses the refrigerant temperature at the evaporator outlet. Also referred to as a thermostatic expansion valve. The valve maintains superheat around a setpoint. The terms TEV and TXV are used interchangeably.

Thermal-electric expansion valve (TEEV or THEV): An expansion valve in which a thermistor senses liquid line temperature and adjusts the orifice by changing the current flow through a bimetal needle. The terms TEEV and THEV are used interchangeably.

Total solids: Refers to the total amount of both dissolved and suspended solids contained in water.

Transformer phasing: Making sure that parallel transformers are operating in phase.

Trim: External controls and accessories attached to the boiler itself, such as sight glasses and water feeder controls.

Turndown ratio: In steam systems, the ratio of downstream pressure to upstream pressure, usually applied to steam pressure-reducing valves. More than a 50 percent turndown ratio is generally considered a large pressure reduction.

Unbalanced flame: A flame that is not centered in the combustion chamber opening of an oil furnace. This can be caused by low oil pressure or a defective nozzle.

Unit cooler: A packaged unit containing evaporator coil, metering device, and evaporator fan(s) that serves as the evaporator section for a refrigeration appliance.

Venturi tube: A short tube with flaring ends and a constricted throat. It is used for measuring flow velocity by measurement of the throat pressure, which decreases as the velocity increases.

Viscosity: The resistance of a fluid to flow. Viscosity of a fluid can vary, depending on the temperature of the fluid. When applied to refrigerant oil, viscosity must be such that the oil will flow and provide proper lubrication over the wide range of temperatures encountered in refrigeration and air conditioning systems. See *pour point*.

Water hammer: A condition that occurs when hot steam comes into contact with cooled condensate, builds pressure, and pushes the water through the line at high speeds, slamming into valves and other devices. Water hammer also occurs in domestic water systems. If the cause is not corrected, water hammer can damage the system.

Watertube boiler: A boiler where the heated water is contained inside the tubes, with the flue gases on the outside.

Zeotrope: A blended refrigerant that never mixes chemically. As a result, it evaporates and condenses over a temperature range, called the glide.

Zone damper: A damper used to control airflow to a zone in a zoned comfort system.

Index

Index

A

Abbreviations, 11.12
Accidents
    burns, 2.40, 5.5
    electrical shock, 10.13
    eye damage, 5.30, 7.39, 12.12
    suffocation, 1.2
Accumulator, 1.20, 4.14, 11.5, 11.6
Accutrol™ metering device, 3.7
Acid
    cleaners, 8.23
    corrosion by, 8.4
    safety, 2.40
    in the system, 1.21, 1.22, 2.31, 6.19, 8.5, 8.7
Acid rain, 7.4, 7.6
Actuator, 5.21, 12.13, 12.14, 12.23
Additives, 1.19, 1.22, 4.34
Adhesive, filter, 7.38
AFUE. *See* Annual fuel utilization efficiency
Air
    as an insulator, 6.19–6.20
    balance in ventilator system, 12.20
    corrosion by, 8.4
    excess (EA), 7.7
    infiltration of outside, 4.17
    primary and secondary, 5.36
    solubility in water, 5.13
    in the system, 1.21, 2.31, 2.33, 5.12–5.13, 5.15–5.16, 6.6
Air cleaners, electronic (EAC), 7.33–7.35, 12.8–12.12
Air conditioners. *See* Cooling units
Air handling unit, 5.35, 5.38, 12.7
Air intake, 5.31, 6.18
Air-water induction unit, 5.36, 5.37
Alarm system, 1.3
Alcohols, 8.10
Algae, 8.4, 8.5, 8.8, 12.6
Alkalinity, 8.3, 8.7, 8.9, 8.30
Alkylbenzene, 1.19, 1.28–1.29, 1.33
AlphaSan®, 4.34
Aluminum, 4.10, 4.22, 4.30, 11.19
American Boiler Manufacturers Association, 8.11, 8.24
American National Standards Institute (ANSI), 1.8, 5.20
American Society of Agricultural Engineers, 8.3
American Society of Heating, Refrigeration, and Air
    Conditioning Engineers (ASHRAE), 1.7–1.8, 7.36
Amides, 8.10
Ammeter, 2.39, 4.14, 7.5, 7.16, 7.25, 7.32

Ammonia, 1.2, 1.3, 1.4
Amperage
    condenser fan, 7.30
    full-load (FLA), 2.18, 2.33
    heat pump, 7.32
    rated-load (RLA), 2.18, 2.33–2.34
Animals, 7.2–7.3, 10.23
Annual fuel utilization efficiency (AFUE), 6.7, 7.7, 7.15
ANSI. *See* American National Standards Institute
Antifoaming agents, 8.10
Apartments, 6.7
Aquastat, 10.15–10.16, 12.4
ASHRAE. *See* American Society of Heating, Refrigeration,
    and Air Conditioning Engineers
Auger, 4.25, 4.26
Autotransformer, 2.28
Azeotrope, 1.10, 1.11, 1.33
    near, 1.12, 1.33, 4.6

B

Backflow, 1.20
Backflow prevention, 2.10, 2.12
Backwashing procedure, 8.13, 8.14, 8.30
Bacteria and bacterial disease, 7.3, 8.4, 8.5, 8.6, 8.8, 8.20
Baffles, wind, 11.28
Ballast, 3.17, 4.34, 9.7
Baseboard units, 5.22–5.23
Bats, excrement hazards, 7.3
Bearings, 2.10, 5.14, 5.15
Belimo Pressure Independent Characterized Control
    Valve™, 5.22
Bellows, 3.11, 6.27. *See also* Diaphragm
Bellows seal, 2.8
Benzotriazole, 8.7
Bicarbonates, 8.2, 8.3, 8.16
Bimetal components
    disc, 2.21, 2.22, 2.23, 2.25
    needle, 3.15
    switch, 6.13, 10.3
    thermal vent damper, 7.12
    thermostatic trap, 6.21, 6.24, 6.27
Biological films, 4.34
Biological growth
    control, 4.34, 7.33, 7.38–7.39, 8.8, 12.12
    in cooling towers, 5.33, 8.5, 8.8
    in humidifiers, 7.33, 12.6
    overview, 5.31, 8.4, 8.5, 8.8
Birds, excrement hazards, 7.3

Black box, 9.3, 9.23
Blackout, 9.6
Bleed-off, 5.33, 6.9, 8.7, 8.18, 8.30
Bleed orifice, 2.16
Blends, binary and ternary refrigerant, 1.10, 1.11, 1.33, 1.34, 4.39
Blowdown
    boiler, 6.39, 6.40, 8.9, 8.10
    controller, 8.19
    definition, 6.9, 6.44
    procedure, 6.40, 8.24
    separator, 6.40, 8.19
Blowers
    cooling system, 4.11, 5.35
    furnace, 7.5, 7.8, 7.16, 9.12, 10.2
    heat pump, 7.32
    humidifier, 12.7
    for zoned system, 12.17
Blowthrough, 6.26
Body, of a valve, 6.14
Bohp. See Boiler horsepower
Boiler horsepower (Bohp), 5.10, 5.47
Boilers
    firetube and watertube, 5.7–5.8, 5.9, 5.47, 6.7
    high-pressure, 5.5, 6.7
    low-pressure, 5.4–5.5, 6.7
    monitoring, 8.25, 8.26
    priming and carryover, 8.10–8.11
    safety, 5.10–5.11, 6.39
    steam
        blowdown procedure, 6.9, 6.40
        operating/safety controls, 5.4, 6.8–6.13, 6.35
        overview, 6.6–6.7
        skimming procedure, 6.39–6.40, 6.44, 8.24–8.25
        and the steam cycle, 6.5
        water level, 6.10–6.12, 6.28, 6.39
        water treatment, 6.41, 8.24–8.25
    water
        cast-iron, 5.5, 5.6–5.7
        commercial hot-water, 5.4–5.10
        copper-finned tube, 5.5–5.6, 5.11
        dry-base, wet-leg, and wet-base, 5.7
        in dual-temperature systems, 5.36, 5.37, 5.38, 5.39, 8.8
        electric, 5.10
        electrode, 5.10
        high-temperature, 5.5
        low-temperature, 5.4–5.5
        operating/safety controls, 5.10–5.11
        Scotch Marine, 5.8, 5.9
        steel firetube and watertube, 5.7–5.8, 5.9, 5.47
        steel vertical tubeless, 5.8–5.9
    water treatment, 6.41, 8.9–8.11, 8.19, 8.24–8.25
Borax, 8.8
Borescope, fiber optic, 7.6, 7.15
Bourdon tube, 4.27–4.28
Brazing, 2.33, 2.40
Breeching, 5.8
Brine, 3.20, 5.24, 8.17
British thermal unit (Btu), 4.17, 6.3, 10.10
Bromine, 1.4
Bronze, 1.3
Broom, do not use, 7.2
Brownout, 9.6
Bubble point, 1.10, 1.11, 1.33
Bucket, trap, 6.20, 6.24, 6.26–6.27
Building
    insulation, 12.7–12.8
    management system, 8.18

plenum as a dump zone, 12.19
    ventilation, 5.31, 8.20, 12.13
Bulb, sensing
    boiler high-pressure limit control, 6.12
    economizer, 12.15–12.16
    heat pump thermostat, 11.18
    humidifier, 12.4
    thermal expansion valve, 3.15, 3.17, 3.19–3.20, 3.21
Burners
    boiler, 5.4, 5.7, 5.8
    conversion, 5.4
    dismantling and removal, 7.4, 7.8
    duel-fuel, 5.4
    flame-retention, 10.27–10.28, 10.31
    gas, combustion test equipment, 7.7
    gun-type, 10.9, 10.24
    LP, 7.10
    oil
        adjustment, 7.20, 10.26
        flame-retention, 10.27–10.28
        maintenance, 7.18–7.19, 7.21–7.24
        nozzle and spray pattern, 7.16, 7.20, 10.3, 10.9, 10.10, 10.25
        overfired, 10.10, 10.24
        supply lines, 10.18
        test equipment, 7.17
        troubleshooting, 7.17, 10.8–10.10, 10.27–10.28
        typical operation, 10.2–10.3
Burnout, 1.21, 2.19, 2.31, 2.40–2.41, 2.45, 7.30
Butane, 1.5, 5.4
Bypass
    hot gas line, 2.15–2.17, 2.35
    pressure relief in zoned system, 12.18–12.19
    in steam system, 6.23

C
Cabinet, fixture, 4.11, 4.21, 4.35
Cable, 4.18, 9.4, 9.6
Cad cell (photocell), 7.21, 10.11, 10.12–10.14, 10.31
Calcium, 8.2, 8.3, 8.7, 8.10. See also Water, hardness
Calcium bicarbonate, 8.3, 8.16, 8.17
Calcium carbonate, 8.3, 8.5
Calcium sulfate, 8.5
Calculator, charging, 1.14
Campuses, 6.7
Capacity control, 2.13–2.18, 2.45, 5.27
Capillary tubes
    copper, 2.26
    cutting, 1.18, 2.40, 3.5, 3.6
    definition, 3.26
    length and refrigerant flow, 3.4
    overview, 3.5–3.6, 4.12
Caps, 1.20, 2.9, 2.16, 4.30
Carbon, 2.31, 2.41
Carbon dioxide
    to clean gas burner orifice, 7.10
    corrosion by, 8.4, 8.5, 8.9
    to determine combustion efficiency, 7.22–7.23, 10.25
    pollution control requirements, 7.6
    sensors used with economizer, 12.13
    in steam, 6.19–6.20
    testing for, 10.4, 10.23–10.24
    in water supply, 8.3
Carbon dioxide tester/sensor, 7.4, 7.7, 7.16, 10.24, 12.13
Carbon monoxide (CO), 7.4, 7.6, 7.7, 7.14, 7.15, 7.24
Carbon monoxide tester, 7.6, 7.16, 7.24
Carryover, 8.10–8.11
Casing, pump, 5.14, 5.15
Cavitation, 5.15–5.16, 5.47

CDC. *See* Centers for Disease Control and Prevention
Cells, charged collector, 7.34–7.35, 12.9
Centers for Disease Control and Prevention (CDC), 7.2, 7.3
Centrifugal action, 2.13, 8.11, 8.15
Certifications, 1.7, 1.14
CFC. *See* Chlorofluorocarbons
CFC-11, 1.4, 1.27
CFC-12, 1.4, 1.5, 1.6, 1.19, 1.27, 4.25
CFC-502, 1.4
Changeout. *See* Replacement
Charging charts, 1.14
Charts
    charging, 1.14
    ice maker energized parts, 4.37
    operating pressure of cooling unit, 7.28, 7.29
    pipe sizing, 6.36
    pressure-temp. (P-T), 1.10, 1.11, 1.12, 4.13
    resistance *vs.* temp., 2.21
    smoke, 10.4, 10.26
    troubleshooting
        electronic air cleaner, 12.12
        heat pump, 11.34–11.41
        oil furnace, 10.36–10.38
        reach-in freezer, 4.33–4.34
Chemicals, water treatment, 8.18, 8.25
Chillers
    absorption liquid, 3.2, 5.29–5.30
    centrifugal liquid, 3.2, 5.27–5.27
    in dual-temp. systems, 5.36, 5.37, 5.38, 5.39, 8.8
    operating/safety control, 5.30–5.31
    reciprocating and scroll liquid, 5.26–5.27
    screw liquid, 5.28–5.29
    water, 5.24, 5.26–5.31
Chimney, 8.9, 10.22–10.27
Chloride, 8.2
Chlorine, 1.2, 1.4, 1.5, 8.8
Chlorofluorocarbons (CFC), 1.5, 1.6, 1.33, 5.27, 5.28. *See also*
    specific CFC entries
Chromates, 8.7, 8.8
Circuit
    control, field wiring for, 11.18, 11.33
    conventional *vs.* microprocessor, 9.3–9.4
Circuit board
    gas furnace, 9.10
    grounding, 9.8
    handling guidelines, 12.22
    heat pump, 9.15
    troubleshooting, 9.4–9.9, 12.22
    zoned system, 12.22
Circuit breaker, 2.27, 11.18
Circuit diagrams and schematics
    bypass timer wiring diagram, 4.29
    compressor, 2.23, 2.25, 2.27
    cubed-ice machine wiring diagram, 4.36
    economizer, 12.15
    electronic control system, 9.14
    freezer wiring diagram, 4.32
    heat pump, 11.8, 11.9, 11.10, 11.11, 11.15
    humidifier, 12.3, 12.4
    HVAC control system, 9.14
    oil furnace, 10.5–10.8
    primary control with an aquastat, 10.17
    time delay, 10.16
    zoned system, 12.17
Clean Air Act of 1990, 1.14
Client, communication with, 4.39, 12.22
Clutch, 2.5
CO. *See* Carbon monoxide
Coal, 8.13, 8.14, 12.9

Cock, gauge (try), 6.9, 6.10
Coils
    condenser, in cooling unit, 7.26
    evaporator, 7.25, 8.23
    hot water heating, 5.23
    refrigerant leak at the, 1.15
    in steam system, 6.5, 6.18, 6.19
Collection leg. *See* Drip leg
Colloidal substance, 8.10, 8.30
Color codes, 1.8–1.9, 1.10, 11.19
Combustion
    effects of incomplete, 8.9
    efficiency
        and carbon dioxide, 7.22–7.23, 10.25
        definition, 7.43
        electronic combustion analyzer, 10.27
        gas furnace, 7.4, 7.7
        oil furnace, 10.4, 10.25
    fresh air from recovery ventilator, 12.20
Comfort systems
    cooling applications, 4.10, 4.11, 4.14, 8.7
    heating applications, 6.16, 6.38
    zoned, 9.12, 9.23, 12.16–12.19
Commercial applications
    air conditioners, 2.28
    compressors, 2.12, 2.13
    hydronic systems. *See* Hydronic systems
    operation at 50% peak load, 2.14
    refrigeration systems. *See* Retail refrigeration systems
Communication with the client, 4.39, 12.22
Compatibility
    of oil and refrigerant, 1.2, 1.18–1.19, 1.27
    of thermostat with equipment, 9.13, 11.21, 12.22
Compensation, rate-of-rise, 2.21, 2.22
Complaints, handling, 4.39, 12.22
Compound, 1.4, 1.33
Compression, 2.2, 2.31, 2.45, 4.8
Compressors
    in air-cooled condensing units, 4.8, 4.9
    burnout, 2.19, 2.31, 2.40–2.41, 7.30
    capacity control, 2.13–2.18
    causes of failure, 2.10, 2.29–2.33, 2.38–2.39
    centrifugal, 2.5, 2.13
    checkout following failure, 2.36–2.39
    in chilled-water cooling system, 5.26, 5.28–5.29
    comparison chart, 2.5
    compliant scroll, 2.12
    components, 1.20
    in cooling unit, 7.27
    electrical system, 2.33–2.35
    lockout protection, 2.26–2.27
    lubricating systems, 1.19–1.20
    motor. *See* Motors, compressor
    nameplate, 2.33–2.34, 2.40
    oil. *See* Oil, compressor
    oil-free, 2.9
    open, hermetic, and semi-hermetic, 2.4, 2.5, 2.9, 4.9, 5.27
    operation checks, 2.38–2.39
    overload protection, 2.19–2.28
    overview, 2.2
    positive-displacement, 2.17, 2.45
    reciprocating, 2.5, 2.6–2.10
    replacement, 2.2, 2.39–2.41
    in retail refrigeration systems, 4.8, 4.9, 4.15, 4.16, 4.31,
        4.37
    role of the, 2.2–2.3
    rotary, 2.5, 2.10–2.11
    safety. *See* Safety, compressor
    screw, 2.5, 2.12–2.13, 2.17

Compressors (*continued*)
  scroll
    capacity control, 2.16, 2.17–2.18
    energy efficiency, 4.10
    overview, 2.5, 2.12, 2.45
    in retail refrigeration systems, 4.9
  shutdown, 2.30
  speed control, 2.17
  types, 2.4–2.13
  use of multiple, 2.14–2.15
  for zoned systems, 12.17
Computers, 9.2. *See also* Microprocessors; Software
Concentration, 8.2, 8.30
Condensation
  in compressor, 2.36
  in cooling unit, 4.11, 7.25
  electric defrost system, 4.19
  in fuel tank, 7.18
  in gas furnace, 7.15
  latent heat of, 6.3, 6.4, 6.44
  in steam systems, 6.6, 6.19, 6.27–6.30, 6.38
Condensers
  air-cooled condensing units, 4.8, 4.9
  in chilled-water cooling systems, 5.24–5.25, 5.26, 5.27,
    5.41
  in cooling unit, 7.26
  evaporative, 5.31, 5.33–5.35
  fin-and-tube, 4.10
  in ice maker, 4.35
  inspection, 2.36, 4.30
  refrigerant in the, 1.2, 3.2
  in retail refrigeration systems, 4.9–4.10, 4.15, 4.16, 4.35
  role of the, 2.2, 2.3
Confined spaces, 1.3
Contaminants in the system
  acid, 1.21, 1.22, 2.31–2.32, 2.33, 8.5, 8.7
  air. *See* Air, in the system
  biological growth control, 4.34, 7.33, 7.38–7.39, 8.8, 12.12
  compressor, 2.29, 2.30–2.33
  with cooling tower, 5.32, 5.33, 8.5, 8.6, 8.7, 8.8
  decrease in dielectric strength, 1.18
  dirt and debris, 1.21, 2.10, 2.31–2.33, 8.5, 8.7
  elimination, 2.33. *See also* Filters
  foam, 6.39, 6.41, 8.9, 8.10
  moisture, 1.17, 1.21, 1.22, 1.23, 2.31, 2.33
  refrigeration system, 1.21–1.23
  rotary vane compressor, 2.10
  scale. *See* Mineral deposits
  sludge, 2.31, 5.5, 6.9, 6.10, 8.10
  steam system, 6.9, 6.10, 6.16, 6.41
  within valve, 6.23
  ventilators, 12.20
  in water, 8.2–8.4
  wax precipitate from oil, 1.18, 3.22
Controllers
  aquastat, 10.15–10.16
  blowdown, 8.19
  speed, 2.17, 11.23–11.24
Convector, 5.22
Convenience stores, 4.10
Converter, heat exchanger used as a, 6.17
Cookout (severe burnout), 2.40, 2.41
Coolant, secondary, 5.24, 5.47
Coolers
  closed-circuit, 5.32
  reach-in, 4.21–4.22
  unit (unitary coils), 4.10, 4.11, 4.45
  walk-in, 4.8, 4.10, 4.11, 4.22–4.23

Cooling process
  free, 12.13, 12.27
  mechanical, 12.13, 12.27
Cooling systems
  chilled water, 5.24–5.36, 5.41
  troubleshooting electrical controls, 9.12–9.15
Cooling towers, 5.31–5.33, 5.34, 8.5, 8.6, 8.7
Cooling units, 3.13, 7.24–7.31
Copper
  boiler fins, 5.5–5.6
  capillary tubes, 2.26, 3.5, 4.10
  corrosion protection, 8.7, 8.8
  piping, 1.3
  plating, 2.31
  sensing bulb strap, 3.19
  wire, 11.18, 11.19
Copper oxide, 2.32–2.33
Corrosion
  air, 8.4
  from biological growth, 8.5, 8.6, 8.8
  in boiler, 8.9
  in closed water system, 8.8
  control, 8.7, 8.8, 8.9
  galvanic, 8.4
  in gas burner, 7.14
  in open system, 8.4
  overview of types, 8.4, 8.9
  by refrigerant, 1.3
  in steam system, 6.21, 6.23, 6.41
  uniform, by acid, 8.4
Corrosive atmosphere, 3.19
CPR. *See* Regulators, crankcase pressure
Crankcase, 2.6, 2.30, 2.34, 4.14–4.15
Crankshaft, compressor, 1.19, 1.20, 2.8, 2.9, 2.10
Cryptococcosis, 7.3
Cutoff control, low-water and high-water, 5.10–5.11, 6.8,
  6.10–6.12
Cutters, tubing, 1.18, 2.40, 3.5
Cycles of concentration, 8.3–8.4, 8.30
Cylinders
  compressor, 2.10, 2.11, 2.15, 2.29, 2.36
  nitrogen, 1.16, 1.17
  refrigerant, 1.8–1.10

D
Dampers
  barometric, 12.18
  in chilled-water cooling system terminal, 5.35, 5.36
  in economizer, 12.13, 12.14
  motorized, 12.18
  outside air, planned maintenance, 7.39
  regulator in oil furnace, 10.9, 10.11
  in steam boiler, 6.12
  troubleshooting, 12.22–12.23
  vent in gas furnace, 7.11, 7.12
  for zoned system, 9.12, 9.23, 12.17, 12.18, 12.22–12.23
DCV. *See* Ventilation, demand control
Deaerator, 6.29–6.30, 8.9, 8.17–8.18
Deck, cooling tower, 5.32
Defrost systems
  electric, 4.18–4.29
  harvest period/mode, 4.13, 4.24, 4.37, 4.38
  heat pump cycle, 7.31, 9.15, 11.5, 11.7, 11.10, 11.28, 11.29
  hot-gas, 4.14, 4.20
  off-cycle, 4.17–4.18
  overview, 4.17
  retail, 4.17–4.20
  timed, 4.18, 4.20, 11.28, 11.29

Degreasers, 8.23
Delay, built-in, 4.28–4.29, 9.4, 9.12, 10.15, 10.16, 11.14
Dew point, 1.10, 1.11, 1.33, 2.14, 4.17
Diagnostics, 9.4, 9.12, 9.18, 9.23
Diameter
    inside (ID), 3.6, 7.10
    outside (OD), 3.19, 10.18
Diaphragm
    in expansion tank, 5.11
    in pressure-reducing valve, 6.14
    seal, 2.8
    in thermal expansion valve, 3.11, 3.12–3.13, 3.14, 3.19
Dielectric strength, 1.18, 1.33
Dip tube, 4.14
Dirt and debris
    and electronic device failure, 9.4
    in the system, 1.21, 2.10, 2.31–2.33, 8.5, 8.7
Disc, bimetal, 2.21, 2.22, 2.23, 2.25
Disinfectant, for rodent droppings, 7.2
Displacement, compressor, 2.17
Disposal. *See* Waste management
Distributor line, 3.12, 3.26
Distributors, 3.12, 3.17–3.18, 3.19, 3.22, 3.26, 11.24
Documentation
    boiler water log sheet, 8.26
    importance of make, model and serial number, 7.2
    operating pressure of gas in heater, 7.10
    water analysis data sheet, 8.21
Doors, 4.17, 4.21
DOT. *See* U.S. Department of Transportation
Downdraft, 10.22
Draft gauge, 10.4, 10.23
Drain-down, of ice maker, 4.24
Drains
    condensate, 4.18, 6.22, 6.23, 6.38–6.39
    cooling tower water, 5.33, 8.6
    gravity, 1.26
    in ice maker, 4.23, 4.37
    steam coils, 6.19
Drip leg, 6.22, 6.39, 6.44
Drive shaft, 2.10, 2.12, 2.18
Drum, water, 5.8
Dry return, 6.32
Ductwork, 10.26, 12.7, 12.18–12.19
Dump zone, 12.18, 12.27
DX. *See* Evaporators, direct-expansion
Dye, 1.3, 1.16, 8.20

E
EA. *See* Air, excess
EAC. *See* Air cleaners, electronic
ECM. *See* Motors, electronically-commutated
Economizers
    control setup diagram, 9.13
    electronic control, 9.14
    flash, 5.28, 5.47
    function, 9.14, 9.15
    thermal, 5.28, 5.47
    troubleshooting, 12.13–12.16, 12.20–12.23
EER. *See* Energy efficiency ratio
EEV. *See* Valves, electronic expansion
Electrical system. *See also* Circuit diagrams
    compressor
        input power, 2.18, 2.29, 2.37
        motor overload module, 2.23, 2.24
        overview and causes of failure, 2.33–2.35, 2.38
    current imbalance, 2.34–2.35, 2.37
    current measurement. *See* Ammeter
    current monitoring, 2.24, 2.25, 2.27, 9.7

current overload, 2.19
current-sensing devices, 2.18, 2.19, 2.20, 2.21
economizer, 12.15
electronic air cleaner, 7.33–7.34
gas furnace, 7.10
grounding, 9.8
heat pump, 11.5, 11.7–11.16
isolation of components, 9.9
shorts, 9.6
troubleshooting controls
    cooling system, 9.12–9.15
    external causes of failure, 9.4–9.10
    heating system, 9.10–9.12, 10.10–10.16, 10.17
    heat pump, 7.32, 9.13, 9.15–9.17, 11.5, 11.7–11.15, 11.28–11.29
    microprocessor, 9.2–9.4
    standardization issues, 9.18
    test instruments for, 9.18
    ventilator, 12.20
    zoned system, 12.22
Electric power
    for boiler, 5.4
    conversion of Bohp to kilowatts, 5.10
    input, 2.18, 2.33, 2.37, 9.6–9.7
    monitoring, 2.24, 2.25, 2.27
    off. *See* Lockout and tagout, electrical equipment
    poor quality, 9.6–9.7
    uninterruptible power supply, 9.7
Electrodes, 5.10, 7.16, 7.19, 7.20, 10.3, 10.13
Electrolysis, 8.4, 8.30
Electrolyte, 8.4, 8.30
Electromagnetic interference (EMI), 9.7
Electronic combustion analyzer, 10.27
Electronic expansion valves (EEV). *See* Valves, electronic
    expansion
Electrostatic discharge (ESD), 9.7–9.8, 12.22
Embrittlement, 8.9
EMI. *See* Electromagnetic interference
Energy efficiency
    AFUE, 6.7, 7.7, 7.15
    combustion efficiency. *See* Combustion, efficiency
    decreased by condensate, 6.38
    economizer, 12.13
    and electronic expansion valves, 4.14
    ENERGY STAR designation, 6.7
    scroll compressor, 4.10
    steam boiler, 6.7
Energy efficiency ratio (EER), 4.8
Enthalpy, 9.14, 9.15, 9.23, 12.13–12.16
Environmental issues
    chromates, 8.7, 8.8
    global warming, 1.6–1.7
    mercury, 6.13, 11.18–11.19, 12.12
    oil disposal, 1.26
    ozone depletion and refrigerants, 1.2, 1.5–1.7, 1.28–1.29
    water conservation, 8.7
EPA. *See* U.S. Environmental Protection Agency
EPR. *See* Valves, evaporator pressure regulating
Equalizer, pressure, 2.14, 3.12–3.15
Equipment. *See also* Instruments
    improper installation and electrical malfunction, 9.6
    maintenance. *See* Maintenance and repair
    replacement. *See* Replacement
    retail refrigeration systems, 4.20–4.26
    shipping and transport, 9.6
    ultrasound, 6.27
    water treatment, mechanical, 8.11–8.19
    for zoned systems, 12.17
ERV. *See* Ventilators, energy recovery

ESD. *See* Electrostatic discharge
Esters, polymerized, 8.10
Ethylene glycol, 5.5
Evaporation, 8.5
Evaporators
    chilled-water cooling system, 5.25, 5.27, 5.29
    components, 2.2, 8.16
    cooling unit, 7.25
    direct-expansion (DX), 3.2, 3.9, 3.10, 3.26, 5.29
    flooded, 3.2, 3.4, 3.9, 3.26
    humidifier, 7.33
    ice maker, 4.11–4.12
    inspection, 2.36, 4.30
    malfunction, 2.30
    overview, 2.2, 2.3, 8.16
    pressure regulating valves in, 4.15
    refrigerant in the, 1.2, 3.2
    retail refrigeration system, 4.10–4.12, 4.15, 4.16
    tube-in-tube, 5.27
    use of suction line trap, 1.20
Exchanger, heat
    cleaning and inspection, 7.14–7.15
    condensing, in condensing oil furnace, 10.28
    leaks, 1.21, 7.4
    oil furnace, 10.3
    steam system, 6.5–6.6, 6.16–6.18
Expansion devices. *See* Metering devices
Expansion valves. *See* Valves, thermal expansion
Explosion, 1.17, 2.41, 5.4, 7.4, 7.16, 8.10

F

Factories, 5.23, 6.7
Fan coil unit, 5.35
Fans
    condenser, 4.8, 4.11, 7.30
    cooling tower, 5.33, 5.34
    fan coil unit, 5.35
    free-blow propeller-type, 4.11
    gas furnace, 7.5, 7.14
    heat pump, 9.15, 9.17
    induced-draft, 9.12
    unit heater and ventilator, 5.23
Fault message, 9.4, 9.5, 9.23
Feeder, water, 6.10–6.12
Feet of head. *See* Pressure, head
Filters. *See also* Separators; Strainers; Traps
    air, in chilled-water cooling system terminal, 5.35
    backwashing to clean, 8.13, 8.14
    bag-type, 8.14–8.15
    cartridge, 8.12
    charged collector cells, 7.34–7.35, 12.9
    duplex multimedia, 8.13
    electronic air cleaner, 7.33–7.35, 12.8–12.12
    electrostatic, 7.38
    fiberglass, 7.36
    filter-drier
        chilled-water cooling system, 5.28
        compressor, 2.40, 2.41
        installation, 3.20
        metering device, 3.6
        refrigeration system, 1.21, 1.22, 1.23, 2.33
        when working with oils, 1.23
    high-efficiency particulate air (HEPA), 7.2, 7.36
    MERV rating, 7.36
    multimedia, 8.13–8.14
    oil, in oil furnace, 7.17–7.18, 10.26
    oil screen, in compressor, 1.20, 2.9, 2.10
    permanent (washable), 7.37–7.38
    pleated, 7.35, 7.36, 8.12, 8.13
        selection, 7.35, 7.38
        standard media, 7.35–7.37
        ventilator system, 12.20
        water, 4.35, 8.11–8.16
Fire, 7.16
Firebox, 5.7
Firebrick, 5.7, 5.8
Firepot, 10.2
Firetube, 5.7–5.8, 5.9, 5.47, 6.7, 8.10
FLA. *See* Amperage, full-load
Flame
    detector or sensor, 7.21, 7.23, 7.31, 9.13, 10.11
    oil furnace, 7.23, 10.26
    pilot light in gas heater, 7.9–7.10
    retarder, 5.9
    safeguard controls, 5.11
    unbalanced, 7.23, 7.43
Flameout, 7.10
Flammability
    and explosion, 1.17, 1.18, 2.41
    of refrigerant, 1.3, 1.5, 1.6, 1.8
Flash gas, 3.2, 3.10, 3.26, 5.28, 5.29
Flash point, 1.18, 1.33
Flash tank, 6.30
Float
    in boiler low/high-water cutoff control, 5.10–5.11,
        6.10–6.12
    in cooling tower, 5.32, 5.33
    float and thermostatic trap (F and T), 6.21, 6.25, 6.26, 6.27,
        6.33
    in ice machine, 4.26
    in valve, 3.8–3.9
Floc point, 1.18, 1.33
Flood-back, 2.28, 3.2, 3.6, 3.26, 7.27
Flooding and flooded starts, 2.29–2.30, 2.45
Flowers, 4.10
Flow meter, 5.17–5.19, 5.47
Flue
    gas furnace, 7.7, 8.9
    oil furnace, 7.22, 7.24, 8.9, 10.22–10.27
    stack control, 10.31
    stack temp., 10.24–10.25
Fluid leak detection, 1.15, 1.17
Fluorine, 1.2, 1.4, 1.5
Foaming, of oil, 1.19, 1.20, 2.30, 8.10
Fouling, 8.8, 8.9, 8.30
Fractionation, 1.11, 1.33
Freezers
    reach-in, 4.21–4.22, 4.30–4.34
    walk-in, 4.8, 4.10, 4.22–4.23
Friction in piping system, 5.2, 6.6
Frost build up, 2.14, 4.4, 4.11, 4.17, 11.28, 11.29
Fuel
    for boiler, 5.4
    oil, 5.4, 10.10, 10.18, 10.19–10.21, 10.25
Fungi and fungal diseases, 7.3, 8.4
Furnaces
    gas, 7.4–7.15, 9.10–9.12, 9.13
    oil, 7.15–7.24, 10.2–10.8, 10.9. *See also* Troubleshooting, oil
        furnace
            condensing, 10.28
            multipoise, 10.2
    pre-wired, 12.5
    use with a heat pump, 11.7
Fuse, control circuit, 11.18
Fusion, latent heat of, 6.3, 6.44

G

Galvanic corrosion, 8.4

Garage, 5.23
Garnet, filtering media, 8.13, 8.14
Gas heating systems, 7.4–7.15, 9.10–9.12, 9.13, 11.7
Gaskets, 1.27, 2.8, 4.17
Gauge, feeler, 7.12
Genetron® refrigerants, 1.28
Geothermal heat, 5.4
Gland, packing, 5.15
Gland seal, 2.8
Glass
    sight, 1.26, 2.36, 2.37
    water gauge, 6.8, 6.9–6.10
Glide, 1.10, 1.33, 4.6
Global warming, 1.6–1.7, 7.6
Global warming potential (GWP), 1.6, 1.33
Glycol, 5.5
Gpg. See Grains per gallon
Grains per gallon (gpg), 8.2, 8.30
Grease, 1.17, 8.10, 8.23, 9.4
Greenhouse effect, 1.6, 1.7, 1.33. See also Global warming
Grid, in ice cube maker, 4.24
Grooving damage, 8.4, 8.9
Grounding, 9.8
GWP. See Global warming potential

H
Halogens, 1.2, 1.4–1.5, 1.33
Hantavirus pulmonary syndrome (HPS), 7.2–7.3, 7.43
Hardness of water, 8.2, 8.3, 8.9, 8.16–8.17, 8.30
Harmonics, electrical, 9.7
Hartford loop, 6.33
HCFC. See Hydrogenated chlorofluorocarbons
HCFC-22
    availability, 1.17
    boiling point, 3.2
    in chilled-water cooling system, 5.27, 5.29
    components, 1.4, 1.5, 1.8
    cylinder color code, 1.9
    in ice maker, 4.25
    ozone depletion potential, 1.6
    phase out, 1.6, 1.8
    P-T chart, 1.12
    retrofits, 1.27
HCFC-123
    in chilled-water cooling system, 5.27, 5.28
    compatibility with oils, 1.19
    cylinder color code, 1.9
    overview, 1.4
    ozone depletion potential, 1.6
    as retrofit for CFC-11, 1.27
HCFC-401A, 1.27, 11.27
HCFC-401B, 1.27
HCFC-409A, 1.27
HCFC-502, 4.25
Health issues
    allergies and electrostatic filters, 7.38
    with animal droppings, 7.2–7.3
    carbon monoxide poisoning, 7.4, 7.14
    eye damage, 5.30, 7.39, 12.12
    heat sickness, 7.24
    Legionnaires' disease, 5.31, 8.20
    low humidity, 12.6
Heat
    of compression, 4.8
    damage to electrical controls, 9.5, 9.8
    latent, 6.2, 6.3, 6.4, 6.44, 12.20
    sensible, 6.2, 6.3, 6.44, 12.20
Heaters
    baseboard and finned-tube, 5.22–5.23

belly, 5.28
    condensate pan, 4.18, 4.19
    convectors, 5.22
    crankcase, 1.19, 2.29, 2.30, 2.31, 2.36, 11.5
    deaerating, 6.29–6.30, 8.9, 8.17–8.18
    defrost, 4.10
    gas furnace, 7.4–7.15
    immersion resistance, 5.10
    infrared quartz, 4.18, 4.19
    low-wattage electric, 4.31
    radiant electric, 4.18, 4.19
    radiant steam, 6.16
    swimming pool, 11.4
    unit (heating element and fan), 5.23
    water. See Water heaters
Heating systems
    gas, 7.4–7.15, 9.10–9.12, 9.13
    hot-water, 5.4–5.23
    oil, 7.15–7.24, 10.2–10.8, 10.9, 10.28. See also
        Troubleshooting, oil furnace
    zoned, 9.12
Heat pumps
    circuit diagrams, 11.8, 11.9, 11.10, 11.11
    classification, 11.3–11.4
    cooling check, 7.30
    diagnostic aid, 9.19
    electronic controls, 7.32, 9.13, 9.15–9.17, 11.5, 11.7–11.15
    emergency heat sequence, 11.11
    low-pressure switches in, 2.26
    metering device in, 3.8
    operation and cycles, 11.4–11.5
    overview, 2.26
    refrigerant charge, 11.28–11.29
    refrigerants, alternative, 1.29
    rooftop units, 11.28
    servicing procedures, 7.31–7.32
    swimming pool, 11.4
    troubleshooting. See Troubleshooting, heat pump
    use with a fossil-fuel furnace, 11.7
    wiring diagram, 9.16
Heat sickness, 7.24
Heat transfer
    in boiler, 5.5
    in convector, 5.22
    in evaporator, 4.13
    factors which impede, 5.33, 8.9, 11.5
    role in refrigeration process, 3.2
HEPA. See Filters, high-efficiency particulate air
Hermetic (welded hermetic) compressor, 2.4, 2.5
HFC. See Hydrogenated fluorocarbons
HFC-23, 1.6
HFC-32, 1.10
HFC-125, 1.10
HFC-134a
    atmospheric life, 1.7
    compatibility with oils, 1.19
    cylinder color code, 1.9
    in liquid chillers, 5.27, 5.29
    overview, 1.4
    ozone depletion potential, 1.6
    typical refrigeration cycle, 4.4–4.6
HFC-152A, 1.7
HFC-404A, 1.4, 1.9, 4.6–4.8, 4.25, 4.26
HFC-407C, 1.4, 1.9, 1.12, 1.27, 5.29
HFC-410A
    blended, 1.10
    charging into a system, 11.29
    in chilled-water cooling systems, 5.29

HFC-410A (*continued*)
  compatibility with oils, 1.19
  cylinder color code, 1.9, 1.10
  near-azeotrope, 1.12
  overview, 1.4
  replacement but not retrofit for HCFC-22, 1.27
HFC-417A, 1.27
High-fire cycle, 9.12, 9.23
High-intensity discharge (HID), 9.7
Histoplasmosis, 7.3
History, refrigeration systems, 4.2
Honeywell, Genetron® refrigerants, 1.28
Hospitals, 6.7
Housing, equipment, 1.20, 2.9, 2.18
HPS. *See* Hantavirus pulmonary syndrome
HRV. *See* Ventilators, heat recovery
HSI. *See* Ignitor, hot surface
Humidifier, 7.32–7.33, 12.3–12.8
Humidistat, 12.6–12.7
Humidity, relative (RH), 12.6
Humidity level, 4.11, 4.17, 4.20, 9.8
Hunting condition, 3.16–3.17, 3.26
Hydrazine, 8.8
Hydrocarbon, 1.4–1.5, 1.33
Hydrogen, 1.5, 8.3
Hydrogenated chlorofluorocarbons (HCFC), 1.4, 1.5, 1.6,
  1.8, 1.33. *See also specific HCFC entries*
Hydrogenated fluorocarbons (HFC), 1.5, 1.17, 1.33. *See also
  specific HFC entries*
Hydrogen sulfide, 8.3
Hydronic, definition, 5.2
Hydronic systems
  air management, 5.12–5.13
  chilled-water cooling systems, 5.24–5.36, 5.41
  circulating pumps, 5.13–5.16
  dual-temp., 5.36, 5.37, 5.38, 5.39, 8.8
  heating system terminals, 5.22–5.23
  hot-water heating system components, 5.4–5.23. *See also*
    Boilers
  overview, 5.2, 5.47
  piping, 5.36–5.41
  valves in, 5.17–5.22
  vs. forced-air systems, 5.2
  water balancing, 5.41–5.43
  water concept review, 5.2–5.4
  water piping systems, 5.36–5.41
Hygroscopic, 1.19, 1.23, 1.33, 2.31

I

Ice
  cubed, 4.23–4.25, 4.34–4.39
  flaked, 4.25–4.26
  historical use, 4.2
  properties, 6.2
Ice makers/machines, 4.10, 4.11–4.12, 4.13, 4.23–4.26,
  4.34–4.39
Ice merchandisers, 4.21
ICM board, 9.15, 9.17, 11.14
ID. *See* Diameter, inside
IFM. *See* Motors, indoor fan
Ignitor and ignition components
  boiler, 5.4
  gas furnace, 7.10, 7.11, 7.12, 9.12
  hot surface (HSI), 7.12
  oil furnace, 10.14–10.15
Impellers
  centrifugal compressor, 2.13, 5.29
  centrifugal pump, 5.14, 5.15, 5.16
  use in series, 5.29

Induction unit, air-water, 5.36, 5.37
Industrial plants. *See* Factories
Inhibitor, 8.4, 8.30
Inspections
  compressor checkout, 2.36
  condenser, 2.36, 4.30
  evaporator, 2.36, 4.30
  heat exchanger, 7.14–7.15
  reach-in freezer, 4.30
Installation
  cable, 9.6
  compressor, 2.39–2.41
  filter-drier, 3.20
  improper, and electrical malfunction, 9.6
  pressure gauge, 6.16
  pressure relief valve in boiler, 6.13
  reversing valve, 11.27
  sensing bulb, 3.19–3.20
  steam trap, 6.22–6.23
  thermal expansion valve, 3.19, 3.20
  tubing, 2.29, 2.31
Instruments
  ammeter, 2.39, 7.5, 7.16, 7.25, 7.32
  Bourdon tube gauge, 4.28
  carbon dioxide tester/sensor, 7.4, 7.16, 10.24, 12.13
  carbon monoxide tester, 7.4, 7.6, 7.16, 7.24
  charging calculator, 1.14
  current monitoring devices, 2.24, 2.25, 9.7
  draft gauge, 10.4, 10.23
  electronic combustion analyzer, 10.27
  electronic leak detector, 1.3, 1.16
  fiber optic borescope, 7.6, 7.15
  flame detector/sensor, 7.21, 7.23, 9.13
  flow meter, 5.17–5.19, 5.47
  gas burner combustion test, 7.7
  gas leak detector, 7.6
  heat, in three-phase compressor motor, 2.23, 2.24, 2.25
  heating equipment maintenance, 7.5–7.6
  ice thickness sensor, 4.24, 4.38
  low ambient temp. head pressure control, 2.28
  magnehelic(c) gauge, 10.4
  manometer, 5.19, 7.7, 7.10, 7.11, 7.25, 7.26
  milliammeter, 7.5
  millivoltmeter, 7.5
  multimeter (VOM), 7.16, 7.25, 7.32, 9.18, 12.20
  ohmmeter, 7.5, 9.9, 11.18, 11.21
  oil burner combustion test, 7.17
  pH meter, 8.20
  pressure gauge, 5.18, 5.19, 6.8–6.9, 6.16, 10.26
  psychrometer, 12.6
  pyrometer, 6.25, 6.27, 6.28
  solid-state reduced-voltage starter, 2.28
  temp. sensor, 2.37, 4.13, 4.19
  test, troubleshooting, 9.18, 9.19–9.20
  thermistor, 2.21, 3.10, 3.15, 5.30, 9.9–9.10
  thermowell, 6.27, 6.28
  voltmeter, 7.5, 10.15, 12.11
  water gauge, 6.8, 6.9–6.10
  water level sensor, in ice maker, 4.38
Insulation
  building, 12.7–12.8
  electrode, 7.19, 7.20, 10.3
  fuel line, 10.21
  ice block, 4.2
  sensing bulb, 3.19, 3.20
Intercooling, 5.27, 5.28
Iron
  cast, boiler, 5.5, 5.6–5.7, 6.7

piping, 1.4
    in water supply, 8.2
Isolation container, 9.9

J

Joints, problems at the, 1.15, 2.32–2.33, 9.6
Junction box, 9.6, 10.16

K

Kerosene, 7.22, 10.20
Klixon (external line break), 2.20

L

Lamps, ultraviolet, 7.38–7.39, 12.12
Latent heat
    of condensation, 6.3, 6.4, 6.44
    definition, 6.2, 6.44
    of fusion, 6.3, 6.44
    transfer by ventilator, 12.20
    of vaporization, 5.31, 5.47, 6.3, 6.44
Leaks
    boiler, 5.5
    compressor, 2.8, 2.39
    from damaged tubes, 2.26, 3.5
    detector, 7.6
    expansion tank, 5.11
    gas furnace, 7.14
    heat exchanger, 1.21, 7.4, 7.14
    heat pump, 11.28–11.29
    oil furnace, 10.22, 10.24
    refrigerant, 1.3, 1.11, 1.14–1.17, 11.28–11.29
    steam system, 6.27, 6.34
Leak test, 7.4
Legionella bacteria, 5.31, 8.20
Lightning, 9.6
Lights, 4.31, 4.34, 7.33, 9.6
Line duty device, 2.18, 2.19, 2.20, 2.21, 2.25, 2.45
Lint, 2.32
Liquid petroleum (LP), 1.5, 5.4, 7.4, 7.10
Lithium bromide, 5.29–5.30, 5.47
Litmus paper, 8.20
Load
    design (maximum cooling), 3.4, 3.26
    low, causes and problems from, 3.22
    peak, in chilled-water cooling system, 5.41
    refrigeration, in Btus, 4.17
    variable, 3.8
Lockout and tagout
    compressor, 2.26–2.27
    cooling unit, 7.24
    electrical equipment, 2.36, 2.39
    heating appliances, 7.4, 7.12, 7.16
Low-fire cycle, 9.12, 9.23
LP. *See* Liquid petroleum
Lubrication system
    overview, 1.17
    pressure (force-feed), 2.10, 2.45
    splash, 1.20, 2.8–2.9, 2.45
Lubriplate®, 1.24–1.25
Lugs, 5.20, 9.6

M

Magnehelic© gauge, 10.4
Magnesium, 8.2, 8.3, 8.10, 8.17. *See also* Water, hardness
Magnesium bicarbonate, 8.16
Maintenance and repair. *See also* Replacement;
    Troubleshooting
    boiler, 5.5, 5.10, 8.24–8.25
    capillary tube, 3.6
    compressor, 2.26, 2.29–2.39
    condenser cleaning, 4.10

cooling tower and open recirculating system, 8.20, 8.22–8.23
    cubed-ice machine, 4.34–4.35
    electronic air cleaner, 12.9–12.10
    flame sensor, 9.13
    importance of make, model, and serial number, 7.2
    metering device, 3.8, 3.21–3.22
    planned. *See* Planned maintenance
    reach-in freezer, 4.30–4.34
    refrigerant leak, 1.11
    steam trap, 6.23–6.25
    tubing, 1.21
    valve adjustment, 3.12, 3.21, 4.14–4.15
    ventilator system, 12.20
Manifold, 2.14, 2.16, 2.26, 2.37
Manometer, 5.19, 7.7, 7.10, 7.11, 7.25, 7.26
Material safety data sheet (MSDS), 1.23, 1.24–1.25, 5.30, 8.25
Maximum continuous current (MCC), 2.33
MCC. *See* Maximum continuous current
Meningitis, 7.3
Mercury, 6.12, 6.13, 11.18–11.19, 12.12, 12.21
MERV. *See* Minimum efficiency reporting value
Metal fatigue, 8.9
Metal particles, 1.21, 2.32
Metering devices
    advances, 11.25
    basic operation, 3.2–3.4
    in cooling unit, 7.28–7.29
    distributors, 3.17–3.18
    expansion valves. *See* Valves, expansion
    fixed-orifice, 3.4–3.8, 3.26, 4.12, 7.28–7.29, 11.25
    hunting condition, 3.16–3.17, 3.26
    individual feeder tube, 3.7
    location in system, 3.3
    malfunction, 2.30, 2.31
    modulating, 4.12
    overview, 3.2, 4.12
    problems, 3.21–3.22
    in retail refrigeration system, 4.12–4.14
    role of, 2.2
    TXV. *See* Valves, thermal expansion
Methyl chloride, 1.2
Methyl orange, 8.3
Microprocessors
    chiller, 5.30
    controls, 9.2–9.4, 9.18. *See also* Troubleshooting, electronic controls
    cooling system, 9.14
    definition, 9.23
    gas furnace, 7.12, 9.12
    handling guidelines, 12.22
    ice maker, 4.39
    thermostat, 7.8, 12.21
    use in troubleshooting, 4.39, 9.20
Milliammeter, 7.5
Milligrams per liter (mg/l), 8.2, 8.30
Millivoltmeter, 7.5
Mineral deposits
    in boiler, 6.41, 8.9–8.10
    checking for, 8.23
    in closed water system, 8.9
    common, 8.2, 8.10
    in cooling tower, 5.33
    in humidifier, 7.32–7.33
    in ice machine, 4.24, 4.34
    overview, 8.4–8.5
    prevention, 8.9, 8.16–8.17
    scale, definition, 8.30
    in steam trap, 6.20

Mineral deposits (*continued*)
    water hardness, 8.2, 8.3, 8.9, 8.16–8.17, 8.30
Mineral oil, 1.17, 1.19, 1.23, 1.28–1.29
Minimum efficiency reporting value (MERV), 7.36
Miscibility, 1.17, 1.33
Mixtures, 1.5, 1.33
Moisture
    in the air, 12.20. *See also* Humidifier
    condensation of airborne, 4.11
    in the fuel tank, 7.18, 10.26
    in the system, 1.17, 1.21, 1.22, 1.23, 2.31, 2.33, 3.21
Mold, 7.3
Molybdate, 8.7
Monitoring
    boiler parameters, 8.25, 8.26
    electrical current, 2.24, 2.25, 2.27, 9.7
    total dissolved solids, 8.19
    water treatment, 8.2, 8.18
Motors
    burner, 10.2, 10.27
    compressor
        burnout, 1.21, 2.19, 2.31, 2.40–2.41
        cooling, 2.18
        current imbalance, 2.34–2.35, 2.37
        drive shaft alignment, 2.18
        input power, 2.18, 2.37
        overload protection, 2.19–2.24, 2.25
        reduced-voltage starting, 2.28
        reverse rotation, 2.15
        variable-speed and multiple-speed, 2.17
        voltage ranges, 2.34
    damper, 12.23
    electronically-commutated (ECM), 11.23
    heat pump, 9.15, 11.14
    indoor fan (IFM), 9.17
    pump, centrifugal, 5.14
    seized, 2.39
    stepper, 3.15–3.16
    three-phase, 2.19, 2.23–2.24, 2.34–2.35
    variable-speed, 11.14, 12.17
    windings
        cooling, 2.18, 2.19
        damage, 2.27, 2.31
        dedicated sets on multiple-speed, 2.17
        location, 2.22
        part-winding, 2.28
MSDS. *See* Material safety data sheet
Multimeter (VOM), 7.16, 7.25, 7.32, 9.18, 12.20

N

Nameplate, compressor, 2.33–2.34, 2.40
National Center for Infectious Diseases (NCID), 7.2
National Electrical Code® (*NEC®*), 2.18, 9.4, 9.5
National Institute of Occupational Safety and Health
    (NIOSH), 7.3
Natural gas, 5.4
NCID. *See* National Center for Infectious Diseases
NEC®. *See National Electrical Code®*
Needle, bimetal, 3.15
Net positive suction head required (NPSHR), 5.15
NIOSH. *See* National Institute of Occupational Safety and
    Health
Nitrate, 8.2
Nitric oxide, 7.4
Nitrogen, 1.16, 1.17, 2.33, 2.41, 3.19, 8.3
Nitrogen dioxide, 7.4
Nitrogen oxides (NO_x), 7.4, 7.6, 7.7
Noise (electrical), 9.6, 9.7
Noise (sound)

air in the system, 5.12, 6.19
    ballast hum, 4.34
    broken suspension spring or bent line, 2.39
    compressor knocks, 2.29, 2.37
    pump cavitation, 5.16
    quieter equipment, 2.11, 12.17
    scroll compressor, 2.12
    water hammer, 6.19, 6.20, 6.21, 6.25, 6.32, 12.6
Nozzle, oil burner, 7.16, 7.20, 10.3, 10.9, 10.10, 10.25
NPSHR. *See* Net positive suction head required
Nylobraid tube, 4.26

O

Occupational Safety and Health Administration (OSHA),
    8.25
OD. *See* Diameter, outside
Odor, 1.15, 7.14, 8.8, 10.22
ODP. *See* Ozone depletion potential
Ohmmeter, 7.5, 9.9, 11.18, 11.21
Oil
    additives, 1.19, 1.23
    breakdown, 2.24, 2.35
    compressor
        addition, removal, and correct level, 1.23, 1.26, 1.27,
            2.10, 2.14
        compatibility with refrigerant, 1.2, 1.18–1.19, 1.27
        draining, 1.21
        free of water, 1.17
        -to-refrigerants ratio, 2.30
    disposal, 1.26, 7.19
    and electronic device failure, 9.4
    fuel, 5.4, 10.10, 10.18, 10.19–10.21, 10.25
    handling guidelines, 1.23–1.26
    leaks, 1.14
    mineral, 1.17, 1.19, 1.23, 1.28–1.29
    MSDS, 1.23, 1.24–1.25
    No. 2, 7.22
    overview, 1.17
    pressure protection, 2.25–2.26
    properties, 1.2, 1.18–1.19
    and the refrigeration system, 1.19–1.23
    sample collection, 1.22
    sludge, 2.31
    synthetic, 1.19, 1.27, 1.28–1.29
    testing, 1.22
    types, 1.19
    volume, 1.27
Oil heating systems, 7.15–7.24, 10.2–10.8, 10.9, 10.28, 11.7.
    See also Troubleshooting, oil furnace
Open-drive compressor, 2.4, 2.5
Orifice, definition, 3.26
Ornithosis, 7.3
OSA. *See* Ventilation, outside air
Overfire, 10.10, 10.24
Overheating, 2.8, 2.19–2.24, 2.35
Overload protection, compressors, 2.19–2.28
Oxygen, 7.7, 7.10, 8.3, 8.4, 8.8, 8.9
Ozone depletion, 1.2, 1.6
Ozone depletion potential (ODP), 1.6, 1.34

P

Packaged (assembled) units
    air-cooled chiller, 5.24, 5.25
    boiler, 5.7
    chilled-water cooling system, 5.27, 5.29
    defrost controls, 11.28
    evaporator, 4.10
    heat pump electronic controls, 9.13, 9.16, 11.14, 11.15
    room air conditioner, 2.10, 2.11
    terminal air conditioner (PTAC), 2.10, 2.11

PAG. *See* Polyalkylene glycol
Parallel-flow system, 6.32
Parrot fever, 7.3
Parts per million (ppm), 8.2, 8.30
Permagum, 7.25, 7.26
Permit, confined space, 1.3
Personal protective equipment
    chemicals, water treatment, 8.25
    cleaning and sanitizing products, 4.35
    compressor, 1.18, 2.35, 2.37, 2.40
    conductive footwear, 9.8
    filter replacement, 7.37
    lithium bromide, 5.30
    oil, 1.23, 1.26
    rodent droppings in work area, 7.2
    soot, 7.16
pH, 8.3, 8.7, 8.9, 8.18, 8.20, 8.30
Phase change, 4.4, 4.6, 4.8, 4.45, 6.2–6.3
Phasing, transformer, 11.22–11.23, 11.33
Phenolphthalein, 8.3
pH meter, 8.20
Phosgene gas, 2.41
Phosphates, 8.7, 8.9, 8.10
Phosphonate, 8.7
Photocell. *See* Cad cell
Pilot duty device, 2.18, 2.19, 2.20, 2.45
Pilot light, 5.4, 7.9–7.10
Pins, in timer, 4.20
Piping
    buried, 2.29
    cooling unit, 7.27
    corrosion, 1.3
    distributor lines, 3.12
    inspection, 2.36
    oil, in oil furnace, 7.18, 10.9
    refrigeration system, 1.20–1.21, 2.14, 2.29
    sizing, for steam, 6.35–6.39
    two independent, with use of multiple condensers,
        2.14–2.15
    to use with ammonia, 1.4
Piping systems
    commercial hydronic systems, 5.36–5.41
    flow rate through, 5.2, 5.17–5.19, 6.36–6.37
    four-pipe, 5.39, 5.40
    friction, 5.2, 6.6
    oil furnace, 10.18–10.21
    one-pipe, 6.31–6.33, 10.18–10.19
    primary-secondary, 5.39–5.41
    rooftop units, 10.21
    schematic layout, how to prepare, 5.42
    steam, 6.30–6.35, 6.39
    three-pipe, 5.38–5.39
    two-pipe, 5.36–5.38, 6.33–6.34, 10.19–10.20
Pistons
    in fixed-orifice metering device, 3.7, 3.8, 4.12, 11.25
    reciprocating compressor, 2.6–2.7, 2.10
    role in cylinder unloading, 2.15, 2.16
    in thermal expansion valve, 3.12
Pitting damage, 8.4, 8.6, 8.9
Planned maintenance (PM)
    bird and bat excrement hazards, 7.3
    cooling units, 7.24–7.31
    fossil-fuel heating appliances, 7.4–7.24
    gas heating systems, 7.4–7.15
    heat pumps, 7.31–7.32
    HVAC accessories, 7.32–7.39
    oil heating systems, 7.15–7.24, 10.17
    overview, 7.2
    rodent hantavirus hazard, 7.2–7.3, 7.43

    steam trap, 6.23
Plates
    access, in boiler, 5.8
    scroll, 2.12, 2.17–2.18
    valve, 2.7, 2.8, 2.16
Plenum, as a dump zone, 12.19
PM. *See* Planned maintenance
POE. *See* Polyolester
Polyalkylene glycol (PAG), 1.19, 1.23, 1.34
Polymers, 8.7, 8.10
Polyolester (POE), 1.19, 1.23, 1.34
Polyvinyl ether (PVE), 1.19, 1.34
Ports, pump, 5.14, 5.15
Pour point, of oil, 1.18, 1.34
Power. *See* Electric power
PPM. *See* Parts per million
Pressure
    balancing with Hartford loop, 6.33
    condensing, 3.4
    control in retail refrigeration systems, 4.27–4.28
    control in steam boilers, 6.12–6.13
    effects on air solubility in water, 5.13
    equalizers, 3.12–3.15
    equilibrium in automatic expansion valve, 3.10
    equilibrium in thermal expansion valve, 3.14, 3.15
    gauge *vs.* atmospheric, 6.25
    head, 2.28, 3.13, 5.2–5.3, 5.4, 5.41, 5.47
    pounds per square inch (psi), conversions, 5.3
    of refrigerant, 4.5, 4.7, 4.14–4.17
    relationship with temp. in steam systems, 6.4–6.5
    relief in zoned systems, 12.18–12.19
    static, 5.4, 5.47
    through distributor, 3.18
    of water, 5.2, 5.3, 5.4
Pressure drop
    capillary tube designed to match, 3.5, 3.6
    in low-temp. refrigeration systems, 4.8
    in medium-temp. refrigeration systems, 4.6
    at metering device, 3.2, 3.3, 3.4
    overview, 5.2, 5.47, 6.44
    pump cavitation, 5.16
    static, 7.25, 7.26, 7.43
    steam line, 6.37–6.38
    and steam pipe sizing, 6.36, 6.37
    in steam system, 6.5
    thermal expansion valve, 3.12
Pressure gauge, 5.18, 5.19, 6.8–6.9, 6.16
Pressurestat, 4.27–4.28
Pressure-temperature (P-T) chart, 1.10, 1.11, 1.12–1.14
Pressuretrol, 6.12
Primary control, 10.17, 10.31
Priming, 8.10–8.11
Produce, 4.11
Propane, 1.5, 5.4
Propylene glycol, 5.5
Protozoa, 8.4
Psittacosis, 7.3
Psychrometer, 12.6
PTAC. *See* Packaged (assembled) units, terminal air
        conditioner
Pulp and paper mills, 6.2
Pumps
    for automatic chemical feed system, 8.18
    centrifugal, 5.13–5.16
    chilled-water, 5.24
    circulating, 5.13–5.16, 5.35
    condensate, 6.6, 6.28, 6.31, 6.33, 7.38
    double-suction centrifugal, 5.16
    geothermal heat, 5.4

Pumps (*continued*)
　heat. *See* Heat pumps
　multi-stage, 5.16, 5.17
　oil
　　boost, 10.20
　　in compressor, 1.20, 1.26, 2.9, 2.10, 2.26
　　in furnace, 10.26
　positive-displacement, 5.13
　vacuum, 6.29, 6.33–6.34, 6.35
　water, 4.37, 4.38, 5.2, 5.3
Pump shaft, 5.14, 5.15
Purge unit, 5.28
PVE. *See* Polyvinyl ether
Pyrometer, 6.25, 6.27, 6.28

R
R- refrigerants, 1.8, 1.12, 1.28–1.29
R-22, 1.8, 1.13
R-134*a*, 2.8
R-410A, 1.12, 1.13, 2.31
Rabies, 7.3
Radio frequency interference (RFI), 9.7, 9.23
Reactor, 2.28
Receivers
　condensate, in steam system, 6.6, 6.28–6.29, 6.33
　in refrigeration system, 1.20, 2.3
　in retail refrigeration systems, 4.14, 4.15, 4.16
Recycling, oil, 1.26
Refrigerants
　alternative, 1.28–1.29, 2.8
　breakdown, 2.24, 2.35
　characteristics, 1.2–1.3, 1.27
　in chilled-water cooling systems, 5.24–5.25, 5.27, 5.28
　common, 1.4
　composition, 1.10–1.14
　in cooling unit, 7.26, 7.27–7.28, 7.29–7.30
　definition, 1.2
　flow
　　regulation. *See* Metering devices
　　through evaporator, 3.11
　　through heat pump, 11.5, 11.6
　　through the system, overview, 2.2, 2.3, 3.2, 3.3, 3.6, 4.3.
　　　See also Refrigeration cycle
　history, 1.2
　identification, 1.7–1.10
　leaks, 1.3, 1.11, 1.14–1.17, 11.28–11.29
　low-temp., alternative, 1.29, 2.8
　newer, and higher pressures, 11.27
　phase change to vapor, 4.4
　reclaimed, 1.4, 1.8
　in sensing bulb, 3.15
　slugging, 2.8, 2.29, 2.30, 3.2, 3.5, 3.6
　structure, 1.4–1.7
　system conversion, 1.26–1.29
　-to-oil ratio, 2.30
Refrigeration cycle
　low-temp., 4.6–4.8
　medium-temp., 4.4–4.6
　off-cycle flow, 2.14, 2.27, 2.45
　overview, 4.4
　short, 2.14, 2.27, 2.45
Refrigeration systems
　charge and recharge, 1.11, 11.28–11.29
　contaminants. *See* Contaminants in the system
　conversion (retrofitting), 1.1, 1.23, 1.26–1.29
　cycle. *See* Refrigeration cycle
　evacuation or purging, 1.21, 1.23, 2.33, 2.40, 2.41, 3.19
　low-temp., 1.29, 2.8, 3.20, 4.6–4.8
　mechanical, 4.2–4.17

　and oil, 1.19–1.23
　operation checks, 2.36–2.39
　overview of components, 4.8–4.17
　overview of refrigerant flow through, 2.2, 2.3, 4.3
　piping. *See* Piping, refrigeration system
　pressurization for leak detection, 1.16–1.17
　retail. *See* Retail refrigeration systems
　split, 2.31
Refuse-burning facilities, 5.30
Regulators
　crankcase pressure (CPR), 2.34, 4.14–4.15
　draft, in oil furnace, 10.9, 10.11
　evaporator pressure, 4.15
　oil pressure, in compressor, 1.20, 2.9, 2.10
　retail refrigeration system pressure, 4.14–4.17
　steam system pressure, 6.14
Relays
　control, 12.15
　heat, 9.15
　lockout, 2.26, 2.27
　magnetic, 11.24
　time delay, 4.28–4.29, 9.4, 9.12, 10.15, 10.16
Release, deminimus, 1.14
Replacement
　cartridge filter, 8.13
　circuit board, 9.4
　compressor, 2.2, 2.39–2.41
　resistive defrost heater, 4.18
　valves, 3.18–3.21, 6.13, 11.26
Reset, 2.19, 2.23, 2.38, 7.16, 10.12, 10.14
Resistance, electrical
　and the lockout relay, 2.27
　oil furnace cad cell, 10.13
　testing, 9.18
　of thermistor, 2.21, 2.23, 9.9, 11.22
　of water in electric boiler, 5.10
Resistor, primary, 2.28, 10.13
Resource Conservation and Recovery Act (RCRA), 1.26
Respirator, HEPA, 7.2, 7.8
Restaurants, 4.30
Retail refrigeration systems
　controls, 4.26–4.30
　defrost, 4.17–4.20
　equipment and fixtures, 4.20–4.26
　mechanical, 4.2–4.17
　overview, 4.2
　troubleshooting, 4.30–4.39
Retail store, 4.30
Retrofitting, 1.1, 1.23, 1.26–1.29
RFI. *See* Radio frequency interference
RH. *See* Humidity, relative
Rings
　check for failure, 2.39
　compression, 2.7
　O-, 1.27, 2.8, 3.7
　piston, 2.7
　slinger, 4.11, 4.45
　wear, in pump, 5.15
Riser, 1.20
RLA. *See* Amperage, rated-load
Rodents, hantavirus hazard, 7.2–7.3, 7.43
Rods, connecting, 1.19, 1.20, 2.6, 2.9, 2.10
Rotary seal, 2.8
Rotor, 2.10, 2.11, 2.12–2.13, 5.28–5.29

S
Safety
　acid, 2.40
　air intake health risk warning, 5.31

blowdown valve, 6.9
boiler, 5.10–5.11, 6.39
chemicals, water treatment, 8.25
compressor, 1.18, 2.35, 2.37, 2.40
cooling unit, 7.24–7.25
gas cylinder, 1.9, 1.17
gas furnace, 7.4–7.5
gear. *See* Personal protective equipment
hantavirus, 7.2–7.3, 7.43
HVAC accessories, 7.32
ice machine sanitation, 4.34
lithium bromide, 5.30
oil furnace, 7.16, 10.3
oil handling, 1.23–1.26
refrigerant toxicity, 1.2–1.3
scroll compressor, 2.12
steam, 5.5
water balancing, 5.42
Safety drop time, 7.11, 7.43
Sag (brownout), 9.6
Sand, filtering media, 8.13, 8.14
Saturation vapor point, 6.3
Scale, 8.30. *See also* Mineral deposits
Screen, oil, 1.20, 2.9, 2.10
Screws, adjustment, 2.16, 4.38
Scroll, 2.12
Sea-going vessels, 5.30
Seals, 2.8, 5.15
Seat, of a valve, 6.14
Sensible heat, 6.2, 6.3, 6.44, 12.20
Separators
    air, 5.12–5.13
    blowdown, 6.40, 8.19
    centrifugal, for water, 8.15–8.16
    inline, for steam (entrainment), 8.11
Service calls, 2.38, 4.30, 4.39, 7.15, 12.22
Short cycling, 2.14, 2.27, 2.45
Shroud, impeller, 5.15
Shutdown
    chiller, 5.30
    compressor, 2.30
    ice maker, 4.24, 4.35, 4.37
    retail refrigeration systems, 4.27, 4.28
Sight glass, 1.26, 2.36, 2.37
Silica, 8.2
Silicate, 8.7
Silt, 8.7
Silver, to control biological growth, 4.34
Simple Green®, 7.35
Single phasing, 2.19, 2.21, 2.34–2.35, 2.45
Siphons, 1.26, 6.8, 6.9
Slide rule, fire/combustion efficiency, 7.7, 7.22
Slimes, 8.8
Sludge, 2.31, 5.5, 6.9, 6.10, 8.10
Slugging, 2.8, 2.29, 2.30, 3.2, 3.5, 3.6
Smog, 7.4
Smoke chart, 10.4
Smoke spot test kit, 7.22
Snubber, 6.8
Soda, caustic, 8.9
Soda ash, 8.9, 8.23
Sodium, 8.2, 8.16–8.17
Sodium carbonate, 8.10
Sodium nitrite, 8.8
Sodium phosphate, 8.9, 8.10
Sodium silicate, 8.9
Softeners, water, 8.16–8.17
Software, 4.13, 5.15, 9.18
Soldering, 3.19, 9.6

Solenoid, pulsing, 3.16
Solids
    dissolved, 8.2, 8.3, 8.11, 8.30
    suspended, 8.3, 8.5, 8.11, 8.30
    total, 8.3, 8.11, 8.30
    total dissolved (TDS), 8.3–8.4, 8.19, 8.24
Soot
    and electronic device failure, 9.4
    in oil furnace flue, 9.4, 10.23
    -removal chemicals, 7.18
    in the system, 2.31, 7.4, 7.16, 7.21, 8.9
Specific heat, 6.3, 6.44
Spike, 9.6
Stability, 1.2, 1.18
Standardization of electrical systems, 9.5, 9.18
Starter, solid-state reduced-voltage, 2.28
Startup
    boiler, 5.10
    chiller, 5.30
    compressor, 2.26, 2.29–2.30
    fixed-orifice steam trap, 6.22
    heat pump, 11.5
    ice maker, 4.37
    refrigeration system, 4.14
    steam system, 6.19, 6.29, 6.38
Static electricity, 9.7–9.8, 12.22
Steam
    contaminants from carryover, 8.10–8.11
    dry, 6.3, 6.44
    flash, 6.9, 6.22, 6.26, 6.27, 6.30–6.31, 6.44
    pressure-temp. relationships, 6.4–6.5
    properties and advantages, 6.2
    saturated, 6.3, 6.4–6.5, 6.44
    as source of heat for absorption chiller, 5.30
    superheated, 5.5, 5.47, 6.4, 6.44
Steam systems
    boiler blowdown and skimming, 6.39–6.40, 6.44,
        8.24–8.25
    boilers, boiler controls, and accessories, 6.6–6.13
    boiler water treatment, 6.41
    condensate return/feedwater system components, 6.6,
        6.27–6.30, 6.39
    cycle principles of operation, 6.5–6.6
    flash tanks, 6.30
    fundamentals and properties of water, 6.2–6.4
    heat exchangers/converters, 6.16–6.18
    overview, 6.2
    piping, 6.30–6.35, 6.39
    pressure-temp. relationships, 6.4–6.5
    terminals, 6.16, 6.18, 6.19
    traps and strainers, 6.18–6.27
    valves, 6.13–6.16
Steel, 1.4, 5.5, 5.7–5.9, 5.10
Stem, of a valve, 6.14
Strainers. *See also* Filters; Traps
    bag type, 8.15
    burner nozzle, 10.9, 10.10
    at capillary tube input, 3.6
    in compressor, 2.16, 2.31
    overview, 8.12
    in steam system, 6.16, 6.23, 6.25, 6.26
    valve, 6.16
Subcooling, 1.12, 2.2, 5.27–5.28, 6.3, 6.4, 6.44, 7.29–7.30
Suction bypass and cutoff unloading, 2.15, 2.16
Suction line
    compressor, 2.14, 2.24–2.25, 2.30, 2.37
    heat pump troubleshooting, 11.29
    installation of sensing bulb on, 3.19
    location in system, 1.21, 2.3

Suction line (*continued*)
    sizing and pitch, 1.20
    thermometer on, 2.37
    trapped, 1.21
    and verification of hunting condition, 3.17
Sulfate, 8.2
Sulfur dioxide, 1.2, 7.4
Sulfuric acid, 8.7, 8.9
Sump, 2.14, 5.32
Superheat
    acceptable values, 3.21, 4.8
    baseline reading, 3.21
    calculation, 1.12, 1.14, 3.21
    checking, 11.29
    definition, 1.12, 3.26, 6.4, 6.44
    effects of outdoor temp., 3.4
    and electronic expansion valve, 3.11, 3.12, 4.13
    and sensible heat, 6.2, 6.3
    and temp. glide, 1.10
Surge (electrical), 9.6
Surge chamber, 3.2, 3.10, 3.26, 5.28
Surge protector, 9.7
Swimming pool, heater, 11.4
Switches
    bimetal, 6.13, 10.3
    bin, in ice maker, 4.38
    boiler cutoff control, 5.10–5.11, 6.11
    dip, 11.14, 11.15
    fan, 4.31
    float, 1.10
    limit, 7.13, 10.3, 10.16
    oil furnace, 10.28
    pressure, 2.24–2.26, 4.27–4.28
    water flow, 5.11
    zone, 12.17
Symbols, 11.13

T
Tanks
    brine, 3.20
    condensate receiver, 6.6, 6.28–6.29, 6.33
    expansion and compression, 5.11–5.12
    flash, 6.30
    fuel, moisture in, 7.18, 10.26
    oil storage, 10.18, 10.19, 10.20, 10.21
Tape, heat, 4.18
TDS. *See* Solids, total dissolved
Technician, refrigerant, 1.7
TEEV. *See* Valves, thermal-electric
Teflon®, 8.23
Temperature
    air duct, 10.26–10.27
    and bacterial growth, 8.5
    box, 4.4, 4.6, 4.18, 4.21
    control, 5.21, 5.36, 6.16, 10.3
    effects on air solubility in water, 5.13
    glide chart, 1.10
    heat pump discharge air, 11.18
    indoor, 4.20, 4.31
    oil furnace, 10.3
    outdoor
        and chilled-water cooling system parameters, 5.41
        during cooling check on heat pump, 7.30
        and cooling/heating systems, 5.37–5.38, 9.14–9.15
        and ice makers, 4.23
        low, 2.28, 3.13, 4.10, 4.23, 5.5, 6.18
    of refrigerant
        acceptable superheat, 3.21
        before and after metering device, 3.2

    effects on viscosity, 1.18
    saturation, 4.4, 4.8
    values in low-temp. refrigeration system, **4.7**
    values in medium-temp. refrigeration system, **4.5**
    vs. resistance chart, 2.21
    relationship with electrical resistance, 2.21, 9.9, **11.22**
    relationship with pressure in steam systems, 6.4–6.5
    stack, 10.24–10.25
    of steam system, 6.16
    vs. resistance chart, in thermistor, 11.22
    wastewater, 8.24
Terminals
    air conditioner, 2.10, 2.11
    air-water induction, 5.36, 5.37
    chilled-water cooling system, 5.25–5.36
    compressor, 2.37, 2.39, 7.25
    electronic air cleaner, 12.11
    equipment, 9.4
    fan coil, 5.35
    heating system, 5.22–5.23
    steam system, 6.16, 6.18, 6.19
    unit, 5.35
Testing
    acid/moisture test kit for, 2.31, 2.32, 2.41
    ball valve with magnet, 11.25–11.26
    boiler low-water cutoff control, 5.11
    carbon dioxide, 10.4, 10.23–10.24, 12.13
    combustion efficiency kit for, 10.4
    compressor operation checks, 2.38–2.39
    cooling unit, 7.26
    gas burner combustion, 7.7
    heating appliance leak, 7.4
    motor, for faulty overload protector, 2.21
    oil, 1.22, 2.32, 2.41
    oil burner combustion, 7.17
    refrigerant leak, 1.3
    refrigeration system operation checks, 2.36–2.38
    smoke, 7.22, 10.4, 10.23–10.24, 10.26
    steam trap, 6.27
    troubleshooting test instruments, 9.18, 9.19–9.20
    valve, solenoid, 11.24
    for voltage imbalance, 2.35, 2.37
    water, 8.19–8.20, 8.21
TEV. *See* Valves, thermal expansion
Thermal expansion, 1.10
Thermal expansion valves (TVE; TXV). *See* Valves, **thermal** expansion
Thermistor
    in chiller, 5.30
    electrical resistance, 2.21, 9.9
    in electronic expansion valves, 3.10, 3.15
    heat pump, 11.21–11.22
    overview, 9.9–9.10
    programmable thermostat, 12.21
Thermocouple, 7.10
Thermometers
    air duct, 10.26–10.27
    calibration, 7.13
    dial, 7.7
    measurement points for reversing valve, 11.26–**11.27**
    outdoor air temp. sensor, 12.15
    stack, 10.25
    suction line placement, 2.37
    to test cooling unit, 7.28, 7.30–7.31
    to test gas furnace, 7.12–7.13
    to test oil furnace, 7.23–7.24, 10.4
Thermopile, 7.10
Thermostats
    call-for-cooling signal, 2.14, 12.14

call-for-heat signal, 7.13, 7.23, 10.16
compatibility with equipment, 9.13, 11.21, 12.22
compressor, 2.27
crankcase heater, 2.30
cycle rate adjustment, 7.8
digital room, 11.19
discharge line, 4.9
economizer, 12.13, 12.14, 12.21–12.22
electric defrost system, 4.19
heat pump, 7.32, 9.15, 11.18–11.22
with humidistat, 12.6
motor, overload, 2.21, 2.22
oil furnace, 10.16
programmable, 7.8, 12.21
remote-bulb, 4.27
retail refrigeration system, 4.19, 4.26–4.27
simulation of, for testing, 11.21, 11.28
steam system, 6.16
troubleshooting, 11.18–11.22, 12.21–12.22
Thermowell, 6.27, 6.28
THEV. *See* Valves, thermal-electric
Timers, 4.18, 4.19, 4.20, 4.31
Tin whiskers, 9.6
Tolyltriazole, 8.7
Tools
    electrode setting, 7.20
    for gas furnace maintenance, 7.5–7.6
    nozzle wrench, 7.16, 7.20, 10.10
    for oil furnace maintenance, 7.16
    for planned maintenance, 7.32
Top dead-center, 2.7
Torch, 1.16, 1.18, 2.40
Totaltest® test tube, 2.32
Towers, cooling, 5.31–5.33, 5.34, 8.5, 8.6, 8.20, 8.22–8.23
Trailers, refrigerated, 2.5
Training, from manufacturers, 11.14, 12.22
Transducer, 2.26, 4.13, 5.30
Transformer, 2.24, 2.25, 7.16, 10.2, 11.18, 11.22–11.23
Transient, 9.6
Traps. *See also* Filters; Strainers
    drain line, 6.38–6.39
    fixed-orifice, 6.22
    float and thermostatic (F and T), 6.21, 6.25, 6.26, 6.27, 6.33
    mechanical, 6.20, 6.24
    in steam system, 6.6, 6.14, 6.18–6.27, 6.38–6.39
    suction line, 1.20, 1.21, 2.30, 3.19–3.20, 4.14
    thermodynamic, 6.21–6.22, 6.25, 6.27
    thermostatic, 6.20–6.21, 6.24–6.25, 6.27, 6.33
Trim, 6.7, 6.44
Troubleshooting
    aids, 12.3. *See also* Troubleshooting, charts
    air cleaner, electronic, 12.8–12.12
    approach, 4.39, 12.2–12.3, 12.22
    bypass pressure relief, 12.18–12.19
    charts, 4.33–4.34, 10.36–10.38, 11.34–11.41
    compressor, 2.38
    cubed-ice machine, 4.34, 4.35–4.39
    damper, 9.12, 9.23, 12.18, 12.22–12.23
    distributor, 3.22
    economizer, 12.13–12.16, 12.20–12.23
    electronic controls
        circuit board, 9.4–9.9
        cooling system, 9.12–9.15
        external causes of failure, 9.4–9.10
        heating system, 9.10–9.12
        heating system, oil, 10.10–10.16, 10.17
        heat pump, 7.32, 9.13, 9.15–9.17, 11.5, 11.7–11.15
        microprocessor, 9.2–9.4
        test instruments for, 9.18, 9.19–9.20

thermistor, 9.9–9.10
energy recovery ventilator, 12.19–12.20
gas heating, 9.10–9.12, 9.13
heat pump
    chart, 11.34–11.41
    control transformer phasing, 11.22–11.23, 11.33
    defrost control, 7.31, 9.15, 11.28, 11.29
    electronic controls, 7.32, 9.13, 9.15–9.17, 11.5, 11.7–11.15
    operation, 11.4–11.5
    overview, 11.2–11.4, 11.16–11.17
    refrigerant charge, 11.28–11.29
    speed controller, 11.23–11.24
    system procedures and diagrams, 11.16–11.29
    thermostats, 11.18–11.22
    valves, 7.31, 11.24–11.27
humidifier, 12.3–12.8
importance of logical thinking, 4.39
lamp, ultraviolet, 12.12
metering device, 3.21–3.22
microprocessor display, sample, 9.20
oil furnace
    burner, 10.8–10.10
    common problems, 10.28
    condensing oil furnace, 10.28
    electronic controls, 10.10–10.16, 10.17
    overview, 10.2
    summary, 10.32
    system, 10.16–10.28, 10.32–10.38
    typical operation, 10.2–10.8
retail refrigeration system, 4.30–4.39
steam system, 6.37
steam trap, 6.25–6.27
thermostat, 11.18–11.22, 12.21–12.22
zoned control system, 12.16–12.19, 12.20–12.23
Tube, cap. *See* Capillary tubes
Tubing, 1.18, 1.21, 2.29, 2.31, 2.40
Turndown ratio, 6.14, 6.44
TXV. *See* Valves, thermal expansion

U

UL. *See* Underwriters' Laboratory
Ultrasound equipment, 6.27
Ultraviolet light. *See* Lamps, ultraviolet
Underwriters' Laboratory (UL), 2.33, 9.4
Uninterruptible power supply (UPS), 9.7
Unloader, hydraulic, 2.15, 2.16
Unloading, cylinder, 2.15
Up-feed system, 6.32
UPS. *See* Uninterruptible power supply
U.S. Department of Transportation (DOT), 1.10
U.S. Environmental Protection Agency (EPA)
    ban of chromates, 8.7, 8.8
    Clean Air Act, 1.14
    ENERGY STAR designation, 6.7
    Office of Solid Waste, 1.26
    refrigerants, 1.6, 1.7
    Resource Conservation and Recovery Act, 1.26
    underground fuel tanks, 10.18

V

Vacuum breaker, 6.6, 6.35, 10.21
Vacuum cleaner, do not use, 7.2
Vacuum-return system, 6.33–6.35
Valves
    automatic expansion, 3.9–3.10
    balancing, 5.17–5.19
    bi-flow, 3.16
    blowdown (skimming), 6.39–6.40, 8.24, 8.25
    blowoff, 6.8, 6.23

Valves (continued)
  butterfly, 5.19–5.20
  capacity control, 2.16
  check
    in bi-flow thermal expansion valve, 3.16
    in compressor, 2.16
    construction, 11.25
    in heat pump, 7.31, 11.5, 11.24–11.26
    in steam boiler, 6.8
    in steam condensate receiver, 6.28
    in steam trap, 6.23
  circuit balancing, 5.36
  in commercial hydronic system, 5.17–5.22
  direct-acting, 11.24, 11.33
  discharge, 2.6, 2.10, 2.11, 2.16
  drain (blowdown), 6.8, 6.9, 6.10
  drain, self-actuating, 6.19
  electronic expansion (EEV), 3.15–3.16, 3.26, 4.13–4.14
  evaporator pressure regulating (EPR), 4.15, 4.16
  expansion, 3.8–3.17
  failure, 2.39, 6.16
  gate, 6.14
  globe or ball, 6.10, 6.14, 6.23, 11.25–11.26
  harvest, in ice maker, 4.38
  high-side float, 3.8–3.9
  intake slide, 2.17
  low-side float, 3.9
  manual expansion, 3.8
  multi-purpose, 5.17
  pilot gas, 7.11
  pilot-operated, 11.24, 11.33
  poppet, 2.16
  pressure reducing, 6.13–6.16, 6.35–6.36
  pressure relief, 1.10, 4.9, 4.14–4.15, 5.12, 6.12–6.13
  reed (flapper), 2.7
  reversing, 7.31, 11.5, 11.20, 11.26–11.27
  ring, 2.7, 2.8
  service, 1.15, 2.31
  setpoint adjustment, 4.14–4.15
  shutoff, 4.14, 6.23
  solenoid
    in air conditioner, 11.24
    in compressor hot gas bypass line, 2.15, 2.16
    in compressor suction line, 2.30
    cooling tower bleed-off, 8.18
    in electronic expansion valve, 4.14
    function, 11.24
    in heat pump, 7.31, 11.20, 11.24
    in humidifier, 7.33
    in oil furnace, 10.15
    in retail refrigeration system, 4.28, 4.29–4.30
    with a sensing bulb in brine tank, 3.20
    in steam boiler, 6.11
  in steam system, 6.13–6.16
  suction, 2.6, 2.15, 2.16
  thermal-electric (TEEV; THEV), 3.10, 3.15, 3.16, 3.26
  thermal expansion (TEV; TXV)
    in compressor, 2.30
    in cooling unit, 7.29
    definition, 3.26
    in heat pump, 11.25
    in metering device, 3.5
    overview, 3.10–3.15
    replacement, 3.18–3.21
    in retail refrigeration system, 4.12–4.13
  thermostatic, 6.16
  two-flange, 5.20
  two-way and three-way, 5.21
  use in series, 6.14–6.15

  wafer lug, 5.20
Vane, 2.10, 2.11, 2.17, 5.15
Vaporization, 4.4, 4.6, 4.8
  latent heat of, 5.31, 5.47, 6.3, 6.44
Vapor point, 2.2
Varnish, 2.31
Vehicle, air-conditioned, 2.5
Velocity sizing, 6.36–6.37
Ventilation
  building, 5.31, 8.20, 12.13
  compressor motor, 2.18
  demand control (DCV), 12.13
  outside air (OSA), 7.39
  work area, 1.3
Ventilators
  energy recovery (ERV), 12.19–12.20
  heat recovery (HRV), 12.19–12.20
  unit, 5.23, 5.35–5.36
Vents
  boiler, 5.4
  crankcase relief, 2.6
  oil furnace, 10.22, 10.23
  in steam flash tank, 6.30, 6.33
  in steam piping, 6.31, 6.33
  in steam trap, 6.20, 6.25
Venturi tube, 5.18, 5.19, 5.47
Vibration, 2.18, 5.16, 9.6, 12.21
Viral disease, 7.3
Viscosity, 1.17, 1.18, 1.34, 5.5
Voltage
  circuit board, 9.4
  compressor motor, 2.28, 2.34
  cooling unit, 7.27
  electronic air cleaner, 7.33–7.34
  imbalance, 2.34, 2.37
  leg-to-let, 2.19
  oil furnace, 10.11–10.12
  and speed controller, 11.23–11.24
Voltage drop, 2.18, 7.10
Voltmeter, 7.5, 10.15, 12.11
Volume
  clearance, 2.7, 2.45
  oil, 1.27
  refrigerant, 3.4–3.5, 3.22

W
Warehouse, 5.23
Waste management
  mercury, 6.13, 11.19, 12.12
  oil, 1.26, 7.19
  oil storage tank, 10.19
  rat nests and droppings, 7.3
  wastewater temp., 8.24
Water
  air solubility in, 5.13
  changing states, 6.2
  conservation, 8.7
  contamination by. See Moisture
  conversions, 5.2, 5.3
  hardness, 8.2, 8.3, 8.9, 8.16–8.17, 8.30
  hot-water heating systems. See Hydronic systems
  humidifier, 12.7, 12.8
  potable (drinking), 6.18, 8.2
  pressure, 5.2, 5.4
  problems with untreated. See Biological growth;
    Corrosion; Mineral deposits; Solids, suspended
  properties, 5.2, 6.2–6.4
  softeners, 8.16–8.17
  substances in, 8.2–8.4

testing, 8.19–8.20, 8.21
treatment, in boiler, 6.41
used as refrigerant under vacuum, 5.30
Water balancing, 5.41–5.43
Water cooler, 3.20
Water hammer
    from boiler, 8.10
    definition and cause, 6.19, 6.20, 6.44, 12.27
    from humidifier, 12.6
    and steam traps, 6.19, 6.20, 6.21, 6.25
    susceptibility of one-pipe systems, 6.20, 6.32
Water heaters, 5.4–5.23, 6.16, 6.18. *See also* Boilers
Water heating systems, commercial, 5.4–5.23
Water supply
    automatic feed, 8.18
    for boilers, 6.10, 8.9, 8.10
    for ice makers, 4.23, 4.24, 4.35
    untreated, problems with, 8.4–8.5
Water treatment
    chemical feed systems, automatic, 8.18
    chemical safety, 8.25
    in closed recirculating systems, 8.8–8.9
    mechanical equipment, 8.11–8.19
    in open recirculating systems, 8.5–8.8, 8.20, 8.22–8.23
    overview, 8.2
    procedures and guidelines, 8.19–8.25, 8.26
    in steam boilers and systems, 6.41, 8.9–8.11
    water characteristics and analysis, 8.2–8.4
Watertube, 5.7–5.8, 5.9, 5.47, 6.7
Weather
    freezing conditions, 5.5, 6.18. *See also* Temperature,
        outdoor, low
    lightning and thunder storms, 2.27, 9.6
    wind, 11.28
Wire
    aluminum to copper connections, 11.19
    copper, 18 AWG, 11.18
    environmental conditions which damage, 9.4–9.5
    ionizing, 7.34, 7.35, 7.43, 12.9
    tin whiskers, 9.6
    tungsten, 12.9
Wiring
    control circuit field, 11.18, 11.33
    diagrams
        bypass timer, 4.29
        cubed-ice machine, 4.36
        freezer, 4.32
        gas furnace, 9.11
        heat pump, 9.16
        heat pump thermostat, 11.20
        oil furnace, 10.6, 10.8, 10.12
    factory, 11.18
    heat pump thermostat, 11.19, 11.20
Work area
    animal droppings in the, 7.2–7.3
    cleanliness, 7.15, 7.37
    conditions which overload electronic air cleaners, 12.10
    effects of excess heat, 7.24, 9.5, 9.8
    humidity and static electricity, 9.8
    in the public eye, 4.30
    ventilation, 1.3
Wrench, nozzle, 7.16, 7.20, 10.10
Wrist strap, grounding, 9.8
Wye-delta connection, 2.28

Z
Zeolite, 8.16–8.17
Zeotrope, 1.10–1.11, 1.34
Zinc, 8.7
Zone, dump, 12.18, 12.27
Zoned control systems
    comfort system, 9.12, 9.23, 12.16–12.19
    equipment for, 12.17
    troubleshooting, 12.16–12.19, 12.20–12.23
    typical system, 12.16–12.17